Designing Sorghum Genome for NextGen Agriculture

With increasing global food demand and changing climatic conditions, sorghum is an important crop for sustainable agriculture. Its inherent resilience to drought, heat, and poor soils makes it a promising solution for enhancing food security and climate adaptation. To further harness its potential, *Designing Sorghum Genome for NextGen Agriculture*, authored by leading researchers and experts, explores the latest advancements aimed at improving sorghum's resilience and productivity. This comprehensive resource provides cutting-edge insights into how modern scientific tools and methods are being leveraged to design the sorghum genome for enhanced yield, stress tolerance, and nutritional value.

This book takes a multidisciplinary approach, integrating diverse strategies to enhance sorghum's resilience. It highlights current achievements and outlines future directions to harness sorghum's genetic potential as a climate-resilient crop for next-generation agriculture.

Key Features:

- Examines sorghum's global agricultural significance and its role in addressing future climatic challenges.
- Covers genomics, pan-genomics, mutagenesis, and targeted gene editing for sorghum improvement.
- Highlights physiological, molecular, and microbiome-based strategies to enhance sorghum's abiotic stress tolerance.
- Explores the synergy between genomic selection and phenomics for accelerated genetic gain.
- Discusses omics approaches and systems biology for advancing sorghum research.
- Showcases artificial intelligence and crop simulation modeling as tools for sorghum breeding for enhanced stress tolerance and yield potential.

This book serves as a valuable resource for researchers, breeders, agronomists, and students working in sorghum genetics, genomics, crop improvement, and sustainable agriculture. By bridging fundamental research with translational breeding strategies, it offers a forward-looking perspective on leveraging genomics for the future of sorghum production.

Dinesh Kumar Saini is a Research Associate in the Department of Plant and Soil Science at Texas Tech University, specializing in crop breeding and genomics. He holds a master's degree in Genetics and Plant Breeding from Chaudhary Charan Singh University, India. During his Ph.D. at Punjab Agricultural University, he received prestigious fellowships, including DST-INSPIRE and ICAR-SRF, along with multiple academic honors, including the Chancellor and Vice-Chancellor Medals. His research focuses on the physiological, molecular, and histological mechanisms driving climate resilience and yield potential in grain sorghum. He has co-authored over 55 peer-reviewed articles, 20 book chapters, and numerous abstracts. He actively reviews for reputed journals, serves on editorial boards, and is a member of several professional societies.

S. V. Krishna Jagadish is a Professor and Thornton Distinguished Chair of Plant and Soil Science at Texas Tech University. He earned a master's degree in Agronomy from the University of Agricultural Sciences, India, and a Ph.D. in Crop Physiology from the University of Reading, U.K. He previously held positions at the International Rice Research Institute, Philippines, and later became a Professor at Kansas State University. Dr. Jagadish has received numerous awards, including the Kansas State Gamma Sigma Delta Outstanding Research Award and International Fellow of the Indian Society of Plant Physiology. With over 200 peer-reviewed publications, 16,500+ citations, an H-index of 73, and $25 million in grant funding, his research focuses on abiotic stress tolerance in sorghum, forage systems, water conservation, and soil health, promoting sustainability in the U.S. Southern High Plains.

Depiction and Designing of Plant Genomes

Chittaranjan Kole and Phullara Kole

Series Editor

The Cannabis Genome: A New Era in Genetics, Genomics, and Breeding
Edited by Angel Fernandez i Marti and Jaime Prohens

Designing Sorghum Genome for NextGen Agriculture
Edited by Dinesh Kumar Saini and S. V. Krishna Jagadish

For more information, please visit our website: https://www.routledge.com/Depiction-and-Designing-of-Plant-Genomes/book-series/DDPG

Designing Sorghum Genome for NextGen Agriculture

Edited by
Dinesh Kumar Saini and S. V. Krishna Jagadish

CRC Press
Taylor & Francis Group
Boca Raton London New York

CRC Press is an imprint of the
Taylor & Francis Group, an **informa** business

Designed cover image: Dinesh Kumar Saini

First edition published 2026
by CRC Press
2385 NW Executive Center Drive, Suite 320, Boca Raton FL 33431

and by CRC Press
4 Park Square, Milton Park, Abingdon, Oxon, OX14 4RN

CRC Press is an imprint of Taylor & Francis Group, LLC

ISBN: 978-1-041-03406-3 (hbk)
ISBN: 978-1-041-03414-8 (pbk)
ISBN: 978-1-003-62371-7 (ebk)

DOI: 10.1201/9781003623717

Typeset in Times
by Newgen Publishing UK

Dedication

This book series is dedicated to my brothers and sisters –

Bansi Badan Kole
Niranjan Kole
Bharati Nandi Nee Kole
Arati Kundu Nee Kole
Ram Ranjan Kole

Contents

Foreword

Sorghum (*Sorghum bicolor* (L.) Moench) is a vital crop known for its resilience to harsh conditions. It ranks fifth among the world's most-produced grains, following corn, wheat, rice, and barley. Believed to have originated in Africa around 8,000 B.C., sorghum remains a major crop in Africa but is now primarily produced in the United States. It serves multiple purposes, including as a food staple for over 500 million people in more than 30 countries, animal feed, and a source for the growing biofuel industry. Its gluten-free nature also makes it a valuable option for those with celiac disease.

The book, *Designing Sorghum Genome for NextGen Agriculture*, serves as a timely and transformative resource that bridges the gap between traditional breeding methods and modern molecular innovations. It represents a collective endeavor by some of the great minds in sorghum research, offering readers a comprehensive exploration of advancements and innovations. The chapters in this book offer a broad view of the topic, covering global production trends and the need for genome design to enhance crop resilience. The chapters explore important topics like molecular mapping, pan-genomics, mutagenesis, root biology, and omics approaches, highlighting their role in yield improvement, stress tolerance, and nutrient use efficiency. The book also discusses modern tools such as speed breeding, advanced bioinformatics, genomic selection, high-throughput phenotyping, artificial intelligence, and crop modeling showing how they can help tackle current and future challenges in sorghum breeding and genetics.

This book is more than a compilation of knowledge. It is a visionary guide that identifies emerging opportunities and lays the foundation for future research. What sets it apart is its emphasis on integrating traditional breeding with translational breeding strategies and tools. By bridging fundamental research with practical applications, it serves as a vital resource for plant scientists, geneticists, agronomists, and policymakers dedicated to advancing sorghum improvement. The insights presented here will not only drive the development of superior sorghum hybrids/varieties but also inspire further research and collaborations to enhance productivity and sustainability. As agriculture continues to evolve, the knowledge compiled in this volume will be instrumental in guiding future innovations in sorghum breeding and genetics.

Editors Dr. Dinesh Kumar Saini and Prof. S.V. Krishna Jagadish have curated an exceptional collection of contributions from 65 experts across eight countries. Their vision and dedication have brought together a wealth of expertise, making this book an invaluable resource for researchers, students, and professionals alike. More than a record of progress, this book is a testament to the power of scientific collaboration and innovation. It celebrates sorghum's rich history while offering a roadmap for unlocking its full potential in modern agriculture.

Glen Ritchie, Ph.D.
Professor of Crop Physiology | J.A. Love Endowed Chair | PSS Department Chair

Preface to the Series

Genome sequencing of the model plant Arabidopsis in 2000 followed by the crop plant rice in 2002 flagged genome sequencing in an array of plants. As of March 5, 2024, 4604 genomes of 1482 different plant species have been sequenced, and the number has continued to grow exponentially. The list includes several crop plants as well as some model plants. Sequencing of the nuclear genomes was followed by sequencing of extranuclear genomes such as mitochondria and chloroplast and by several studies on functional genomics. However, detailed reports on them are scattered in the literature including research articles and reviews. Therefore, a comprehensive compilation of deliberations on the sequences with enumeration of the sequences, annotation, assembly, gene families, and comparative genomics of these useful genomes, and elucidation of functional genomics including transcriptomics, metabolomics, and proteomics in these higher plants is timely.

A number of higher plants provide the basic F5 – food, feed, fuel, fiber and furniture – needs; and also, nutraceuticals, beverages, luxury, and aesthetic items. Genetic improvement in these plants with regard to their utilities was accomplished in the last century mainly by traditional breeding. The advent of molecular breeding and genetic engineering in the last two decades of the last century strengthened the breeding strategies in economic plants. Advances in plant genomics have ultimately revolutionized the concepts and strategies of crop improvement. Employment of high-throughput genotyping and phenotyping together with other genomics-assisted breeding tools like gene discovery, allele mining, genome-wide association studies, genomic selection, and genome-wide prediction, etc., have facilitated accelerated and precise plant breeding. The recent emergence of the genome editing technique has enormously supplemented molecular and genomics tools, making it possible to design and tailor genomes according to need. However, success stories of next-generation breeding techniques need to be compiled for easy access of readers.

This present series, entitled Depiction and Designing of Plant Genomes, includes a number of books, each dedicated to the deliberations on genomic attributes and/or genetic improvement of a higher plant, and hopefully will serve as handy compendium for readers.

The books of this series have been edited by highly reputed scientists and I remain thankful for their invaluable academic efforts and cooperation. I am also thankful to the publisher and specifically to its Senior Commissioning Editor (Chemistry / Agricultural Sciences), Dr. Renu Upadhyay, and its Editorial Assistant, Dr. Renu Upadhyay, for their outstanding and timely help and cordial co-operation since the inception of this book series.

Finally, I must express my thanks to my wife Phullara and our children, Sourav and Devleena, for assisting me in editing the books of this series and for patiently bearing with me as I devoted time that should have been theirs to the work of editing this series.

Kolkata **Chittaranjan Kole**

Special Acknowledgement to
Phullara Kole
For her outstanding assistance in editing this book series

Preface

Sorghum (*Sorghum bicolor* (L.) Moench) is a resilient cereal crop extensively cultivated in arid and semi-arid regions due to its remarkable ability to withstand drought, heat, and poor soil conditions. It serves as a staple food for millions and as a key resource for animal feed, biofuel, and industrial applications. Its high nutritional value, gluten-free nature, and adaptability to diverse agro-climatic conditions enhance its global significance. Different types of sorghum serve distinct purposes: grain sorghum supports human consumption and animal feed; forage sorghum provides silage and grazing material; sweet sorghum is ideal for syrup and ethanol production; biomass sorghum is cultivated for bioenergy; dual-purpose sorghum offers opportunities for food and fodder; broomcorn sorghum is harvested for brush-making; and waxy sorghum caters to specialty food and industrial applications. Sorghum's significance extends beyond food security in developing nations, playing an expanding role in industrial and bioenergy sectors, while its adaptability and low input requirements position it as a critical crop for addressing climate change. Modern breeding and biotechnological innovations continue to unlock its potential across food, feed, and energy domains, with this book focusing primarily on grain sorghum due to its pivotal role in food security and sustainable agriculture.

This edited volume, *Designing Sorghum Genome for NextGen Agriculture* (a part of the *Depiction and Designing of Plant Genomes* series), brings together leading experts in sorghum research to present a comprehensive overview of recent advances, emerging technologies, and future directions in sorghum science. The book begins with Chapter 1, highlighting sorghum's global significance, production trends, and the pressing need for genome design to adapt to future climatic challenges. Chapter 2 explores advanced tools to harness genetic diversity, while Chapter 3 investigates the role of bioinformatics in facilitating sorghum omics research. Chapters 4 and 5 focus on molecular mapping advancements and next-generation sequencing's impact on sorghum genomics and pangenomics. Chapter 6 examines mutagenesis, transgenesis, and gene editing, while Chapter 7 discusses genomic selection integrated with phenomics for accelerated genetic gains. Chapters 8–10 cover hybrid breeding, omics approaches, and physiological strategies for next-generation sorghum improvement. Root biology's role in enhancing resilience is explored in Chapters 11 and 12, covering root-phenes engineering and microbiome redesign. Chapters 13–15 discuss emerging technologies, from omics-based seed composition modulation to high-throughput phenotyping, artificial intelligence, and crop simulation modeling to address climate change.

This book takes a forward-looking approach to sorghum research, making it a valuable resource for researchers, students, and professionals. It brings together recent advancements, encourages collaboration, and supports the development of better sorghum varieties for the future. *Designing Sorghum Genome for NextGen Agriculture* includes 15 chapters written by 65 scientists from 31 institutes and universities across eight countries, including the United States, Australia, India, Sweden, the United Kingdom, Pakistan, China, and Turkey. We sincerely thank all contributors for their valuable insights and efforts in creating this book.

Dinesh Kumar Saini
S.V. Krishna Jagadish

Acknowledgements

The completion of this book would not have been possible without the support and contributions of many individuals and organizations. We sincerely thank the United Sorghum Checkoff Program (USCP) for their generous funding and dedication to sorghum research and education, including their support to Prof. S.V. Krishna Jagadish's program at Texas Tech University, and the senior editor of this book Dr. Dinesh Kumar Saini. USCP plays a crucial role in enhancing the profitability of U.S. sorghum producers by focusing on productivity, demand, and value improvements. We also appreciate the dedication of researchers, industry experts, and collaborators who continue to advance sorghum and other essential crops worldwide. Their efforts benefit both farmers and the agricultural community. Special thanks to the authors and reviewers whose insights and expertise have enriched this book. Their contributions have helped create a valuable resource for researchers, students, and professionals. Finally, we extend our heartfelt appreciation to our colleagues and mentors for their guidance and encouragement throughout this endeavor. Their unwavering support has been instrumental in bringing this book to fruition.

Contributors

Kibrom B. Abreha is a researcher in Plant Breeding at the Swedish University of Agricultural Sciences, Uppsala, Sweden. His research focuses on sorghum resistance to anthracnose, drought, and striga, and he develops genomic tools for stress tolerance and disease resistance in diverse crops, including sorghum, finger millet, and oilseeds.

Preeti Adhana is a Ph.D. research scholar at Chaudhary Charan Singh University, Meerut, India. Her research spans environmental biology and crop improvement, with recent work focusing on advancing sorghum breeding through genetic diversity and modern biotechnological tools. She also has interests in toxicology and wildlife conservation.

Praveenkumar Alagappan is a Ph.D. student at Punjab Agricultural University, India, with research interests in improving quality traits in wheat through wide hybridization, morphological characterization, stress resilience, and genetic enhancement. His work integrates traditional evaluation and molecular tools to identify elite genotypes, contributing to crop improvement and sustainable production systems. He is also a recipient of the Prime Minister Fellowship.

Josef Aranjo is a computer science undergraduate at Kansas State University, USA, specializing in machine learning and DevOps. As a student researcher, he has contributed to agricultural AI by annotating image datasets, evaluating model performance, and supporting tool development. Josef also holds AWS certifications and brings hands-on experience in design, cloud services, and customer relations.

Fatma Azgın is Researcher in the Biology Department at Çukurova University, Turkey. Her interdisciplinary research spans plant fertilization, environmental science, nutrition, plant protection, and bioinformatics. With a strong foundation in microbiology and molecular biology, she focuses on advancing sustainable agricultural practices and plant health through molecular insights.

Aliva Bakshi is a PhD student in computer science at Kansas State University, USA, specializing in computer vision. Her research spans object detection, segmentation, and behavioral video analytics with applications in agriculture, veterinary science, and neuroscience. With prior experience in NLP and formal verification, she brings interdisciplinary expertise to real-world AI solutions.

Gangaiah Bandla, Principal Scientist at ICAR–IIMR, India, is a renowned agronomist with over 25 years of experience in natural resource management and sustainable crop production systems. His pioneering work on conservation practices, mulching technologies, and climate-resilient agriculture has shaped policies and restoration efforts in island and mainland ecosystems alike.

Mayank Pratap Singh Bangari is a Ph.D. student at Texas Tech University, USA, working with the TeCSIS Research Team. His research focuses on dissecting the morphological, anatomical, and biomechanical traits influencing stalk strength under water-limited conditions, with a broader interest in stress physiology, endophyte interactions, and photosynthetic efficiency in plants.

Kirti Bardhan, Assistant Professor at Navsari Agricultural University, India, teaches and researches abiotic stress physiology in crops. A former visiting scientist at Texas Tech University, USA, he focuses on root plasticity and drought resilience, integrating plant physiology with root traits, and has coauthored several impactful studies on stress tolerance.

Md Abdullah Al Bari is an assistant professor at New Mexico State University, USA. He previously worked as a postdoctoral researcher at Kansas State University. His expertise spans plant breeding, quantitative genetics, genomics, and phenomics. He has contributed extensively to genetic dissection and genomic prediction in legumes, cereals, and vegetables to enhance stress resilience and yield-related traits.

Scott R. Bean is a research chemist at USDA-ARS, USA, specializing in sorghum grain composition. His work explores how genetic, environmental, and management factors influence grain quality in sorghum, with a focus on proteins and their functional applications in food, bio-based products, and ethanol. He integrates fundamental chemistry with applied grain science.

Geeta Boken is a researcher at Chaudhary Charan Singh Haryana Agricultural University, Hisar, India. She specializes in plant biotechnology and genome engineering, focusing on genetic improvement for abiotic stress tolerance. Her work includes marker-assisted selection, RNA interference, and molecular breeding for different crops.

Arshvir Kaur Boparai is a DST WISE–KIRAN Fellow at Punjab Agricultural University. Her research focuses on chapati quality, resistant starch in wheat, and evaluating genotypes for yield and quality traits under stress conditions. She is dedicated to advancing agricultural sustainability through field-based performance assessment and trait-based selection.

Doina Caragea is a professor of computer science at Kansas State University, USA, and a leading expert in machine learning and data mining. Her research addresses data-intensive challenges in bioinformatics, security, and recommender systems, with a focus on semi-supervised learning and domain adaptation. She has published over 100 scholarly articles.

Deepti Chaturvedi has a Ph.D. from Ch. Charan Singh University, Meerut, India. Her expertise lies in plant genomics, bioinformatics, and genetic resistance to plant-parasitic nematodes. She has contributed to QTL mapping, GWAS, and database development, advancing molecular breeding and disease resistance strategies in crop improvement programs.

Deepika Cheruku, Scientist at ICAR–IIMR, India, specializes in genetics and plant breeding. Her work focuses on the genetic enhancement of Kodo millet and sorghum improvement. A Ph.D. graduate from IARI, New Delhi, she has contributed to both molecular and applied breeding research, including association mapping and end-use trait selection.

Seva Nayak D., Senior Scientist at ICAR–IIMR, India, specializes in plant stress physiology and millet genetic resource management. With prior contributions to wheat varietal development and genetic stock registration, his research advances pre-breeding and abiotic stress tolerance in millets. He received the "Scientist of the Year" award in 2022.

Bandana Devi is a researcher at Banaras Hindu University, India, with expertise in plant immunity and stress biology. Her work focuses on intergenerational defense priming, microbial interactions, and nanoparticle-mediated strategies to enhance plant resilience. She has coauthored several studies on sustainable plant protection under environmental stress conditions.

Pallavi Dhiman is a Ph.D. student at Texas Tech University, USA, specializing in plant tissue culture, genome editing, molecular biology, and genomics in sorghum. Her research includes gene function analysis, stress response mechanisms, and crop engineering strategies, with a growing record of publications in plant science and biotechnology.

Sadık Dinçer, Head of the Biology Department at Çukurova University, Turkey, conducts cutting-edge research on stress-response networks using integrative omics. He develops innovative strategies for microbial ecology, bacteria-plant interactions, carbon and nitrogen mineralization, microbial fertilizers, sustainable agriculture, and increasing crop durability.

Yves Emendack is a research plant physiologist at USDA-ARS, USA. His expertise lies in agro-morphological and physiological trait characterization to enhance abiotic stress tolerance in crops. Trained in Germany, Dr. Emendack focuses on improving crop resilience through detailed stress-response research in millet, sorghum, and wheat.

Terry Felderhoff is a research assistant professor at Kansas State University, USA, focusing on sorghum genetics and molecular breeding. His research targets stress tolerance traits and trait-marker development. He integrates genomic tools into breeding programs, bridging basic and applied science to accelerate sorghum improvement for both public and commercial use.

Harsimardeep Singh Gill is a postdoctoral research associate at the University of Minnesota, USA. His research focuses on integrating genomics, phenomics, and deep learning to improve disease resistance, grain quality, and stress resilience in cereals. He has contributed extensively to advancing predictive breeding through multi-trait and multi-environment genomic selection.

Ponnaiah Govintharaj is a consultant at the International Crops Research Institute for the Semi-Arid Tropics (ICRISAT), India. His research focuses on breeding high-yielding forage lines, hybrids, and open pollinated varieties with improved forage quality, incorporating traits like photosensitivity, brown midrib (bmr) mutation, and enhanced lodging tolerance.

Karthika Gunasekaran is a postdoctoral research fellow at the International Crops Research Institute for the Semi-Arid Tropics (ICRISAT), India. Her research focuses on stress physiology and genetic improvement in pearl millet and sorghum, particularly salinity, drought tolerance, and disease resistance using marker-assisted breeding, physiological phenotyping, and trait introgression strategies.

Chad Hayes is a research geneticist at USDA-ARS and adjunct professor affiliated with Texas A&M and Kansas State University, USA. His work focuses on developing cold- and drought-tolerant sorghum germplasm and high-yielding hybrids. He has released numerous lines, including R.LBK1 and R.LBK2 with sugarcane aphid resistance.

Vinay Hegde is an assistant professor at M S Ramaiah Institute, Bengaluru. His research explores plant responses to abiotic stresses using phenomics, thermal imaging, and molecular approaches. He works on desiccation stress memory, root phenotyping, and biostimulant applications, contributing to crop resilience and sustainable agriculture.

S.V. Krishna Jagadish is Professor and Thornton Distinguished Chair at Texas Tech University, USA. With global research experience, he focuses on abiotic stress tolerance, soil health, and sustainable agriculture. He has published over 200 peer-reviewed articles and secured major research funding to address climate-resilient crop production.

Krishna Kumar Jangid is an associate researcher at the National Institute of Abiotic Stress Management, India. His research focuses on crop phenotyping and physiology under drought, heat, and waterlogging stress. He leverages trait-based selection and phenomics for enhancing abiotic stress tolerance in different crops.

Rintu Jha is a Ph.D. scholar at the Chinese Academy of Agricultural Sciences, China. His research focuses on improving nutritional quality and stress tolerance in different crops. With a strong publication record, his work integrates association mapping, transcriptomics, and omics approaches to address global food and nutrition challenges.

Uday Chand Jha is a senior scientist at ICAR-IIPR, Kanpur, India. He works on improving resilience to abiotic stresses through breeding and genomics. His research integrates physiological, molecular, and genetic approaches, with a strong publication record focused on enhancing stress tolerance, yield stability, and nutritional traits in diverse plant systems.

Yinping Jiao is an assistant professor at Texas Tech University, USA, specializing in plant genomics and trait genetics. Her research enhances crop adaptation and breeding through genome-scale analyses. She has published in top journals including *Nature*, *Nature Genetics*, and *The Plant Cell*, with landmark contributions in sorghum and maize.

Sirisha Kaniganti is a crop transformation and genome editing specialist at Rothamsted Research, UK. Her work focuses on developing *Striga* resistance in sorghum using CRISPR/Cas9 technology. She has published extensively on genome editing, plant hormone signaling, and tissue culture, advancing biotechnological solutions for crop improvement.

Jatinder Paul Kaur is a research fellow at Punjab Agricultural University, India. Her expertise lies in crop breeding and genetic improvement, with a strong focus on enhancing grain quality and yield. She contributes to the development of resilient and high-performing cultivars through innovative breeding strategies and trait selection.

Pradeep Kumar is a Ph.D. student at Texas Tech University and USDA-ARS, Lubbock, TX, USA, specializing in sorghum breeding and improvement. His research integrates advanced phenotyping and genetic tools to enhance crop architecture and stress resilience. He has contributed to studies spanning trait mapping and breeding innovations in cereals.

Pradeep Kumar is a scientific officer in the Pearl Millet Breeding program at ICRISAT, India. He specializes in restorer inbred line development and holds an M.Phil. in genetics and plant breeding. A DST Inspire Fellow and university gold medalist, he has authored a book and multiple book chapters.

Ritesh Kumar is a postdoctoral fellow at the Boyce Thompson Institute, USA, with expertise in plant molecular biology and stress physiology. He previously served as a postdoctoral researcher at Kansas State University, USA. His research focuses on heat stress tolerance, root architecture, and reproductive resilience, bridging molecular insights with practical crop improvement strategies.

Zhiyuan Liu is a Ph.D. student at Texas Tech University, USA, specializing in plant molecular genetics and data analysis. With prior teaching experience in laboratory-based plant science, his research integrates genomics, physiology, and cross-disciplinary insights to explore genetic mechanisms underlying plant development and stress responses.

Mukesh Kumar Mehla is a postdoctoral research associate at Texas A&M University, USA, and previously held the same position at Texas Tech University. His work focuses on irrigation management, crop-hydrological modeling, and agricultural sustainability, with expertise in soil and water conservation, remote sensing, and GIS.

Anuradha Narala, Scientist at ICAR–IIMR, India, specializes in agricultural economics with expertise in impact analysis, price forecasting, and gender studies in agriculture. With prior experience at CICR, Nagpur, she has developed forecasting models and conducted socioeconomic assessments to support informed decision-making in cotton and millet systems.

Priyanka Negi is a Ph.D. scholar at the National Institute of Abiotic Stress Management, India. Her research focuses on genetic improvement and agronomic optimization of different crops. She also contributes to sustainable farming innovations, including urban agriculture and nature-based techniques, enhancing resilience and productivity in climate-smart agriculture.

Senthilkumar Palanisamy is an associate professor in the Department of Genetic Engineering at SRM Institute of Science and Technology, India. His research covers genetics, biotechnology, and bioinformatics, with a focus on plant defense responses and molecular mechanisms in crops. He actively contributes to advancements in crop improvement and stress biology.

Gunvant B. Patil is an assistant professor at Texas Tech University, USA, with expertise in genetic engineering, translational genomics, and quantitative genetics. His research focuses on developing genotype-independent transformation systems and targeted genome editing approaches to accelerate trait discovery and crop improvement under abiotic stress conditions.

Aliza Pradhan is a scientist at the National Institute of Abiotic Stress Management, India. Her research focuses on sustainable land management, nature-based solutions, and resource-efficient farming. She integrates agronomic, ecological, and phenomic strategies to improve soil health, carbon sequestration, and system resilience under changing climatic conditions.

Swaraj Pramanik is a Ph.D. student in computer science at Kansas State University, USA, with a research focus on computer vision, deep learning, few-shot learning, and transfer learning. His work contributes to cutting-edge AI applications in agriculture, including automated phenotyping, flower and pod counting, and deep learning tools for crop trait assessment.

Saksham Pundir has a Ph.D. from botany at Ch. Charan Singh University, Meerut, India. Her research focuses on molecular genetics, QTL mapping, and plant–pathogen interactions. She also works on bioinformatics and epigenetics, contributing to studies on trait improvement and resistance mechanisms using advanced genetic tools and statistical approaches.

Jagadish Rane is a crop physiologist and director at ICAR-CIAH, India. He specializes in abiotic stress physiology, phenomics, and climate-resilient agriculture. With an exceptional publication record across high-impact journals, his work advances precision phenotyping, plant-environment interactions, and sustainable strategies for improving crop productivity under challenging conditions.

Asad Riaz is a Ph.D. student at The University of Queensland, Australia. His research explores molecular and physiological responses to abiotic and biotic stresses, with expertise in plant secondary metabolites, nutrient-use efficiency, and stress signaling. He has published widely in plant genomics, transcriptomics, and genome editing applications.

Swarna Ronanki, Scientist at ICAR–IIMR, India, holds a Ph.D. in agronomy from IARI, New Delhi. Her research focuses on conservation agriculture, nutrient management, and improving resource-use efficiency. She has significantly contributed to sustainable production systems and rabi sorghum improvement through both field research and simulation modeling.

Srividhya S., Scientist at ICAR–IIMR, India, specializes in plant stress physiology, postharvest biology, and crop phenotyping. With academic excellence recognized by multiple gold medals from TNAU, she integrates her deep understanding of plant responses to abiotic stress to advance physiological research and improve crop productivity and resilience.

Mumtaz Ali Saand is the Chairman of the Department of Botany at Shah Abdul Latif University, Pakistan. His research integrates bioinformatics, transcriptomics, proteomics, and multi-omics approaches to investigate various biotic and abiotic stresses in crops. He has coauthored several highly cited studies across different crop species.

Dinesh Kumar Saini is a research associate at Texas Tech University, USA, specializing in crop breeding and genomics. His research focuses on physiological, molecular, and histological mechanisms enhancing climate resilience in sorghum. He has authored over 55 articles and serves as a reviewer and editorial board member for scientific journals.

Muhammad Sajjad is a tenured associate professor at COMSATS University Islamabad and a member of NCGE. He specializes in cereal genetics, genomics, and genome editing, with over 90 SCI publications. His research enhances yield, stress tolerance, and quality traits, supported by national and international fellowships and leadership roles.

Himanshu Sharma, Research Associate at Punjab Agricultural University, specializes in postharvest physiology, nutritional biochemistry, and plant secondary metabolism. His interdisciplinary research spans from food safety and pectin extraction to molecular insights into anthocyanin biosynthesis, biofortification, and stress mitigation strategies, contributing to innovations in plant health, food quality, and nutritional security.

Akanksha Singh is a researcher at CSJM Kanpur University, India, with expertise in microbiology. Her work focuses on host pathogen interactions and genomic strategies to enhance plant tolerance to disease. She has coauthored several studies on sustainable plant protection under environmental stress conditions.

Gurjeet Singh is a postdoctoral research associate at Texas A&M University, USA. He focuses on integrating advanced and conventional breeding strategies for multi-trait improvement. He earned his Ph.D. from Punjab Agricultural University, India, and has authored several peer-reviewed publications in the field of plant breeding and genetics.

Prashant Singh is an assistant professor at Banaras Hindu University, India. His research spans plant–pathogen interactions, innate immunity, and transgenerational defense priming. He also contributes to genomics and pan-genomics research, applying next-generation sequencing to uncover genetic diversity and resilience traits in crop improvement.

Vikas Kumar Singh has a Ph.D. from Ch. Charan Singh University, Meerut, India, and a DST-INSPIRE fellow. His research focuses on plant–pathogen interactions, bioinformatics, genomics, transcriptomics, and epigenomics. He also works on root phenotyping and nutrient use efficiency, contributing to advancements in molecular breeding and data-driven crop improvement strategies.

Impa Somayanda is a research assistant professor at Texas Tech University, USA. A crop physiologist by training, her research focuses on abiotic stress responses, grain quality, and micronutrient dynamics across cereals, including sorghum. She has authored more than 40 peer-reviewed articles and previously held positions at Kansas State University and IRRI, Philippines.

Javaria Tabassum is an assistant professor at the University of the Punjab, Lahore, Pakistan. She specializes in crop genetics, genome editing, and crop improvement using genome editing. She has published extensively in high-impact journals and authored several book chapters, with expertise in molecular biology, bioinformatics, and plant biotechnology.

Mohammad Jafar Tanin is a visiting scholar in the Division of Plant Science and Technology at the University of Missouri, USA. His research focuses on improving biotic and abiotic stress tolerance and grain quality in crops through QTL mapping, transcriptomics, and meta-analyses. He has coauthored impactful publications in leading plant science journals. He is also a recipient of the IIE-SRF fellowship.

Niranjan Ravindra Thakur is a Ph.D. research scholar at ICRISAT, specializing in genetics and plant breeding. He has expertise in GWAS, biofortification, and seed quality analysis across diverse crops. With multiple publications and awards, he is committed to advancing agricultural science through innovative breeding and genomic tools.

Shivani Upadhyay is a Ph.D. student at Punjab Agricultural University, India, supported by the Prime Minister's Fellowship. Her research focuses on crop breeding and improvement, specifically the development of climate-resilient wheat germplasm with enhanced tolerance to biotic and abiotic stresses. She aims to improve agronomic and quality traits through innovative breeding strategies to support food security and agricultural resilience.

Chaoxin Wang is an assistant professor of computer science at Dominican University in Illinois, USA, specializing in bioinformatics, computer vision, and database systems. His work integrates deep learning with plant phenotyping, including sorghum stomatal trait analysis and root anatomical imaging. He also contributes to therapeutic target databases, advancing both agricultural and biomedical data science.

Ali Yetgin is an R&D specialist at Toros Agri Industry and Trade Co. Inc., Turkey, with over a decade of experience in molecular biology, biotechnology, and microbiology. He leads project on microbial fertilizers and soil microbiome interactions and work national and international projects focused on biostimulants and sustainable nutrient management.

Abbreviations

2DE	Two-Dimensional Gel Electrophoresis
ABA	Abscisic Acid
AFLP	Amplified Fragment Length Polymorphism
AFM	Arbuscular Mycorrhizal Fungi
AI	Artificial Intelligence
APSIM	Agricultural Production Systems Simulator
ARF	Auxin Response Factor
AWC	Available Water Capacity
BAP	Bioenergy Association Panel
BBM	Baby Boom
BC-NAM	Backcross Nested Association Mapping Population
BME	Bayesian Multi-Environment
BP	Breeding Population
BPH	Better Parent Heterosis
BRR	Bayesian Ridge Regression
CHA	Chemical Hybridizing Agent
CIM	Composite Interval Mapping
CMS	Cytoplasmic Male Sterility
CNN	Convolutional Neural Network
CNV	Copy Number Variation
COG	Cost-Of-Goods
CRISPR/Cas9	Clustered Regularly Interspaced Palindromic Repeats//CRISPR-Associated Protein 9
CSM	Crop Simulation Model
CV	Cross-Validation
CWR	Crop Wild Relative
DArT	Diversity Array Technology
ddRAD	Double Digest Restriction-Site Associated DNA
DGE	Differential Gene Expression and Digital Gene Expression
DH	Doubled Haploid
DNA	Deoxyribonucleic Acid
DREB	Dehydration-Responsive Element-Binding
DSSAT	Decision Support System for Agrotechnology Transfer
EMS	Ethyl Methanesulfonate
EST	Expressed Sequence Tag
FAO	Food And Agriculture Organization
FAOSTAT	Food And Agriculture Organization Corporate Statistical Database
FISH	Fluorescence In Situ Hybridization
FPN	Feature Pyramid Network
FTIR	Fourier Transform Infrared Spectroscopy
GA	Gibberellic Acid
GAB	Genomics-Assisted Breeding
GAN	Generative Adversarial Network
GBLUP	Genomic Best Linear Unbiased Prediction
GBS	Genotyping-By-Sequencing
GCA	General Combining Ability
GCM	General Circulation Model

GC-MS	Gas Chromatography-Mass Spectrometry
GD	Genetic Distance
GDD	Growing Degree Day
GEBV	Genomic Estimated Breeding Value
GEI	Genotype-Environment Interaction
GELAN	Generalized Efficient Layer Aggregation Network
GM	Genetically Modified
GO	Gene Ontology
GP	Gene Pool
GRF	Growth Regulatory Factors
GRIN	Germplasm Resources Information Network
GS	Genomic Selection
GWAS	Genome Wide Association Study
HDR	Homology-Directed Repair
HSP	Heat Shock Protein
HTP	High Throughput Phenotyping
HTPP	High-Throughput Plant Phenotyping
HTS	High-Throughput Sequencing
ICIM	Inclusive Composite Interval Mapping
ICP	Inductively Coupled Plasma
ICRISAT	International Crops Research Institute for The Semi-Arid Tropics
InDel	Insertion Deletion
IoU	Intersection-Over-Union
IPCC	Intergovernmental Panel on Climate Change
IPOP	Integrative Plant Multi-Omics Platform
IWUE	Irrigation Water Use Efficiency
LAI	Leaf Area Index
LC-MS	Liquid Chromatography-Mass Spectrometry
LD	Linkage Disequilibrium
LiDAR	Light Detection and Ranging
MAE	Mean Absolute Error
MAGIC	Multi-parental Advanced Generation Inter-Cross
MALDI-TOF	Matrix-Assisted Laser Desorption Ionization-Time of Flight
MAPE	Mean Absolute Percentage Error
MAS	Marker Assisted Selection
MIM	Multiple Interval Mapping
ML	Machine Learning
MMEJ	Microhomology-Mediated End Joining
MMT	Million Metric Tons
MPH	Mid-Parent Heterosis
MS	Mass Spectrometry
MSI	Mass Spectrometry Imaging
MTA	Marker-Trait Association
NAM	Nested Association Mapping
NBPGR	National Bureau of Plant Genetic Resources
NGS	Next Generation Sequencing
NHEJ	Non-Homologous End Joining
NMR	Nuclear Magnetic Resonance
NMS	Nuclear Male Sterility
NPGS	National Plant Germplasm System

NUE	Nitrogen Use Efficiency
ONT	Oxford Nanopore Technologies
PAV	Presence Absence Variation
PCR	Polymerase Chain Reaction
PEG	Polyethylene Glycol
PGI	Programmable Gradient Information
PHG	Practical Haplotype Graph
PUE	Phosphorus Use Efficiency
QTL	Quantitative Trait Loci
RADSeq	Restriction Site-Associated DNA Sequencing
RAPD	Random Amplified Polymorphic DNA
RCP	Representative Concentration Pathway
RFLP	Restriction Fragment Length Polymorphism
RGB	Red Green Blue
RIL	Recombinant Inbred Line
RMSE	Root Mean Squared Error
RNA	Ribonucleic Acid
RNA-Seq	RNA Sequencing
ROI	Regions Of Interest
ROS	Reactive Oxygen Species
RPN	Region Proposal Network
RSA	Root System Architecture
SA	Salicylic Acid
SAGE	Serial Analysis of Gene Expression
SAP	Sorghum Association Panel
SCA	Specific Combining Ability and Stomatal Complex Area
SD	Stomatal Density
SDI	Subsurface Drip Irrigation
SG	Stay Green
SIM	Simple Interval Mapping
SL	Strigolactone
SNP	Single Nucleotide Polymorphism
SSD	Single Seed Descent
SSR	Simple Sequence Repeat
SV	Structural Variation
T2T	Telomere-To-Telomere
TALEN	Transcription Activator-Like Effector Nuclease
TAPIS	Transcriptome Analysis Pipeline for Isoform Sequencing
TBV	True Breeding Value
TE	Transposable Element
TFSMA	Trifluoromethanesulfonamide
TGS	Third Generation Sequencing
TILLING	Targeting Induced Local Lesions in Genomes
TP	Training Population
UAS	Uncrewed Aircraft System
UAV	Unmanned Aerial Vehicle
USDA	United States Department of Agriculture
USDA-FAS	USDA Foreign Agricultural Service

WGCNA	Weighted Gene Co-Expression Network Analysis
WUE	Water Use Efficiency
XRF	X-Ray Fluorescence
YOLO	You Only Look Once
ZFN	Zinc Finger Nuclease

1 Global Importance, Production Trends, and the Need for Designing the Sorghum Genome to Address Future Climatic Conditions

Mukesh Kumar Mehla[#], Senthilkumar Palanisamy[#], Dinesh Kumar Saini, and S.V. Krishna Jagadish

1.1 INTRODUCTION

Sorghum is the fifth most important cereal crop globally and is widely regarded as a climate-smart crop. Its versatility makes it a crucial component of human nutrition, animal feed, biofuel production, and livestock systems, collectively enhancing agricultural economics and rural food security (Tonapi et al., 2020). Thriving in hot, dry environments, sorghum demonstrates remarkable adaptability to various biotic and abiotic stresses (Hao et al., 2021; Huang, 2018; Zhang et al., 2019). Despite its global presence, only ten countries account for 80% of the cropped area, underscoring its concentrated cultivation (Dalton & Hodjo, 2020; Figure 1.1). Sorghum plays a vital role for smallholder producers, especially in semi-arid regions of Africa, where its economic impact is substantial. Cultivated across over 42 million hectares in approximately 66 countries, sorghum supports local economies and creates employment opportunities (Tonapi et al., 2020). The United States, India, and several African countries dominate global sorghum trade and market dynamics, strengthening its importance in global food systems (Tonapi et al., 2020; Figure 1.1)

From a nutritional standpoint, sorghum is rich in protein, fiber, and essential minerals like iron and phosphorus. Its gluten-free nature and high antioxidant content makes it a preferred choice for individuals with dietary restrictions and health concerns (Visarada & Aruna, 2019). Promoting sorghum production and consumption is critical for addressing the growing global demand for nutritious food, driven by population growth and threats to food security from climate change. To strengthen the crop's resilience, measures such as optimized fertilizer use, efficient irrigation practices, and the development of climate-resilient varieties are essential (Khalifa & Eltahir, 2023). As agriculture faces increasing challenges from climate change, sorghum emerges as a resilient crop due to its tolerance to elevated temperatures, water scarcity, and even excessive rainfall (Visarada & Aruna, 2019). Its deep root systems confer drought resilience, while adaptive root responses help mitigate waterlogging. The crop's low input requirements, such as minimal reliance on fertilizers and pesticides, further highlight its sustainability.

[#] These authors contributed equally to this work.

DOI: 10.1201/9781003623717-1

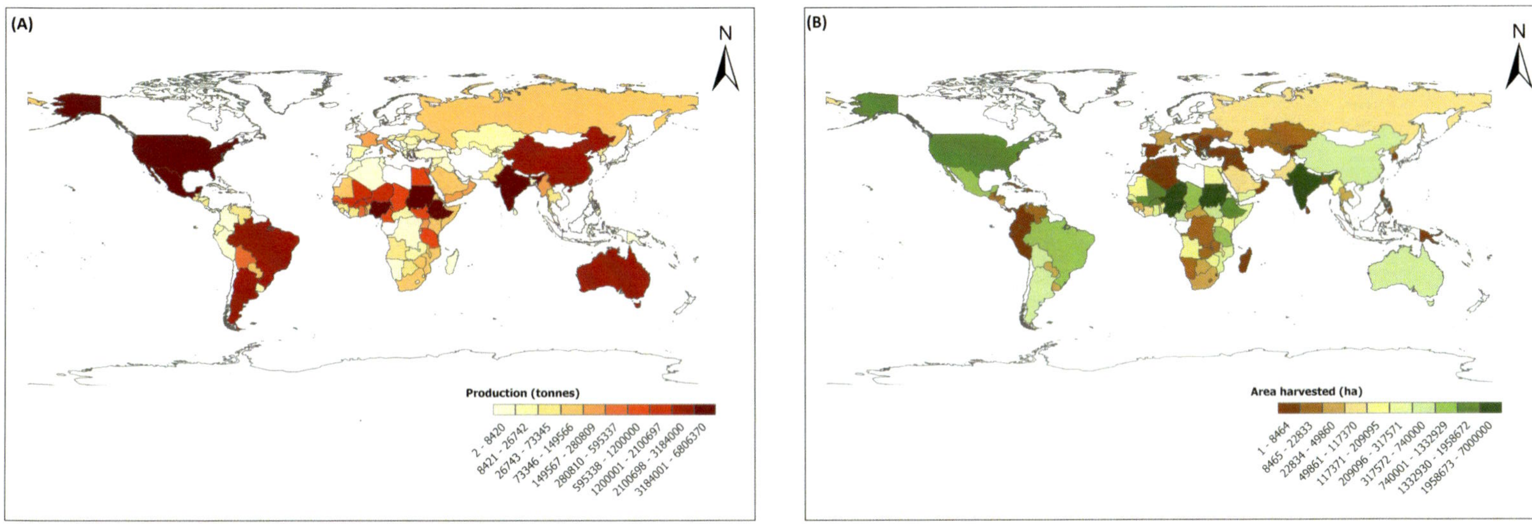

FIGURE 1.1 Geographical distribution of (a) global sorghum production and (b) harvested areas (Source: FAOSTAT, 2023). The map highlights major sorghum-producing regions worldwide, illustrating the spatial variation in cultivation intensity and harvested areas.

While sorghum is well-adapted to harsh climatic conditions, it still faces significant challenges that impact its productivity. Prolonged drought significantly impacts sorghum growth, often leading to premature plant death, reduced seed size, and lower grain yields. Severe drought during flowering and grain-filling stages can result in complete yield loss (Assefa et al., 2010; Derese et al., 2018). For instance, in Eritrea, drought alone can cause total yield loss, while biotic stress accounts for up to 60% of additional losses (Abraha et al., 2015). Water scarcity disrupts panicle development and seed viability, while also negatively affecting morphological traits and metabolic processes (Bibi et al., 2012; Ndlovu et al., 2021).

Excess rainfall, leading to waterlogging, depletes soil oxygen, stunts plant growth, and reduces nutrient absorption during key developmental stages (Zhang et al., 2023). Waterlogged conditions during the vegetative and reproductive phases can reduce shoot and root biomass by 10–15% (Sharma et al., 2019). Studies in Brazil and Pakistan report yield losses of up to 70–80% under waterlogged conditions compared to non-stressed plants (Queiroz et al., 2019). Sorghum can tolerate a wide temperature range, from 7°C to 38°C, but extreme heat or cold can severely disrupt its productivity (Butchee et al., 2012). High temperatures during flowering cause pollen abortion, reduced stigma receptivity, and poor grain filling, leading to yield losses exceeding 60% when daily temperatures exceed 40°C (Jagadish, 2020; Prasad et al., 2015). Low temperatures delay flowering and cause pollen sterility, further compounding losses (Emendack et al., 2021).

Strong winds impact sorghum by accelerating evapotranspiration and causing physical damage to crops, particularly through sand abrasion (Ondiko & Recha, 2022). While moderate wind aids pollen dispersion, excessive wind can result in pollen loss, reducing grain set and yields. Humidity also plays a dual role; low humidity exacerbates water loss and limits photosynthesis, while elevated humidity fosters fungal growth, further reducing yields (Sharma et al., 2019). Sorghum's global average yield is approximately 2.5 tonnes per hectare, with significant regional disparities. For example, the United States achieves average yields of 4.1 tonnes per hectare, whereas sub-Saharan Africa and South Asia often experience yields as low as 0.9 tonnes per hectare due to harsh climatic and resource limitations (Khalifa & Eltahir, 2023). Addressing its production challenges through innovative breeding, improved agronomic practices, and enhanced resource management remains essential to maximize its contribution to global food systems.

Advances in genomics provide an unprecedented opportunity to develop climate-resilient sorghum varieties. The availability of high-quality reference genomes, next-generation sequencing technologies, trait mapping, and genome-wide association studies, enables researchers to identify key genes and genetic pathways involved in stress tolerance and adaptation (Paterson, 2008; Bailey-Serres et al., 2019). Additionally, tools such as marker-assisted selection (MAS) and genomic selection (GS) accelerate the breeding process, making it possible to develop cultivars with improved traits and enhanced nutrient use efficiency (Hao et al., 2021; Somegowda et al., 2023). Multi-omics and gene-editing technologies, such as CRISPR/Cas9, further enhance the potential of sorghum genome design. These approaches enable precise modifications in gene expression, allowing researchers to fine-tune traits that contribute to yield stability under extreme climatic conditions (Parikh et al., 2021; Mukherjee et al., 2024). By integrating traditional breeding with cutting-edge genomic tools, the development of sorghum varieties suited for future climates can be significantly expedited.

This chapter deals with the global importance of sorghum, its challenges in the face of climate change, and its adaptability as a climate-smart crop. Sorghum's role in food security, its nutritional value, and economic significance are covered. The chapter also highlights the impact of different stresses on sorghum production and the need for sustainable practices. The role of genomic resources in developing climate-resilient sorghum varieties is also discussed, along with future research directions for improving sorghum production. A summary of the topics covered in the following chapters of the book is also provided.

1.2 GLOBAL IMPORTANCE OF SORGHUM

1.2.1 Historical and Current Trends in Global Sorghum Production

Sorghum ranks as the fifth most produced grain globally, after maize, barley, rice, and wheat (Figure 1.2a). Sorghum has a long history, originating in Africa about 3000 BC before being exported to India and the Americas (Joshi et al., 2021). It is a significant food crop in over 66 countries, with the U.S. being the largest producer followed by Nigeria, Sudan, Mexico, and India (Figure 1.2b). The resilience and nutritional value of sorghum have made it a highly favored crop for cultivation, serving as an important alternative in the face of climate change due to its ability to thrive in arid climates (Tonapi et al., 2020). The U.S. primarily cultivates sorghum for livestock feed and biofuel; however, it is rapidly gaining popularity as a food grain for dishes like hot cereals, baby foods, and grain bowls. The sugar-laden canes of sweet sorghum varieties serve as the main component for conversion into sorghum syrup, a traditional Southern American sweetener (Visarada & Aruna, 2019). Sorghum is a preferred choice for gluten-free diets among 500 million or more people who rely on it (Rather et al., 2023). This versatile grain may be prepared in several ways and is rich in protein, iron, and other essential nutrients. Sorghum has many health benefits, including its capacity to diminish inflammation and cancer risk, facilitate weight loss, and is safe for those with celiac disease.

Sorghum serves not only as a food source for people but also as livestock feed and a constituent in biofuels. Sorghum is also a green feed and biofuel crop due to its drought endurance and ability to flourish in different soil and climatic conditions. Cattle and poultry industries primarily use sorghum for its nutritional benefits, which are comparable to those of wheat and with a higher protein content than maize (Visarada & Aruna, 2019). A hard kernel enhances resistance to damage and disease, although it makes digestion for animals difficult. In the U.S., around 46% of the sorghum harvested is used for animal feed. Sorghum is an important biofuel crop that can be used for producing ethanol with one metric ton of sorghum generating 402 liters of ethanol (Sorghum, 2011).

Global aggregate production statistics for sorghum crop during the period of 1961–2022 are presented in Figure 1.3. Sorghum production has been in a rising trend after dropping in the early 1990s (Dalton & Hodjo, 2020). Sorghum production has grown substantially in sub-Saharan Africa but reduced in Asia and the Americas, particularly in India and the U.S. (Prasad et al., 2021).

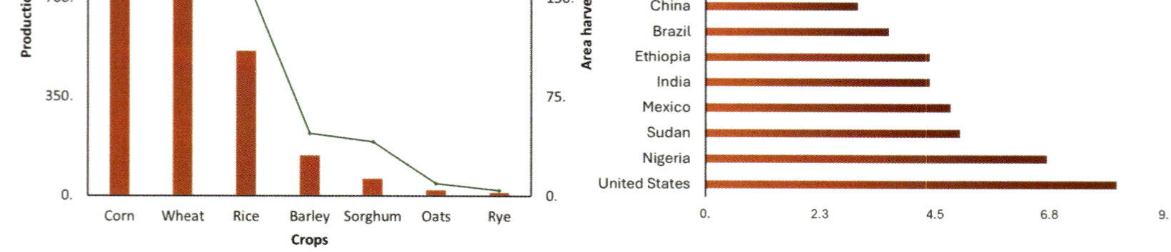

FIGURE 1.2 (a) Global production of major cereal crops, highlighting their agricultural significance (Source: FAO and USDA, 2024). (b) Top sorghum-producing countries, emphasizing their role in global sorghum output (Source: FAO and USDA, 2024).

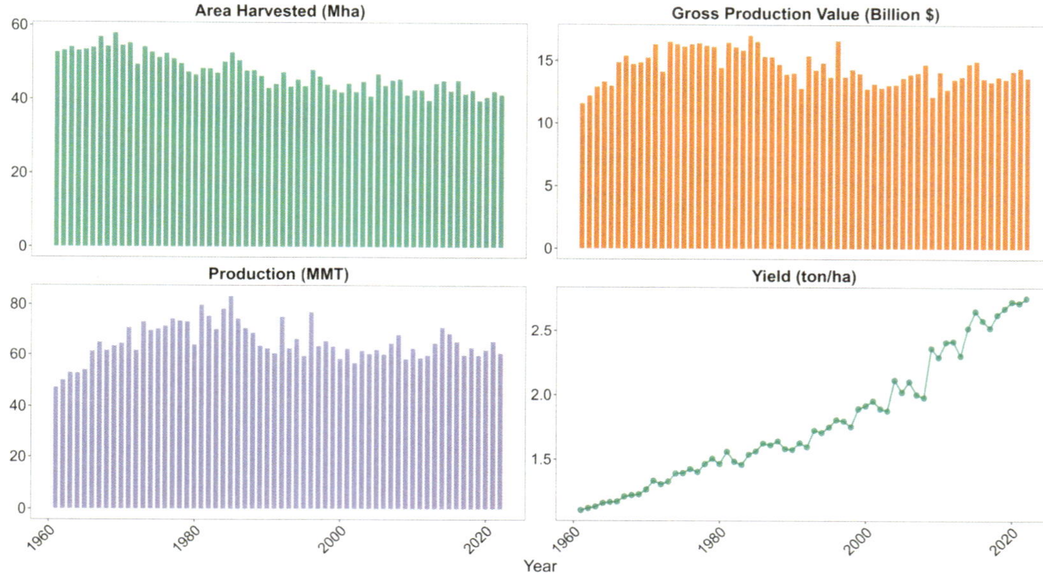

FIGURE 1.3 Global trends in sorghum production from 1961 to 2022, including area harvested (Mha), production (MMT), gross production value (billion $), and yield (ton/ha) (Source: FAOSTAT, 2023).

There has been a significant decline in sorghum yield in Sudan (55%) and considerable decrease in harvested areas (77%) in India during the period of 1961–2020 (Khalifa & Eltahir, 2023).

This has split the sorghum usage and production scenario, with some nations producing extra for the global market and others using it as a staple food crop. Sorghum has a comparatively lower average yield (2.5 tonnes/ha) worldwide than other major cereal crops, and its genetic yield gain has been much slower than that of other principal cereal crops (Khalifa & Eltahir, 2023). Although this fact may first seem alarming, it also underscores the possibility for substantial improvements through targeted breeding programs (Baloch et al., 2023; Leon et al., 2024; Mayor et al., 2023). The advancement of sorghum farming has been very slow, with average yield advances of just 0.01 tons per hectare annually in Argentina and the U.S. (Pfeiffer et al., 2019). Conversely, other major crops such maize and soybeans have substantially profited from research and development efforts, resulting in notable advancements. Yield of sorghum and other staple crops may decline in the future owing to intensified competition for limited agricultural resources resulting from the prevailing trend in biofuel production (Damerau et al., 2016). Therefore, we should adopt better resource management approaches and streamline research and development efforts with focus on climate resilient crops like sorghum, to ensure food and life security under the threat of climate change.

Sorghum is a crucial crop for smallholder producers in developing countries and its economic impact is substantial in such regions. On a global scale, sorghum exports are being controlled by the U.S., Argentina and Australia (Figure 1.4a). Breaking the trade barriers that limit sorghum grains from being supplied by high-production to high-demand countries is essential for sustained sorghum markets and international trade.

In the 2023–2024 cropping season, U.S. sorghum production reached 8 million metric tons (MMT), followed by Nigeria (6.7 MMT), Sudan (5.2 MMT), India (4.8 MMT), and Niger (3.5 MMT). The importance of the crop in several geographical regions is shown by Figure 1.4b. Sorghum production also strengthens local economies and provides employment opportunities. In 2022, global exports were valued at over $2.24 billion, with the U.S. leading, followed by Australia ($737 million), Argentina ($469 million), France ($52.6 million), and Ukraine ($17.2 million).

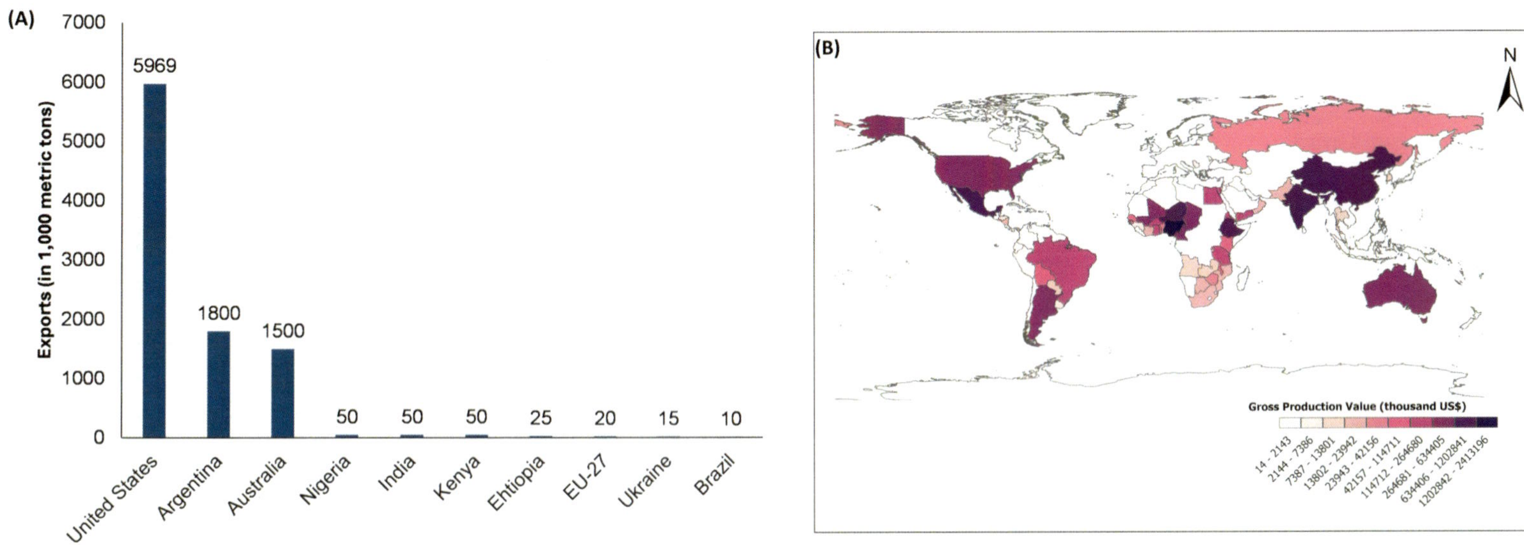

FIGURE 1.4 (a) Global sorghum exports from leading exporting countries, highlighting their contribution to international trade (Source: IndexMundi, 2024). (b) Gross production value of sorghum worldwide, reflecting its economic significance (Source: FAOSTAT, 2023).

These numbers indicate that sorghum is significant not just for human/animal sustenance but also for economic stability and the facilitation of trade.

1.2.2 Sorghum Production Dynamics and Regional Shifts

Over the past few decades sorghum consumption and production have largely remained stable, despite continental and regional shifts (Dalton & Hodjo, 2020). For millennia, people have harvested sorghum, a multifaceted grain known for its longevity and storage capabilities. A significant portion of grain consumption in the U.S. is attributed to animal feed, whereas a smaller but significant fraction is used for ethanol generation. Ethanol production, as a largely renewable energy source in conjunction with petrol, has been on rise since 2000 (Graham-Rowe, 2011). Ethanol production utilizes 30–35% of domestic sorghum crop in the U.S. Bio-butanol, a renewable fuel derived from sweet sorghum bagasse, is found to be compatible with existing pipelines and exhibits reduced engine corrosion, making it quite popular (Jeya et al., 2022). This shift to biofuels has affected production and usage of sorghum grain in the U.S. significantly. Concerns over meat quality and animal welfare have initiated debates on the ethical implications and health consequences of feeding grains to herbivorous animals (Rosegrant & Msangi, 2014). Despite these reservations, sorghum's adoption and production have steadily increased, highlighting its contribution to both food and energy production systems. However, this growth is not without its challenges or side effects. The increased sorghum production for biofuel production is exerting pressure on food production due to limited resources such as water and land (Damerau et al., 2016; Graham-Rowe, 2011). Some research indicates that the shift to biofuels significantly contributed to the substantial rise in food prices (Graham-Rowe, 2011). Individuals already experiencing financial difficulties in developing countries are likely to be most vulnerable to this surge in food prices (Khalifa & Eltahir, 2023; Rosegrant & Msangi, 2014).

South Asia and sub-Saharan Africa, mostly low-income regions, have persistently low sorghum yields unlike developed countries that show higher average yields. Sorghum is mostly grown for human use in these areas, unlike its prevalent application as animal feed and biofuel in developed nations (Tonapi et al., 2020). Nigeria, Sudan, and Ethiopia would gain significantly from a 29.5% rise in sorghum production if the yield gap were eliminated to achieve the highest local yields (1.3 t/ha) in 31 sub-Saharan African countries (Khalifa & Eltahir, 2023). The poor agricultural yields in sub-Saharan Africa, particularly for sorghum, pose a challenge to food security while also presenting an opportunity for enhanced resource management (Figure 1.5).

Sorghum encounters several biotic and abiotic stresses during its crop cycle, including pests, diseases, heat stress, and soil moisture deficits. The food security of regions reliant on sorghum as a primary crop is jeopardized due to these pressures, which diminish productivity and heighten the risk of crop failure. Despite the changes in the agricultural landscape caused by climate change, there is optimism for the sustainability of sorghum cultivation via enhanced agronomic practices, advanced technology, and efficient resource management.

1.2.3 Nutritional Value and Uses of Sorghum

Sorghum is a versatile grain used for food, animal feed, and bioenergy. Its adaptability and nutrient-rich profile, including B vitamins, magnesium, iron, and antioxidants, make it essential in many regions worldwide. These elements contribute to various health benefits, including supporting metabolic processes, enhancing energy production, and reducing oxidative stress (Tanwar et al., 2023). Sorghum is increasingly valued for its diverse applications, driven by its nutritional benefits, versatility, and sustainability. In human diets, it can be prepared as whole grain, similar to rice or quinoa, for salads and pilafs, while milled sorghum flour serves as a

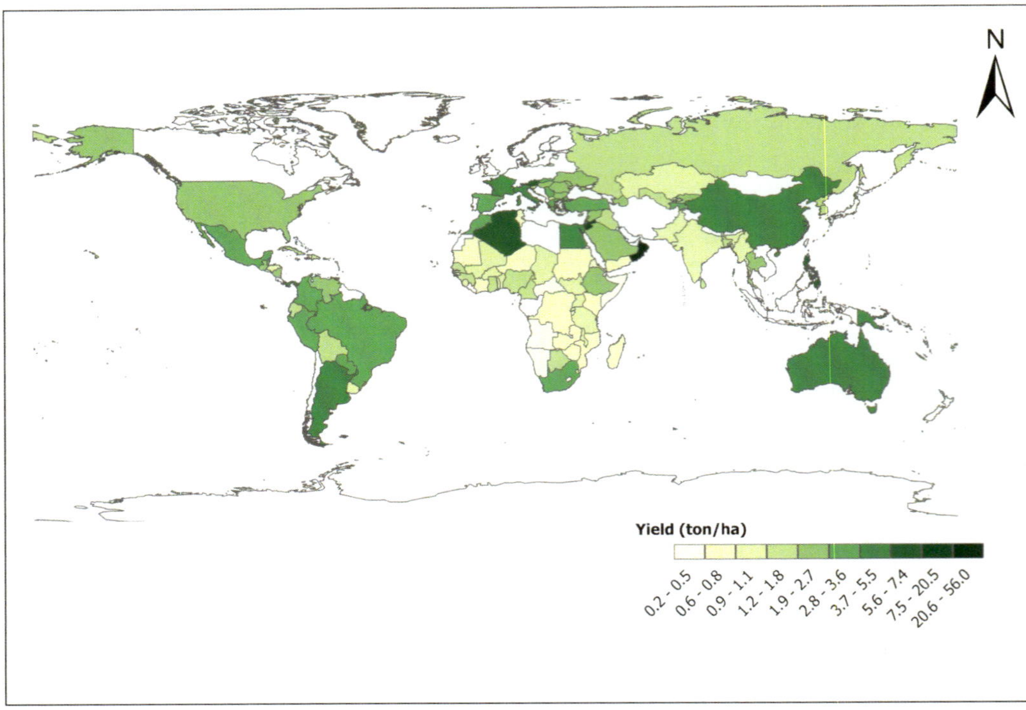

FIGURE 1.5 Global variation in average sorghum yield in 2022, highlighting regional differences in productivity (Source: FAOSTAT, 2023).

gluten-free alternative to wheat in baking (Tanwar et al., 2023). Sorghum syrup, a natural sweetener, is used in processed foods and as a substitute for molasses in recipes. This adaptability has led to its incorporation into over 350 product lines in the U.S. consumer food industry (www.sorghumcheckoff.com/sorghum-101/).

Beyond its dietary uses, sorghum is a primary feed grain for livestock, including beef, dairy, poultry, and swine industries (McCuistion et al., 2019). Its grain, stalks, and leaves are integral to various feed formulations, especially in drought-prone regions where maize is less viable. Sorghum's comparable nutritional profile and lower cost, combined with improved digestibility through fine grinding and the advantages of low-tannin varieties, make it a valuable feed crop (Cruz et al., 2024). Sorghum has also emerged as a sustainable option for bioenergy production, with its stalks and leaves processed into biofuels like ethanol and bio-butanol (Rao et al., 2019). Its drought resistance and ability to grow on marginal lands minimize competition with food crops, positioning it as an ideal biofuel source. Bioenergy sorghum hybrids are being studied for their potential to enhance soil health, carbon sequestration, and nitrogen use efficiency, making them environmentally sustainable options for renewable energy systems (Dou et al., 2014; Shahandeh et al., 2016; Maw et al., 2017).

In addition, sorghum supports various industrial applications. Its fibers contribute to building materials and agricultural tools, while grains and byproducts are used in biodegradable plastics and eco-friendly materials (Qi et al., 2019). As traditional food markets for sorghum decline, innovative uses like pet food production, where its high fiber content and low glycemic index are appealing, are gaining traction (Aruna & Visarada, 2019). Overall, sorghum's nutritional value, adaptability, and diverse applications make it an increasingly important crop for global food security and sustainable agriculture.

1.2.4 Economic, Cultural, and Policy Issues in Sorghum Production

Sorghum production faces a myriad of economic, cultural, and policy-related issues that significantly influence its cultivation, market dynamics, and societal roles across different regions. Economic factors play a crucial role in shaping sorghum production. Key issues include market prices, production costs, subsidies, and trade policies. The cost of production for sorghum often includes expenses related to inputs such as seeds, fertilizers, labor, and water management. Fluctuating market prices can impact decisions of producers regarding cultivation or sale, as producers aim to maximize profitability while minimizing risks related to price volatility (Mundia et al., 2019). Additionally, production subsidies provided by governments can significantly affect the economic viability of sorghum farming by lower production costs and encouraging higher outputs. However, these subsidies can also distort market prices, leading to challenges in international competitiveness for producers in regions reliant on subsidies (Hansen, 2024).

Culturally, sorghum holds substantial significance in many regions, especially in Africa and parts of Asia (Westengen et al., 2014; Erramouspe, 2021). It is deeply embedded in traditional diets and plays a critical role in local customs and social practices. Sorghum is often used to prepare staple foods and beverages, such as porridge and beer, highlighting its importance in daily life and cultural celebrations (Erramouspe, 2021; Sorghum checkoff, 2022). In African communities, sorghum has historically played a vital role in sustenance, serving as a "safety net" for households facing food insecurity (Erramouspe, 2021). Additionally, in some regions, sorghum is regarded as a "cultural keystone species" due to its deep integration into livelihood practices, traditional cuisine, medicine, and the fulfillment of material needs (Garibaldi & Turner, 2004). These deep cultural ties to sorghum varieties contribute to resistance against adopting newer, potentially higher-yielding hybrids, as many communities prioritize traditional crops for both their nutritional benefits and their cultural importance (Wendmu et al., 2022).

Government policies significantly influence sorghum production, encompassing aspects like research funding, regulations on crop management, and trade agreements. Programs such as the United Sorghum Checkoff aim to promote research and dissemination of knowledge to enhance agricultural practices and market opportunities for sorghum producers (United Sorghum Board, 2024). However, disparities in policy implementation across regions can create uneven advantages for certain producers, often favoring large-scale operations over smallholder producers (Mundia et al., 2019). Furthermore, trade policies impact sorghum exports and imports, as countries regulate markets to protect domestic agriculture while aiming for competitive participation in global trade (Trupo, 2022).

1.3 SORGHUM AS A CLIMATE-SMART COMMODITY

Sorghum is recognized as a climate-smart commodity due to its unique attributes that contribute to water conservation, environmental sustainability, and resilience to climate change. These qualities make it an attractive option for producers and policymakers looking to mitigate the impacts of climate variability. The climate-smart status of sorghum is underscored by its remarkable drought tolerance and ability to thrive in arid conditions with minimal resource requirements compared to other cereals (Liaqat et al., 2024). This resilience is critical as climate change increases the frequency and severity of droughts in many farming regions. Sorghum's adaptation to heat and its ability to grow in poorer soils further enhance its viability as a crop in challenging environments. In addition to its drought and heat resistance, sorghum has a deep root system that promotes soil carbon sequestration (Wright & Hons, 2005). This feature can help lower the carbon intensity of farming practices by capturing carbon dioxide and storing it underground over extended periods. Consequently, integrating sorghum into crop rotations has been shown to positively impact both carbon capture and soil health, supporting carbon-neutral objectives (Bean, 2024).

Sorghum requires significantly less water compared to other cereal crops, using one-third less water than maize (Schlegel et al., 2016). This attribute is particularly advantageous in water-scarce regions where sustainable irrigation practices are essential. The cultivation of sorghum therefore aligns with global water conservation efforts and contributes to reduced competition for dwindling water supplies. Moreover, sorghum farming often involves fewer tillage practices (Musafiri et al., 2022). Reduced tillage limits soil disruption, minimizing greenhouse gas emissions associated with soil aeration. This reduction in tillage not only aids in maintaining soil structure and health but also decreases energy use in farming practices, contributing to lower overall emissions.

Beyond its environmental advantages, sorghum offers economic benefits to producers (Sirany et al., 2022). As soil health is improved through practices associated with sorghum cultivation, producers may experience increased yields and lower input costs over time. This economic viability makes sorghum an attractive and competitive choice for producers, as it can be sold at prices comparable to or even higher than those of other crops, thereby enhancing income potential of producers. Additionally, with the increasing focus on climate-smart agriculture, sorghum producers can participate in carbon credit markets, enhancing their revenue streams by adopting sustainable practices recognized by ecological markets (Williams et al., 2002). Thus, the adoption of sorghum can improve not only environmental outcomes but also the financial stability of producers in changing agricultural landscapes.

Sorghum plays a critical role in food security, particularly in developing regions vulnerable to climate change (Hossain et al., 2022). Its high nutritional value, including protein, fiber, and essential minerals, positions it as an important staple in the diets of many communities (Hossain et al., 2022). By promoting sorghum cultivation, regions can diversify food sources and reduce reliance on less climate-resilient crops, bolstering local food security amidst a changing climate. This multifaceted potential of sorghum, encompassing environmental, economic, and nutritional aspects, ever more solidifies its status as a climate-smart commodity. As the agricultural sector seeks solutions to cope with climate challenges, sorghum stands out as a model of sustainability and resilience.

1.3.1 Water Use in Sorghum Production

Sorghum is a highly water-efficient crop that thrives in arid and semi-arid environments. Unlike other crops, which require nearly 12 inches of water to produce the first bushel, grain sorghum needs only six inches of water, whether from soil moisture, rainfall, or irrigation, to yield its initial bushel (www.sorghumcheckoff.com/our-farmers/grain-production/water-use-and-irrigation/). The water use of sorghum is significantly influenced by its developmental stage, environmental conditions, and soil characteristics. The typical irrigation requirements for a medium to late maturing cultivar (110 to 130 days) vary from 450 to 650 mm (Assefa et al., 2010; FAO, 2002). Daily water requirements vary greatly depending on the growth stage and climatic conditions (Figure 1.6). During the initial stages of growth, the irrigation requirements are just 1 to 2.5 mm per day, escalating to a peak of 7 to 10 mm per day at the boot stage (Krieg, 1983; McWilliams, 2003). Booting, commencing 35–40 days post-emergence, is a critical developmental period for sorghum and requires the highest water intake (New, 2004).

Grain sorghum is known for superior drought tolerance and lower water usage when compared to other cereal crops; yet its exposure to drought leads to a reduction in crop yield. Water stress during the vegetative stage may decrease yield by over 36%, while water stress during the reproductive stage might diminish production by more than 55% (Assefa et al., 2010). Sorghum mostly extracts moisture from the upper 0 to 1.65 meters and roots can go as deep as 2.5 meters, making substantial water reserves in rootzone essential for successful crop production (Moroke et al., 2005). Sorghum yields can be susceptible to climatic conditions, so irrigating at 75% of the available water capacity (AWC) can optimize irrigation water use efficiency (IWUE). In

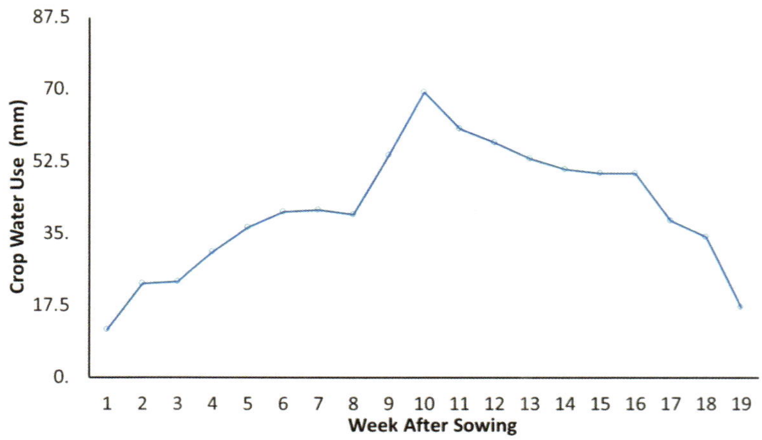

FIGURE 1.6 Weekly water consumption of the sorghum crop, illustrating its water use patterns during different growth stages.

normal cold and wet conditions, starting irrigation at 50% AWC and switching off at 85% AWC, and in warm-dry weather conditions, beginning at 60% and terminating at 100% AWC, would optimize IWUE (Kothari et al., 2019). Enhanced yield and water conservation can be attained by effective management practices and strategies (Stone & Schlegel, 2006). Soil moisture sensors and probes are valuable instruments for farmers, enabling precise monitoring of soil moisture levels and the assessment of irrigation needs (Hardie, 2020). This approach promotes robust root development and abundant yields by preventing both over- and under-irrigation. Deficit irrigation may be used to enhance water saving while maintaining consistent yields (Araya et al., 2016; Bell et al., 2018). This entails the selective application of water at less critical developmental stages. The use of subsurface drip irrigation (SDI) and other modern irrigation techniques minimize water loss attributable to evaporation and optimizing water use (Lamm et al., 2021). Furthermore, cultural practices such as no-till or reduced tillage can enhance soil moisture retention and reduce evaporation.

Sorghum's extensive root system, superior osmotic adjustment ability, and stress tolerance allow it to endure water shortages and other extreme environmental conditions (Krupa et al., 2017). Research has revealed that many sorghum cultivars exhibit noticeable advantages in water-use, production capabilities, and physiological characteristics, highlighting the need of studying the drought responses of diverse hybrids (Assefa et al., 2010; Krupa et al., 2017). Osmotic adjustments, stomatal conductance, leaf water potential, and other agronomic factors such as changes in root structure contribute to this adaptability. Supplying water according to the crop's developmental stages is crucial, but assessing water requirements is complex due to variable environmental factors. Farmers can enhance water-use by changing planting dates and adopting better irrigation strategies by identifying the developmental stages most susceptible to water stress. The gap between anticipated and actual yields in arid circumstances can be diminished by the use of drought-resistant hybrids, optimizing irrigation and fertilizer application, adopting management practices such as no-till agriculture and cover crop. Sorghum is an important global crop due to its ability to thrive in water limited conditions and prospers in rained agriculture. Apart from its reputation for drought resilience, sorghum performs well with irrigation, making it an economically viable choice for irrigated systems in regions with limited water resources. Due to its adaptability, sorghum has the ability to increase agricultural resilience in areas with limited water resources, which in turn might help ensure the world's food security.

1.4 CURRENT CHALLENGES IN SORGHUM PRODUCTION

Sorghum, a vital food and fodder crop in tropical and semi-arid Africa, plays a critical role in food security for nations like Sudan, Nigeria, and Ethiopia. Despite its diverse uses, including grain, forage, and industrial applications (e.g., syrup, sugar, and textiles), its productivity remains low, averaging ~2.5 t/ha (Mundia et al., 2019; Hao et al., 2021; Khalifa & Eltahir, 2023). This low productivity is due to a combination of challenges that affect both yield and quality. One key challenge is limited access to quality seeds. Many farmers rely on traditional practices, which, while locally adaptive, often result in poor yields. In Zimbabwe, nearly 80% of sorghum and pearl millet seeds are exchanged through non-monetary village seed markets, further limiting access to improved varieties (Hlatshwayo et al., 2021).

Adding to these challenges are sorghum diseases caused by fungi, bacteria, viruses, nematodes, and parasitic plants, which vary in severity from mild to devastating (Little & Perumal, 2019). Some pathogens, such as those causing zonate leaf spot, sooty stripe, and bacterial leaf diseases, are cosmopolitan and typically result in minimal losses. However, regional epidemics can still occur (Little & Perumal, 2019; Anitha et al., 2020). In contrast, diseases like anthracnose, sorghum downy mildew, and ergot have caused severe epidemics, while others, including Fusarium stalk rot, charcoal rot, and grain mold, persistently reduce yields and require ongoing research (Little et al., 2011; Tesso et al., 2012). These diseases, along with parasitic weeds like *Striga hermontica*, which aggressively compete for nutrients and moisture, significantly impact yields and contribute to economic losses (Ahmed et al., 2024).

Abiotic stresses, such as drought, salinity, and extreme weather, also severely limit crop growth and yield (Zheng et al., 2023). In addition, industrial activities that contaminate soils with heavy metals pose further threats to food security and human health. Drought affects 50% of global arid and semi-arid land, reducing photosynthesis, yield, and nutrient uptake, particularly during critical growth stages (Zika & Erb, 2009; Abreha et al., 2022). Soil salinization affects 20% of irrigated land, and projections suggest that 50% of arable land may be salinized by 2050 (Zheng et al., 2023). Both stresses disrupt plant growth and require costly mitigation strategies, such as cultivating stress-tolerant crops, to enhance food security.

While sorghum is known for its drought tolerance, significant yield losses still occur under drought, particularly in regions like sub-Saharan Africa and the U.S., where rainfed agriculture is prevalent (Abreha et al., 2022). Drought not only reduces yields but also worsens protein digestibility, increasing malnutrition risks in food-insecure areas (Impa et al., 2019; Liaqat et al., 2024). Global climate change exacerbates this issue by increasing temperature extremes, which further reduces yields (Abreha et al., 2022). Further, industrial contamination with heavy metals, such as cadmium and lead, generates reactive oxygen species that stunt plant growth and endanger human health through accumulation in the food chain (Yilmaz et al., 2017). Moreover, harvesting and post-harvest processes also present significant challenges (Arama et al., 2014). High moisture content complicates mechanical processing, leading to losses during harvesting and transportation. Inadequate storage facilities and processing technologies exacerbate these issues, resulting in considerable post-harvest grain losses and reducing farmers' incomes (Adeyeye, 2017). Market access remains another major obstacle, as much of the sorghum is consumed at the household level, preventing farmers from engaging with profitable markets, specifically in under-developed and developing countries (Duff et al., 2019). Industrial demand fluctuations, coupled with limited local processing facilities, hinder market diversification and income opportunities. To address these challenges, integrated strategies are needed to enhance production, reduce losses, and create more stable market opportunities for sorghum growers.

1.5 DESIGNING THE SORGHUM GENOME FOR FUTURE CLIMATE CONDITIONS

Designing the sorghum genome for future climate conditions is essential to address the challenges posed by climate change and other stress factors that threaten sorghum production. Increasing drought frequency, temperature extremes, and various biotic and abiotic stresses compromise yield and sustainability. Genomics-assisted breeding approaches can expedite the development of sorghum cultivars with traits like drought tolerance, heat resistance, and improved nutrient efficiency. These modifications are vital for maintaining high yields in a warming world and ensuring the crop's resilience to climate-induced stresses. Additionally, genomics has enabled the identification of genes for disease, pest, and environmental stress resistance, enhancing sorghum's adaptability. It can also improve seed quality, reduce reliance on traditional seed exchange, and address post-harvest losses by developing varieties with better storage and uniform maturity. Ultimately, leveraging the sorghum genome will help create more resilient, high-yielding varieties, improving food security and farmer incomes in sorghum-dependent regions.

1.5.1 Sorghum Germplasm and Genomic Resources

Sorghum gene banks worldwide hold over 240,000 accessions, including major collections at ICRISAT (India, over 39,000 accessions), USDA-NPGS (U.S.A., 45,000 accessions), ICS-CAAS (China, 18,263 accessions), and NBPGR (India, 20,221 accessions) (Xin et al., 2021). Although cultivated sorghum constitutes 98.3% of these collections, redundancies and gaps, including geographical gaps in regions like Burkina Faso and Nigeria, and taxonomic gaps, persist and reduce their efficiency (Upadhyaya et al., 2016, 2017a, 2017b). Efforts to conserve wild relatives face challenges due to habitat loss driven by urbanization, modern cultivation, and climate change. Core collections, representing 10% of total accessions, enhance germplasm evaluation and utilization. ICRISAT and USDA-NPGS have established core and mini-core collections, such as the ICRISAT mini-core of 242 accessions, selected based on morpho-agronomic traits (Upadhyaya & Ortiz, 2001; Upadhyaya et al., 2009). These collections have been instrumental in identifying accessions with desirable traits, including disease resistance and adaptive traits (Cuevas et al., 2017, 2018). Regional evaluations have proven efficient for adaptive trait identification. For instance, Ethiopian germplasm has been extensively studied, revealing high genetic diversity and rare variations (Cuevas et al., 2017; Girma et al., 2020). Screening efforts in Senegal and Sudan have identified accessions resistant to anthracnose and grain mold, highlighting the value of targeted germplasm evaluations (Cuevas & Prom, 2020). These studies underline the potential of geographically targeted core collections for breeding programs.

Genome sequencing is a crucial resource for understanding gene functions and facilitating genomic studies. The first sorghum genome sequence, completed in the inbred line BTx623 over 12 years ago, has served as a reference for identifying genetic variations across sorghum accessions (Paterson et al., 2009). Subsequent sequencing of sweet sorghums and grain sorghum genotypes revealed over one million SNPs, nearly 100,000 Indels, and other variations, highlighting differences in starch and sucrose metabolism genes between sweet and grain sorghums (Zheng et al., 2011). Ten years later, a high-quality reference genome (version 3) added 29.6 Mb of new data, increased annotated genes by 24%, and significantly reduced sequencing errors (McCormick et al., 2018). Advanced technologies also enabled chromosome-level assemblies for sorghum lines like Tx430 and sweet sorghum Rio, showing structural similarity despite agronomic differences (Cooper et al., 2019). A pan-genome constructed from *de novo* assemblies of 13 diverse sorghum lines identified 44,079 gene families, with 64% showing presence/absence variation (Tao et al., 2021). Analysis of 354 accessions revealed nearly two million SNPs and 3.9 million InDels (Ruperao et al., 2021). This pan-genome, consisting of core, shell, and cloud genes, highlights extensive genetic variation and provides a valuable resource for sorghum improvement.

Linkage and association mapping are essential for studying genetic traits, leveraging natural variation for ecophysiological relevance, agronomic utility, and evolutionary insights. Over the past 25 years, linkage mapping in sorghum has uncovered the genetic basis of key traits, including agronomic, morphological, physiological and stress tolerance/resistance traits (Mace et al., 2019).

Recently, more sorghum RIL families have been made publicly available via NPGS-GRIN under "SORGHUM-GENSTOCKS". Notable examples include the BTx623 x IS3620C family, developed from a cross between the reference line BTx623 and the divergent Nigerian guinea margaritiferum line IS3620C, offering wide genetic diversity and high-density GBS markers (Burow et al., 2011; Kong et al., 2018). Additionally, three RIL families for cold tolerance, derived from crosses of U.S. cold-susceptible and Chinese cold-tolerant lines, are available along with GBS markers, segregating for traits like pigmentation, flowering time, and plant architecture (Marla et al., 2019).

The advent of high-density genome-wide markers has shifted genetic mapping in crops from linkage mappings to genome-wide association studies (GWAS). In sorghum, GWAS resources like the Sorghum Association Panel (SAP) and the Bioenergy Association Panel (BAP) are available through GRIN, providing diverse germplasm and extensive genomic data for studying global diversity and trait associations. These panels include temperate-adapted and tropical accessions, along with specialized collections like sweet sorghum and landraces with georeferenced data (Boyles et al., 2019; Brenton et al., 2016). Given the tradeoffs between linkage and association mapping, there has been increasing interest in combining the strengths of both approaches. Multi-parent mapping strategies, such as nested association mapping (NAM), backcross NAM, and multi-parent advanced generation intercross (MAGIC), have been developed and utilized in sorghum (Hao et al., 2021). NAM uses multiple RIL families with diverse founders but a common parent and has been successfully applied in sorghum with RTx430 as the common parent, comprising over 2,200 RILs (Perumal et al., 2021). This resource has mapped traits like flowering time, height, and inflorescence morphology (Perumal et al., 2021). It has also shown greater power than GWAS for mapping adaptive traits (Bouchet et al., 2017; Olatoye et al., 2020). Similarly, a carbon-partitioning NAM population was developed using 11 diverse, male founder accessions, each crossed with a recurrent female parent "Grassl" (Kumar et al., 2022). MAGIC, which increases recombination and mapping resolution using advanced intercrosses, aims to balance allele frequencies and enhance QTL detection compared to GWAS (Singh & Shrivastav, 2023). One sorghum MAGIC resource has been described and can be requested from the developers (Ongom & Ejeta, 2018). The backcross NAM (BC-NAM) approach, which recovers lines similar to the elite parent, has advantages in agronomic relevance (Patil et al., 2024).

Recent advances in sorghum mapping resources have facilitated genetic studies, but two major gaps remain. First, while many QTLs and candidate genes have been identified, causative variants for natural variation in sorghum have only been found for a few traits (Casto et al., 2019; Zou et al., 2020). Future efforts should focus on identifying these variants, perhaps through whole-genome resequencing of NAM parents and SAP. Second, genetic mapping has not consistently led to advances in sorghum improvement. Future mapping resources may have a greater impact on improvement if they focus on BCNAM or similar approaches, linking diversity with elite genetic backgrounds.

1.6 INTRODUCTION TO THE STRUCTURE AND THEMES OF THE BOOK

The book *Designing Sorghum Genome for NextGen Agriculture* covers a wide array of innovative topics essential for advancing sorghum breeding and improving agricultural productivity. It begins by exploring how harnessing genetic diversity, through advanced tools and speed breeding techniques, can accelerate the development of high-yielding and stress-resistant sorghum varieties (*Chapter 2*). The following chapter focuses on the role of bioinformatics, emphasizing the tools and resources available for sorghum omics, which include genomic, transcriptomic, proteomic, and metabolomic data analysis to support research and breeding (*Chapter 3*). The book also delves into advancements in molecular mapping, specifically how these techniques are applied to improve sorghum genome design and facilitate the identification of important genetic traits (*Chapter 4*).

Additionally, it discusses sorghum genomics and pan-genomics in the next-generation sequencing era, highlighting how these technologies uncover genetic variations and improve our understanding of sorghum diversity (*Chapter 5*). The chapter on genetic modification strategies examines mutagenesis, transgenesis, and CRISPR/Cas-based gene editing to enhance sorghum traits such as yield, quality, and stress tolerance (*Chapter 6*). Another key chapter integrates genomic and phenomic data, explaining how predictive tools can accelerate genetic gain in sorghum improvement (*Chapter 7*). The book continues by discussing hybrid breeding strategies that improve sorghum grain yield, stress resilience, and quality attributes (*Chapter 8*). It also highlights how omics and systems biology approaches can develop climate-resilient sorghum varieties capable of withstanding extreme weather events and other environmental stresses (*Chapter 9*). The physiological mechanisms that enable sorghum to cope with abiotic stresses like drought and heat are explored, as well as how these mechanisms can be used in breeding climate-resilient varieties (*Chapter 10*). The subsequent chapter focuses on molecular engineering of root phenotypes, demonstrating how modifications can enhance sorghum's resilience to abiotic stresses (*Chapter 11*). The redesigning of the sorghum root microbiome is another engaging topic, emphasizing how this can improve water and nutrient use efficiency (*Chapter 12*). Omics tools are used to understand and modify sorghum seed composition, providing insights into the development of varieties with improved nutritional and health benefits (*Chapter 13*). The integration of high-throughput phenotyping technologies and artificial intelligence is also explored, showing how these tools can accelerate the identification of desirable traits in sorghum and speed up breeding cycles (*Chapter 14*). Finally, the book examines how crop simulation modeling can help address climate change challenges in sorghum production, guiding future breeding efforts to mitigate climate-related issues (*Chapter 15*).

ACKNOWLEDGMENT

The authors are grateful for financial support from "Love Tito's", the philanthropic arm of Tito's Handmade Vodka, supporting Dr. Mehla's hire to the Davis College Water Center headed by Dr Jagadish at Texas Tech University.

REFERENCES

Abraha, T., Githiri, S. M., Kasili, R., Araia, W., & Nyende, A. B. (2015). Genetic variation among sorghum (*Sorghum bicolor* L. Moench) landraces from Eritrea under post-flowering drought stress conditions. *American Journal of Plant Sciences*, *6*(9), 1410–1424.

Abreha, K. B., Enyew, M., Carlsson, A. S., Vetukuri, R. R., Feyissa, T., Motlhaodi, T., . . . Geleta, M. (2022). Sorghum in dryland: Morphological, physiological, and molecular responses of sorghum under drought stress. *Planta*, *255*, 1–23.

Adeyeye, S. A. O. (2017). The role of food processing and appropriate storage technologies in ensuring food security and food availability in Africa. *Nutrition & Food Science*, *47*(1), 122–139.

Ahmed, A., Elsafy, M., Zhourghane, A., Abdalla, A. A., Abreha, K. B., Geleta, M., . . . Abdelhalim, T. S. (2024). Exploring the genetic variability of Sudanese wild sorghum (*Sorghum bicolor* (L.) moench) germplasm for post-attachment striga hermonthica resistance mechanisms using single sequence repeat (SSR) primers. *Genetic Resources and Crop Evolution*, *72*, 4583–4595.

Anitha, K., Das, I. K., Holajjer, P., Sivaraj, N., Reddy, C. R., & Balijepalli, S. B. (2020). Sorghum diseases: Diagnosis and management. In *Sorghum in the 21st century: Food–fodder–feed–fuel for a rapidly changing world*, 565–619. Springer, Singapore.

Arama, P., Ochola, S. O., Kange, A. M., Cheruiyot, E. K., & Ogendo, J. O. (2014). Pre-and post harvest factors affecting sorghum production (*Sorghum bicolor* L. Moench) among smallholder farming communities. *International Journal of Agronomy and Agricultural Research*, *5*(4), 40–47.

Araya, A., Kisekka, I., & Holman, J. (2016). Evaluating deficit irrigation management strategies for grain sorghum using AquaCrop. *Irrigation Science*, *34*, 465–481.

Aruna, C., & Visarada, K. B. R. S. (2019). Other industrial uses of sorghum. In C. Aruna, K. B. R. S. Visarada, B. V. Bhat, & V. A. Tonapi (Eds.), *Breeding sorghum for diverse end uses* (pp. 271–292). Cambridge, UK: Woodhead Publishing.

Assefa, Y., Staggenborg, S. A., & Prasad, V. P. (2010). Grain sorghum water requirement and responses to drought stress: A review. *Crop Management*, *9*(1), 1–11.

Bailey-Serres, J., Parker, J. E., Ainsworth, E. A., Oldroyd, G. E., & Schroeder, J. I. (2019). Genetic strategies for improving crop yields. *Nature*, *575*(7781), 109–118.

Baloch, F. S., Altaf, M. T., Liaqat, W., Bedir, M., Nadeem, M. A., Cömertpay, G., . . . Sun, H. J. (2023). Recent advancements in the breeding of sorghum crop: Current status and future strategies for marker-assisted breeding. *Frontiers in Genetics*, *14*, Art. 1150616.

Bean, B. (2024). *Sorghum in 2025: Low seed costs, resilience and crop rotation benefits*. www.sorghumcheckoff.com/agronomy-insights/sorghum-in-2025-low-seed-costs-resilience-and-crop-rotation-benefits/#

Bell, J. M., Schwartz, R., McInnes, K. J., Howell, T., & Morgan, C. L. (2018). Deficit irrigation effects on yield and yield components of grain sorghum. *Agricultural Water Management*, *203*, 289–296.

Bibi, A., Sadaqat, H. A., Tahir, M. H. N., & Akram, H. M. (2012). Screening of sorghum (*Sorghum bicolor* var Moench) for drought tolerance at seedling stage in polyethylene glycol. *Journal of Animal and Plant Sciences*, *22*(3), 671–678.

Bouchet, S., Olatoye, M. O., Marla, S. R., Perumal, R., Tesso, T., Yu, J., . . . Morris, G. P. (2017). Increased power to dissect adaptive traits in global sorghum diversity using a nested association mapping population. *Genetics*, *206*(2), 573–585.

Boyles, R. E., Brenton, Z. W., & Kresovich, S. (2019). Genetic and genomic resources of sorghum to connect genotype with phenotype in contrasting environments. *The Plant Journal*, *97*(1), 19–39.

Brenton, Z. W., Cooper, E. A., Myers, M. T., Boyles, R. E., Shakoor, N., Zielinski, K. J., . . . Kresovich, S. (2016). A genomic resource for the development, improvement, and exploitation of sorghum for bioenergy. *Genetics*, *204*(1), 21–33.

Burow, G., Burke, J. J., Xin, Z., & Franks, C. D. (2011). Genetic dissection of early-season cold tolerance in sorghum (*Sorghum bicolor* (L.) Moench). *Molecular Breeding*, *28*, 391–402.

Butchee, K., Arnall, D. B., Sutradhar, A., Godsey, C., Zhang, H., & Penn, C. (2012). Determining critical soil pH for grain sorghum production. *International Journal of Agronomy*, *2012*(1), Art. 130254.

Casto, A. L., Mattison, A. J., Olson, S. N., Thakran, M., Rooney, W. L., & Mullet, J. E. (2019). Maturity2, a novel regulator of flowering time in *Sorghum bicolor*, increases expression of *SbPRR37* and *SbCO* in long days delaying flowering. *PLoS One*, *14*(4), Art. e0212154.

Cooper, E. A., Brenton, Z. W., Flinn, B. S., Jenkins, J., Shu, S., Flowers, D., . . . Kresovich, S. (2019). A new reference genome for *Sorghum bicolor* reveals high levels of sequence similarity between sweet and grain genotypes: Implications for the genetics of sugar metabolism. *BMC Genomics*, *20*, 1–13.

Cruz, A., Saini, D. K., Aviles, D., Norris, A., & Jagadish, S. V. K. (2024). Sorghum and pearl millet as sustainable alternative forage options for water limited environments: Opportunities and challenges. *Advances in Agronomy*, *189*, 137–192.

Cuevas, H. E., & Prom, L. K. (2020). Evaluation of genetic diversity, agronomic traits, and anthracnose resistance in the NPGS Sudan Sorghum Core collection. *BMC Genomics*, *21*, 1–15.

Cuevas, H. E., Prom, L. K., & Rosa-Valentin, G. (2018). Population structure of the NPGS Senegalese sorghum collection and its evaluation to identify new disease resistant genes. *PLoS One*, *13*(2), Art. e0191877.

Cuevas, H. E., Rosa-Valentin, G., Hayes, C. M., Rooney, W. L., & Hoffmann, L. (2017). Genomic characterization of a core set of the USDA-NPGS Ethiopian sorghum germplasm collection: Implications for germplasm conservation, evaluation, and utilization in crop improvement. *BMC Genomics*, *18*, 1–17.

Dalton, T. J., & Hodjo, M. (2020). Trends in global production, consumption, and utilization of Sorghum. In *Sorghum in the 21st Century: Food–Fodder–Feed–Fuel for a Rapidly Changing World*, 3–15.

Damerau, K., Patt, A. G., & Van Vliet, O. P. (2016). Water saving potentials and possible trade-offs for future food and energy supply. *Global Environmental Change*, *39*, 15–25.

Derese, S. A., Shimelis, H., Mwadzingeni, L., & Laing, M. (2018). Agro-morphological characterisation and selection of sorghum landraces. *Acta Agriculturae Scandinavica, Section B—Soil & Plant Science*, *68*(7), 585–595.

Dou, F., Wight, J. P., Wilson, L. T., Storlien, J. O., & Hons, F. M. (2014). Simulation of biomass yield and soil organic carbon under bioenergy sorghum production. *PLoS One*, *9*(12), Art. e115598.

Duff, J., Bice, D., Hoeffner, I., & Weinheimer, J. (2019). The sorghum industry and its market perspective. *Sorghum: A State of the Art and Future Perspectives*, *58*, 503–514.

Emendack, Y., Sanchez, J., Hayes, C., Nesbitt, M., Laza, H., & Burke, J. (2021). Seed-to-seed early-season cold resiliency in sorghum. *Scientific Reports*, *11*(1), 7801.

Erramouspe, H. (2021). The global significance of Sorghum – Africa. *National Sorghum Producers*. https://sorghumgrowers.com/magazine/the-global-signficance-of-sorghum-africa/

FAO (2002). *Crop water management* [Online]. Rome, Italy: AGLW Water Management Group, United Nations FAO.

FAO and USDA. (2024). Worldwide production of grain in 2023/24, by type (in MMT). *Statista*. Accessed September 27, 2024, from www.statista.com/statistics/263977/world-grain-production-by-type/

FAOSTAT. (2023). *Crops and livestock products*. Food and Agriculture Organization of the United Nations. Accessed September 27, 2024, from www.fao.org/faostat/en/#data/QCL

Garibaldi, A., & Turner, N. (2004). Cultural keystone species: Implications for ecological conservation and restoration. *Ecology and Society*, *9*(3).

Girma, G., Nida, H., Tirfessa, A., Lule, D., Bejiga, T., Seyoum, A., . . . Mengiste, T. (2020). A comprehensive phenotypic and genomic characterization of Ethiopian sorghum germplasm defines core collection and reveals rich genetic potential in adaptive traits. *The Plant Genome*, *13*(3), Art. e20055.

Graham-Rowe, D. (2011). Agriculture: Beyond food versus fuel. *Nature*, *474*(7352), S6–S8.

Hansen, J. (2024). *USDA ERS – China's Agricultural Policies Reshape Global Sorghum Trade*. Retrieved from www.ers.usda.gov/amber-waves/2015/june/china-s-agricultural-policies-reshape-global-sorghum-trade/

Hao, H., Li, Z., Leng, C., Lu, C., Luo, H., Liu, Y., . . . Jing, H. C. (2021). Sorghum breeding in the genomic era: Opportunities and challenges. *Theoretical and Applied Genetics*, *134*, 1899–1924.

Hardie, M. (2020). Review of novel and emerging proximal soil moisture sensors for use in agriculture. *Sensors*, *20*(23), 6934.

Hlatshwayo, S. I., Modi, A. T., Hlahla, S., Ngidi, M., & Mabhaudhi, T. (2021). Usefulness of seed systems for reviving smallholder agriculture: A South African perspective. *African Journal of Food, Agriculture, Nutrition and Development*, *21*(2), 17581–17603.

Hossain, M. S., Islam, M. N., Rahman, M. M., Mostofa, M. G., & Khan, M. A. R. (2022). Sorghum: A prospective crop for climatic vulnerability, food and nutritional security. *Journal of Agriculture and Food Research*, *8*, Art. 100300.

Huang, R. D. (2018). Research progress on plant tolerance to soil salinity and alkalinity in sorghum. *Journal of Integrative Agriculture*, *17*(4), 739–746.

Impa, S. M., Perumal, R., Bean, S. R., Sunoj, V. J., & Jagadish, S. V. K. (2019). Water deficit and heat stress induced alterations in grain physico-chemical characteristics and micronutrient composition in field grown grain sorghum. *Journal of Cereal Science*, *86*, 124–131.

IndexMundi. (2024). Sorghum exports worldwide in 2023, by leading country (in 1,000 metric tons). *Statista*. Accessed September 27, 2024, from www.statista.com/statistics/1134655/global-sorghum-export-by-country/

Jagadish, S. V. K. (2020). Heat stress during flowering in cereals–effects and adaptation strategies. *New Phytologist*, *226*(6), 1567–1572.

Jeya, G., Dhanalakshmi, R., Anbarasu, M., Vinitha, V., & Sivamurugan, V. (2022). Techno-economic analysis of butanol biosynthesis. In H. Li, S. Saravanamurugan, A. Pandey, & S. Elumalai (Eds.), *Biomass, biofuels, biochemicals* (pp. 75–94). Amsterdam, The Netherlands: Elsevier.

Joshi, R. P., Jain, A. K., Malhotra, N., & Kumari, M. (2021). Origin, domestication, and spread. In *Millets and pseudo cereals* (pp. 33–38). Cambridge, MA: Woodhead Publishing.

Khalifa, M., & Eltahir, E. A. (2023). Assessment of global sorghum production, tolerance, and climate risk. *Frontiers in Sustainable Food Systems*, *7*, Art. 1184373.

Kong, W., Kim, C., Zhang, D., Guo, H., Tan, X., Jin, H., . . . Paterson, A. H. (2018). Genotyping by sequencing of 393 *Sorghum bicolor* BTx623× IS3620C recombinant inbred lines improves sensitivity and resolution of QTL detection. *G3: Genes, Genomes, Genetics*, *8*(8), 2563–2572.

Kothari, K., Ale, S., Bordovsky, J. P., Thorp, K. R., Porter, D. O., & Munster, C. L. (2019). Simulation of efficient irrigation management strategies for grain sorghum production over different climate variability classes. *Agricultural Systems*, *170*, 49–62.

Krieg, D. R. (1983). Sorghum (Chapter 11). In I. D. Teare & M. M. Peet (Eds.), *Crop water relations*. Chichester: John Wiley.

Krupa, K. N., Dalawai, N., Shashidhar, H. E., Harinikumar, K. M., Manojkumar, H. B., Bharani, S., & Turaidar, V. (2017). Mechanisms of drought tolerance in Sorghum: A review. *Indian Journal of Pure & Applied Biosciences*, *5*(4), 221–237.

Kumar, N., Brenton, Z., Myers, M. T., Boyles, R. E., Sapkota, S., Boatwright, J. L., . . . Kresovich, S. (2022). Registration of the sorghum carbon-partitioning nested association mapping (CP-NAM) population. *Journal of Plant Registrations*, *16*(3), 656–663.

Lamm, F. R., Colaizzi, P. D., Sorensen, R. B., Bordovsky, J. P., Dougherty, M., Balkcom, K., . . . Peters, R. T. (2021). A 2020 vision of subsurface drip irrigation in the US. *Transactions of the ASABE*, *64*(4), 1319–1343.

Leon, F., Harvey, L., & Rooney, W. L. (2024). Historical multi-environmental trials reveal trends in hybrid grain sorghum performance from 1970 to 2021 in Texas. *Crop Science*, *64*(6), 3014–3027.

Liaqat, W., Altaf, M. T., Barutçular, C., Mohamed, H. I., Ahmad, H., Jan, M. F., & Khan, E. H. (2024). Sorghum: A star crop to combat abiotic stresses, food insecurity, and hunger under a changing climate: A Review. *Journal of Soil Science and Plant Nutrition*, *24*(1), 74–101.

Little, C. R., & Perumal, R. (2019). The biology and control of sorghum diseases. *Sorghum: A State of the Art and Future Perspetives*, *58*, 297–346.

Little, C. R., Perumal, R., Tesso, T. T., Prom, L. K., Odvody, G. N., & Magill, C. W. (2011). Sorghum pathology and biotechnology-A fungal disease perspective: Part I. Grain mold, head smut, and ergot. *European Journal of Plant Science and Biotechnology*, *6*(Special Issue 1), 10–30.

Mace, E., Innes, D., Hunt, C., Wang, X., Tao, Y., Baxter, J., . . . Jordan, D. (2019). The Sorghum QTL Atlas: A powerful tool for trait dissection, comparative genomics and crop improvement. *Theoretical and Applied Genetics*, *132*, 751–766.

Marla, S. R., Burow, G., Chopra, R., Hayes, C., Olatoye, M. O., Felderhoff, T., . . . Morris, G. P. (2019). Genetic architecture of chilling tolerance in sorghum dissected with a nested association mapping population. *G3: Genes, Genomes, Genetics*, *9*(12), 4045–4057.

Maw, M. J., Houx III, J. H., & Fritschi, F. B. (2017). Nitrogen use efficiency and yield response of high biomass sorghum in the lower Midwest. *Agronomy Journal*, *109*(1), 115–121.

Mayor, L., Demarco, P., Lira, S., Fang, Y., Abadie, T., Gambin, B. L., . . . Messina, C. (2023). Retrospective study in US commercial sorghum breeding: I. Genetic gain in relation to relative maturity. *Crop Science*, *63*(2), 501–510.

McCormick, R. F., Truong, S. K., Sreedasyam, A., Jenkins, J., Shu, S., Sims, D., . . . Mullet, J. E. (2018). The *Sorghum bicolor* reference genome: Improved assembly, gene annotations, a transcriptome atlas, and signatures of genome organization. *The Plant Journal*, *93*(2), 338–354.

McCuistion, K. C., Selle, P. H., Liu, S. Y., & Goodband, R. D. (2019). Sorghum as a feed grain for animal production. In John R.N. Taylor and K.G. Duodu (Eds.), *Sorghum and millets* (pp. 355–391). Amsterdam, The Netherlands: Elsevier.

McWilliams, D. (2003). *Drought strategies for corn and grain sorghum*. Las Cruces, NM: New Mexico State University, Cooperative Extension Service, College of Agriculture and Home Economics.

Moroke, T. S., Schwartz, R. C., Brown, K. W., & Juo, A. S. R. (2005). Soil water depletion and root distribution of three dryland crops. *Soil Science Society of America Journal*, *69*(1), 197–205.

Mukherjee, A., Maheshwari, U., Sharma, V., Sharma, A., & Kumar, S. (2024). Functional insight into multi-omics-based interventions for climatic resilience in sorghum (*Sorghum bicolor*): A nutritionally rich cereal crop. *Planta*, *259*(4), 91.

Mundia, C. W., Secchi, S., Akamani, K., & Wang, G. (2019). A regional comparison of factors affecting global sorghum production: The case of North America, Asia and Africa's Sahel. *Sustainability*, *11*(7), 2135.

Musafiri, C. M., Kiboi, M., Macharia, J., Ng'etich, O. K., Okoti, M., Mulianga, B., . . . Ngetich, F. K. (2022). Does the adoption of minimum tillage improve sorghum yield among smallholders in Kenya? A counterfactual analysis. *Soil and Tillage Research*, *223*, 105473.

Ndlovu, E., Van Staden, J., & Maphosa, M. (2021). Morpho-physiological effects of moisture, heat and combined stresses on *Sorghum bicolor* [Moench (L.)] and its acclimation mechanisms. *Plant Stress*, *2*, Art. 100018.

New, L. (2004). Grain sorghum irrigation. In *Texas FARMER Collection*. Texas Cooperative Extension Report B-6152:6–04, Texas.

Olatoye, M. O., Marla, S. R., Hu, Z., Bouchet, S., Perumal, R., & Morris, G. P. (2020). Dissecting adaptive traits with nested association mapping: Genetic architecture of inflorescence morphology in sorghum. *G3: Genes, Genomes, Genetics*, *10*(5), 1785–1796.

Ondiko, J. H., & Recha, C. W. (2022). Global sorghum production constraints: A Review. *Annals of Arid Zone*, *61*(1), 1–12.

Ongom, P. O., & Ejeta, G. (2018). Mating design and genetic structure of a multi-parent advanced generation intercross (MAGIC) population of sorghum (*Sorghum bicolor* (L.) Moench). *G3: Genes, Genomes, Genetics*, *8*(1), 331–341.

Parikh, A., Brant, E. J., Baloglu, M. C., & Altpeter, F. (2021). CRISPR/Cas-mediated genome editing in sorghum—recent progress, challenges and prospects. *In Vitro Cellular & Developmental Biology-Plant*, *57*, 1–11.

Paterson, A. H. (2008). Genomics of sorghum. *International Journal of Plant Genomics*, *2008*(1), Art. 362451.

Paterson, A. H., Bowers, J. E., Bruggmann, R., Dubchak, I., Grimwood, J., Gundlach, H., . . . Rokhsar, D. S. (2009). The *Sorghum bicolor* genome and the diversification of grasses. *Nature*, *457*(7229), 551–556.

Patil, N. Y., Hoffmann Jr, L., Winans, N., Perumal, R., Hayes, C., Emendack, Y., . . . Rooney, W. L. (2024). Registration of sorghum backcross-nested association mapping (BC-NAM) families in a BTx623 or RTx436 background. *Journal of Plant Registrations*, *18*(1), 204–219.

Perumal, R., Tesso, T. T., Morris, G. P., Jagadish, S. V. K., Little, C. R., Bean, S. R., . . . Tuinstra, M. R. (2021). Registration of the sorghum nested association mapping (NAM) population in RTx430 background. *Journal of Plant Registrations*, *15*(2), 395–402.

Pfeiffer, B. K., Pietsch, D., Schnell, R. W., & Rooney, W. L. (2019). Long-term selection in hybrid sorghum breeding programs. *Crop Science*, *59*(1), 150–164.

Prasad, P. V. V., Djanaguiraman, M., Perumal, R., & Ciampitti, I. A. (2015). Impact of high temperature stress on floret fertility and individual grain weight of grain sorghum: Sensitive stages and thresholds for temperature and duration. *Frontiers in Plant Science*, *6*, 820.

Prasad, V. R., Govindaraj, M., Djanaguiraman, M., Djalovic, I., Shailani, A., Rawat, N., . . . Prasad, P. V. (2021). Drought and high temperature stress in sorghum: Physiological, genetic, and molecular insights and breeding approaches. *International Journal of Molecular Sciences*, *22*(18), 9826.

Qi, G., Li, N., Sun, X. S., & Wang, D. (2019). Overview of sorghum industrial utilization. *Sorghum: A State of the Art and Future Perspectives*, *58*, 463–476.

Queiroz, M. S., Oliveira, C. E., Steiner, F., Zuffo, A. M., Zoz, T., Vendruscolo, E. P., . . . Menis, F. T. (2019). Drought stresses on seed germination and early growth of maize and sorghum. *Journal of Agricultural Science*, *11*(2), 310.

Rao, P. S., Vinutha, K. S., Kumar, G. A., Chiranjeevi, T., Uma, A., Lal, P., . . . Jose, S. (2019). Sorghum: A multipurpose bioenergy crop. *Sorghum: A State of the Art and Future Perspectives*, *58*, 399–424.

Rather, M. A., Thakur, R., Hoque, M., Das, R. S., Miki, K. S. L., Teixeira-Costa, B. E., . . . Gupta, A. K. (2023). Sorghum (*Sorghum bicolor*). In *Nutri-cereals: Nutraceutical and techno-functional potential*. Boca Raton, Florida: CRC Press.

Rosegrant, M. W., & Msangi, S. (2014). Consensus and contention in the food-versus-fuel debate. *Annual Review of Environment and Resources*, *39*(1), 271–294.

Ruperao, P., Thirunavukkarasu, N., Gandham, P., Selvanayagam, S., Govindaraj, M., Nebie, B., . . . Rathore, A. (2021). Sorghum pan-genome explores the functional utility for genomic-assisted breeding to accelerate the genetic gain. *Frontiers in Plant Science*, *12*, Art. 666342.

Schlegel, A. J., Assefa, Y., O'Brien, D., Lamm, F. R., Haag, L. A., & Stone, L. R. (2016). Comparison of corn, grain sorghum, soybean, and sunflower under limited irrigation. *Agronomy Journal*, *108*(2), 670–679.

Shahandeh, H., Hons, F. M., Storlien, J. O., & Wight, J. P. (2016). Long-term bioenergy sorghum harvest strategy and soil quality. *Aims Energy*, *4*(4), 633–657.

Sharma, O. P., Kannan, N., Cook, S., Pokhrel, B. K., & McKenzie, C. (2019). Analysis of the effects of high precipitation in Texas on rainfed sorghum yields. *Water*, *11*(9), 1920.

Singh, S., & Shrivastav, S. P. (2023). MAGIC populations improved traits and provided different accomplishments. *International Journal of Environment and Climate Change*, *13*(9), 2174–2182.

Sirany, T., Tadele, E., Aregahegn, H., & Wale, D. (2022). Economic potentials and use dynamics of sorghum food system in Ethiopia: Its implications to resolve food deficit. *Advances in Agriculture*, *2022*(1), Art. 4580643.

Somegowda, V. K., Reddy, S. D., Gaddameedi, A., Kiranmayee, K. U., Naravula, J., Kishor, P. K., & Penna, S. (2023). Genomics breeding approaches for developing *Sorghum bicolor* lines with stress resilience and other agronomic traits. *Current Plant Biology*, *37*, Art. 100314.

Sorghum. (2011, July 21). *10 things you may not know about sorghum*. National Sorghum Producers. https://sorghumgrowers.com/2011/07/21/ten-things-you-may-not-know-about-sorghum/

Sorghum Checkoff. (2022). *Around the Globe*. www.sorghumcheckoff.com/sorghum-101/around-the-globe/

Stone, L. R., & Schlegel, A. J. (2006). Yield–water supply relationships of grain sorghum and winter wheat. *Agronomy Journal*, *98*(5), 1359–1366.

Tanwar, R., Panghal, A., Chaudhary, G., Kumari, A., & Chhikara, N. (2023). Nutritional, phytochemical and functional potential of sorghum: A review. *Food Chemistry Advances*, *3*, Art. 100501.

Tao, Y., Luo, H., Xu, J., Cruickshank, A., Zhao, X., Teng, F., . . . Mace, E. (2021). Extensive variation within the pan-genome of cultivated and wild sorghum. *Nature Plants*, *7*(6), 766–773.

Tesso, T. T., Perumal, R., Little, C. R., Adeyanju, A., Radwan, G. L., Prom, L. K., & Magill, C. W. (2012). Sorghum pathology and biotechnology—a fungal disease perspective: Part II. Anthracnose, stalk rot, and downy mildew. *European Journal of Plant Science and Biotechnology*, *6*, 31–44.

Tonapi, V. A., Talwar, H. S., Are, A. K., Bhat, B. V., Reddy, C. R., & Dalton, T. J. (Eds.). (2020). *Sorghum in the 21st century: Food, fodder, feed, fuel for a rapidly changing world*. Singapore: Springer.

Trupo, P. (2022). *U.S. sorghum prices rally with China's return to the market*. International Agricultural Trade Report. https://fas.usda.gov/data/us-sorghum-prices-rally-china-s-return-market

United Sorghum Board | Agricultural Marketing Service. (2024). usda.gov. www.ams.usda.gov/rules-regulations/research-promotion/sorghum

Upadhyaya, H. D., & Ortiz, R. (2001). A mini core subset for capturing diversity and promoting utilization of chickpea genetic resources in crop improvement. *Theoretical and Applied Genetics*, *102*, 1292–1298.

Upadhyaya, H. D., Pundir, R. P. S., Dwivedi, S. L., Gowda, C. L. L., Reddy, V. G., & Singh, S. (2009). Developing a mini core collection of sorghum for diversified utilization of germplasm. *Crop Science*, *49*(5), 1769–1780.

Upadhyaya, H. D., Reddy, K. N., Vetriventhan, M., Ahmed, M. I., Krishna, G. M., Reddy, M. T., & Singh, S. K. (2017a). Sorghum germplasm from West and Central Africa maintained in the ICRISAT genebank: Status, gaps, and diversity. *The Crop Journal*, *5*(6), 518–532.

Upadhyaya, H. D., Reddy, K. N., Vetriventhan, M., Gumma, M. K., Ahmed, M. I., Reddy, M. T., & Singh, S. K. (2017b). Status, genetic diversity and gaps in sorghum germplasm from South Asia conserved at ICRISAT genebank. *Plant Genetic Resources*, *15*(6), 527–538.

Upadhyaya, H. D., Vetriventhan, M., & Deshpande, S. (2016). Sorghum germplasm resources characterization and trait mapping. In *The Sorghum genome* (pp. 77–94). Cham: Springer.

Visarada, K. B. R. S., & Aruna, C. (2019). Chapter 1 – Sorghum: A bundle of opportunities in the 21st century. In C. Aruna, K. B. R. S. Visarada, B. V. Bhat, & V. A. Tonapi (Eds.), *Breeding Sorghum for diverse end uses* (pp. 1–14). Amsterdam, The Netherlands: Elsevier.

Wendmu, T. A., Cuni-Sanchez, A., Abebe, H. T., de Boer, H. J., Abera, F. A., & Westengen, O. T. (2022). Cultural effects on sorghum varieties grown, traits preferred, and seed management practices in northern Ethiopia. *Economic Botany*, *76*(3), 233–249.

Westengen, O. T., Okongo, M. A., Onek, L., Berg, T., Upadhyaya, H., Birkeland, S., . . . Brysting, A. K. (2014). Ethnolinguistic structuring of sorghum genetic diversity in Africa and the role of local seed systems. *Proceedings of the National Academy of Sciences*, *111*(39), 14100–14105.

Williams, J. R., Nelson, R. G., Aller, T. D., Claassen, M. M., & Rice, C. W. (2002). Derived carbon credit values for carbon sequestration: Do CO_2 emissions from production inputs matter?. American Agricultural Economics Association Meetings, Long Beach, CA, July 28–31.

Wright, A. L., & Hons, F. M. (2005). Carbon and nitrogen sequestration and soil aggregation under sorghum cropping sequences. *Biology and Fertility of Soils*, *41*, 95–100.

Xin, Z., Wang, M., Cuevas, H. E., Chen, J., Harrison, M., Pugh, N. A., & Morris, G. (2021). Sorghum genetic, genomic, and breeding resources. *Planta*, *254*(6), 114.

Yilmaz, S. H., Kaplan, M., Temizgul, R., & Yilmaz, S. (2017). Antioxidant enzyme response of sorghum plant upon exposure to aluminum, chromium and lead heavy metals. *Turkish Journal of Biochemistry*, *42*(4), 503–512.

Zhang, R., Yue, Z., Chen, X., Huang, R., Zhou, Y., & Cao, X. (2023). Effects of waterlogging at different growth stages on the photosynthetic characteristics and grain yield of sorghum (*Sorghum bicolor* L.). *Scientific Reports*, *13*(1), 7212.

Zhang, R., Zhou, Y., Yue, Z., Chen, X., Cao, X., Ai, X., . . . Xing, Y. (2019). The leaf-air temperature difference reflects the variation in water status and photosynthesis of sorghum under waterlogged conditions. *PLoS One*, *14*(7), Art. e0219209.

Zheng, H., Dang, Y., Diao, X., & Sui, N. (2023). Molecular mechanisms of stress resistance in sorghum: Implications for crop improvement strategies. *Journal of Integrative Agriculture, 23*(3), 741–768.

Zheng, L. Y., Guo, X. S., He, B., Sun, L. J., Peng, Y., Dong, S. S., . . . Jing, H. C. (2011). Genome-wide patterns of genetic variation in sweet and grain sorghum (*Sorghum bicolor*). *Genome Biology*, *12*, 1–15.

Zika, M., & Erb, K. H. (2009). The global loss of net primary production resulting from human-induced soil degradation in drylands. *Ecological Economics*, *69*(2), 310–318.

Zou, G., Zhai, G., Yan, S., Li, S., Zhou, L., Ding, Y., . . . Tao, Y. (2020). Sorghum *qTGW1a* encodes a G-protein subunit and acts as a negative regulator of grain size. *Journal of Experimental Botany*, *71*(18), 5389–5401.

2 Advancing Sorghum Improvement by Harnessing Genetic Diversity Through Advanced Tools and Speed Breeding

Ponnaiah Govintharaj, Karthika Gunasekaran, Pradeep Kumar, Niranjan Ravindra Thakur, Geeta Boken, and Preeti Adhana

2.1 INTRODUCTION

Sorghum (*Sorghum bicolor* L. Moench) is a climate-resilient staple food crop, essential for millions of people in arid and semi-arid regions globally. Known for its resilience to abiotic stresses such as drought, heat, and salinity (Karthika & Govintharaj, 2022; Somegowda et al., 2023), sorghum thrives in marginal lands, where its ability to grow under low-input conditions helps avoid direct competition with other food crops. As a diploid species (2n = 20) with a genome size of approximately 730 Mbp (Obilana, 2004; Paterson et al., 2009), sorghum ranks as the fifth most cultivated crop worldwide, valued for its use in food, feed, forage, and biofuel production. Its status as a C_4 crop endows it with high photosynthetic efficiency, exceptional water-use efficiency, and tolerance to abiotic stresses, making it a crucial crop for food security in the face of climate change (Karthika & Govintharaj, 2022).

Sorghum's origin date back approximately 7,500 years in the Sahara region, where it was initially used as a food source (Venkateswaran et al., 2019). Evidence of domesticated and wild sorghum, dating back to 3500–1500 B.C., has been discovered in the Kassala region (Beldados & Costantini, 2011). Sorghum is broadly categorized into four major groups based on utilization: grain sorghum, forage sorghum, biomass sorghum, and sweet sorghum (www.sorghumcheckoff.com/sorghum-101/what-is-sorghum/). The crop is rich in proteins, fats, carbohydrates, fiber, and essential minerals (Chadalavada et al., 2022; Ingle et al., 2023; Jadhav et al., 2024) and is highly suitable for gluten-free diets (Mathur et al., 2017). Additionally, it serves as an excellent source of health-promoting bioactive compounds (Stefoska-Needham et al., 2015).

The United States, Nigeria, India, Mexico, Brazil, Ethiopia, Sudan, China, Argentina, and Australia are the top ten sorghum-producing countries, collectively contributing approximately 74% of global sorghum production (USDA-FAS, 2024). Of this production, about 47.5% is consumed as food, while 34.7% is used as animal feed. Despite its importance, global sorghum yields remain low compared to other cereals, underscoring the need to explore and incorporate yield-enhancing genes from sorghum's crop wild relatives (CWRs) (Thakur et al., 2024a; Sasaki & Antonio, 2009).

CWRs, including wild progenitors, landraces, and closely related taxa, are reservoirs of valuable genetic diversity that can be harnessed to improve sorghum varieties. However, most studies

DOI: 10.1201/9781003623717-2

have focused on genetic variability within cultivated species, with limited exploration of CWRs, especially those in the secondary and tertiary gene pools. This gap is increasingly concerning as urbanization, rising global temperatures, uneven rainfall patterns, salinity issues, and the emergence of new pests and diseases exacerbate the challenges faced by agriculture (Karthika & Govintharaj, 2022). To address these challenges, breeders must develop superior cultivars capable of thriving in warmer, drier, and more variable climates by leveraging the genetic diversity of CWRs. Recent advancements in molecular breeding, genomic technologies, and CRISPR-based gene editing have enabled the efficient utilization of beneficial genes from CWRs, offering significant potential to enhance sorghum productivity and resilience in the face of changing global conditions. This chapter explores the role of CWRs and cutting-edge biotechnological tools in advancing sorghum breeding.

2.2 TAXONOMY AND CLASSIFICATION

The genus sorghum comprises 25 species, which are further divided into five subgenera or sections: Eu-Sorghum, Chaetosorghum, Heterosorghum, Para-Sorghum, and Stiposorghum (Garber, 1950). **(i) Section Eu-Sorghum**: this section includes *Sorghum bicolor* subsp. bicolor, *Sorghum halepense* (L.) Pers. (Johnsongrass), *Sorghum propinquum* (Kunth) Hitchc., Sorghum × almum Parodi, Sorghum × drummondii (Steud.) Millsp. & Chase, and *Sorghum arundinaceum* (Desv.) Stapf., which is the progenitor of *Sorghum bicolor* (Harlan & de Wet, 1971; Doggett, 1988). All *Sorghum bicolor* subsp. bicolor are diploid (2n = 2x = 20), annual in growth habit, and classified into five primary races: bicolor, guinea, caudatum, kafir, and durra, with an additional ten intermediate races recognized (Harlan & de Wet, 1972). **(ii) Section Chaetosorghum**: the section *Chaetosorghum* contains *Sorghum macrospermum* (E. D. Garber), an annual species endemic to the Northern Territory of Australia (Price et al., 2005a; Lazarides et al., 1991). **(iii) Section Heterosorghum**: the section *Heterosorghum* includes *Sorghum laxiflorum* (F. M. Bailey), an annual species native to Australia and New Guinea (Price et al., 2005a). **(iv) Section Para-Sorghum**: the section *Para-Sorghum* comprises *Sorghum grande* Lazarides, *Sorghum leiocladum* (Hack.) C. E. Hubb., *Sorghum matarankense* E. D. Garber & Snyder, *Sorghum nitidum* (Vahl) Pers., *Sorghum purpureosericeum* (Hochst. ex A. Rich.) Schweinf. & Asch., *Sorghum timorense* (Kunth) Büse, and *Sorghum versicolor* Andersson. Most of these species are native to Australia, with the exceptions of *Sorghum purpureosericeum* and *Sorghum versicolor*. Species such as *Sorghum grande*, *Sorghum leiocladum*, and *Sorghum matarankense* are more prevalent (Lazarides et al., 1991). Among these, *Sorghum nitidum* and *Sorghum leiocladum* exhibit perennial growth habits. **(v) Section Stiposorghum**: the section *Stiposorghum* includes *Sorghum amplum* (Lazarides), *Sorghum brachypodum* (Lazarides), *Sorghum angustum* (S. T. Blake), *Sorghum intrans* (F. Muell. ex Benth.), *Sorghum ecarinatum* (Lazarides), *Sorghum bulbosum* (Lazarides), *Sorghum plumosum* (R. Br.) (P. Beauv.), *Sorghum stipoideum* (Ewart & Jean White) (C. A. Gardner & C. E. Hubb.), *Sorghum interjectum* (Lazarides), and *Sorghum exstans* (Lazarides). All these species are native to Australia. Among them, *Sorghum interjectum* and *Sorghum plumosum* are perennial in nature (Lazarides et al., 1991).

2.3 GENETIC RESOURCES AND MAPPING POPULATIONS

The global conservation of sorghum germplasm comprises 259,595 accessions, including 2,465 wild and weedy relatives, preserved across 135 genebanks worldwide (www.croptrust.org/work/projects/global-crop-conservation-strategies/sorghum/). The International Crops Research Institute for the Semi-Arid Tropics (ICRISAT) holds the world's largest collection of sorghum germplasm, comprising 42,416 cultivated and 453 wild accessions from 112 countries (Vetriventhan et al., 2024). Of these, approximately 60% of wild accessions are from the primary gene pool (GP1), with the secondary (GP2) and tertiary gene pools (GP3) contributing fewer accessions (Vetriventhan et al., 2024).

The screening of wild and weedy relatives has led to the identification of numerous trait-specific donors that have proven valuable for sorghum breeding. These include traits related to grain yield, such as grains per panicle and seed size (Abreha et al., 2022), as well as physiological traits like higher photosystem II functionality, increased photosynthetic rate, and better relative water content under stress (Cowan et al., 2020). Additionally, traits related to drought tolerance, such as stay-green (Borrell et al., 2014), as well as chlorophyll content, green leaf retention at maturity, and yield, have also been identified as key breeding targets (Ochieng et al., 2021). Morphological traits such as leaf rolling, drying, root length, plant height, and flag leaf width (Techale et al., 2014), as well as reduced dhurrin content in the above-ground biomass (Cowan et al., 2020), have further expanded the potential of wild relatives in breeding programs.

However, while the use of wild sorghum germplasm in breeding programs has introduced beneficial traits such as photoperiod insensitivity and short stature (Stephens et al., 1967), it has also resulted in a reduction in genetic diversity for key traits like plant height and maturity. This reduction has become a concern for continued breeding success (Morris et al., 2013; Thurber et al., 2013). To address the decline in genetic diversity, several innovative mapping populations have been developed. Biparental mapping populations allow for targeted studies of specific traits, while association mapping panels facilitates broader genetic analyses for diverse traits. Additionally, multiparent populations, such as Nested Association Mapping (NAM), Multiparent Advanced Generation Intercross (MAGIC), and Back Cross-Nested Association Mapping (BC-NAM), have been developed to enhance the understanding of trait genetics and to promote their application in crop improvement (Narkhede et al., 2022; Guden et al., 2023; Kumar et al., 2023; Perumal et al., 2021; Winans et al., 2023). These advances in genetic resource utilization and mapping populations offer promising avenues for improving sorghum's genetic diversity, adaptability, and overall productivity.

2.4 SORGHUM IMPROVEMENT BY CLASSICAL APPROACHES

2.4.1 Wide Crosses

Crop wild relatives (CWR) serve as valuable sources of untapped genetic variability that can be leveraged for crop improvement. However, gene transfer between CWRs and cultivated sorghum species is often hindered by pre- and post-zygotic reproductive barriers. These barriers are primarily caused by pollen-pistil incompatibilities, differences in genome size and chromosome morphology, and embryo abortions (Garber & Snyder, 1951; Price et al., 2005b). Despite these challenges, various techniques have been developed to overcome these reproductive barriers, including chromosome doubling, bridge species, embryo rescue, and pollen grain irradiation (Kumari et al., 2016).

Several successful breeding attempts have been made to transfer genes from wild relatives to cultivated species. For instance, the embryo rescue method was successfully employed to recover progenies from the crosses between *Sorghum bicolor* and *S. macrospermum* (Price et al., 2005b). In another study, Kuhlman et al. (2010) developed a homozygous *S. bicolor* line carrying the recessive *iap* (inhibition of alien pollen) gene, which allows pollen tubes to grow and complete fertilization when pollen grains from GP3 are involved. Using this line, hybrids were successfully developed between *S. bicolor* and *S. macrospermum* (Kuhlman et al., 2010), as well as with *Saccharum* spp. (Hodnett et al., 2010). Other successful breeding attempts have focused on biomass-related traits, such as biomass yield, overwintering, and antioxidant traits, using a panel developed from crossing *Sorghum bicolor* × *S. halepense* (Habyarimana et al., 2018, 2019, 2020).

The high cross-compatibility within GP1 and between GP1 and the GP2 has enabled the development of many cultivars with useful traits that have been made available to sorghum producers. For instance, traits such as higher grain yield have been transferred from *Sorghum bicolor* subsp. *verticilliflorum* (Cox et al., 1984; Jordan et al., 2004) and *S. propinquum* (Wooten, 2001), while taller plant height and early maturity have been introduced from *S. propinquum* (Wooten, 2001). Additionally, genes for senescence and allelopathic properties have been transferred from *S. propinquum* (Feltus et al., 2006) and *S. bicolor* nothosubsp. *Drummondii* (Viaene & Abawi, 1998; Baerson et al., 2008), respectively. Furthermore, perennialism has been successfully introgressed from *S. halepense* into GP1 (Cox et al., 2002; Dweikat, 2005). Numerous other traits useful for breeding programs have also been identified in sorghum wild relatives, as summarized in Table 2.1. For instance, *Sorghum leiocladum* has shown potential for cold tolerance, likely due to its adaptation to temperate regions of New South Wales and Victoria, Australia. Similarly, species from GP3, such as *S. angustum*, *S. amplum*, and *S. bulbosum*, have demonstrated resistance to biotic stresses, including resistance to egg laying by sorghum midge (Sharma & Franzmann, 2001).

In addition to wide crosses within the *Sorghum* genus, several attempts have been made to introgress genes from closely related genera through intergeneric hybridization. *Saccharum* is a close relative of *Sorghum*, and intergeneric hybridization between *Saccharum* and *Sorghum* has been attempted, although with limited success (Venkatraman & Thomas, 1932; Subramonian, 1991). Nevertheless, clones resulting from the hybridization of *Saccharum officinarum* × *S. bicolor* have been recommended for biomass production in Japan (Terajima et al., 2007). Additionally, hybridization attempts between *sorghum* and other crops, such as *Miscanthus* spp., *Erianthus* spp., maize, and *Pennisetum*, have been made, but no viable plants were recovered (Ohadi et al., 2017).

2.4.2 Heterotic Grouping and Hybrid Breeding

Genetic similarity, or genetic distance (GD), is a key indicator for determining heterotic groups or pools within sorghum germplasm and genotypes. Heterotic groups are typically formed based on combining ability, pedigree, geographical origin, and molecular markers (Melchinger, 1999). Efforts to identify heterotic groups in sorghum have been ongoing. For instance, Menz et al. (2004) categorized sorghum germplasm into heterotic pools such as Zera-Zera, Kafir, Kafir-Milo, Durra, and Feterita using DNA-based molecular markers. Numerous studies have shown a positive correlation between GD of hybrid parents and grain yield heterosis in sorghum (Gabriel, 2005; Jordan et al., 2003). In Ethiopian sorghum breeding programs, Mindaye et al. (2015) identified six heterotic groups based on geographical origin, agro-ecological adaptation, and genetic differentiation in 184 genotypes.

Sorghum hybrid breeding typically involves the use of seed (B-lines) and restorer (R-lines) parental lines, which are often treated as two separate heterotic pools, though these groups may not always be genetically distinct (Menz et al., 2004; Mindaye et al., 2015). A study of 160 sorghum parental lines genotyped with 29,649 Genotyping-by-Sequencing (GBS) SNPs identified four heterotic groups based on pedigree and maintainer/restorer reactions (Silva et al., 2020). Similarly, K-means clustering of 435 sorghum inbred parental lines from Texas A&M using GBS revealed the formation of B-lines, R-lines, and forage lines, with minimal divergence among them (Crozier et al., 2020). In another study, Verma et al. (2022) identified two heterotic groups in rabi sorghum breeding materials based on maintainer and restorer reactions. More recently, Begna et al. (2024) classified sorghum lines into two heterotic groups (A and B) based on specific combining ability effects, and into three heterotic groups based on general combining ability effects, highlighting the potential for developing superior hybrids from broad-based and suitable parents.

2.4.3 Doubled Haploid Technology

The doubled haploid (DH) approach is one of the most promising methods for rapidly developing inbred (completely homozygous) parental lines, outperforming classical breeding techniques in terms of speed. This method involves various processes for haploid induction, such as androgenesis (microspore or anther culture), gynogenesis (ovary or unfertilized ovule culture), wide hybridization

TABLE 2.1
Crop wild relatives and their gene pool for trait improvement in sorghum

Biological status/ Species name	Traits	Reference
Primary gene pool		
S. propinquum	Increased grain yield, increased height, and earliness of development	Wooten (2001)
	Senescence	Feltus et al. (2006)
	Rhizomatousness, tillering and regrowth	Paterson et al. (1995)
	Rhizomatousness and overwintering	Washburn et al. (2013); Jessup et al. (2017)
	Rhizomatousness and vegetative branching	Kong et al. (2014, 2015)
S. bicolor subsp. *verticilliflorum*	Increased grain yield	Cox et al. (1984); Jordan et al. (2004)
	Ability to grow in drought conditions, seeds with tolerance to high temperatures, high yield, parasite resistance	Bramel-Cox and Cox (1988); Rich et al. (2004)
S. bicolor nothosubsp. Drummondii	Allelopathic properties, resistance to ergot and nematodes	Mojtahedi et al. (1993); Tsukiboshi et al. (1998); Viaene and Abawi (1998); Baerson et al. (2008)
Secondary gene pool		
S. halepense	Perennialism	Cox et al. (2002); Dweikat (2005)
	Resistance to green bug, chinch bug, and sorghum shoot fly	Nwanze et al. (1995); Dweikat, (2005)
Tertiary gene pool		
S. angustum	Resistance to egg laying by sorghum midge	Sharma and Franzmann, (2001)
S. amplum	Resistance to egg laying by sorghum midge	Sharma and Franzmann, (2001)
S. bulbosum	Resistance to egg laying by sorghum midge	Sharma and Franzmann, (2001)
S. macrospermum	Insect and disease resistance, higher growth rate and an insignificant aboveground dhurrin content under drought conditions	Kuhlman et al. (2008); Cowan et al. (2020)
S. brachypodum	Higher growth rate and an insignificant aboveground dhurrin content under drought conditions	Cowan et al. (2020)
S. exstans	Resistance to shoot fly	Kamala et al. (2009)
S. stipoideum	Resistance to shoot fly	Kamala et al. (2009)
S. matarankense	Resistance to shoot fly	Kamala et al. (2009)
S. leiocladum	Cold tolerance	Fiedler et al. (2016)

(elimination of chromosomes from one parent in a cross), and parthenogenesis (embryo development from a haploid egg cell or embryo sac).

Sorghum is particularly challenging for DH production, especially when using immature inflorescences as explants in tissue culture. The success rate of sorghum tissue culture is often limited by factors such as poor callus induction, low regeneration rates, phenolic compounds from somatic cells, and the occurrence of albino plantlets (Liang et al., 1997). To date, three methods including androgenesis, gynogenesis, and wide hybridization have been utilized in crop breeding to produce haploids. Kumaravadivel and Sree Rangasamy (1994) reported DH sorghum from anther culture using mid- to late uninucleate pollen, although the regeneration rate was low, at just 7.6%. Furthermore, anther culture in sorghum is genotype-dependent (Dora & Polumahanthi, 2011; Arulselvi & Krishnaveni, 2009; Bidmeshkipour et al., 2007).

Genotypes such as grain sorghum C401, C2-97, and Tx403 (Can et al., 1998; Sato et al., 2004), as well as sweet sorghum varieties R9188, Ramada, and Wray (Raghuwanshi & Birch, 2010), have shown higher responses to tissue culture. However, there have been no reports of DH production using gynogenesis or wide hybridization methods. More recently, two haploid inducer lines, SMHI01 and SMHI02D, were identified from a global collection of 4,000 sorghum germplasm accessions. These inducer lines exhibited a haploid generation frequency of about 1–2% (Hussain & Franks, 2019), marking a significant step forward in the production of DH sorghum.

2.4.4 Apomixis

Apomixis is an asexual reproductive mechanism through seeds, which is of great interest to breeders because it allows for the fixation and propagation of heterozygous genotypes, thereby exploiting heterosis or hybrid vigor. This mechanism reduces the time required to generate new commercial cultivars and decreases seed production costs. However, apomixis is a complex phenomenon that involves a combination of factors, including autonomous endospermogenesis (or the formation of pseudogamous endosperm), parthenogenesis of the egg cell, and the ability to form unreduced embryo sacs (Barcaccia & Albertini, 2013). These embryo sacs can either be aposporous (developing from somatic cells near the macrospore mother cell or meiotic embryo sac) or diplosporous (arising from disturbances in meiosis) (Barcaccia & Albertini, 2013; Fiaz et al., 2021). Studies have shown that progenies derived from crosses between apomictic lines and sexual lines in sorghum result in reduced apomictic seed formation, with the ability to transmit facultative apomixis characteristics differing in each case (Tang et al., 1980).

Several elite breeding lines and accessions have been used by researchers to study the apomixis mechanism in sorghum. Among these, the R473 breeding line has been extensively utilized to express varying levels of apomixis (Murty, 1993). Numerous reports have identified different apomixis mechanisms, such as aposporous or diplosporous embryo sacs (Hanna et al., 1970; Rao et al., 1978; Wu ShuBiao et al., 1994; Elkonin et al., 1995, 2012; Junai et al., 2004; Carman et al., 2011) and the autonomous development of both embryo and endosperm, as observed through cytological methods (Rao et al., 1978; Wu ShuBiao et al., 1994; Elkonin et al., 1995, 2012; Junai et al., 2004). A study by Elkonin et al. (2012) found that the AS-1a line exhibited apomictic properties only under drought and hot conditions, with no expression under moderate temperature and adequate moisture. More recently, the grain sorghum line AS-3, used in apomictic studies, has shown apomictic properties, including apospory, parthenogenesis, and autonomous endospermogenesis (Belyaeva et al., 2021).

2.5 SORGHUM IMPROVEMENT BY MODERN APPROACHES

Recent advances in genomic research, including the sequencing of the BTx623 reference genome, have provided critical tools for sorghum breeding (Somegowda et al., 2023). This section delves into the potential role of modern breeding approaches, such as quantitative trait locus (QTL) mapping, genome-wide association studies (GWAS), marker-assisted selection (MAS), genomic selection (GS), and transgenic technologies, in improving sorghum traits.

2.5.1 QTL Mapping

Quantitative trait locus (QTL) mapping is a fundamental approach for dissecting complex quantitative traits in plants, including sorghum (Choudhary et al., 2024). It identifies genomic regions associated with specific traits, facilitating their incorporation into breeding programs (Somegowda et al., 2023). The availability of the BTx623 reference genome has enabled the construction of high-density genetic linkage maps, significantly enhancing QTL mapping (Paterson et al., 2009). Early efforts using restriction fragment length polymorphism (RFLP) markers provided a broad overview of the sorghum genome but lacked resolution (Tao et al., 1998). The introduction of single nucleotide polymorphism (SNP) markers and reduced representation sequencing (i.e., RAD sequencing) has revolutionized QTL mapping by offering higher accuracy and throughput (Yonemaru et al., 2009).

For instance, a high-density map containing 3,710 SNPs was used to identify QTLs for traits such as plant height and flowering time (Kajiya-Kanegae et al., 2020). QTLs associated with drought resilience have been extensively studied in sorghum, with the stay-green trait, maintaining leaf functionality under drought stress, mapped to several loci (Mace et al., 2013). Similarly, QTLs for salinity tolerance, shoot fly resistance, and grain yield have been identified, demonstrating the versatility of this approach (Adeyanju et al., 2015). One challenge in QTL mapping is the limited transferability of identified QTLs across different populations due to genotype-environment interactions (Mwamahonje et al., 2021). Meta-analysis techniques address this issue by integrating data from diverse populations and environments to identify stable and transferable meta-QTLs (Somegowda et al., 2023; Karnatam et al., 2023). These advancements provide a robust framework for marker-assisted breeding and offer valuable insights for enhancing sorghum traits in diverse environments.

2.5.2 Genome-wide Association Mapping

Genome-wide association studies (GWAS) are a powerful tool for identifying marker-trait associations (MTAs) by leveraging the natural genetic diversity within sorghum populations. Unlike traditional QTL mapping, GWAS does not require the development of bi-parental populations, making it a cost-effective and efficient approach for trait discovery (Narkhede et al., 2022; Saini et al., 2022). In a study involving Ethiopian landraces, GWAS identified MTAs for drought resilience using 108,107 SNP markers distributed across the sorghum genome (Wondimu et al., 2021). Another GWAS analysis, using a sorghum association panel with 268,289 SNPs, identified loci on chromosomes 1 and 8 associated with grain mold resistance (Cuevas et al., 2019). Similarly, MTAs for anthracnose resistance were detected on chromosomes 4 and 10, contributing to the development of resistant cultivars (Adeyanju et al., 2015).

GWAS has also played a key role in understanding grain quality traits. For instance, loci associated with starch and protein content in sorghum were identified, providing valuable insights into the genetic basis of these nutritional traits (Mace et al., 2013). More recently, Thakur et al. (2024b) identified genomic regions related to grain nutrient traits such as iron and zinc, which are essential for developing nutrient-dense sorghum varieties. The integration of GWAS with genomic prediction models has further enhanced the efficiency of trait selection, particularly for complex traits with low heritability. This approach allows for the identification of promising genetic regions and accelerates breeding efforts to improve sorghum for various agronomic and nutritional traits (Somegowda et al., 2023).

2.5.3 Marker-assisted Selection

Over the past two decades, approximately 150 QTL and GWAS studies in sorghum have provided extensive data on the genetic basis of quantitative traits, identifying over 6,000 QTLs/MTAs, including their genomic locations, allelic effects, and epistatic interactions (https://aussorgm.org.au/sorghum-qtl-atlas/). Despite this wealth of information, its full potential has not been realized in sorghum improvement programs due to the heterogeneous nature of these studies, which has led to varying reliability in the identified QTLs. Marker-assisted selection (MAS) combines molecular markers with phenotypic data to expedite the breeding of improved cultivars, particularly for traits with complex genetic architectures (Ingle et al., 2023).

SSR markers, known for their high polymorphism and reproducibility, have been widely used in sorghum breeding. For instance, QTLs for shoot fly resistance located on SBI-01, SBI-07, and SBI-10 were introgressed using SSR markers into elite post-rainy season varieties, significantly enhancing both resistance and grain yield (Adeyanju et al., 2015). Similarly, stay-green QTLs were incorporated into B35-derived lines, resulting in superior drought resilience (Mace et al., 2013). The advent of next-generation genotyping platforms, such as GBS and KASP, has further revolutionized MAS by enabling high-throughput screening of SNP markers (Somegowda et al., 2023). These technologies allow the simultaneous selection of multiple traits, such as salinity tolerance and pest resistance, within a single breeding cycle (Cuevas et al., 2019). MAS has also shown promise in enhancing sorghum's bioenergy potential. For instance, low-lignin lines containing the *bmr6* allele have been developed to improve biomass digestibility for ethanol production (Reddy et al., 2024; Somegowda et al., 2023). These examples underscore the versatility of MAS in addressing diverse breeding objectives.

2.5.4 Genomic Selection

Recent advancements in NGS technologies have significantly enhanced breeding programs, enabling rapid genetic gains. While MAS has been successful for traits controlled by a few major genes (Lande & Thompson, 1990), most agronomically and economically important traits in crops are controlled by the cumulative effects of minor genes (Riedelsheimer et al., 2012), where MAS is less effective. To address this, genomic selection (GS) predicts breeding values by utilizing genome-wide markers across all chromosomes. GS accelerates genetic gain by reducing costs and breeding cycle time (Hickey et al., 2014), as well as lowering labor and field phenotyping costs compared to traditional breeding methods (Bhat et al., 2016).

Although GS studies in sorghum have been limited, most focus on training models with designed populations such as testcross hybrids and natural populations. These studies have shown that GS is both effective and economical (Yu et al., 2016; de Oliveira et al., 2018; Fernandes et al., 2018; Hunt et al., 2018; Velazco et al., 2019; Habyarimana et al., 2020; Dos Santos et al., 2020; Sapkota et al., 2020). Among various GS models, the rrBLUP model has proven to be the most promising in sorghum (Yu et al., 2016; Dos Santos et al., 2020). Incorporating genotype-by-environment interaction (GEI) effects into genomic prediction models has been shown to improve selection efficiency and hybrid deployment (Hunt et al., 2018). Multi-trait GS has also been found to be more effective than single-trait GS in sorghum (Habyarimana et al., 2020).

Fonseca et al. (2021) compared kernel-based Bayesian GBLUP models (GB) with classical GCA-SCA-based models (BRR) in grain sorghum and found higher prediction accuracies for both models, with the GB model showing greater accuracy when GEI effects were included. Crozier et al. (2023) confirmed that GEI effects positively influence the prediction accuracy of key traits in sorghum. In another study, higher prediction accuracy for agronomic traits was observed in 602 grain sorghum hybrids when both parents were included in the training set for the GBLUP model, with accuracy dropping by 12–90% when only the female tester parent was included (Sapkota et al., 2023). Increasing the training population size also improved prediction accuracy for all traits studied in 204 grain sorghum hybrids (Maulana et al., 2023).

More recently, Charles et al. (2024) evaluated 250 sorghum breeding lines under irrigated and water-stressed conditions using Bayes B and Bayes ridge regression (BRR) models. They found higher prediction accuracy within environmental conditions than across environments, with prediction accuracy ranging from 0.38 for total stem number to 0.71 for total soluble solids. Additionally, Winans et al. (2024) applied both genotypic and environmental data to predict hybrid performance in various U.S. sorghum-producing regions. Their study highlighted that enviromic-assisted genomic prediction can enhance accuracy for unobserved environments, supporting breeding programs with sparse hybrid trials.

2.5.5 Transgenesis

Transgenic technologies offer significant opportunities to introduce novel traits into sorghum, helping to address various challenges such as abiotic stress, pest resistance, and nutritional deficiencies. For abiotic stress tolerance, transgenic sorghum lines have shown enhanced drought resilience. For instance, overexpression of drought-responsive genes such as *sbNAC9* improved drought tolerance by enhancing the scavenging ability of reactive oxygen species and activating stress-responsive genes (Jin et al., 2023). Additionally, proline over-accumulation has been shown to alleviate salt stress and protect photosynthetic and antioxidant enzyme activities in transgenic sorghum (Reddy et al., 2015).

In terms of pest resistance, Bt sorghum expressing *Cry1Ac* toxins has proven effective in controlling stem borers, with up to 60% reduction in leaf damage observed five days after infestation (Girijashankar et al., 2005). Moreover, transgenic approaches have enhanced resistance to fungal pathogens through the expression of antifungal proteins and enzymes (Akosambo-Ayoo et al., 2011). Nutritional enhancement is another important focus of transgenic research. Biofortified sorghum lines with increased iron and zinc content have been developed to address micronutrient deficiencies, thus contributing to the global effort to combat hidden hunger (Lipkie et al., 2013). These advancements demonstrate the potential of transgenic technologies not only to improve sorghum's resilience to environmental stress and pests but also to contribute to global food security and nutritional improvements.

2.5.6 Genome Editing

Genome editing is a revolutionary breeding technique that enables sorghum researchers to make targeted and specific modifications in the crop's DNA. These techniques provide a precise, efficient, and effective way to identify, alter, and transfer novel genes, ultimately improving sorghum production and productivity (Balakrishna et al., 2020). Several target genes in sorghum have been edited using CRISPR/Cas9 technology through methods like Agrobacterium-mediated transformation and biolistic bombardment (Balakrishna et al., 2020).

For instance, the kafirin-encoding *K1C* genes have been targeted to enhance grain protein digestibility and lysine content (Li et al., 2018a). Additionally, positive homozygous mutants were achieved in T_1 generations by editing the cinnamyl alcohol dehydrogenase (*CAD*) and phytoene desaturase (*PDS*) genes (Liu et al., 2019). Moreover, the knockout of betaine aldehyde dehydrogenase 2 (*SbBADH2*) using CRISPR/Cas9 resulted in an aromatic smell in both seeds and leaves (Zhang et al., 2022). Recently, the *SbBADH2* gene has been targeted for fragrant traits in sorghum using genome editing technologies (Cheng et al., 2024).

Furthermore, a multiplex CRISPR/Cas9 strategy can be effectively applied in sorghum, enabling the simultaneous editing of sequences in cis-regulatory regions and upstream open reading frames (uORFs) of genes associated with key agronomic traits such as shoot architecture, flower and seed production, day-length sensitivity, and nutrient content. This approach holds significant potential to accelerate the domestication of wild species, enhancing traits that are critical for crop improvement (Li et al., 2018b). By enabling precise editing of multiple genes, genome editing technologies offer exciting opportunities to enhance sorghum productivity, providing breeders with the ability to target specific traits with greater efficiency and speed.

2.6 EMERGING APPROACHES FOR SORGHUM IMPROVEMENT

2.6.1 Haplotype-assisted Breeding

Recent advancements in NGS technologies, such as GBS, whole genome sequencing (WGS), whole-genome resequencing (WGRS), restriction site-associated DNA sequencing (RAD-seq), and fixed SNP arrays, have significantly facilitated the genotyping of larger germplasm collections for GS and GWAS (Zargar et al., 2015; Annicchiarico et al., 2017). These technologies have greatly improved our ability to explore genetic variation in crops like sorghum. However, the use of SNPs identified in these studies has some limitations due to their biallelic nature, the occurrence of rare alleles, and high levels of linkage drag (Wray et al., 2013; Voss-Fels & Snowdon, 2016). As a result, candidate genomic regions identified in GWAS often do not correspond directly to the causative loci but rather to loci linked to a gene or regulatory element that influences the trait of interest (Mackay et al., 2009; Korte & Farlow, 2013).

To overcome these limitations, haplotypes have emerged as a more effective approach. A haplotype is a group of DNA variants or markers that are inherited together on the same chromosome, often reflecting a more comprehensive genetic profile than individual SNPs. Recent work has introduced practical haplotype graphs (PHGs) in sorghum, which integrate WGS and marker-based technologies to better capture genetic variability (Jensen et al., 2020). These PHGs provide a more refined understanding of genetic variation, enhancing the resolution of candidate genomic regions linked to key traits. For instance, Thakur et al. (2024b) identified haplotypes associated with grain Fe and zinc Zn content in sorghum using genetically diverse lines and SNPs. These findings provide valuable insights that can serve as the foundation for biofortification breeding programs aimed at improving the nutritional quality of sorghum.

2.6.2 Speed Breeding

The development and release of cultivars for commercial cultivation typically takes 8 to 12 years through traditional plant breeding methods (Shimelis & Laing, 2012). In recent years, speed breeding has gained popularity among researchers as a strategy to reduce the breeding cycle time by manipulating factors such as light intensity, temperature, photoperiod, nutrition, soil moisture, and planting density. For instance, Yang et al. (2014) demonstrated that sorghum genotypes (BTx642/Tx7000 RIL population and parental lines) flowered earlier under long-day (LD, 14/10 hr light/dark) or short-day (SD, 10/14 hr light/dark) conditions at a temperature of 30°C during the day and 23°C at night. It was also observed that temperatures above the optimum (33°C) negatively affected pollen viability, leading to sterility in sorghum (Singh et al., 2015).

Further studies have shown that sorghum genotypes flowered nine days earlier under well-watered conditions compared to drought-stressed conditions (Munamava & Riddoch, 2001). In another study, Jones and Johnson (1991) reported that crop development stages and grain yield were unaffected by higher planting densities (four to eight plants/m^2) in sorghum. Similarly, Villar et al. (1989) found a reduction in the number of flowering days from 59 to 50 under high-density planting. The use of automated glasshouses and growth chambers is also instrumental in accelerating breeding cycles. By manipulating factors such as plot size, temperature, irrigation, photoperiod, and employing *in vitro* rescue techniques, breeding cycles can be significantly shortened. With such management practices, including small plots, restricted watering, regular lighting, and embryo rescue, sorghum can be grown for up to six generations per year (Ghanim et al., 2014; Indriatama, 2020).

A study by Rizal et al. (2014) found that selfing in monoculm plants combined with the embryo rescue technique reduced the breeding cycle from 17 weeks to 11 weeks, enabling faster cultivar development. More recently, Ghanim et al. (2023) optimized and validated a sorghum crop protocol by manipulating pot size, watering, temperature, photoperiod, and employing the immature embryo rescue technique. Their findings indicated that under optimized protocols with limited rooting areas, the seed-to-seed or embryo rescue cycle could be completed in two to three months, compared to the five to six months required under normal field conditions. This optimization allows for four to six breeding cycles per year, significantly accelerating cultivar development (Ghanim et al., 2023).

2.6.3 Pan-genomics Assisted Breeding

Recent advancements in sequencing technologies have made it more feasible to sequence a wider range of species, particularly as sequencing costs have decreased. Previously, molecular markers and gene cloning studies revealed variations between individuals of the same species, but due to low-quality genome assemblies, it was challenging to distinguish between biological variation and errors in assembly and annotation. To date, three sorghum pan-genomes have been reported (Jensen et al., 2020; Ruperao et al., 2021; Tao et al., 2021), which have potential to significantly advance crop improvement through genome-assisted breeding and gene discovery.

Pan-genomics-assisted breeding involves the comprehensive analysis of genetic diversity across a species by examining both core and accessory genes in various genotypes. This approach enables the identification of novel alleles associated with important traits such as drought tolerance, pest resistance, and nutritional enhancement. In sorghum breeding, pan-genomics allows breeders to access a broader pool of genetic variation, improving the identification of genes that control complex traits. It also enhances genomic prediction models, increasing the accuracy of selecting superior genotypes. By incorporating wild relatives and landraces, pan-genomics can accelerate the domestication of new varieties, leading to more resilient and productive crops. When combined with tools like GWAS and GS, pan-genomics offers a powerful strategy to address climate change challenges and improve crop performance.

2.7 CONCLUSION

Sorghum is a highly versatile crop with a variety of end uses, and it is widely cultivated in hot and drought-prone environments, particularly by smallholder farmers in arid and semi-arid regions across the globe. Despite its widespread cultivation, sorghum has relatively lower productivity compared to other cereal crops. The wild relatives of sorghum harbor many valuable traits that could be leveraged to address current and future climate challenges. However, gene transfer between wild species and cultivated sorghum has been limited due to a lack of detailed information on trait variability and the presence of reproductive barriers, both pre- and post-zygotic. Recent advancements in genomic technologies, including genome sequencing and marker-assisted selection, have facilitated the introgression of beneficial genes from wild relatives into cultivated sorghum. These developments have led to improvements in sorghum productivity, resilience, and nutritional value, offering promising prospects for the future of sorghum breeding. As genomic tools continue to evolve, they will likely accelerate the incorporation of beneficial traits, enhancing sorghum's adaptability to changing climate conditions and contributing to global food security.

REFERENCES

Abreha, K. B., Enyew, M., Carlsson, A. S., Vetukuri, R. R., Feyissa, T., Motlhaodi, T., . . . Geleta, M. (2022). Sorghum in dryland: Morphological, physiological, and molecular responses of sorghum under drought stress. *Planta*, *255*, 1–23.

Adeyanju, A., Little, C., Yu, J., & Tesso, T. (2015). Genome-wide association study on resistance to stalk rot diseases in grain sorghum. *G3: Genes, Genomes, Genetics*, *5*(6), 1165–1175.

Akosambo-Ayoo, L. M., Bader, M., Loerz, H., & Becker, D. (2011). Transgenic sorghum (*Sorghum bicolor* L. Moench) developed by transformation with chitinase and chitosanase genes from *Trichoderma harzi*anum expresses tolerance to anthracnose. *African Journal of Biotechnology*, *10*(19), 3659–3670.

Annicchiarico, P., Nazzicari, N., Pecetti, L., Romani, M., Ferrari, B., Wei, Y., & Brummer, E. C. (2017). GBS-based genomic selection for pea grain yield under severe terminal drought. *The Plant Genome*, *10*(2), 1–13.

Arulselvi, P. I., & Krishnaveni, S. (2009). Effect of hormones, explants and genotypes in in vitro culturing of sorghum. *Journal of Biochemical Technology*, *1*(4), 96–103.

Baerson, S. R., Dayan, F. E., Rimando, A. M., Nanayakkara, N. D., Liu, C. J., Schroder, J., . . . Duke, S. O. (2008). A functional genomics investigation of allelochemical biosynthesis in *Sorghum bicolor* root hairs. *Journal of Biological Chemistry*, *283*(6), 3231–3247.

Balakrishna, D., Singode, A., Narasimham, D., & Bhat, B. V. (2020). Current status and future prospects of genetic transformation and gene editing in sorghum. In Tonapi, V.A., Talwar, H.S., Are, A.K., Bhat, B.V., Reddy, C.R., Dalton, T.J. (eds.), *Sorghum in the 21st century: Food–fodder–feed–fuel for a rapidly changing world* (pp. 511–535). Singapore: Springer.

Barcaccia, G., & Albertini, E. (2013). Apomixis in plant reproduction: A novel perspective on an old dilemma. *Plant Reproduction*, *26*, 159–179.

Begna, T., Birhan, T., & Tadesse, T. (2024). Estimation of standard heterosis and heterotic grouping of elite sorghum [*Sorghum bicolor* (L) Moench] inbred lines under moisture stress conditions. *Ecological Genetics and Genomics*, *31*, Art. 100242.

Beldados, A., & Costantini, L. (2011). Sorghum exploitation at Kassala and its environs, Northeastern Sudan in the second and first millennium BC. *Nyame Akuma*, *75*, 33–39.

Belyaeva, E. V., Elkonin, L. A., Vladimirova, A. A., & Panin, V. M. (2021). Manifestation of apomictic potentials in the line AS-3 of *Sorghum bicolor* (L.) Moench. *Planta*, *254*(2), 37.

Bhat, J. A., Ali, S., Salgotra, R. K., Mir, Z. A., Dutta, S., Jadon, V., . . . Prabhu, K. (2016). Genomic selection in the era of next generation sequencing for complex traits in plant breeding. *Frontiers in Genetics*, *7*, 221.

Bidmeshkipour, A., Thengane, R. J., Bahagvat, M. D., Ghafari, S. M., & Rao, V. S. (2007). Production of haploid wheat via maize pollination. *Journal of Sciences, Islamic Republic of Iran*, *18*, 5–11

Borrell, A. K., Mullet, J. E., George-Jaeggli, B., van Oosterom, E. J., Hammer, G. L., Klein, P. E., & Jordan, D. R. (2014). Drought adaptation of stay-green sorghum is associated with canopy development, leaf anatomy, root growth, and water uptake. *Journal of Experimental Botany*, *65*(21), 6251–6263.

Bramel-Cox, P. J., & Cox, T. S. (1988, December). Use of wild germplasm in sorghum improvement. In *Proceedings of 43rd Annual Corn and Sorghum Research Conference*, USA (Washington DC: American Seed Trade Association) (pp. 13–25).

Can, N. D., Nakamura, S., Haryanto, T. A. D., & Yoshida, T. (1998). Effects of physiological status of parent plants and culture medium composition on the anther culture of sorghum. *Plant Production Science*, *1*(3), 211–215.

Carman, J. G., Jamison, M., Elliott, E., Dwivedi, K. K., & Naumova, T. N. (2011). Apospory appears to accelerate onset of meiosis and sexual embryo sac formation in sorghum ovules. *BMC Plant Biology*, *11*, 1–13.

Chadalavada, K., Guna, K., Kumari, B. R., & Kumar, T. S. (2022). Drought stress in sorghum: Impact on grain quality. In Shanker, A. K., Shanker, C., Anand, A., Maheswari, M. (eds.), *Climate change and crop stress* (pp. 113–134). Cambridge, MA: Academic Press.

Charles, J. R., Dorval, M. D., Durone, J. B., Ferrão, L. F. V., Amadeu, R. R., Munoz, P. R., . . . Pressoir, G. (2024). Genomic prediction of sweet sorghum agronomic performance under drought and irrigated environments in Haiti. *Crop Science, 64*, 1595–1607.

Cheng, Z., Li, K., Liu, H., Wei, X., Yin, T., Xing, X., . . . Sui, Y. (2024). Establishment of a genome-editing system to create fragrant germplasm in sweet sorghum. *aBIOTECH*, *5*, 502–506.

Choudhary, M., Kumar, P., Prakash, N. R., Gupta, M., Sheoran, S., Kumar, P., . . . Saini, D. K. (2024). QTLomics-based genetic dissection of key traits in major cereals: Challenges and prospects. In Saini, D.K. and Kole, C. (eds.), *Omics and System Biology Approaches for Delivering Better Cereals* (pp. 71–110). Boca Raton: CRC Press.

Cowan, M. F., Blomstedt, C. K., Norton, S. L., Henry, R. J., Møller, B. L., & Gleadow, R. (2020). Crop wild relatives as a genetic resource for generating low-cyanide, drought-tolerant Sorghum. *Environmental and Experimental Botany*, *169*, Art. 103884.

Cox, T. S., Bender, M., Picone, C., Tassel, D. V., Holland, J. B., Brummer, E. C., . . . Jackson, W. (2002). Breeding perennial grain crops. *Critical Reviews in Plant Sciences*, *21*(2), 59–91.

Cox, T. S., House, L. R., & Frey, K. J. (1984). Potential of wild germplasm for increasing yield of grain sorghum. *Euphytica*, *33*, 673–684.

Crozier, D., Hoffmann Jr, L., Klein, P. E., Klein, R. R., & Rooney, W. L. (2020). Predicting heterosis in grain sorghum hybrids using sequence-based genetic similarity estimates. *Journal of Crop Improvement*, *34*(5), 600–617.

Crozier, D., Leon, F., Fonseca, J. M., Klein, P. E., Klein, R. R., & Rooney, W. L. (2023). Inbred phenotypic data and non-additive effects can enhance genomic prediction models for hybrid grain sorghum. *Crop Science*, *63*(3), 1183–1196.

Cuevas, H. E., Fermin-Pérez, R. A., Prom, L. K., Cooper, E. A., Bean, S., & Rooney, W. L. (2019). Genome-wide association mapping of grain mold resistance in the US sorghum association panel. *The Plant Genome*, *12*(2), Art. 180070.

de Oliveira, A. A., Pastina, M. M., de Souza, V. F., da Costa Parrella, R. A., Noda, R. W., Simeone, M. L. F., . . . Margarido, G. R. A. (2018). Genomic prediction applied to high-biomass sorghum for bioenergy production. *Molecular Breeding*, *38*, 1–16.

Doggett, H. (1988). *Sorghum* (2nd ed.). New York: Wiley.

Dora, S. V. V. N., & Polumahanthi, S. (2011). Efficient callus induction protocol for *Sorghum bicolor*. *Asian Journal of Plant Science & Research, 4*, 14–21.

Dos Santos, J. P., Fernandes, S. B., McCoy, S., Lozano, R., Brown, P. J., Leakey, A. D., . . . Gore, M. A. (2020). Novel bayesian networks for genomic prediction of developmental traits in biomass sorghum. *G3: Genes, Genomes, Genetics*, *10*(2), 769–781.

Dweikat, I. (2005). A diploid, interspecific, fertile hybrid from cultivated sorghum, *Sorghum bicolor*, and the common Johnsongrass weed *Sorghum halepense*. *Molecular Breeding*, *16*, 93–101.

Elkonin, L. A., Belyaeva, E. V., & Fadeeva, I. Y. (2012). Expression of the apomictic potential and selection for apomixis in sorghum line AS-1a. *Russian Journal of Genetics*, *48*(1), 32–40.

Elkonin, L. A., Enaleeva, N. K., Tsvetova, M. I., Belyaeva, E. V., & Ishin, A. G. (1995). Partially fertile line with apospory obtained from tissue culture of male sterile plant of sorghum (*Sorghum bicolor* L. Moench). *Annals of botany*, *76*(4), 359–364.

Feltus, F. A., Hart, G. E., Schertz, K. F., Casa, A. M., Kresovich, S., Abraham, S., . . . Paterson, A. H. (2006). Alignment of genetic maps and QTLs between inter-and intra-specific sorghum populations. *Theoretical and Applied Genetics*, *112*, 1295–1305.

Fernandes, S. B., Dias, K. O., Ferreira, D. F., & Brown, P. J. (2018). Efficiency of multi-trait, indirect, and trait-assisted genomic selection for improvement of biomass sorghum. *Theoretical and Applied Genetics*, *131*, 747–755.

Fiaz, S., Wang, X., Younas, A., Alharthi, B., Riaz, A., & Ali, H. (2021). Apomixis and strategies to induce apomixis to preserve hybrid vigor for multiple generations. *GM Crops & Food*, *12*(1), 57–70.

Fiedler, K., Bekele, W. A., Matschegewski, C., Snowdon, R., Wieckhorst, S., Zacharias, A., & Uptmoor, R. (2016). Cold tolerance during juvenile development in sorghum: A comparative analysis by genome wide association and linkage mapping. *Plant Breeding*, *135*(5), 598–606.

Fonseca, J. M., Klein, P. E., Crossa, J., Pacheco, A., Perez-Rodriguez, P., Ramasamy, P., . . . Rooney, W. L. (2021). Assessing combining abilities, genomic data, and genotype × environment interactions to predict hybrid grain sorghum performance. *The Plant Genome*, *14*(3), Art. e20127.

Gabriel, K. (2005). *A study of heterotic relationships in sorghum*. College Station, TX, USA: Texas A&M University.

Garber, E. D. (1950). Cytotaxonomic studies in the genus *Sorghum*. Berkeley, CA: University of California Publications in Botany.

Garber, E. D., & Snyder, L. A. (1951). Cytotaxonomic studies in the genus *Sorghum*. II. Two new species from Australia. *Madroño*, *11*(1), 6–10.

Ghanim, A. M. A., Ali, A. B., Sen, A., Ingelbrecht, I., & Sivasankar, S. (2023). Rapid cycling and generation advancement for accelerated mutation breeding in sorghum. In *Mutation breeding and efficiency enhancing technologies for resistance to striga in cereals* (pp. 161–170). Berlin: Springer.

Ghanim, A. M. A., Bado, S., Den, A., Ahmed, M. S., Tassoob-Shirazi, F., & Amos, E. K. (2014). *Techniques in accelerating plant mutation breeding in wheat and sorghum*. Vienna, Austria: VISCEA.

Girijashankar, V., Sharma, H. C., Sharma, K. K., Swathisree, V., Prasad, L. S., Bhat, B. V., . . . Seetharama, N. (2005). Development of transgenic sorghum for insect resistance against the spotted stem borer (*Chilo partellus*). *Plant Cell Reports*, *24*, 513–522.

Guden, B., Yol, E., Erdurmus, C., Lucas, S. J., & Uzun, B. (2023). Construction of a high-density genetic linkage map and QTL mapping for bioenergy-related traits in sweet sorghum [*Sorghum bicolor* (L.) Moench]. *Frontiers in Plant Science*, *14*, Art. 1081931.

Habyarimana, E., Dall'Agata, M., De Franceschi, P., & Baloch, F. S. (2019). Genome-wide association mapping of total antioxidant capacity, phenols, tannins, and flavonoids in a panel of *Sorghum bicolor* and *S. bicolor* × *S. halepense* populations using multi-locus models. *PLoS One*, *14*(12), Art. e0225979.

Habyarimana, E., Lopez-Cruz, M., & Baloch, F. S. (2020). Genomic selection for optimum index with dry biomass yield, dry mass fraction of fresh material, and plant height in biomass sorghum. *Genes*, *11*(1), 61.

Habyarimana, E., Lorenzoni, C., Redaelli, R., Alfieri, M., Amaducci, S., & Cox, S. (2018). Towards a perennial biomass sorghum crop: A comparative investigation of biomass yields and overwintering of *Sorghum bicolor* x *S. halepense* lines relative to long term *S. bicolor* trials in northern Italy. *Biomass and Bioenergy*, *111*, 187–195.

Hanna, W. W., Schertz, K. F., & Bashaw, E. C. (1970). Apospory in *Sorghum bicolor* (L.) Moench. *Science*, *170*(3955), 338–339.

Harlan, J. R., & de Wet, J. M. (1971). Toward a rational classification of cultivated plants. *Taxon*, *20*(4), 509–517.

Harlan, J. R., & de Wet, J. M. (1972). A simplified classification of cultivated sorghum 1. *Crop Science*, *12*(2), 172–176.

Hickey, J. M., Dreisigacker, S., Crossa, J., Hearne, S., Babu, R., Prasanna, B. M., . . . Gorjanc, G. (2014). Evaluation of genomic selection training population designs and genotyping strategies in plant breeding programs using simulation. *Crop Science*, *54*(4), 1476–1488.

Hodnett, G. L., Hale, A. L., Packer, D. J., Stelly, D. M., Da Silva, J., & Rooney, W. L. (2010). Elimination of a reproductive barrier facilitates intergeneric hybridization of *Sorghum bicolor* and *Saccharum*. *Crop science*, *50*(4), 1188–1195.

Hunt, C. H., van Eeuwijk, F. A., Mace, E. S., Hayes, B. J., & Jordan, D. R. (2018). Development of genomic prediction in sorghum. *Crop Science*, *58*(2), 690–700.

Hussain, T., & Franks, C. (2019). Discovery of sorghum haploid induction system. In *Sorghum: Methods and protocols* (pp. 49–59). New York, NY: Springer New York.

Indriatama, W. M. (2020). Accelerated purification of sorghum mutant line by using rapid cycling methods. *Journal of Physics: Conference Series*, *1436*(1), 012028.

Ingle, K., Thakur, N., Moharil, M. P., Suprasanna, P., Awio, B., Narkhede, G., . . . Abdi, G. (2023). Current status and future prospects of molecular marker assisted selection (MAS) in millets. In Pudake, R.N., Solanke, A.U., and Kole, C. (eds.), *Nutriomics of Millet Crops*. Boca Raton, FL: CRC Press.

Jadhav, Y., Thakur, N. R., Ingle, K. P., & Ceasar, S. A. (2024). The role of phenomics and genomics in delineating the genetic basis of complex traits in millets. *Physiologia Plantarum*, *176*(3), Art. e14349.

Jensen, S. E., Charles, J. R., Muleta, K., Bradbury, P. J., Casstevens, T., Deshpande, S. P., . . . Ramstein, G. P. (2020). A sorghum practical haplotype graph facilitates genome-wide imputation and cost-effective genomic prediction. *The Plant Genome*, *13*(1), Art. e20009.

Jessup, R. W., Klein, R. R., Burson, B. L., Murray, S. C., Washburn, J. D., Heitholt, J. J., & Foster, J. L. (2017). Registration of perennial *Sorghum bicolor* × *S. propinquum* line PSH12TX09. *Journal of Plant Registrations*, *11*(1), 76–79.

Jin, X., Zheng, Y., Wang, J., Chen, W., Yang, Z., Chen, Y., . . . Sun, B. (2023). SbNAC9 improves drought tolerance by enhancing scavenging ability of reactive oxygen species and activating stress-responsive genes of sorghum. *International Journal of Molecular Sciences*, *24*(3), 2401.

Jones, O. R., & Johnson, G. L. (1991). Row width and plant density effects on Texas High Plains sorghum. *Journal of Production Agriculture*, *4*(4), 613–621.

Jordan, D., Tao, Y., Godwin, I., Henzell, R., Cooper, M., & McIntyre, C. (2003). Prediction of hybrid performance in grain sorghum using RFLP markers. *Theoretical and Applied Genetics*, *106*, 559–567.

Jordan, J., Butler, D., Henzell, B., Drenth, J., & McIntyre, L. (2004, September). Diversification of Australian sorghum using wild relatives. In *New directions for a diverse planet: Proceedings of the 4th International Crop Science Congress, Brisbane, Australia* (Vol. 26).

Junai, P., Fuyao, Z., Guimei, C., Qingjun, C., Zhihong, D., & Yixian, Z. (2004). A study on the properties of autonomous seed setting and embryology in sorghum apomictic line 2083. *Zuo wu xue bao*, *30*(7), 714–718.

Kajiya-Kanegae, H., Takanashi, H., Fujimoto, M., Ishimori, M., Ohnishi, N., Wacera W, F., . . . Sakamoto, W. (2020). RAD-seq-based high-density linkage map construction and QTL mapping of biomass-related traits in sorghum using the Japanese landrace Takakibi NOG. *Plant and Cell Physiology*, *61*(7), 1262–1272.

Kamala, V., Sharma, H. C., Manohar Rao, D., Varaprasad, K. S., & Bramel, P. J. (2009). Wild relatives of sorghum as sources of resistance to sorghum shoot fly, *Atherigona soccata*. *Plant Breeding*, *128*(2), 137–142.

Karnatam, K. S., Chhabra, G., Saini, D. K., Singh, R., Kaur, G., Praba, U. P., . . . Vikal, Y. (2023). Genome-wide meta-analysis of QTLs associated with root traits and implications for maize breeding. *International Journal of Molecular Sciences*, *24*(7), 6135.

Karthika, G., & Govintharaj, P. (2022). Breeding climate-resilience crops for future agriculture. In Shanker, A.K., Shanker, C., Anand, A., and Maheswari, M. (eds.), *Climate change and crop stress* (pp. 1–32). Cambridge, MA: Academic Press.

Kong, W., Guo, H., Goff, V. H., Lee, T. H., Kim, C., & Paterson, A. H. (2014). Genetic analysis of vegetative branching in sorghum. *Theoretical and Applied Genetics*, *127*, 2387–2403.

Kong, W., Kim, C., Goff, V. H., Zhang, D., & Paterson, A. H. (2015). Genetic analysis of rhizomatousness and its relationship with vegetative branching of recombinant inbred lines of *Sorghum bicolor* × *S. propinquum*. *American Journal of Botany*, *102*(5), 718–724.

Korte, A., & Farlow, A. (2013). The advantages and limitations of trait analysis with GWAS: A review. *Plant Methods*, *9*, 1–9.

Kuhlman, L. C., Burson, B. L., Klein, P. E., Klein, R. R., Stelly, D. M., Price, H. J., & Rooney, W. L. (2008). Genetic recombination in *Sorghum bicolor* × *S. macrospermum* interspecific hybrids. *Genome*, *51*(9), 749–756.

Kuhlman, L. C., Burson, B. L., Stelly, D. M., Klein, P. E., Klein, R. R., Price, H. J., & Rooney, W. L. (2010). Early-generation germplasm introgression from *Sorghum macrospermum* into sorghum (*S. bicolor*). *Genome*, *53*(6), 419–429.

Kumar, N., Boatwright, J. L., Brenton, Z. W., Sapkota, S., Ballén-Taborda, C., Myers, M. T., . . . Boyles, R. E. (2023). Development and characterization of a sorghum multi-parent advanced generation inter-cross (MAGIC) population for capturing diversity among seed parent gene pool. *G3: Genes, Genomes, Genetics*, *13*(4), Art. jkad037.

Kumaravadivel, N. S. R. S., & Sree Rangasamy, S. R. (1994). Plant regeneration from sorghum anther cultures and field evaluation of progeny. *Plant Cell Reports*, *13*, 286–290.

Kumari, P., Pahuja, S. K., Arya, S., & Patil, J. V. (2016). Sorghum. In: Singh, M., Kumar, S. (Eds.), *Broadening the genetic base of grain cereals*. New Delhi, India: Springer.

Lande, R., & Thompson, R. (1990). Efficiency of marker-assisted selection in the improvement of quantitative traits. *Genetics*, *124*(3), 743–756.

Lazarides, M., Hacker, J. B., & Andrew, M. H. (1991). Taxonomy, cytology and ecology of indigenous Australian sorghums (Sorghum Moench: Andropogoneae: Poaceae). *Australian Systematic Botany*, *4*(4), 591–635.

Li, A., Jia, S., Yobi, A., Ge, Z., Sato, S. J., Zhang, C., . . . Holding, D. R. (2018a). Editing of an alpha-kafirin gene family increases, digestibility and protein quality in sorghum. *Plant Physiology*, *177*(4), 1425–1438.

Li, T., Yang, X., Yu, Y., Si, X., Zhai, X., Zhang, H., . . . Xu, C. (2018b). Domestication of wild tomato is accelerated by genome editing. *Nature Biotechnology*, *36*(12), 1160–1163.

Liang, G. H., Gu, X., Yue, G., Shi, Z. S., & Kofoid, K. D. (1997). Haploidy in sorghum. In *In vitro haploid production in higher plants: Volume 4—Cereals* (pp. 149–161). Dordrecht, The Netherlands: Springer.

Lipkie, T. E., De Moura, F. F., Zhao, Z. Y., Albertsen, M. C., Che, P., Glassman, K., & Ferruzzi, M. G. (2013). Bioaccessibility of carotenoids from transgenic provitamin A biofortified sorghum. *Journal of Agricultural and Food Chemistry*, *61*(24), 5764–5771.

Liu, G., Li, J., & Godwin, I. D. (2019). Genome editing by CRISPR/Cas9 in sorghum through biolistic bombardment. In Zhao, Z. U. and Dahlberg, J. (eds.), *Sorghum: Methods and protocols* (vol 1931, pp. 169–183). New York, NY: Humana Press.

Mace, E. S., Tai, S., Gilding, E. K., Li, Y., Prentis, P. J., Bian, L., . . . Wang, J. (2013). Whole-genome sequencing reveals untapped genetic potential in Africa's indigenous cereal crop sorghum. *Nature Communications*, *4*(1), 2320.

Mackay, T. F., Stone, E. A., & Ayroles, J. F. (2009). The genetics of quantitative traits: Challenges and prospects. *Nature Reviews Genetics*, *10*(8), 565–577.

Mathur, S., Umakanth, A. V., Tonapi, V. A., Sharma, R., & Sharma, M. K. (2017). Sweet sorghum as biofuel feedstock: Recent advances and available resources. *Biotechnology for Biofuels*, *10*, 1–19.

Maulana, F., Perumal, R., Serba, D. D., & Tesso, T. (2023). Genomic prediction of hybrid performance in grain sorghum (*Sorghum bicolor* L.). *Frontiers in Plant Science*, *14*, Art. 1139896.

Melchinger, A. E. (1999). Genetic diversity and heterosis. In Coors, J.G. and Pandey, S. (eds.), *Genetics and exploitation of heterosis in crops* (pp. 99–118). Madison, WI: American Society of Agronomy, Inc. Crop Science Society of America, Inc.

Menz, M. A., Klein, R. R., Unruh, N. C., Rooney, W. L., Klein, P. E., & Mullet, J. E. (2004). Genetic diversity of public inbreds of sorghum determined by mapped AFLP and SSR markers. *Crop Science*, *44*(4), 1236–1244.

Mindaye, T. T., Mace, E. S., Godwin, I. D., & Jordan, D. R. (2015). Genetic differentiation analysis for the identification of complementary parental pools for sorghum hybrid breeding in Ethiopia. *Theoretical and Applied Genetics*, *128*, 1765–1775.

Mojtahedi, H., Santo, G. S., & Ingham, R. E. (1993). Suppression of Meloidogyne chitwoodi with sudangrass cultivars as green manure. *Journal of Nematology*, *25*(2), 303.

Morris, G. P., Ramu, P., Deshpande, S. P., Hash, C. T., Shah, T., Upadhyaya, H. D., . . . Kresovich, S. (2013). Population genomic and genome-wide association studies of agroclimatic traits in sorghum. *Proceedings of the National Academy of Sciences*, *110*(2), 453–458.

Munamava, M., & Riddoch, I. (2001). Response of three sorghum (*Sorghum bicolor* L. Moench) varieties to soil moisture stress at different developmental stages. South *African Journal of Plant and Soil*, *18*(2), 75–79.

Murty, U. R. (1993). Appraisal on the present status of research on apomixis in sorghum. *Current Science*, *64*(5), 315–318.

Mwamahonje, A., Eleblu, J. S. Y., Ofori, K., Feyissa, T., Deshpande, S., Garcia-Oliveira, A. L., . . . Tongoona, P. (2021). Introgression of QTLs for drought tolerance into farmers' preferred sorghum varieties. *Agriculture*, *11*(9), 883.

Narkhede, G. W., Thakur, N. R., Ingle, K. P., & Kasanaboina, K. (2022). Perspectives of genome-wide association studies (GWAS) in plant breeding. In *Exploration in genomics: A contemporary approach*. Agra, India: SR Scientific Publications.

Nwanze, K. F., Seetharama, N., Sharma, H. C., & Stenhouse, J. W. (1995). Biotechnology in pest management: Improving resistance in sorghum to insect pests. *African Crop Science Journal*, *3*(2), 209–215.

Obilana, A. B. (2004). Sorghum breeding research in Africa. In *Sorghum genetic enhancement: Research process, dissemination and impacts* (320 pp, pp. 33, 105). Andhra Pradesh, India: International Crops Research Institute for the Semi-Arid Tropics.

Ochieng, G., Ngugi, K., Wamalwa, L. N., Manyasa, E., Muchira, N., Nyamongo, D., & Odeny, D. A. (2021). Novel sources of drought tolerance from landraces and wild sorghum relatives. *Crop Science*, *61*(1), 104–118.

Ohadi, S., Hodnett, G., Rooney, W., & Bagavathiannan, M. (2017). Gene flow and its consequences in *Sorghum* spp. *Critical Reviews in Plant Sciences*, *36*(5–6), 367–385.

Paterson, A. H., Bowers, J. E., Bruggmann, R., Dubchak, I., Grimwood, J., Gundlach, H., . . . Rokhsar, D. S. (2009). The *Sorghum bicolor* genome and the diversification of grasses. *Nature*, *457*(7229), 551–556.

Paterson, A. H., Schertz, K. F., Lin, Y. R., Liu, S. C., & Chang, Y. L. (1995). The weediness of wild plants: Molecular analysis of genes influencing dispersal and persistence of johnsongrass, Sorghum halepense (L.) Pers. *Proceedings of the National Academy of Sciences*, *92*(13), 6127–6131.

Perumal, R., Tesso, T. T., Morris, G. P., Jagadish, S. V. K., Little, C. R., Bean, S. R., . . . Tuinstra, M. R. (2021). Registration of the sorghum nested association mapping (NAM) population in RTx430 background. *Journal of Plant Registrations*, *15*(2), 395–402.

Price, H. J., Dillon, S. L., Hodnett, G., Rooney, W. L., Ross, L., & Johnston, J. S. (2005a). Genome evolution in the genus *Sorghum* (Poaceae). *Annals of Botany*, *95*(1), 219–227.

Price, H. J., Hodnett, G. L., Burson, B. L., Dillon, S. L., & Rooney, W. L. (2005b). A *Sorghum bicolor* × *S. macrospermum* hybrid recovered by embryo rescue and culture. *Australian Journal of Botany*, *53*(6), 579–582.

Raghuwanshi, A., & Birch, R. G. (2010). Genetic transformation of sweet sorghum. *Plant Cell Reports*, *29*, 997–1005.

Rao, N. G. P., Narayana, L. L., & Reddy, R. N. (1978). Apomixis and its utilization in grain Sorghum-I. Embryology of two apomictic parents. *Caryologia*, *31*(4), 427–433.

Reddy, P. S., Jogeswar, G., Rasineni, G. K., Maheswari, M., Reddy, A. R., Varshney, R. K., & Kishor, P. K. (2015). Proline over-accumulation alleviates salt stress and protects photosynthetic and antioxidant enzyme activities in transgenic sorghum [*Sorghum bicolor* (L.) Moench]. *Plant Physiology and Biochemistry*, *94*, 104–113.

Reddy, Y., Lavanya, G. R., Chavan, S., Gorthy, S., Sargar, P. R., Krishna, K., . . . Habyarimana, E. (2024). Early-generation evaluation of marker-assisted transfer of bmr6 and bmr12 alleles into sweet and high-biomass sorghum. *Journal of Advances in Biology & Biotechnology*, *27*(5), 365–380.

Rich, P. J., Grenier, C., & Ejeta, G. (2004). Striga resistance in the wild relatives of sorghum. *Crop Science*, *44*(6), 2221–2229.

Riedelsheimer, C., Czedik-Eysenberg, A., Grieder, C., Lisec, J., Technow, F., Sulpice, R., . . . Melchinger, A. E. (2012). Genomic and metabolic prediction of complex heterotic traits in hybrid maize. *Nature Genetics*, *44*(2), 217–220.

Rizal, G., Karki, S., Alcasid, M., Montecillo, F., Acebron, K., Larazo, N., . . . Quick, W. P. (2014). Shortening the breeding cycle of sorghum, a model crop for research. *Crop Science*, *54*(2), 520–529.

Ruperao, P., Thirunavukkarasu, N., Gandham, P., Selvanayagam, S., Govindaraj, M., Nebie, B., . . . Rathore, A. (2021). Sorghum pan-genome explores the functional utility for genomic-assisted breeding to accelerate the genetic gain. *Frontiers in Plant Science*, *12*, 666342.

Saini, D. K., Chopra, Y., Singh, J., Sandhu, K. S., Kumar, A., Bazzer, S., & Srivastava, P. (2022). Comprehensive evaluation of mapping complex traits in wheat using genome-wide association studies. *Molecular Breeding*, *42*, 1–52.

Sapkota, S., Boatwright, J. L., Kumar, N., Myers, M., Cox, A., Ackerman, A., . . . Kresovich, S. (2023). Genomic prediction of hybrid performance for agronomic traits in sorghum. *G3: Genes, Genomes, Genetics*, *13*(4), Art. jkac311.

Sapkota, S., Boyles, R., Cooper, E., Brenton, Z., Myers, M., & Kresovich, S. (2020). Impact of sorghum racial structure and diversity on genomic prediction of grain yield components. *Crop Science*, *60*(1), 132–148.

Sasaki, T., & Antonio, B. A. (2009). Sorghum in sequence. *Nature*, *457*(7229), 547–548.

Sato, S., Clemente, T., & Dweikat, I. (2004). Identification of an elite sorghum genotype with high in vitro performance capacity. *In Vitro Cellular & Developmental Biology-Plant*, *40*, 57–60.

Sharma, & Franzmann. (2001). Host-plant preference and oviposition responses of the sorghum midge, *Stenodiplosis sorghicola* (Coquillett) (Dipt., Cecidomyiidae) towards wild relatives of sorghum. *Journal of Applied Entomology*, *125*(3), 109–114.

Shimelis, H., & Laing, M. (2012). Timelines in conventional crop improvement: Pre-breeding and breeding procedures. *Australian Journal of Crop Science*, *6*(11), 1542–1549.

Silva, K. J. D., Pastina, M. M., Guimarães, C. T., Magalhães, J. V., Pimentel, L. D., Schaffert, R. E., . . . Menezes, C. B. D. (2020). Genetic diversity and heterotic grouping of sorghum lines using SNP markers. *Scientia Agricola*, *78*(6), Art. e20200039.

Singh, V., Nguyen, C. T., van Oosterom, E. J., Chapman, S. C., Jordan, D. R., & Hammer, G. L. (2015). Sorghum genotypes differ in high temperature responses for seed set. *Field Crops Research*, *171*, 32–40.

Somegowda, V. K., Reddy, S. D., Gaddameedi, A., Kiranmayee, K. U., Naravula, J., Kishor, P. K., & Penna, S. (2023). Genomics breeding approaches for developing *Sorghum bicolor* lines with stress resilience and other agronomic traits. *Current Plant Biology*, *37*, Art. 100314.

Stefoska-Needham, A., Beck, E. J., Johnson, S. K., & Tapsell, L. C. (2015). Sorghum: An underutilized cereal whole grain with the potential to assist in the prevention of chronic disease. *Food Reviews International*, *31*(4), 401–437.

Stephens, J. C., Miller, F. R., & Rosenow, D. T. (1967). Conversion of alien sorghums to early combine genotypes 1. *Crop Science*, *7*(4), 396–396.

Subramonian, N. (1991). Chromosome behaviour in sugarcane × sorghum hybrids. *Cytologia*, *56*(1), 11–15.

Tang, C. Y., Schertz, K. F., & Bashaw, E. C. (1980). Apomixis in sorghum lines and their F1 progenies. *Botanical Gazette*, *141*(3), 294–299.

Tao, Y., Luo, H., Xu, J., Cruickshank, A., Zhao, X., Teng, F., . . . Mace, E. (2021). Extensive variation within the pan-genome of cultivated and wild sorghum. *Nature Plants*, *7*(6), 766–773.

Tao, Y. Z., Jordan, D. R., Henzell, R. G., & McIntyre, C. L. (1998). Construction of a genetic map in a sorghum recombinant inbred line using probes from different sources and its comparison with other sorghum maps. *Australian Journal of Agricultural Research*, *49*(4), 729–736.

Techale, B., Tamene, Y., & Ferede, T. (2014). Response of wild type sorghum (*Sorghum bicolor* (L.) Moench) accessions to pre-flowering drought stress. *African Journal of Agricultural Research*, *9*(41), 3077–3081.

Terajima, Y., Matsuoka, M., Irei, S., Sakaigaichi, T., Fukuhara, S., Ujihara, K., . . . Sugimoto, A. (2007). Breeding for high-biomass sugarcane and its utilisation in Japan. *Proceedings of International Society of Sugar Cane Technologists*, *26*, 759–763.

Thakur, N. R., Gorthy, S., Vemula, A., Odeny, D. A., Ruperao, P., Sargar, P. R., . . . Habyarimana, E. (2024b). Genome-wide association study and expression of candidate genes for Fe and Zn concentration in sorghum grains. *Scientific Reports*, *14*(1), 12729.

Thakur, N. R., Ingle, K. P., Sargar, P. R., Baraskar, S. S., Kasanaboina, K., Awio, B., . . . Abdi, G. (2024a). Sustainable utilization of wild germplasm resources. In *Sustainable utilization and conservation of plant genetic diversity* (pp. 551–590). Singapore: Springer Nature Singapore.

Thurber, C. S., Ma, J. M., Higgins, R. H., & Brown, P. J. (2013). Retrospective genomic analysis of sorghum adaptation to temperate-zone grain production. *Genome Biology*, *14*, 1–13.

Tsukiboshi, T., Koga, H., Uematsu, T., & Shimanuki, T. (1998). Resistance of sorghum and sudangrass to ergot caused by *Claviceps* sp. and the cultural control of the disease. *Bulletin of The National Grassland Research Institute*, *56*, 28–35.

USDA-FAS. (2024). *Production – Sorghum*. Retrieved from https://fas.usda.gov/data/production/commodity/0459200

Velazco, J. G., Jordan, D. R., Mace, E. S., Hunt, C. H., Malosetti, M., & Van Eeuwijk, F. A. (2019). Genomic prediction of grain yield and drought-adaptation capacity in sorghum is enhanced by multi-trait analysis. *Frontiers in Plant Science*, *10*, 997.

Venkateswaran, K., Elangovan, M., & Sivaraj, N. (2019). Origin, domestication and diffusion of *Sorghum bicolor*. In Aruna, C., Visarada, K.B.R.S., Bhat, B.V., and Tonapi, V.A. (eds.), *Breeding Sorghum for diverse end uses* (pp. 15–31). Amsterdam, The Netherlands: Elsevier.

Venkatraman, T. S., & Thomas, R. (1932). Sugarcane-sorghum hybrids. Part I. General outline and early characters. *Indian Journal of Agricultural Sciences*, *2*, 19–27.

Verma, L. K., Biradar, B. D., & Meena, A. K. (2022). Heterotic grouping in rabi sorghum [*Sorghum bicolor* (L.) Monech] through diallel analysis. *The Pharma Innovation Journal*, *11*(2), 1955–1957.

Vetriventhan, M., Jagadeesh, K., Baraskar, S., Dagunapur, N. R., Srinivas, R., Anilkumar, K., & Singh, K. (2024). Crop wild relatives of sorghum: A novel source of genetic variation for crop improvement. In *Omics and biotechnological approaches for product profile-driven Sorghum improvement* (pp. 145–164). Singapore: Springer Nature.

Viaene, N. M., & Abawi, G. S. (1998). Management of *Meloidogyne hapla* on lettuce in organic soil with sudangrass as a cover crop. *Plant Disease*, *82*(8), 945–952.

Villar, J. L., Maranville, J. W., & Gardner, J. C. (1989). High density sorghum production for late planting in the Central Great Plains. *Journal of Production Agriculture*, *2*(4), 333–338.

Voss-Fels, K., & Snowdon, R. J. (2016). Understanding and utilizing crop genome diversity via high-resolution genotyping. *Plant Biotechnology Journal*, *14*(4), 1086–1094.

Washburn, J. D., Murray, S. C., Burson, B. L., Klein, R. R., & Jessup, R. W. (2013). Targeted mapping of quantitative trait locus regions for rhizomatousness in chromosome SBI-01 and analysis of overwintering in a *Sorghum bicolor* × *S. propinquum* population. *Molecular Breeding*, *31*, 153–162.

Winans, N. D., Fonseca, J. M., Perumal, R., Klein, P. E., Klein, R. R., & Rooney, W. L. (2024). Envirotyping can increase genomic prediction accuracy of new environments in grain sorghum trials depending on mega-environment. *Crop Science, 64*, 2519–2533.

Winans, N. D., Klein, R. R., Fonseca, J. M. O., Klein, P. E., & Rooney, W. L. (2023). Evaluating introgression sorghum germplasm selected at the population level while exploring genomic resources as a screening method. *Plants*, *12*(3), 444.

Wondimu, Z., Dong, H., Paterson, A. H., Worku, W., & Bantte, K. (2021). Genetic diversity, population structure, and selection signature in Ethiopian sorghum [*Sorghum bicolor* L. (Moench)] germplasm. *G3*, *11*(6), Art. jkab087.

Wooten, D. R. (2001). *The use of* Sorghum propinquum *to enhance agronomic traits in sorghum* (Doctoral dissertation, Texas A&M University).

Wray, N. R., Yang, J., Hayes, B. J., Price, A. L., Goddard, M. E., & Visscher, P. M. (2013). Pitfalls of predicting complex traits from SNPs. *Nature Reviews Genetics*, *14*(7), 507–515.

Wu ShuBiao, W. S., Shang YongJin, S. Y., Han XueMei, H. X., Wang JingXue, W. J., Zhang FuYao, Z. F., & Wei YaoMing, W. Y. (1994). Embryological study on apomixis in a sorghum line SSA-1. *Acta Botanica Sinica*, *36*, 833–837.

Yang, S., Weers, B. D., Morishige, D. T., & Mullet, J. E. (2014). CONSTANS is a photoperiod regulated activator of flowering in sorghum. *BMC Plant Biology*, *14*, 1–15.

Yonemaru, J., Ando, T., Mizubayashi, T., Kasuga, S., Matsumoto, T., & Yano, M. (2009). Development of genome-wide simple sequence repeat markers using whole-genome shotgun sequences of sorghum (*Sorghum bicolor* (L.) Moench). *DNA Research*, *16*(3), 187–193.

Yu, X., Li, X., Guo, T., Zhu, C., Wu, Y., Mitchell, S. E., . . . Yu, J. (2016). Genomic prediction contributing to a promising global strategy to turbocharge gene banks. *Nature Plants*, *2*(10), 1–7.

Zargar, S. M., Raatz, B., Sonah, H., MuslimaNazir, Bhat, J. A., Dar, Z. A., . . . Rakwal, R. (2015). Recent advances in molecular marker techniques: Insight into QTL mapping, GWAS and genomic selection in plants. *Journal of Crop Science and Biotechnology*, *18*, 293–308.

Zhang, D., Tang, S., Xie, P., Yang, D., Wu, Y., Cheng, S., . . . Xie, Q. (2022). Creation of fragrant sorghum by CRISPR/Cas9. *Journal of Integrative Plant Biology*, *64*(5), 961–964.

3 Bioinformatics Tools and Resources for Sorghum Omics

Vikas Kumar Singh, Deepti Chaturvedi, Saksham Pundir, Kibrom B. Abreha, and Gurjeet Singh

3.1 INTRODUCTION

Sorghum (*Sorghum bicolor* (L.) Moench) is a globally significant cereal crop, renowned for its adaptability and resilience in drought-prone and semi-arid regions (Hadebe et al., 2017). Amid growing concerns over food security, sustainability, and climate change, research to enhance sorghum's productivity and stress tolerance has intensified (Prasad et al., 2021). As the fifth most important cereal globally, sorghum plays a crucial role in food supplies, particularly in Africa and developing regions. Cultivated in over 100 countries, the top ten producers, the U.S., Sudan, Mexico, Nigeria, India, Niger, Ethiopia, Australia, Brazil, and China, account for approximately 77% of global production (Aruna & Cheruku, 2018). Sorghum faces substantial production challenges from climate stresses, such as drought and heat, as well as biotic stresses, which together can reduce yields by over 60% during critical stages (Ondiko & Recha, 2022). Furthermore, poor soil fertility marked by deficiencies in essential nutrients like nitrogen and phosphorus adds to the crop's productivity constraints (Ondiko & Recha, 2022; Khalifa & Eltahir, 2023). Efforts to enhance sorghum production focus on developing high-yield, climate-resilient varieties suited to local agro-ecological zones. Breeding programs have introduced varieties with improved resistance to biotic and abiotic stresses (Khalifa & Eltahir, 2023).

Omics technologies, including genomics, transcriptomics, proteomics, and metabolomics, have significantly advanced sorghum research by providing insights into molecular processes and enabling efficient genetic improvement. Bioinformatics plays a central role by managing and analyzing the extensive data generated, integrating multi-omics datasets to uncover gene expression patterns, protein interactions, and metabolic pathways crucial for enhancing traits like drought tolerance, nutrient efficiency, and yield. This combination of technologies allows for deeper exploration of biological systems, revealing novel pathways to improve sorghum agronomic performance (Figure 3.1).

Genomics provides a fundamental framework for understanding the genetic structure of sorghum (Boyles et al., 2019). The sequencing of the sorghum genome has opened new possibilities for marker-assisted selection (MAS) and genetic mapping, enabling researchers to identify key genes associated with important agronomic traits, such as drought tolerance and disease resistance (Ibraheem et al., 2018; Madhusudhana, 2019). Modern bioinformatics platforms, such as genome browsers and sequence alignment tools, support the analysis of sorghum's genetic architecture (Mace et al., 2013; Naithani et al., 2023). Genome-wide association studies (GWAS) and quantitative trait loci (QTL) analysis have proven crucial for identifying loci associated with phenotypic traits, enabling breeders to develop new varieties that exhibit enhanced resilience in sorghum (Morris et al., 2013; Li et al., 2018; Elias et al., 2024). The advent of advanced sequencing technologies, including Next-Generation Sequencing (NGS), has greatly accelerated research in sorghum

DOI: 10.1201/9781003623717-3

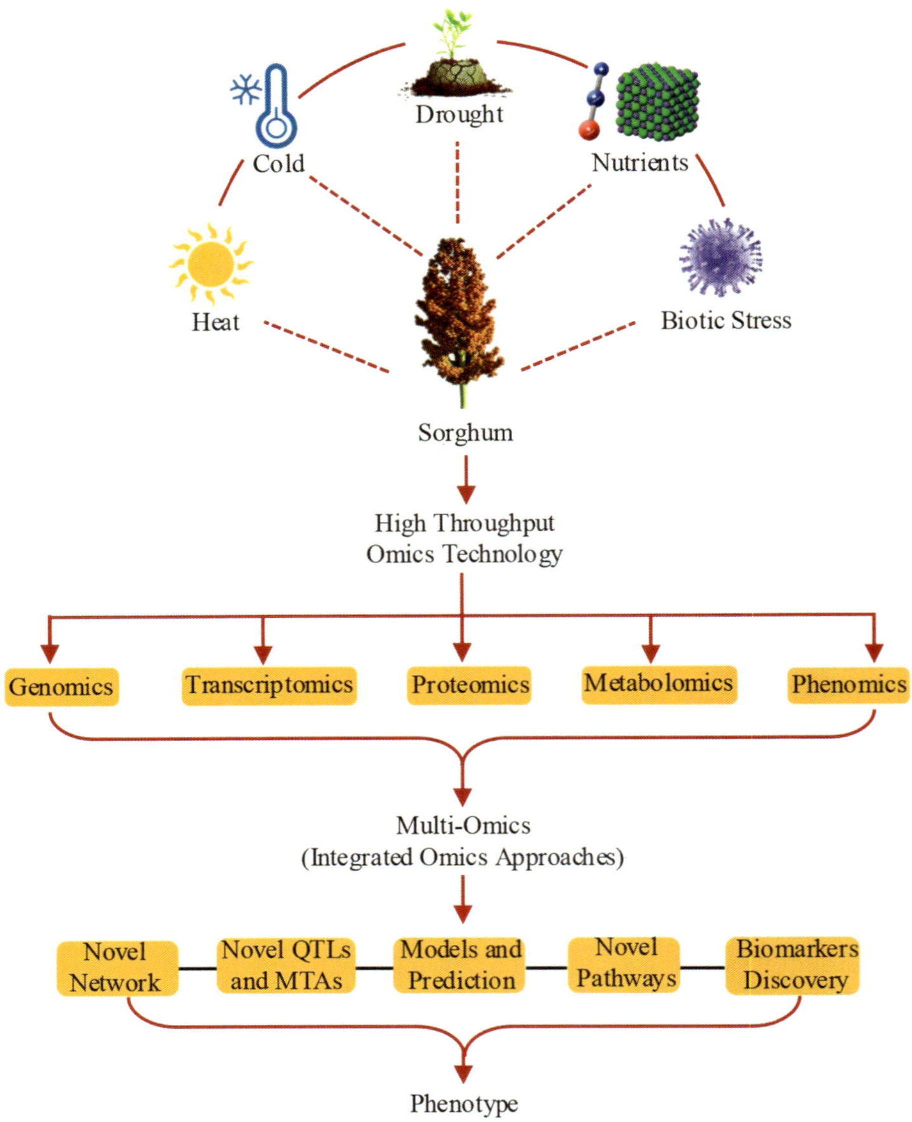

FIGURE 3.1 Illustration of various challenges affecting sorghum production, alongside different omics approaches that can aid in identifying novel candidate regions in the genome, pathways, and interactions regulating trait phenotypes. Additionally, various bioinformatics tools are essential for analyzing the data generated through these omics approaches.

genomics by offering outstanding resolution for identifying genetic variations among different sorghum varieties. This progress facilitates a more thorough and efficient investigation of the rich sorghum genetic diversity (Luo et al., 2014). By leveraging bioinformatics tools, large datasets are processed and analyzed to establish relationship between genotype and phenotype, thereby facilitating precision breeding efforts that aim to mitigate the impact of changing weather conditions (Thriveni et al., 2024).

Transcriptomics focuses on genome-wide gene expression profiling to understand how sorghum responds to environmental stimuli (Ngara et al., 2021; Ibraheem et al., 2018; Varoquaux et al.,

2019). RNA sequencing (RNA-Seq) is the primary technique used to quantify gene expression across tissues, developmental stages, and environmental conditions (Wolf, 2013; Lowe et al., 2017; Monisha & Sowndarya, 2024). Bioinformatic tools ensure sequence quality, transcript assembly, and analysis of large datasets to uncover regulatory networks and gene functions (Niranjan et al., 2024). By examining gene expression under stress conditions like drought or nutrient deficiency, researchers can identify potential targets for genetic modification or breeding (Marla et al., 2017; Sun et al., 2020; Zhou et al., 2022; Puri et al., 2023; Shrestha et al., 2024). Transcriptomic studies reveal how genes collaborate in regulatory pathways, providing insights into sorghum's adaptation to biotic and abiotic stress, essential for breeding climate-resilient varieties.

Proteomics offers a comprehensive understanding by analyzing the protein composition of sorghum (Mukherjee et al., 2024). While genomics and transcriptomics identify gene presence and expression, proteomics focuses on protein production and interactions (Srivastava et al., 2024). Techniques like mass spectrometry (MS), including LC-MS, are used to identify and quantify proteins, examining their expression and modifications under various conditions (Karpievitch et al., 2010; Qu et al., 2017; Cajka, 2024). Bioinformatics tools are crucial for managing complex datasets, protein identification, quantification, and functional annotation, providing insights into post-translational modifications and protein interactions that drive sorghum's physiology. Proteomics complements transcriptomics by validating biological processes and identifying biomarkers related to stress tolerance, nutrient efficiency, and yield (Tu et al., 2023).

Metabolomics investigates metabolites, crucial for cellular functions, and maps pathways related to stress responses and metabolic regulation in sorghum (Tu et al., 2023; Jiao et al., 2023). Techniques like NMR and GC-MS profile metabolites under different conditions (Kisiel et al., 2023). Bioinformatics integrates metabolomic, genomic, and proteomic data, providing insights into biochemical networks regulating growth and development (Tu et al., 2023; Khan et al., 2024). Metabolomics aids in understanding sorghum's adaptation to stress and nutrient limitations, revealing pathways to enhance resilience (Tu et al., 2023; Mukherjee et al., 2024; Singh et al., 2024). Combining omics platforms enables a systems biology approach linking molecular data to phenotypic traits.

The goal of using bioinformatics tools in sorghum omics is to integrate genomics, transcriptomics, proteomics, and metabolomics data to understand the plant's molecular biology. This integration helps develop predictive models, identify gene networks, and discover new targets for crop improvement. For instance, combining transcriptomic and metabolomic data reveals how gene expression changes relate to metabolic shifts that affect stress tolerance. Advanced bioinformatics platforms facilitate pathway analysis, network construction, and machine learning, identifying molecular markers for breeding programs. These tools enhance our understanding of sorghum's molecular biology, crucial for developing resilient varieties to withstand climate change. This chapter discusses available bioinformatics tools and their integration for sorghum improvement.

3.2 TOOLS AND APPROACHES FOR GENOMIC STUDIES IN SORGHUM

For plant breeders, agronomists, molecular biologists, geneticists, and bioinformaticians, centralizing valuable data into an accessible hub is essential for fostering collaboration and advancing germplasm enhancement. The SorghumBase web portal (www.sorghumbase.org) serves as a comprehensive resource for the sorghum research community, offering a wealth of genomic information within a modular framework built on open-source software, ensuring sustainability. Its primary focus is the management of genomic datasets, promoting data standards, open access, and information sharing. A key priority is the stewardship of sorghum reference genomes, which are crucial for accessing and characterizing allelic variation linked to agronomic traits. Utilizing Ensembl and Gramene open-source software (Tello-Ruiz et al., 2021), SorghumBase benefits from over 23 years of development in data models, workflows, visualizations, and APIs. Similarly, the

Sorghum Genome Science Database (SorGSD: https://ngdc.cncb.ac.cn/sorgsd/) has been significantly enhanced, now including SNPs from 289 sorghum lines, mapped to the updated BTx623 genome (v3.1). The updated version also features phenotypic data and panicle images from key accessions. SorGSD integrates genomic variation and phenotypic information while offering online tools for data mining, genome navigation, and analysis (Liu et al., 2021).

The Sorghum QTL Atlas (https://aussorgm.org.au/sorghum-qtl-atlas/) is a high-resolution, open-access platform designed to facilitate the identification of candidate genes in sorghum, maize, and rice. By refining QTL confidence intervals, it provides the most comprehensive and up-to-date sorghum QTL data available, incorporating findings from extensive research since 1995. The database includes predicted gene models associated with QTL confidence intervals across all sorghum genome assembly gene sets, offering insights into gene diversity and signatures of selection in sorghum (Mace et al., 2019). Other widely used databases include Phytozome, PlantGDB, and EnsemblPlants (Upadhyay et al., 2024).

3.2.1 Genome Sequencing and Assembly Tools

The advent of NGS technologies has revolutionized plant genomics in recent years. NGS platforms, including Roche/454, Illumina/Solexa, Ion Torrent, and SOLiD, offer high-throughput capabilities that enable millions of sequencing reactions to occur simultaneously, making them essential for crop improvement, including sorghum (Hu et al., 2021; Reuter et al., 2015). These technologies have significantly enhanced sorghum breeding programs by enabling comprehensive DNA and RNA sequencing, uncovering genetic mechanisms underlying agronomic traits, and aiding in the identification of QTLs linked to biotic and abiotic stress tolerance. Comparative genomics, which utilizes related species with conserved regions, has become a valuable strategy, further benefiting from NGS's capabilities. NGS is crucial for mapping the sorghum genome and identifying QTLs through genome-wide analysis which can be applied in MAS (Varshney et al., 2009).

Third-generation sequencing (TGS) technologies, such as Pacific Biosciences (PacBio), Oxford Nanopore Technologies (ONT), and BioNano Genomics, have taken genome sequencing to the next level by enabling accurate *de novo* assemblies, improving sequencing efficiency, and minimizing biases associated with PCR amplification (Pillai et al., 2017). TGS offers significant advantages over earlier NGS platforms, including long read lengths (over 20kb), high consensus accuracy (greater than 99.99% at 30× coverage), and the ability to detect epigenetic modifications. Notably, ONT sequencing technologies, such as the PromethION, allow for high-throughput analysis of large genomes and provide real-time data processing through cloud-based platforms like EPI2ME (Jain et al., 2016). By leveraging these advanced sequencing platforms, researchers can map the sorghum genome with unprecedented accuracy, identify key QTLs associated with crop yield, and enhance resilience, driving forward sorghum breeding efforts.

High-throughput sequencing (HTS) technologies, particularly TGS, have significantly advanced research by enabling rapid and efficient annotation, prioritization, and navigation of large genomic datasets. In recent years, tools like RefAligner, Canu, and Nanopore Synthetic-Long (NaS) have been developed, focusing on *de novo* and genome-based sequencing analysis. TGS platforms, such as PacBio and MinION, offer long-read lengths from single-molecule templates, providing a distinct advantage in genome assembly, especially for repetitive regions. In contrast, second-generation platforms like Illumina's MiSeq, HiSeq, and NextSeq excel in high throughput and base-calling accuracy but produce shorter reads, complicating assembly in complex genomic regions (Wee et al., 2019). The integration of NGS with MAS facilitates targeted sorghum breeding strategies, enhancing desirable traits. Moreover, advancements in TGS, particularly Oxford Nanopore's long-read capability, have simplified the assembly of the complex sorghum genome. This integrated approach not only streamlines comparative genomics among related species but also deepens our understanding of genetic diversity in sorghum, ultimately contributing to more sustainable agricultural practices.

3.2.2 SNP Detection and Genotyping Tools

Once genetic sequences are obtained, the next step involves using various bioinformatics tools and packages to convert the raw genetic data into a final set of SNP and genotype calls, which typically involves determining the allele at each SNP locus for every individual analyzed. While the detailed analysis methods may vary depending on the sequencing technology used, rigorous validation is essential to ensure accuracy and avoid biases (Shafer et al., 2017). The bioinformatics pipeline for genotype production must be tailored to the specific sequencing technology and library generation method, but all workflows share common objectives: addressing moderate genotyping error rates typical in HTS data, identifying and removing SNPs in paralogous and repetitive regions, and generating a set of unlinked loci.

The first step in these pipelines is the demultiplexing process, where each sequence is assigned to its individual of origin (Torkamaneh et al., 2017). Following this, the sequence reads undergo a preprocessing phase for quality control and filtering, where poor-quality SNP loci or those likely to be remnants are discarded. The processed reads are then aligned or assembled. If a reference genome is available, alignment tools such as TopHat2, BWA, Bowtie2, or STAR are used (Musich et al., 2021). The final step involves discovering polymorphic loci and gathering genotypes for each individual at these loci. Bi-allelic SNPs are identified, and only high-quality, error-free genotypes are selected for downstream analysis. Loci that deviate from Hardy-Weinberg equilibrium, exhibit low call frequencies, or show linkage disequilibrium are filtered out (Trenkel et al., 2022). For sorghum breeding, improving SNP detection and genotyping reliability is crucial, and HTS data analysis enables precise estimation of SNPs and genotype calls.

Several bioinformatics software packages have been developed to accelerate the analysis of reduced-representation sequencing data. While some platforms apply tools and algorithms similar to those used in whole-genome sequencing, others employ specialized algorithms. For instance, Stacks v2, the most widely used software for RADseq, sequentially cleans and filters raw sequence data, builds loci, creates a catalog of loci, matches samples to the catalog, transposes data, and integrates paired-end reads for genotype calling (Rochette et al., 2019). Stacks utilizes a de Bruijn graph assembler to create contigs from paired-end reads, which are then merged with corresponding single-end loci. This allows a Bayesian genotype caller to generate precise SNPs and phase them into long haplotypes, resulting in RAD loci spanning hundreds of base pairs (Rivera-Colón & Catchen, 2022).

Other tools like Ipyrad (Eaton & Overcast, 2020), which accommodates InDel variation, and dDocent (Puritz et al., 2014), which uses Bayesian statistics and FreeBayes for SNP calling, also play key roles in SNP analysis. Additionally, bioinformatics tools like ANGSD (Korneliussen et al., 2014) and polyRAD (Clark et al., 2019) incorporate Bayesian methods to assess posterior probabilities for each allele, incorporating statistical uncertainty in genotype assessments. By filtering low-quality loci through quality control, Stacks v2 ensures accurate SNP identification, while Bayesian techniques in ANGSD improve genotype assessments. This combined approach reveals useful genetic variation in sorghum, which supports breeding programs aimed at improving the crop's performance under diverse environmental conditions (see Table 3.1).

3.3 TOOLS AND APPROACHES FOR TRANSCRIPTOMIC STUDIES IN SORGHUM

Transcriptomics is the large-scale analysis of expressed mRNAs in a cell or plant at a specific moment in time, whether during a particular growth stage or in response to an environmental stimulus, such as biotic stress (e.g., viral, bacterial, and fungal infections) or abiotic stress (e.g., salinity, drought, waterlogging, heavy metals, and extreme temperatures) (Ngara et al., 2021). A wide range of techniques has been used to study transcripts in various crops, including sorghum. These methods include northern

TABLE 3.1
Summary of some key genetic mapping studies conducted in sorghum

Population type (size)	Trait Name	MTAs/QTLs	Remarks	References
Diverse sorghum germplasm (96)	Agronomic traits	42	Focused on major agronomic, root, and stomatal traits	Ramalingam et al. (2023)
Diverse genotypes (390)	Grain yield components	53	Examined associations across the genome for yield, grain number, and components	Boyles et al. (2016)
Diverse genotypes (324)	Agronomic traits	20	Focused on identifying genomic regions and candidate genes related to yield and related traits	Enyew et al. (2022)
Diverse genotypes (324)	Grain yield components	52	Identified genomic regions and candidate genes related to yield traits	Kumar et al. (2023)
Diverse germplasm (NA)	Sugarcane aphid resistance	2	Identified a WRKY transcription factor as a candidate gene	Poosapati et al. (2022)
Diverse germplasm (245)	Forage quality-related traits	42	Utilized multi-locus GWAS to investigate forage quality-related traits	Li et al. (2018)
Diverse genotypes (276)	Aphid resistance	87	Identified three new sources of resistant lines to sorghum aphid	Punnuri et al. (2022)
Three connected populations (207 lines each)	Agronomic and yield related traits under drought stress	105	Identified QTLs related to yield and drought tolerance through multi-environment trials	Techale et al. (2022)
$F_{3:5}$ RILs (177)	Multiple traits including stalk sugar yield	19	Covered all ten sorghum chromosomes with 1991 markers, mapping traits related to stalk sugar yield	Hostetler et al. (2021)
$F_{2:3}$ (248)	Economically important traits	98	Focused on key economically important traits in sorghum through comprehensive mapping	Xiao-ping et al. (2011)
RILs (210)	Protein, starch, and amylose content	27	Aimed to identify stable QTLs for grain quality traits	Ayalew et al. (2022)
RILs (213)	Spikelet-related traits and grain size	43	Reported that *Dw* genes were not strongly involved in the regulation of grain size	Takanashi et al. (2021)
Two populations (401 and 403 F_3 progenies)	Grain morphology and quality traits	51 (P114), 71 (P118)	Identified QTLs with high R^2 corresponding to major genes within mapping populations	Guindo et al. (2019)

blotting, hybridization techniques, expressed sequence tags (ESTs), reverse transcriptase quantitative PCR (RT-qPCR), serial analysis of gene expression (SAGE), RNA microarrays, and RNA-seq (Lowe et al., 2017). Among these, microarrays and RNA-seq have become essential tools for elucidating the differential expression of genes involved in stress responses across multiple plant species.

In sorghum, numerous transcriptomic studies have been conducted to better understand its response to both biotic and abiotic stresses. For instance, Cheng et al. (1993) investigated drought tolerance in the sorghum variety TX430, observing up-regulation of dehydrin mRNA after 22 hours of drought exposure, with expression increasing as the stress duration lengthened. Similarly, Mizuno et al. (2012) used Illumina sequencing on the sorghum variety BTx623 to examine its response to *Bipolaris sorghicola*, identifying 5,617 upregulated and 3,052 downregulated transcripts at seven days post-inoculation (dpi). These studies, among others, contribute to the growing body of transcriptomic research aimed at enhancing sorghum's resilience to various stresses (see Table 3.2).

TABLE 3.2
Summary of transcriptomic studies in sorghum and key findings

Sorghum genotype	Trait	Time point	Technique	Key findings	Reference
BTx623	Fungi; *Bipolaris sorghicola*	7dpi	Illumina sequencing	5,617 transcripts upregulated and 3,052 transcripts downregulated	Mizuno et al. (2012)
S. propinquum	Physiological processes	40 days after sowing (DAS)	Illumina sequencing	1,963 genes in aerial shoot and 599 genes in rhizome	Zhang et al. (2014)
S. propinquum	-	Vegetative growth stage	Microarray	548 genes including 31 and 114 unique genes	Zhang et al. (2014)
R16	Drought & heat	Seedlings	Microarray, qRT-PCR	28585 gene	Johnson et al. (2014)
B35 and R16	Stay-green trait	~45 DAS (booting stage)	Microarray	Upregulation of delta1-pyrroline-5-carboxylate synthase 2 (P5CS2)	Johnson et al. (2015)
M-81E and Roma	Salt	7 DAT (Days after treatment)	Illumina sequencing	Upregulation of 236 in M81/E and 442 genes in Roma	Sui et al. (2015)
BTx623	Drought	Seedling (8d)	Isoform Sequencing	APA of ~11,000 expressed genes and more than 2,100 novel genes	Abdel-Ghany et al. (2020)
Grain cultivar MR43	Defensive compound dhurrin	0, 0.5, 3, 4, 10, 13, 17, 20, 25, 32, 40, 45, 49, 52, 56, 61, 63 and 67 DAA (Days after anthesis)	Illumina HiSeqTM 2000 platform sequencing	Existence of two endogenous dhurrin turnover pathways in sorghum	Nielsen et al. (2016)
Kaoliang and Ji2731	Salt	mature, early senescence (ES), middle senescence (MS) and late senescence (LS)	Illumina sequencing	Up-regulated 2,351 genes, 897 down-regulated and 1,045 genes with either up-down or down-up regulated	Wu et al. (2016)
BTx623 and Niu Sheng Zui	Chilling	10-d old Seedling	RNA-seq	Up-regulation of cold response regulator C-repeat binding factor (CBF) transcription factor and reactive oxygen detoxification, jasmonic acid (JA) biosynthesis, and lipid remodeling phospholipase Da1 (PLDa1) gene	Marla et al. (2017)
Inbred lines; M-81E and Roma	Salt	48 HAT (Hours After Treatment)	RNA-seq	Increase abscisic acid (ABA) in M-81E leaves, and increase concentration of jasmonate changed in roma roots, indole-3-acetic acid (IAA) increased in both genotypes leaves	Yang et al. (2017)
RILs F_7 with CK60 x San Chi San	Low NUE	3-week-old	Illumina sequencing	486 DEGs	Gelli et al. (2017)
623B and HN (Henong 16)	Salt	0, 48, and 72 h	RNA-seq analysis	5,647 DEGs	Cui et al. (2018)

(continued)

TABLE 3.2 (Continued)
Summary of transcriptomic studies in sorghum and key findings

Sorghum genotype	Trait	Time point	Technique	Key findings	Reference
DKS 44–20 (S) & DKS 37–07 (R)	Sugarcane aphid – *Melanaphis sacchari*	2- and 6-week post-emergence	RNA-seq	Up-regulated 2,376 genes in resistant and 985 genes in susceptible genotype	Kiani and Szczepaniec (2018)
Inbred lines (M-81E x Roma)	Salt tolerance	Four leaf stage	Illumina sequencing	2,085 and 3,172 DEGs	Yang et al. (2018)
Tolerant–Keza15 (KZ15) and sensitive – Suiza7 (SZ7)	Low nitrogen	7DAT	RNA-seq	793 DEGs; upregulated 404 DEGs in KZ15 and 762 DEGs in SZ7, downregulated 642 in KZ15 and 1,132 DEGs in SZ7	Yang et al. (2019)
RIO, BTx406 and R9188	Carbon allocation	Flag leaf and flowering		18 275, 19 727 and 19 102 genes were expressed in RIO, BTx406 and R9188, respectively	Li et al. (2019)
SC56 and Tx7000	Drought	13 DAT	Illumina HiSeq	2,407 DEGs	Azzouz-Olden et al. (2020)
M-81E and Roma line	Salt	24-h AT	RNA-seq	126 DEGs in M-81E and 133 in Roma	Sun et al. (2020)
BTx623	Saline-alkali	-	Illumina sequencing	16 DEGs	Ma et al. (2020)
Cultivar Aikang-8	Exogenous MeJA	1h, 2.5h, 4.5h, 6.5h AT	RNA-seq	Increased DEGs in plasma cells with the increase of MeJA concentration	Liu et al. (2021)
RTx2911 (R) and RTx430	Grain mold	0, 24 and 48hpi	RNA-seq	1661 DEGs at 0 hpi 1955 DEGs at 24 hpi	Nida et al. (2021)
SSG 59-3-Tolerant and PC-5	Salt	48 h AT	Illumina sequencing	32,412 DEGs; SSG59-3 (1,284 upregulated; 624 downregulated) and PC-5 (806 upregulated; 515 downregulated)	Punia et al. (2021)
TAM428 (R) and Tx2737 (S)	Sugarcane aphid	0, 96, 216, and 360 hpi	Illumina sequencing	-	Serba et al. (2021)
Mota Maradi andBTx623 (S)	Salt	10 DAT	Illumina sequencing	-	Ukwatta et al. (2021)
RILs; 407B (R) x 7B (S)	Sugarcane aphid	at 4 and 7 dpi	Illumina sequencing	865 DEGs	Zhang et al. (2021)
Hiro-1	Callus regeneration	Whole tissue	Illumina sequencing	6,000 DEGs	Zhou et al. (2021)
RIO and SN005	Salt	1, 2, 6 & 12hAT	Illumina sequencing	5,823 DEGs in RIO and 5,658 in SN005	Chen et al. (2022)

TABLE 3.2 (Continued)
Summary of transcriptomic studies in sorghum and key findings

Sorghum genotype	Trait	Time point	Technique	Key findings	Reference
Dorado, Giza-113, Shandaweel-6, Hybrid-306	Drought	15 DAT	RNA-seq	2 DEGs	Hassan et al. (2022)
Liaonian3 and Liaoza10	Starch accumulation patterns	3-, 18-, and 30-days post pollination	Illumina sequencing	6,634 DEGs between 3 and 18d, and 779 DEGs between 18 and 30d in Liaoza10. 6,768 DEGs between 3 and 18d, and 7,630 DEGs between 18 and 30d in Liaonian3	Ke et al. (2022)
HN and GZ	Salt	0, 24, 48, and 72h AT	Illumina sequencing	26,553 genes	Ren et al. (2022)
Liaotian No. 1	Drought and salt	48h AT	Illumina sequencing	1,897 DEGs; drought, 3761; salinity, 6,285; combined stress	Wang and Wei (2022)
cultivar Tx430	Forage yield	At 2wk interval until 10wk after planting	RNA-seq	5,456 DEGs between stage1_stage2, 1,246 DEGs between stage2_stage3, & 1,379 DEGs between stages 3_stage 4	Wang et al. (2022)
LRNK1and LR2381	Salt	72h	Illumina sequencing	10,771 DEGs	Zhang et al. (2022)
Z6, Z27, and HC4	Seed quality	At harvesting	Illumina sequencing	3,054 DEGs	Zhou et al. (2022)
PI329435, PI510924, PI585347, and PI585348	Grain carotenoids	14, 28, or 42 days after pollination (DAP)	Illumina sequencing	21,236 DEGs	Cruet-Burgos and Rhodes (2023)
IT124115 and IT124115M	Salt	48hAT	Illumina sequencing	582: Mu200/MuCT, 1062L: WT200/WtCT, 510 DEGs: Mu200/WT200	Jeon et al. (2023)
Inbred line "407B"	Cadmium	3DAT	Illumina sequencing	2,683 DEGs (1,812 up- and 871 down-regulated)	Jiao et al. (2023)
ICSV219, TCSV361, Yidumi and Jinhui75	Mesocotyl Elongation	8d old seedling	Illumina sequencing	7,541 DEGs between Yidumi_ICSV219, 6,895 DEGs between Yidumi_TCSV361, 6,042 DEGs between Jinhui75_ICSV219 & 6,093 DEGs between Jinhui75_TCSV361	Ju et al. (2023)
ICSR-56 (S) and CSV 15	Salt	24hAT	Illumina sequencing	70 and 162 DEGs to SS and SSN and 112 and 26 DEGs to ST and STN	Karumanchi et al. (2023)
var. BTx623	Seed germination	6h imb., 6, 10, 14, 18, 22hAG	Illumina sequencing	20,603 genes	Liu et al. (2023)
RTx430, SC265, SC1345	Sugarcane aphid	0, 6, 24, and 48 hpi	Illumina sequencing	26,794 DEGs	Puri et al. (2023)

(*continued*)

TABLE 3.2 (Continued)
Summary of transcriptomic studies in sorghum and key findings

Sorghum genotype	Trait	Time point	Technique	Key findings	Reference
Accession 59	Mucilage secretion	12h with mucilage secretion	Illumina sequencing	4,461 DEGs	Xu et al. (2023)
Alto2	Cadmium	7DAT	Illumina sequencing	14,564 DEGs	Zhou et al. (2023)
cv. Wray	Stem sucrose	74-DAE	RNA-seq	8,242 DEGs	Fu et al. (2024)
'BTx623	Nutrition synthesis	1-9, 6-25 & 10-25 dpa	RNA-seq	21,971 DEGs	Khan et al. (2024)
Tx2783 (R) and BTx623 (S)	Sugarcane aphid	0, 3, 6, 9, and 12 dpi	Illumina sequencing	Highest, 3,533 DEGs (6dpi) in Tx2783 and 6,747 DEGs (9dpi) in BTx623	Shrestha et al. (2024)
TX430	Drought	22 h	Northern blotting	Up-regulation of the dehydrin mRNA with increase in stress duration.	Cheng et al. (1993)
P954035 and P721N	Drought	Seedlings and/or mature plants	Northern blotting	Up-regulation of the dehydrin mRNA in both seedlings and mature plants.	Wood and Goldsbrough (1997)
BTx623	Drought	3 and 27 h	Microarray	~2,200 DEGs	Buchanan et al. (2005)
BTx623	Drought	27 h	RNA-seq	-	Dugas et al. (2011)
IS22330 and IS20351	Drought	Seedlings	RNA-seq	-	Fracasso et al. (2016)
South African landrace-LR6	Drought	4, 6 & 7 days	Microarray	-	Devnarain et al. (2019)
XGL-1	Drought	7 DAT	RNA-seq	3,687 DEGs in leaves and 4,635 DEGs in roots	Zhang et al. (2019)
BTx623 and SC56 Tx7000 and PI-482662	Drought	1 and 6 h AT	RNA-seq	180 DEGs	Abdel-Grany et al. (2020)
SC56 Tx7000	Drought	Anthesis for 13 DAT	RNA-seq	2,407 DEGs	Azzouz-Olden et al. (2020)
RTx430, BTx642	Drought	17 weeks from seedlings	RNA-seq	-	Varoquaux et al. (2019)
M35-1 and C43	Drought	30 DAS	Illumina sequencing	-	Katiyar et al. (2015)
Arfa Gadamak-tolerant, N98, Atlas	Drought	10 days & 21days	qRT-PCR targeting 8 microRNA	-	Hamza et al. (2016)

Additionally, in response to the recent emergence of the sugarcane aphid as a significant pest of sorghum, Shrestha et al. (2024) conducted Illumina sequencing on both resistant (Tx2783) and susceptible (BTx623) sorghum varieties at 0, 3, 6, 9, and 12 dpi. They identified the highest number of differentially expressed genes (DEGs) at six dpi (3,533 DEGs) and nine dpi (6,747 DEGs), providing valuable insights into the molecular mechanisms underlying sorghum's defense against this pest. These transcriptomic analyses are crucial for improving sorghum's pest and stress resistance, ultimately supporting more resilient agricultural practices.

3.4 TOOLS AND APPROACHES FOR PROTEOMIC AND METABOLOMIC STUDIES IN SORGHUM

The phenotype of plants results from a complex interplay of interactions among various molecular components, including DNA, RNA, proteins, and metabolites, which collectively regulate biochemical and physiological acclimation to environmental stimuli (Guidetti-Gonzalez et al., 2015; Ibraheem et al., 2018). The proteome, which represents the complete set of proteins in an organism, is highly dynamic and subject to change in response to environmental challenges, such as biotic and abiotic stresses. Proteins are responsible for a myriad of cellular events, and the study of protein structure and function, referred to as proteomics, seeks to understand how proteins perform their functions, interact with other molecules, and contribute to the plant's response to various environmental stimuli.

In sorghum, several proteomic and metabolomic studies have been conducted to explore the plant's response to biotic, abiotic, and nutrient stresses (Table 3.3). For instance, Roy et al. (2016) employed MALDI-TOF-TOF MS to identify 33 differentially expressed proteins in leaves of the inbred sorghum line BTx623 after three days of cadmium stress. In a follow-up study, Roy et al. (2017) used the same technology to identify 422 proteins in roots under copper stress at five days

TABLE 3.3
Summary of proteomics and metabolomics studies in sorghum

Source	Plant tissue sampled	Environmental Condition	Time Point	Technique	Summary of Findings	Reference
Inbred line BTx623	Leaves	Cd Stress	3DAT	MALDI TOF/ TOFMS	33 differentially expressed proteins	Roy et al. (2016)
BTx623, RTx430, IS3620C, BTx642, Tx7000, SC56, SC170, M35-1, 100M, Rio, N598	Leaves	-	4wk old	GC/LC MS	1,181 metabolites	Turner et al. (2016)
MR43	Grain	-	0, 0.5, 3, 4, 10, 13, 17, 20, 25, 32, 40, 45, 49, 52, 56, 61, 63 and 67 DAA	LC-MS/MS		Nielsen et al. (2016)
S. bicolor	Roots	Cu Stress	5DAT	MALDI-TOF-TOFMS	422 proteins	Roy et al. (2017)
BTx623 & SC56	Roots	-	21 d old	GC- and UPLC-MS	1,304 compounds	Miller et al. (2019)
Amazi Mhlophe, NS 5511, NS 5655	4-leaf stage	*Colletotrichum sublineolum*	0, 3, 5, 7, and 9 dpi	LC-ESI QTOF-MS	1,536 in ESI positive and 2,759 in ESI negative datasets	Tugizimana et al. (2019)

(continued)

TABLE 3.3 (Continued)
Summary of proteomics and metabolomics studies in sorghum

Source	Plant tissue sampled	Environmental Condition	Time Point	Technique	Summary of Findings	Reference
Albino mutant sbe6-a1	Leaves	-	-	Q Exactive plus Orbitrap mass spectrometer	4,233 proteins	Zhu et al. (2019)
RIO, BTx406 and R9188	Internode	-	Flag leaf stage, 100% flowering, 10, 15 and 30 DAF	UHPLC-MS/MS, UHPLC (HILIC)-MS/MS and GC-MS	507 biochemicals, of which 300 were known metabolites	Li et al. (2019)
SA1441-tolerant & ICSB338 (S)	Roots	Drought	0, 4, 8 and 12 DAT	LC-MS	1,169 and 1,043 proteins in ICSB338 and SA1441, respectively	Goche et al. (2020)
BTx623	Leaves	Saline-alkali		LC-MS/MS	657 metabolites	Ma et al. (2020)
S. bicolor	Seedling, advanced vegetative (booting) and ripening	Different field condition	25, 75, 110 DAS	NMR	Presence of Dhurrin and p-glucosyloxy-2-hydroxyphenylacetic acid	Mandrone et al. (2021)
Swarna (susceptible) and ICSV700, IS2205	Leaves	*Chilo partellus*	3h and 24h	LC-MS/MS	967 proteins	Tamhane et al. (2021)
SC265	Leaves	Sugarcane aphid	1 and 7 dpi	LC-MS/MS	4,211 proteins	Grover et al. (2022)
347 genotypes of SAP	Leaves	Low nitrogen	65 DAS	LC-MS/MS	3,496 metabolites	Grzybowski et al. (2022)
S4 (Heibeinuoza5) and S4-1 (Qianjinchui) (S) and T33 (Theis) and T14 (Rio) (R)	Leaves	Drought	24h	nano LC-MS/MS	3,927 proteins	Li et al. (2022)
HN and GZ	Root	Salt	0, 24, 48, and 72h AT	UPLC-MS/MS	58 secondary metabolites and 246 proteins	Ren et al. (2022)
LRNK1-tolerant and LR2381 (S)	Leaves	Salt	72h	LC-MS	548 and 824 metabolites	Zhang et al. (2022)
Z6, Z27, and HC4	Seed	-	At harvesting	LC-ESI MS/MS	651 metabolites	Zhou et al. (2022)
Inbred line "407B"	Roots	Cadmium	3DAT	UPLC-MS/MS	1,555 metabolites	Jiao et al. (2023)

TABLE 3.3 (Continued)
Summary of proteomics and metabolomics studies in sorghum

Source	Plant tissue sampled	Environmental Condition	Time Point	Technique	Summary of Findings	Reference
var. BTx623	Germinating grain; embryo, scutellum, and aleurone	-	6h imb, 6, 10, 14, 18, 22h AG	LC-MS/ MS	Presence of dhurrin	Liu et al. (2023)
Chuanhongliang No. 9 and Dongzajiao	Grain	-	75d	UHPLC-Q-TOF-MS/ MS	267 metabolites	Li et al. (2024)
BTx623	Whole seed, endosperm and embryo	-	5, 10, 15, 20 and 25 dpa	SEM	2,073 metabolites	Khan et al. (2024)

after treatment (DAT). Moreover, metabolomic analyses using LC-MS/MS have identified key metabolites in response to a range of stresses, including saline-alkali, drought, sugarcane aphid infestation, low nitrogen, and salt (Goche et al., 2020; Tamhane et al., 2021; Grover et al., 2022; Li et al., 2022; Ren et al., 2022).

Metabolomics enables the identification of a broad spectrum of metabolic compounds, including sugars, fatty acids, amino acids, hormones, alkaloids, and phenolic compounds, that are produced, degraded, or transformed in response to biotic and abiotic stresses in plants (Singh et al., 2024). These metabolites are chemically diverse, varying in size, polarity, and hydrophobicity. The primary objective of metabolomics is to provide a comprehensive profile of both qualitative and quantitative metabolites in plants. Primary metabolites, such as amino acids, nucleotides, lipids, and carbohydrates, are essential for fundamental processes like photosynthesis, respiration, and solute transport. In contrast, secondary metabolites, including terpenes, phenolics, and alkaloids, are primarily produced in response to specific environmental conditions, playing key roles in plant defense and adaptation.

Proteomics and metabolomics are powerful omics technologies used for identifying proteins and metabolites, employing a variety of techniques. Common methods in both fields include (i) Liquid chromatography–tandem mass spectrometry (LC–MS/MS), a widely used platform for identifying proteins and metabolites; (ii) Gas chromatography–mass spectrometry (GC–MS), which is primarily used in metabolomics; (iii) Nuclear magnetic resonance (NMR) spectroscopy, which is particularly useful for analyzing polar metabolites in biofluids; (iv) Two-dimensional gel electrophoresis (2DE), combined with mass spectrometry (MS), a technique used in proteomics; (v) Shotgun proteomics, an approach for protein identification; (vi) Capillary electrophoresis (CE), a separation method often coupled with MS; (vii) Fluorescence-based methods; (viii) Raman-based techniques; and (viii) Direct mass spectrometry imaging (MSI), applied in both proteomics and metabolomics. Historically, gel-based proteomics has been the most frequently used method, involving protein separation via one- or two-dimensional gel electrophoresis or differential in-gel electrophoresis, followed by quantification, digestion of protein spots, and mass spectrometric identification. Alternatively, gel-free analysis involves proteolytic digestion of protein mixtures, followed by separation using liquid chromatography and mass spectrometry-based identification, with multidimensional protein identification technology often used for peptide separation (Tan et al., 2017; Labuschagne, 2018). While

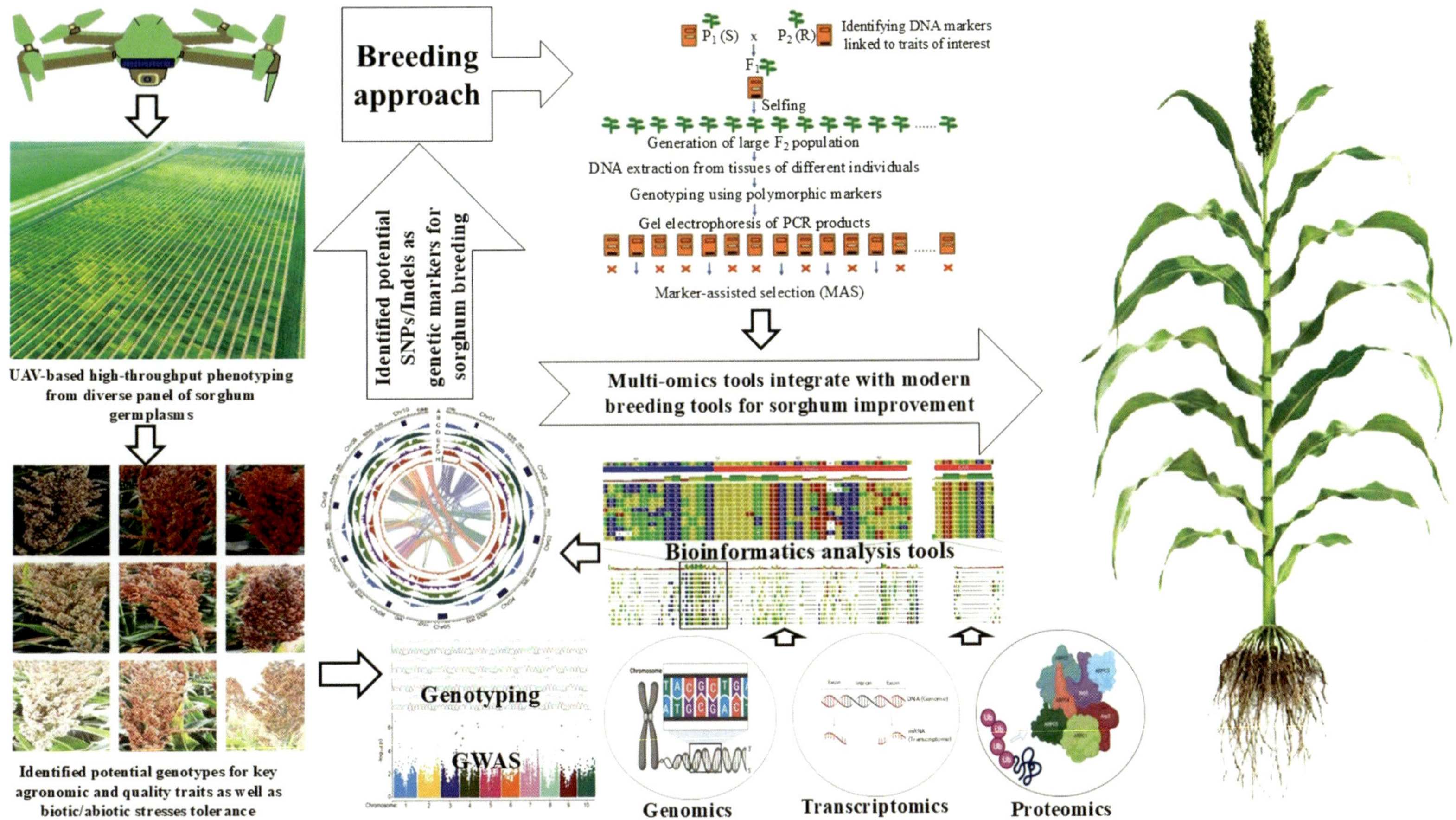

FIGURE 3.2 Leveraging bioinformatics resources to advance sorghum improvement by integrating high-throughput phenotyping tools and multi-omics datasets with modern breeding strategies.

NMR offers high selectivity and non-destructiveness, its low sensitivity is a limitation. In contrast, MS offers an excellent combination of selectivity, sensitivity, and precision, making it widely used for analyzing complex samples (Guidetti-Gonzalez et al., 2015). More recently, gel-free protein separation techniques and advanced proteomic methods, such as MudPIT, ICAT, TMT, and iTRAQ, have become increasingly common in plant development and metabolic studies, particularly in relation to adaptation to biotic and abiotic stresses (Matros et al., 2011).

Additionally, gene ontology-related databases, such as Gene Ontology (GO), UniProt, and PANTHER, provide valuable insights into the biological processes, molecular functions, and cellular components of proteins. However, compiling comprehensive databases that cover all the data from proteomic and metabolomic studies of a species under specific environmental conditions remains a significant challenge. This highlights the need for continued research in systems biology to address these data gaps.

Plants' ability to resist or tolerate climate changes and survive under biotic and abiotic stresses is enabled by the reprogramming of their transcriptome, proteome, and metabolome (Tu et al., 2023). These "omics" technologies provide insights into the genes, gene products, and pathways that regulate phenotypic traits of agronomic importance, thereby facilitating the development of platforms for screening and genetic improvement strategies (Langridge & Fleury, 2011; Guidetti-Gonzalez et al., 2015). However, plant proteomic studies are often conducted across different tissues and genotypes under various biological conditions, leading to considerable variability that complicates comparative analyses.

3.5 INTEGRATION OF OMICS DATA

3.5.1 Multi-omics Data Integration Strategies

Multi-omics approaches, including genomics, transcriptomics, proteomics, metabolomics, and others, generate comprehensive data essential for understanding plant physiological processes (Ashraf et al., 2022; Zhou et al., 2022). Integrating these datasets, supported by advancements in NGS technologies and mass spectrometry, enhances sorghum breeding for stress tolerance (Fondi & Liò, 2015). This integrated approach provides a systems-level view, enabling the identification of key regulatory networks, biomarkers, and candidate genes for targeted breeding (Saini et al., 2024a, 2024b). By combining multi-omics data with advanced bioinformatics and machine learning algorithms, predictive modeling and precision breeding for stress-tolerant crops are improved. Recent studies have applied multi-omics techniques to investigate salt stress tolerance in sorghum. For instance, analysis of two sorghum cultivars, Henong16 and Gaoliangzhe, under salt stress revealed significant differences in gene expression, metabolites, and protein abundance, with flavonoid biosynthesis playing a key role in salinity adaptation (Ren et al., 2022). Another study on drought stress in cv. Hongyingzi and cv. Kangsi identified 156 lipid compounds, with key lipids showing significant changes. Drought stress reduced the unsaturation indices of certain lipids, while transcriptome analysis identified lipid remodeling genes, including *PLDα1*, *PLDδ*, and *PLA2*. Integration of data via WGCNA and PPI analysis highlighted candidate genes, which could aid in developing drought-tolerant sorghum lines (Xu et al., 2022). Such insights are critical for breeding programs aimed at developing more resilient sorghum varieties.

Despite its potential, integrating multi-omics data poses challenges, particularly in finding effective computational tools and methodologies for analysis. The sheer volume and complexity of data generated across different omics layers can complicate the integration process. Additionally, researchers must develop strategies to ensure that data from distinct omics platforms can be compared and analyzed cohesively. Several computational tools and platforms have been developed to facilitate multi-omics analysis. For instance, the IPOP (Integrative Plant multi-Omics Platform) is designed specifically for integrating various types of omics data across different plant species, including sorghum (Huang et al., 2023). It provides functionalities for comparative analysis and evolutionary genomics, assisting researchers in understanding how multi-omics integration can enhance crop improvement strategies.

3.5.2 Visualization Techniques for Complex Biological Data

The life sciences are increasingly reliant on data science, where visualization plays a crucial role in interpreting complex data generated by advanced experimental techniques such as 3D genomics, spatial transcriptomics, 3D proteomics, epiproteomics, high-throughput imaging, and metagenomics. Data visualization involves not only analysis, design, and rendering but also observation and cognitive interpretation (O'Donoghue, 2021). Visualization tools can be broadly categorized into three types: (i) Spreadsheets: widely used tools like Microsoft Excel and Google Sheets that offer basic visualizations such as bar charts, line graphs, and scatter plots; (ii) Data Visualization Software: specialized tools like Tableau, QlikView, and Power BI that provide advanced capabilities such as interactive dashboards, heat maps, and network diagrams; and (iii) Programming Libraries: tools like Matplotlib, ggplot2, and D3.js, which allow for custom visualizations with greater flexibility but require more technical expertise.

3.6 APPLICATIONS AND FUTURE DIRECTIONS

The genetic diversity of sorghum has been significantly enhanced through genome sequencing of several inbred lines, including the reference BTx623 (v1.0) (Paterson et al., 2009) and its subsequent versions (v3.1 and v5.1) (McCormick et al., 2018), along with RTx430 (Deschamps et al., 2018), Rio (Cooper et al., 2020), and Ji2731 (Tao et al., 2021). These genomes, available on Phytozome at the Joint Genome Institute (JGI), have enabled the identification of key genetic variations such as SNPs, InDels, structural variations, and copy-number variations. Recent advancements include high-quality telomere-to-telomere (T2T) genome assemblies for BTx623, the Chinese inbred line Ji2055 (Wei et al., 2024), and two Chinese Baijiu brewing landraces (HYZ and HDN) (Bao et al., 2024), achieved through ultra-long ONT, PacBio HiFi reads, and Hi-C techniques. These assemblies have resolved gaps in centromere and telomere regions and corrected assembly inaccuracies, marking significant progress in genome assembly quality. Moving forward, sorghum genomics research should prioritize developing bioinformatics resources to increase access to genome assemblies and genetic data. The integration of multi-omics data, supported by enhanced analytical tools and collaboration platforms, will accelerate breeding efforts and drive the discovery of important traits, ultimately improving agricultural yield and resilience. Figure 3.2 illustrates the integration of high-throughput phenotyping tools and multi-omics datasets with modern breeding strategies, showcasing how bioinformatics resources can be leveraged to advance sorghum improvement.

3.7 CONCLUSION

This chapter highlights the critical role of bioinformatics in advancing sorghum research, leveraging omics technologies to enable comprehensive genomic and molecular analyses. Sorghum, recognized for its significant agricultural value, thrives in challenging environments and plays a crucial role in food security and sustainability. The chapter systematically reviews key bioinformatics tools applied to sorghum omics, including genomic sequencing, transcriptomic analysis, proteomic profiling, and metabolomic studies. It begins by emphasizing sorghum's resilience across diverse climatic zones and its applications in food, feed, and bioenergy. The discussion further explores specialized bioinformatics platforms and the integration of multi-omics data, underscoring the importance of advanced visualization techniques for interpreting complex biological information. By critically evaluating the practical implications of bioinformatics, the chapter demonstrates its essential contribution to improving sorghum's productivity and resilience, which are key to sustainable agricultural practices and addressing global food security challenges. This chapter serves as a foundational resource for researchers and practitioners, promoting innovation and agricultural advancement in the near future.

REFERENCES

Abdel-Ghany, S. E., Ullah, F., Ben-Hur, A., & Reddy, A. S. (2020). Transcriptome analysis of drought-resistant and drought-sensitive sorghum (*Sorghum bicolor*) genotypes in response to PEG-induced drought stress. *International Journal of Molecular Sciences*, *21*(3), 772.

Aruna, C., Visarada, K. B. R. S., Bhat, B. V., & Tonapi, V. A. (Eds.). (2018). *Breeding sorghum for diverse end uses*. Amsterdam, The Netherlands: Elsevier.

Ashraf, U., Mahmood, S., Shahid, N., Imran, M., Siddique, M., & Abrar, M. (2022). Multi-omics approaches for strategic improvements of crops under changing climatic conditions. In *Principles and practices of OMICS and genome editing for crop improvement* (pp. 57–92). Cham, Switzerland: Springer International Publishing.

Ayalew, H., Peiris, S., Chiluwal, A., Kumar, R., Tiwari, M., Ostmeyer, T., . . . Jagadish, S. V. K. (2022). Stable sorghum grain quality QTL were identified using SC35× RTx430 mapping population. *The Plant Genome*, *15*(3), Art. e20227.

Azzouz-Olden, F., Hunt, A. G., & Dinkins, R. (2020). Transcriptome analysis of drought-tolerant sorghum genotype SC56 in response to water stress reveals an oxidative stress defense strategy. *Molecular Biology Reports*, *47*(5), 3291–3303.

Bao, J., Zhang, H., Wang, F., Li, L., Zhu, X., Xu, J., . . . Zhu, Y. (2024). Telomere-to-telomere genome assemblies of two Chinese Baijiu-brewing sorghum landraces. *Plant Communications, 5*(6), 1–4.

Boyles, R. E., Brenton, Z. W., & Kresovich, S. (2019). Genetic and genomic resources of sorghum to connect genotype with phenotype in contrasting environments. *The Plant Journal*, *97*(1), 19–39.

Boyles, R. E., Cooper, E. A., Myers, M. T., Brenton, Z., Rauh, B. L., Morris, G. P., & Kresovich, S. (2016). Genome-wide association studies of grain yield components in diverse sorghum germplasm. *The Plant Genome*, *9*(2), 1–17.

Buchanan, C. D., Lim, S., Salzman, R. A., Kagiampakis, I., Morishige, D. T., Weers, B. D., . . . Mullet, J. E. (2005). *Sorghum bicolor*'s transcriptome response to dehydration, high salinity and ABA. *Plant Molecular Biology*, *58*, 699–720.

Cajka, T. (2024). Liquid chromatography–mass spectrometry-based metabolomics approaches for foodomics research. *Current Opinion in Food Science*, *58*, Art. 101201.

Chandrashekar, K., Niranjan, V., Vishal, A., & Setlur, A. S. (2024). Integration of artificial intelligence, machine learning and deep learning techniques in genomics: Review on computational perspectives for NGS analysis of DNA and RNA seq data. *Current Bioinformatics*, *19*(9), 825–844.

Chen, C., Shang, X., Sun, M., Tang, S., Khan, A., Zhang, D., . . . Xie, Q. (2022). Comparative transcriptome analysis of two sweet sorghum genotypes with different salt tolerance abilities to reveal the mechanism of salt tolerance. *International Journal of Molecular Sciences*, *23*(4), 2272.

Cheng, Y., Weng, J., Joshi, C. P., & Nguyen, H. T. (1993). Dehydration Stress-Induced Changes in Translatable RNAs in Sorghum. *Crop Science*, *33*(6), 1397–1400.

Clark, L. V., Lipka, A. E., & Sacks, E. J. (2019). polyRAD: Genotype calling with uncertainty from sequencing data in polyploids and diploids. *G3: Genes, Genomes, Genetics*, *9*(3), 663–673.

Cooper, Z. J. (2020). The initial farming population of the Northern Rio Grande: A multidisciplinary analysis. *Journal of Anthropological Research*, *76*(4), 439–479.

Cruet-Burgos, C., & Rhodes, D. H. (2023). Unraveling transcriptomics of sorghum grain carotenoids: A step forward for biofortification. *BMC Genomics*, *24*(1), 233.

Cui, J., Ren, G., Qiao, H., Xiang, X., Huang, L., & Chang, J. (2018). Comparative transcriptome analysis of seedling stage of two sorghum cultivars under salt stress. *Journal of Plant Growth Regulation*, *37*, 986–998.

Deschamps, S., Zhang, Y., Llaca, V., Ye, L., Sanyal, A., King, M., . . . Lin, H. (2018). A chromosome-scale assembly of the sorghum genome using nanopore sequencing and optical mapping. *Nature Communications*, *9*(1), 4844.

Devnarain, N., Crampton, B. G., Olivier, N., Van der Westhuyzen, C., Becker, J. V., & O'Kennedy, M. M. (2019). Transcriptomic analysis of a *Sorghum bicolor* landrace identifies a role for beta-alanine betaine biosynthesis in drought tolerance. South *African Journal of Botany*, *127*, 244–255.

Dugas, D. V., Monaco, M. K., Olson, A., Klein, R. R., Kumari, S., Ware, D., & Klein, P. E. (2011). Functional annotation of the transcriptome of *Sorghum bicolor* in response to osmotic stress and abscisic acid. *BMC Genomics*, *12*, 1–21.

Eaton, D. A., & Overcast, I. (2020). ipyrad: Interactive assembly and analysis of RADseq datasets. *Bioinformatics*, *36*(8), 2592–2594.

Elias, M., Chere, D., Lule, D., Serba, D., Tirfessa, A., Gelmesa, D., . . . Menamo, T. M. (2024). Multi-locus genome-wide association study reveal genomic regions underlying root system architecture traits in Ethiopian sorghum germplasm. *The Plant Genome*, *17*(2), Art. e20436.

Enyew, M., Feyissa, T., Carlsson, A. S., Tesfaye, K., Hammenhag, C., Seyoum, A., & Geleta, M. (2022). Genome-wide analyses using multi-locus models revealed marker-trait associations for major agronomic traits in *Sorghum bicolor*. *Frontiers in Plant Science*, *13*, Art. 999692.

Fondi, M., & Liò, P. (2015). Multi-omics and metabolic modelling pipelines: Challenges and tools for systems microbiology. *Microbiological Research*, *171*, 52–64.

Fracasso, A., Trindade, L. M., & Amaducci, S. (2016). Drought stress tolerance strategies revealed by RNA-Seq in two sorghum genotypes with contrasting WUE. *BMC Plant Biology*, *16*, 1–18.

Fu, J., McKinley, B., James, B., Chrisler, W., Markillie, L. M., Gaffrey, M. J., . . . Marshall-Colon, A. (2024). Cell-type-specific transcriptomics uncovers spatial regulatory networks in bioenergy sorghum stems. *The Plant Journal*, *118*(5), 1668–1688.

Gelli, M., Konda, A. R., Liu, K., Zhang, C., Clemente, T. E., Holding, D. R., & Dweikat, I. M. (2017). Validation of QTL mapping and transcriptome profiling for identification of candidate genes associated with nitrogen stress tolerance in sorghum. *BMC Plant Biology*, *17*, 1–18.

Goche, T., Shargie, N. G., Cummins, I., Brown, A. P., Chivasa, S., & Ngara, R. (2020). Comparative physiological and root proteome analyses of two sorghum varieties responding to water limitation. *Scientific Reports*, *10*(1), 11835.

Grover, S., Cardona, J. B., Zogli, P., Alvarez, S., Naldrett, M. J., Sattler, S. E., & Louis, J. (2022). Reprogramming of sorghum proteome in response to sugarcane aphid infestation. *Plant Science*, *320*, Art. 111289.

Grzybowski, M. W., Zwiener, M., Jin, H., Wijewardane, N. K., Atefi, A., Naldrett, M. J., . . . Schnable, J. C. (2022). Variation in morpho-physiological and metabolic responses to low nitrogen stress across the sorghum association panel. *BMC Plant Biology*, *22*(1), 433.

Guidetti-Gonzalez, S., Veneziano Labate, M. T., de Santana Borges, J., Frasson Budzinski, I. G., Marques, F. G., Regiani, T., . . . Labate, C. A. (2015). Proteomics and metabolomics as large-scale phenotyping tools. In Fritsche-Neto, R. and Borém, A. (eds.), *Phenomics: How next-generation phenotyping is revolutionizing plant breeding* (pp. 125–139). Cham: Springer.

Guindo, D., Teme, N., Vaksmann, M., Doumbia, M., Vilmus, I., Guitton, B., . . . Rami, J. F. (2019). Quantitative trait loci for sorghum grain morphology and quality traits: Toward breeding for a traditional food preparation of West-Africa. *Journal of cereal Science*, *85*, 256–272.

Hadebe, S. T., Modi, A. T., & Mabhaudhi, T. (2017). Drought tolerance and water use of cereal crops: A focus on sorghum as a food security crop in sub-Saharan Africa. *Journal of Agronomy and Crop Science*, *203*(3), 177–191.

Hamza, N. B., Sharma, N., Tripathi, A., & Sanan-Mishra, N. (2016). MicroRNA expression profiles in response to drought stress in *Sorghum bicolor*. *Gene Expression Patterns*, *20*(2), 88–98.

Hassan, A. M., Ahmed, M. F., & Rashed, M. A. (2022). Performance and transcriptomic analysis of *Sorghum bicolor* responding to drought stress. *SABRAO Journal of Breeding and Genetics*, *54*(4) 814–825.

Hostetler, A. N., Govindarajulu, R., & Hawkins, J. S. (2021). QTL mapping in an interspecific sorghum population uncovers candidate regulators of salinity tolerance. *Plant Stress*, *2*, Art. 100024.

Hu, T., Chitnis, N., Monos, D., & Dinh, A. (2021). Next-generation sequencing technologies: An overview. *Human Immunology*, *82*(11), 801–811.

Huang, W., Hu, X., Ren, Y., Song, M., Ma, C., & Miao, Z. (2023). IPOP: An integrative plant multi-omics platform for cross-species comparison and evolutionary study. *Molecular Biology and Evolution*, *40*(12), Art. msad248.

Ibraheem, O., Adigun, R. O., & Olatunji, I. T. (2018). Omics technologies in unraveling plant stress responses; using sorghum as a model crop, how far have we gone. *Vegetos*, *31*, 2.

Jain, M., Olsen, H. E., Paten, B., & Akeson, M. (2016). The Oxford Nanopore MinION: Delivery of nanopore sequencing to the genomics community. *Genome Biology*, *17*, 1–11.

Jeon, D., Kim, J. B., Kang, B. C., & Kim, C. (2023). Deciphering the genetic mechanisms of salt tolerance in *Sorghum bicolor* L.: Key genes and SNP associations from comparative transcriptomic analyses. Plants, *12*(14), 2639.

Jiao, Z., Shi, Y., Wang, J., Wang, Z., Zhang, X., Jia, X., . . . Lv, P. (2023). Integration of transcriptome and metabolome analyses reveals sorghum roots responding to cadmium stress through regulation of the flavonoid biosynthesis pathway. *Frontiers in Plant Science*, *14*, Art. 1144265.

Johnson, S. M., Cummins, I., Lim, F. L., Slabas, A. R., & Knight, M. R. (2015). Transcriptomic analysis comparing stay-green and senescent *Sorghum bicolor* lines identifies a role for proline biosynthesis in the stay-green trait. *Journal of Experimental Botany*, *66*(22), 7061–7073.

Johnson, S. M., Lim, F. L., Finkler, A., Fromm, H., Slabas, A. R., & Knight, M. R. (2014). Transcriptomic analysis of *Sorghum bicolor* responding to combined heat and drought stress. *BMC Genomics*, *15*, 1–19.

Ju, L., Lv, N., Yin, F., Niu, H., Yan, H., Wang, Y., . . . Ping, J. (2023). Identification of key genes regulating sorghum mesocotyl elongation through transcriptome analysis. *Genes*, *14*(6), 1215.

Karpievitch, Y. V., Polpitiya, A. D., Anderson, G. A., Smith, R. D., & Dabney, A. R. (2010). Liquid chromatography mass spectrometry-based proteomics: Biological and technological aspects. *Annals of Applied Statistics*, *4*(4), 1797.

Karumanchi, A. R., Sivan, P., Kummari, D., Rajasheker, G., Kumar, S. A., Reddy, P. S., . . . Kishor, P. K. (2023). Root and leaf anatomy, ion accumulation, and transcriptome pattern under salt stress conditions in contrasting genotypes of *Sorghum bicolor*. *Plants*, *12*(13), 2400.

Katiyar, A., Smita, S., Muthusamy, S. K., Chinnusamy, V., Pandey, D. M., & Bansal, K. C. (2015). Identification of novel drought-responsive microRNAs and trans-acting siRNAs from *Sorghum bicolor* (L.) Moench by high-throughput sequencing analysis. *Frontiers in Plant Science*, *6*, 506.

Ke, F., Zhang, K., Li, Z., Wang, J., Zhang, F., Wu, H., . . . Zhu, K. (2022). Transcriptomic analysis of starch accumulation patterns in different glutinous sorghum seeds. *Scientific Reports*, *12*(1), 11133.

Khalifa, M., & Eltahir, E. A. (2023). Assessment of global sorghum production, tolerance, and climate risk. *Frontiers in Sustainable Food Systems*, *7*, Art. 1184373.

Khan, A., Tian, R., Bean, S. R., Yerka, M., & Jiao, Y. (2024). Transcriptome and metabolome analyses reveal regulatory networks associated with nutrition synthesis in sorghum seeds. *Communications Biology*, *7*(1), 841.

Kiani, M., & Szczepaniec, A. (2018). Effects of sugarcane aphid herbivory on transcriptional responses of resistant and susceptible sorghum. *BMC Genomics*, *19*, 1–18.

Kisiel, A., Krzemińska, A., Cembrowska-Lech, D., & Miller, T. (2023). Data science and plant metabolomics. *Metabolites*, *13*(3), 454.

Korneliussen, T. S., Albrechtsen, A., & Nielsen, R. (2014). ANGSD: Analysis of next generation sequencing data. *BMC Bioinformatics*, *15*, 1–13.

Kumar, N., Boatwright, J. L., Sapkota, S., Brenton, Z. W., Ballén-Taborda, C., Myers, M. T., . . . Boyles, R. E. (2023). Discovering useful genetic variation in the seed parent gene pool for sorghum improvement. *Frontiers in Genetics*, *14*, Art. 1221148.

Labuschagne, M. T. (2018). A review of cereal grain proteomics and its potential for sorghum improvement. *Journal of Cereal Science*, *84*, 151–158.

Langridge, P., & Fleury, D. (2011). Making the most of 'omics' for crop breeding. *Trends in Biotechnology*, *29*(1), 33–40.

Li, J., Tang, W., Zhang, Y. W., Chen, K. N., Wang, C., Liu, Y., . . . Wang, L. (2018). Genome-wide association studies for five forage quality-related traits in sorghum (*Sorghum bicolor* L.). *Frontiers in Plant Science*, *9*, 1146.

Li, Y., Ma, Y., Zhu, H., Liu, Y., Pan, S., Chen, X., & Wu, T. (2024). Identifying distinct markers in two Sorghum varieties for baijiu fermentation using untargeted metabolomics and molecular network approaches. *Food Chemistry: X*, *23*, Art. 101646.

Li, Y., Tan, B., Wang, D., Mu, Y., Li, G., Zhang, Z., . . . Zhu, L. (2022). Proteomic analysis revealed different molecular mechanisms of response to PEG stress in drought-sensitive and drought-resistant sorghums. *International Journal of Molecular Sciences*, *23*(21), 13297.

Li, Y., Wang, W., Feng, Y., Tu, M., Wittich, P. E., Bate, N. J., & Messing, J. (2019). Transcriptome and metabolome reveal distinct carbon allocation patterns during internode sugar accumulation in different sorghum genotypes. *Plant Biotechnology Journal*, *17*(2), 472–487.

Liu, H., Micic, N., Miller, S., Crocoll, C., & Bjarnholt, N. (2023). Species-specific dynamics of specialized metabolism in germinating sorghum grain revealed by temporal and tissue-resolved transcriptomics and metabolomics. *Plant Physiology and Biochemistry*, *196*, 807–820.

Liu, Y., Wang, Z., Wu, X., Zhu, J., Luo, H., Tian, D., . . . Jing, H. C. (2021). SorGSD: Updating and expanding the sorghum genome science database with new contents and tools. *Biotechnology for Biofuels*, *14*(1), 165.

Lowe, R., Shirley, N., Bleackley, M., Dolan, S., & Shafee, T. (2017). Transcriptomics technologies. *PLoS Computational Biology*, *13*(5), e1005457.

Luo, H., Mocoeur, A., & Jing, H. C. (2014). Next-generation sequencing technology for genetics and genomics of sorghum. In Luo, H., Mocoeur, A., and Jing, H.C. (eds.), *Genetics, genomics and breeding of sorghum* (pp. 226–250). Boca Raton: CRC Press.

Ma, S., Lv, L., Meng, C., Zhang, C., & Li, Y. (2020). Integrative analysis of the metabolome and transcriptome of *Sorghum bicolor* reveals dynamic changes in flavonoids accumulation under saline–alkali stress. *Journal of Agricultural and Food Chemistry*, *68*(50), 14781–14789.

Mace, E. S., Tai, S., Gilding, E. K., Li, Y., Prentis, P. J., Bian, L., . . . Wang, J. (2013). Whole-genome sequencing reveals untapped genetic potential in Africa's indigenous cereal crop sorghum. *Nature Communications*, *4*(1), 2320.

Mace, E., Innes, D., Hunt, C., Wang, X., Tao, Y., Baxter, J., . . . Jordan, D. (2019). The Sorghum QTL Atlas: A powerful tool for trait dissection, comparative genomics and crop improvement. *Theoretical and Applied Genetics*, *132*, 751–766.

Madhusudhana, R. (2019). Marker-assisted breeding in sorghum. In Aruna, A., Visarada, K.B.R.S., Bhat, B.V., and Tonapi, V.A. (eds.), *Breeding sorghum for diverse end uses* (pp. 93–114). Cambridge, MA: Woodhead Publishing.

Mandrone, M., Chiocchio, I., Barbanti, L., Tomasi, P., Tacchini, M., & Poli, F. (2021). Metabolomic study of sorghum (*Sorghum bicolor*) to interpret plant behavior under variable field conditions in view of smart agriculture applications. *Journal of Agricultural and Food Chemistry*, *69*(3), 1132–1145.

Marla, S. R., Shiva, S., Welti, R., Liu, S., Burke, J. J., & Morris, G. P. (2017). Comparative transcriptome and lipidome analyses reveal molecular chilling responses in chilling-tolerant sorghums. *The Plant Genome*, *10*(3), 1–16.

Matros, A., Kaspar, S., Witzel, K., & Mock, H. P. (2011). Recent progress in liquid chromatography-based separation and label-free quantitative plant proteomics. *Phytochemistry*, *72*(10), 963–974.

McCormick, R. F., Truong, S. K., Sreedasyam, A., Jenkins, J., Shu, S., Sims, D., . . . Mullet, J. E. (2018). The *Sorghum bicolor* reference genome: Improved assembly, gene annotations, a transcriptome atlas, and signatures of genome organization. *The Plant Journal*, *93*(2), 338–354.

Miller, S. B., Heuberger, A. L., Broeckling, C. D., & Jahn, C. E. (2019). Non-targeted metabolomics reveals sorghum rhizosphere-associated exudates are influenced by the belowground interaction of substrate and sorghum genotype. *International Journal of Molecular Sciences*, *20*(2), 431.

Mizuno, H., Kawahigashi, H., Kawahara, Y., Kanamori, H., Ogata, J., Minami, H., . . . Matsumoto, T. (2012). Global transcriptome analysis reveals distinct expression among duplicated genes during sorghum-*Bipolaris sorghicola* interaction. *BMC Plant Biology*, *12*, 1–15.

Monisha, G., & Sowndarya, K. (2024). RNA-SEQ: A High-Revolution View of the Transcriptome. *Biotechnology and Biological Sciences*, *73*, 71–90.

Morris, G. P., Ramu, P., Deshpande, S. P., Hash, C. T., Shah, T., Upadhyaya, H. D., . . . Kresovich, S. (2013). Population genomic and genome-wide association studies of agroclimatic traits in sorghum. *Proceedings of the National Academy of Sciences*, *110*(2), 453–458.

Mukherjee, A., Maheshwari, U., Sharma, V., Sharma, A., & Kumar, S. (2024). Functional insight into multi-omics-based interventions for climatic resilience in sorghum (*Sorghum bicolor*): A nutritionally rich cereal crop. *Planta*, *259*(4), 91.

Musich, R., Cadle-Davidson, L., & Osier, M. V. (2021). Comparison of short-read sequence aligners indicates strengths and weaknesses for biologists to consider. *Frontiers in Plant Science*, *12*, Art. 657240.

Naithani, S., Deng, C. H., Sahu, S. K., & Jaiswal, P. (2023). Exploring pan-genomes: An overview of resources and tools for unraveling structure, function, and evolution of crop genes and genomes. *Biomolecules*, *13*(9), 1403.

Ngara, R., Goche, T., Swanevelder, D. Z., & Chivasa, S. (2021). Sorghum's whole-plant transcriptome and proteome responses to drought stress: A review. *Life*, *11*(7), 704.

Nida, H., Lee, S., Li, Y., & Mengiste, T. (2021). Transcriptome analysis of early stages of sorghum grain mold disease reveals defense regulators and metabolic pathways associated with resistance. *BMC Genomics*, *22*, 1–17.

Nielsen, L. J., Stuart, P., Pičmanová, M., Rasmussen, S., Olsen, C. E., Harholt, J., . . . Bjarnholt, N. (2016). Dhurrin metabolism in the developing grain of *Sorghum bicolor* (L.) Moench investigated by metabolite profiling and novel clustering analyses of time-resolved transcriptomic data. *BMC Genomics*, *17*, 1–24.

O'Donoghue, S. I. (2021). Grand challenges in bioinformatics data visualization. *Frontiers in Bioinformatics*, *1*, Art. 669186.

Ondiko, J. H., & Recha, C. W. (2022). Global Sorghum Production Constraints: A Review. *Annals of Arid Zone*, *61*(1), 1–12.

Paterson, A. H., Bowers, J. E., Bruggmann, R., Dubchak, I., Grimwood, J., Gundlach, H., . . . Rokhsar, D. S. (2009). The *Sorghum bicolor* genome and the diversification of grasses. *Nature*, *457*(7229), 551–556.

Pillai, S., Gopalan, V., & Lam, A. K. Y. (2017). Review of sequencing platforms and their applications in phaeochromocytoma and paragangliomas. *Critical Reviews in Oncology/Hematology*, *116*, 58–67.

Poosapati, S., Poretsky, E., Dressano, K., Ruiz, M., Vazquez, A., Sandoval, E., . . . Huffaker, A. (2022). A sorghum genome-wide association study (GWAS) identifies a WRKY transcription factor as a candidate gene underlying sugarcane aphid (*Melanaphis sacchari*) resistance. *Planta*, *255*(2), 37.

Prasad, V. R., Govindaraj, M., Djanaguiraman, M., Djalovic, I., Shailani, A., Rawat, N., . . . Prasad, P. V. (2021). Drought and high temperature stress in sorghum: Physiological, genetic, and molecular insights and breeding approaches. *International Journal of Molecular Sciences*, *22*(18), 9826.

Punia, H., Tokas, J., Malik, A., Singh, S., Phogat, D. S., Bhuker, A., . . . Sheokand, R. N. (2021). Discerning morpho-physiological and quality traits contributing to salinity tolerance acquisition in sorghum [*Sorghum bicolor* (L.) Moench]. *South African Journal of Botany*, *140*, 409–418.

Punnuri, S. M., Ayele, A. G., Harris-Shultz, K. R., Knoll, J. E., Coffin, A. W., Tadesse, H. K., . . . Wallace, J. G. (2022). Genome-wide association mapping of resistance to the sorghum aphid in *Sorghum bicolor*. *Genomics*, *114*(4), Art. 110408.

Puri, H., Grover, S., Pingault, L., Sattler, S. E., & Louis, J. (2023). Temporal transcriptomic profiling elucidates sorghum defense mechanisms against sugarcane aphids. *BMC Genomics*, *24*(1), 441.

Puritz, J. B., Hollenbeck, C. M., & Gold, J. R. (2014). dDocent: A RADseq, variant-calling pipeline designed for population genomics of non-model organisms. *PeerJ*, *2*, Art. e431.

Qu, M., An, B., Shen, S., Zhang, M., Shen, X., Duan, X., . . . Qu, J. (2017). Qualitative and quantitative characterization of protein biotherapeutics with liquid chromatography mass spectrometry. *Mass Spectrometry Reviews*, *36*(6), 734–754.

Ramalingam, A. P., Mohanavel, W., Kambale, R., Rajagopalan, V. R., Marla, S. R., Prasad, P. V., . . . Perumal, R. (2023). Pilot-scale genome-wide association mapping in diverse sorghum germplasms identified novel genetic loci linked to major agronomic, root and stomatal traits. *Scientific Reports*, *13*(1), 21917.

Ren, G., Yang, P., Cui, J., Gao, Y., Yin, C., Bai, Y., . . . Chang, J. (2022). Multiomics analyses of two sorghum cultivars reveal the molecular mechanism of salt tolerance. *Frontiers in Plant Science*, *13*, Art. 886805.

Reuter, J. A., Spacek, D. V., & Snyder, M. P. (2015). High-throughput sequencing technologies. *Molecular Cell*, *58*(4), 586–597.

Rivera-Colón, A. G., & Catchen, J. (2022). Population genomics analysis with RAD, reprised: Stacks 2. In *Marine genomics: Methods and protocols* (pp. 99–149). New York, NY: Springer US.

Rochette, N. C., Rivera-Colón, A. G., & Catchen, J. M. (2019). Stacks 2: Analytical methods for paired-end sequencing improve RADseq-based population genomics. *Molecular Ecology*, *28*(21), 4737–4754.

Roy, S. K., Cho, S. W., Kwon, S. J., Kamal, A. H. M., Lee, D. G., Sarker, K., . . . Woo, S. H. (2017). Proteome characterization of copper stress responses in the roots of sorghum. *Biometals*, *30*, 765–785.

Roy, S. K., Kwon, S. J., Cho, S. W., Kamal, A. H. M., Kim, S. W., Sarker, K., . . . Woo, S. H. (2016). Leaf proteome characterization in the context of physiological and morphological changes in response to copper stress in sorghum. *Biometals*, *29*, 495–513.

Saini, D. K. (2024a). Establishing the proper links: Network-based techniques for comprehending gene functions in cereals. In Saini, D.K. and Kole, C. (eds.), *Omics and system biology approaches for delivering better cereals* (pp. 401–430). Boca Raton, FL: CRC Press.

Saini, D. K., Kumar, S., Saini, K., & Goyal, S. (2024b). Omics and system biology for developing climate-ready cereals: An overview. In Saini, D.K. and Kole, C. (eds.), *Omics and system biology approaches for delivering better cereals* (pp. 1–25). Boca Raton, FL: CRC Press.

Serba, D. D., Meng, X., Schnable, J., Bashir, E., Michaud, J. P., Vara Prasad, P. V., & Perumal, R. (2021). Comparative transcriptome analysis reveals genetic mechanisms of sugarcane aphid resistance in grain sorghum. *International Journal of Molecular Sciences*, *22*(13), 7129.

Shafer, A. B., Peart, C. R., Tusso, S., Maayan, I., Brelsford, A., Wheat, C. W., & Wolf, J. B. (2017). Bioinformatic processing of RAD-seq data dramatically impacts downstream population genetic inference. *Methods in Ecology and Evolution*, *8*(8), 907–917.

Shrestha, K., Huang, J., Yan, L., Doust, A. N., & Huang, Y. (2024). Integrated transcriptomic and pathway analyses of sorghum plants revealed the molecular mechanisms of host defense against aphids. *Frontiers in Plant Science*, *15*, Art. 1324085.

Singh, G., Kumar, P., Mandal, P., Aziz, D., Rafiq, S., Saini, P., . . . Saini, D. K. (2024). Plant metabolomic profiling to better understand biotic and abiotic stress resistance in cereals. In Saini, D.K. and Kole, C. (eds.), *Omics and system biology approaches for delivering better cereals* (pp. 268–292). Boca Raton, FL: CRC Press.

Srivastava, U., Kanchan, S., Kesheri, M., Gupta, M. K., & Singh, S. (2024). Types of omics data: Genomics, metagenomics, epigenomics, transcriptomics, proteomics, metabolomics, and phenomics. In *Integrative omics* (pp. 13–34). Cambridge, MA: Academic Press.

Sui, N., Yang, Z., Liu, M., & Wang, B. (2015). Identification and transcriptomic profiling of genes involved in increasing sugar content during salt stress in sweet sorghum leaves. *BMC Genomics*, *16*, 1–18.

Sun, X., Zheng, H., Li, J., Liu, L., Zhang, X., & Sui, N. (2020). Comparative transcriptome analysis reveals new lncRNAs responding to salt stress in sweet sorghum. *Frontiers in Bioengineering and Biotechnology*, *8*, 331.

Takanashi, H., Shichijo, M., Sakamoto, L., Kajiya-Kanegae, H., Iwata, H., Sakamoto, W., & Tsutsumi, N. (2021). Genetic dissection of QTLs associated with spikelet-related traits and grain size in sorghum. *Scientific Reports*, *11*(1), 9398.

Tamhane, V. A., Sant, S. S., Jadhav, A. R., War, A. R., Sharma, H. C., Jaleel, A., & Kashikar, A. S. (2021). Label-free quantitative proteomics of *Sorghum bicolor* reveals the proteins strengthening plant defense against insect pest Chilo partellus. *Proteome Science*, *19*, 1–25.

Tan, T., Wu, D., Li, W., Zheng, X., Li, W., & Shan, A. (2017). High specific selectivity and membrane-active mechanism of synthetic cationic hybrid antimicrobial peptides based on the peptide FV7. *International Journal of Molecular Sciences*, *18*(2), 339.

Tao, Y., Luo, H., Xu, J., Cruickshank, A., Zhao, X., Teng, F., . . . Mace, E. (2021). Extensive variation within the pan-genome of cultivated and wild sorghum. *Nature Plants*, *7*(6), 766–773.

Techale, B., Dong, H., Mihrete, G., Aregash, G., Paterson, A. H., & Kassahun, B. (2022). QTL analysis in multiple sorghum mapping populations facilitates dissection of the genetic control of agronomic and yield-related traits in sorghum [*Sorghum bicolor* (Moench)]. *Euphytica*, *218*(3), 24.

Tello-Ruiz, M. K., Naithani, S., Gupta, P., Olson, A., Wei, S., Preece, J., . . . Ware, D. (2021). Gramene 2021: Harnessing the power of comparative genomics and pathways for plant research. *Nucleic Acids Research*, *49*(D1), D1452–D1463.

Thriveni, V., Teotia, J., Hazra, S., Bharti, T., Kumar, M., Lallawmkimi, M. C., & Panwar, D. (2024). A review on integrating bioinformatics tools in modern plant breeding. *Archives of Current Research International*, *24*(9), 293–308.

Torkamaneh, D., Laroche, J., Bastien, M., Abed, A., & Belzile, F. (2017). Fast-GBS: A new pipeline for the efficient and highly accurate calling of SNPs from genotyping-by-sequencing data. *BMC Bioinformatics*, *18*, 1–7.

Trenkel, V. M., Charrier, G., Lorance, P., & Bravington, M. V. (2022). Close-kin mark–recapture abundance estimation: Practical insights and lessons learned. *ICES Journal of Marine Science*, *79*(2), 413–422.

Tu, M., Du, C., Yu, B., Wang, G., Deng, Y., Wang, Y., . . . Li, Y. (2023). Current advances in the molecular regulation of abiotic stress tolerance in sorghum via transcriptomic, proteomic, and metabolomic approaches. *Frontiers in Plant Science*, *14*, Art. 1147328.

Tugizimana, F., Djami-Tchatchou, A. T., Steenkamp, P. A., Piater, L. A., & Dubery, I. A. (2019). Metabolomic analysis of defense-related reprogramming in *Sorghum bicolor* in response to *Colletotrichum sublineolum* infection reveals a functional metabolic web of phenylpropanoid and flavonoid pathways. *Frontiers in Plant Science*, *9*, 1840.

Turner, M. F., Heuberger, A. L., Kirkwood, J. S., Collins, C. C., Wolfrum, E. J., Broeckling, C. D., . . . Jahn, C. E. (2016). Non-targeted metabolomics in diverse sorghum breeding lines indicates primary and secondary metabolite profiles are associated with plant biomass accumulation and photosynthesis. *Frontiers in Plant Science*, *7*, 953.

Ukwatta, J., Pabuayon, I. C. M., Park, J., Chen, J., Chai, X., Zhang, H., . . . Shi, H. (2021). Comparative physiological and transcriptomic analysis reveals salinity tolerance mechanisms in *Sorghum bicolor* (L.) Moench. *Planta*, *254*, 1–20.

Upadhyay, P., Thakur, R., Arora, A., Yadav, S., Thakur, S., Sanu, K., . . . Bisht, A. (2024). Databases: A tool in the bioinformatics armory for cereals research. In Saini, D.K. and Kole, C. (eds.), *Omics and system biology approaches for delivering better cereals* (pp. 26–48). Boca Raton, FL: CRC Press.

Varoquaux, N., Cole, B., Gao, C., Pierroz, G., Baker, C. R., Patel, D., . . . Purdom, E. (2019). Transcriptomic analysis of field-droughted sorghum from seedling to maturity reveals biotic and metabolic responses. *Proceedings of the National Academy of Sciences*, *116*(52), 27124–27132.

Varshney, R. K. (2009). Gene-based marker systems in plants: High throughput approaches for marker discovery and genotyping. In Jain, S. and Brar, D. (eds.), *Molecular techniques in crop improvement* (2nd ed., pp. 119–142). Dordrecht: Springer.

Wang, L., Liu, Y., Gao, L., Yang, X., Zhang, X., Xie, S., . . . Shen, Y. (2022). Identification of candidate forage yield genes in sorghum (*Sorghum bicolor* L.) using integrated genome-wide association studies and RNA-seq. *Frontiers in Plant Science*, *12*, Art. 788433.

Wang, Z., & Wei, Y. (2022). Physiological and transcriptomic analysis of antioxidant mechanisms in sweet sorghum seedling leaves in response to single and combined drought and salinity stress. *Journal of Plant Interactions*, *17*(1), 1006–1016.

Wee, Y., Bhyan, S. B., Liu, Y., Lu, J., Li, X., & Zhao, M. (2019). The bioinformatics tools for the genome assembly and analysis based on third-generation sequencing. *Briefings in Functional Genomics*, *18*(1), 1–12.

Wei, C., Gao, L., Xiao, R., Wang, Y., Chen, B., Zou, W., . . . Tao, Y. (2024). Complete telomere-to-telomere assemblies of two sorghum genomes to guide biological discovery. *iMeta*, *3*(2), 1–5.

Wolf, J. B. (2013). Principles of transcriptome analysis and gene expression quantification: An RNA-seq tutorial. *Molecular Ecology Resources*, *13*(4), 559–572.

Wood, A. J., & Goldsbrough, P. B. (1997). Characterization and expression of dehydrins in water-stressed *Sorghum bicolor*. *Physiologia Plantarum*, *99*(1), 144–152.

Wu, X. Y., Hu, W. J., Luo, H., Xia, Y., Zhao, Y., Wang, L. D., . . . Jing, H. C. (2016). Transcriptome profiling of developmental leaf senescence in sorghum (*Sorghum bicolor*). *Plant Molecular Biology*, *92*, 555–580.

Xiao-ping, L., Jin-feng, Y., Cui-ping, G., & Acharya, S. (2011). Quantitative trait loci analysis of economically important traits in *Sorghum bicolor*× *S. sudanense* hybrid. *Canadian Journal of Plant Science*, *91*(1), 81–90.

Xu, D., Ni, Y., Zhang, X., & Guo, Y. (2022). Multiomic analyses of two sorghum cultivars reveals the change of membrane lipids in their responses to water deficit. *Plant Physiology and Biochemistry*, *176*, 44–56.

Xu, S., Li, X. Q., Guo, H., Wu, X. Y., Wang, N., Liu, Z. Q., . . . Jing, H. C. (2023). Mucilage secretion by aerial roots in sorghum (*Sorghum bicolor*): Sugar profile, genetic diversity, GWAS and transcriptomic analysis. *Plant Molecular Biology*, *112*(6), 309–323.

Yang, G. D., Zhou, Y. F., Huang, R. D., Lin, F., Hu, Z. Y., Hao, Z. Y., . . . Dong, L. D. (2019). Identification of differentially expressed genes of sorghum [*Sorghum bicolor* (L.) Moench] seedlings under nitrogen stress by RNA-SEQ. *Applied Ecology & Environmental Research*, *17*(5), 11525–11536.

Yang, Z., Wang, Y., Wei, X., Zhao, X., Wang, B., & Sui, N. (2017). Transcription profiles of genes related to hormonal regulations under salt stress in sweet sorghum. *Plant Molecular Biology Reporter*, *35*, 586–599.

Yang, Z., Zheng, H., Wei, X., Song, J., Wang, B., & Sui, N. (2018). Transcriptome analysis of sweet Sorghum inbred lines differing in salt tolerance provides novel insights into salt exclusion by roots. *Plant and Soil*, *430*, 423–439.

Zhang, D. F., Zeng, T. R., Liu, X. Y., Gao, C. X., Li, Y. X., Li, C. H., . . . Yu, L. I. (2019). Transcriptomic profiling of sorghum leaves and roots responsive to drought stress at the seedling stage. *Journal of Integrative Agriculture*, *18*(9), 1980–1995.

Zhang, F., Lu, F., Wang, Y., Zhang, Z., Wang, J., Zhang, K., . . . Zhu, K. (2022). Combined transcriptomic and physiological metabolomic analyses elucidate key biological pathways in the response of two sorghum genotypes to salinity stress. *Frontiers in Plant Science*, *13*, Art. 880373.

Zhang, J., Li, W., Lv, P., Yang, X., Xu, W., Ni, X., . . . Wang, W. (2021). Whole-genome resequencing and transcriptome analysis provide insights on aphid-resistant quantitative trait loci/genes in *Sorghum bicolor*. *Plant Breeding*, *140*(4), 618–629.

Zhang, T., Zhao, X., Wang, W., Huang, L., Liu, X., Zong, Y., . . . Li, Z. (2014). Deep transcriptome sequencing of rhizome and aerial-shoot in *Sorghum propinquum*. *Plant Molecular Biology*, *84*, 315–327.

Zhou, C., Wang, S., Zhou, H., Yuan, Z., Zhou, T., Zhang, Y., . . . Zhang, D. (2021). Transcriptome sequencing analysis of sorghum callus with various regeneration capacities. *Planta*, *254*, 1–14.

Zhou, T., Ling, D., He, Q., Wang, P., & Zhu, J. (2023). RNA-Seq Analysis of Aboveground and Underground Parts of Biomass Sorghum Was Performed to Evaluate Its Suitability for Environmental Remediation. *Biomolecules*, *13*(6), 925.

Zhou, W., Zhou, Y., Wang, Z., Li, Y., Liu, H., Cui, F., . . . Shi, Z. (2022). Comparative transcriptome analysis in three sorghum (*Sorghum bicolor*) cultivars reveal genomic basis of differential seed quality. *Plant Biosystems-An International Journal Dealing with all Aspects of Plant Biology*, *156*(1), 232–241.

Zhu, L., Wang, D., Sun, J., Mu, Y., Pu, W., Ma, B., . . . Pan, Y. (2019). Phenotypic and proteomic characteristics of sorghum (*Sorghum bicolor*) albino lethal mutant sbe6-a1. *Plant Physiology and Biochemistry*, *139*, 400–410.

4 Advances in Molecular Mapping in Sorghum Genome Design

Sirisha Kaniganti, Javaria Tabassum, Asad Riaz, and Muhammad Sajjad

4.1 INTRODUCTION

A variety of factors, including continued soil degradation, rising global temperatures, increasing atmospheric CO_2, and more frequent extreme weather events such as heatwaves, flash floods, and prolonged droughts, pose significant threats to global food and nutritional security (Muluneh, 2021). These challenges are particularly pressing in developing countries, where they exacerbate food insecurity and hinder agricultural productivity. Furthermore, a significant portion of land exists as marginal agricultural land, which remains underutilized or abandoned, awaiting the development of crop varieties suited to these less favorable environments. Addressing this issue will require evaluating the vulnerability of key agricultural regions to climate change and developing crop varieties that can withstand these changing environmental conditions (Meena et al., 2021).

Sorghum (*Sorghum bicolor* (L.) Moench), often referred to as the "camel of crops" due to its exceptional stress tolerance, stands as one of the primary crops capable of meeting these challenges. It is cultivated in 93 countries and provides food for 97% of the world's poor population (Meena et al., 2021). Sorghum's ability to thrive under high temperatures, drought, saline soils, and nutrient-deficient environments makes it crucial for agricultural regions facing resource scarcity. Despite advances in traditional breeding methods that have improved stress tolerance, a 66–96% yield gap remains between actual and potential non-irrigated yields (Assefa et al., 2024). This highlights the urgent need for precision breeding techniques, such as molecular mapping, to enhance both productivity and stress resilience. Recent advancements in molecular mapping of the sorghum genome have significantly improved our understanding of this important crop.

Mapping climate-resilient traits remains a significant challenge, but progress has been made with the creation of a reference genome for the elite sorghum line BTx623 (Paterson et al., 2009), which has provided valuable insights into sorghum's genetic architecture. The integration of high-throughput sequencing technologies has further advanced genetic mapping, enabling the development of more precise maps that are crucial for identifying genes associated with desirable traits. Trait mapping, including linkage-based mapping and genome-wide association studies (GWAS), has become an essential tool in identifying the genetic control of key traits. These methods allow researchers to correlate phenotypic variations with genetic markers, pinpointing specific loci associated with critical agronomic traits. Additionally, marker-assisted backcross breeding, which utilizes these identified genomic regions has shown promise in improving sorghum's adaptability to unfavorable environmental conditions (Mwamahonje et al., 2021).

DOI: 10.1201/9781003623717-4

This chapter provides an overview of the latest molecular mapping techniques employed in sorghum genome design, including interval mapping and GWAS. The integration of these methods has led to the identification of key genomic regions linked to yield-related traits and biotic and abiotic stress resistance. The roles of marker-assisted selection (MAS) and genomic selection (GS) in accelerating breeding programs are discussed. Finally, future directions in sorghum genomic research are outlined, with a focus on overcoming challenges related to data management, phenotyping, and the broader implications for food security and sustainability.

4.2 ADVANCEMENTS IN MARKER SYSTEMS IN SORGHUM

Initially, the Restriction Fragment Length Polymorphism (RFLP) marker system was widely used in sorghum (Tao et al., 1998), followed by the development of other marker systems such as Amplified Fragment Length Polymorphism (AFLP), Randomly Amplified Polymorphic DNA (RAPD), Single Sequence Repeats (SSRs), and Diversity Array Technology (DArT) markers (Haussmann et al., 2002; Wu et al., 2006; Mace et al., 2008). These markers have been widely applied for QTL mapping, genetic diversity analysis, and fingerprinting. The availability of the sorghum genome sequence has greatly expanded the potential of SSR molecular markers, which are valuable due to their abundance, high polymorphism, codominance, and compatibility with high-throughput genotyping (HTG). SSRs in sorghum have been identified through genomic library screening, database searches, and cross-species primers. The reference genome of the BTx623, created through whole-genome shotgun sequencing, contains 5,599 non-redundant SSR markers, and computational analysis has predicted an additional 163,943 SSRs, providing enhanced tools for studying genetic diversity and developing improved cultivars (Paterson et al., 2009).

Advancements in sequencing technologies, such as next-generation sequencing (NGS), have significantly enhanced HTG in sorghum. These developments, along with whole-genome marker arrays, have led to the creation of extensive genomic resources (McCormick et al., 2018; Baloch et al., 2023). Techniques like Whole Genome Shotgun Sequencing and Illumina Genome Analyzer Sequencing, along with advanced software, have been used to re-sequence sorghum lines and identify structural variants and SNPs (Nelson et al., 2011; Bekele et al., 2013; Tao et al., 2021). Recent studies have uncovered thousands of SNPs, copy number variations, insertions-deletions (InDels), duplications, and other structural changes (Zhang et al., 2024; Boatwright et al., 2022).

SNPs have gained prominence over SSRs due to their abundance, uniform distribution, and higher accuracy and throughput. Tools like HaploSNPer have facilitated rapid SNP discovery, enabling the identification of high-confidence SNPs across sorghum genomes (Singhal et al., 2011). Genotyping-by-sequencing (GBS) has also proven effective in detecting genetic variations, with studies revealing large numbers of SNPs and InDels from sorghum accessions (Wondimu et al., 2021). High-resolution genotyping and SNP identification using techniques such as RAD-seq have provided comprehensive molecular marker resources that hold great promise for advancing sorghum breeding efforts (Jin et al., 2021). DNA-based molecular markers have transformed sorghum breeding by facilitating the detection of genetic diversity and traits associated with yield potential and stress resilience. Their application has been instrumental in assessing germplasm resources, expediting breeding efforts, and advancing the creation of climate-adaptive sorghum varieties.

4.3 ADVANCES IN THE CONSTRUCTION OF GENETIC MAPS IN SORGHUM

Genetic linkage maps are foundational to QTL analysis (Choudhary et al., 2024), helping identify specific genomic regions associated with important traits like agronomic performance and stress tolerance. These maps, constructed by placing molecular markers relative to one another, have been crucial in advancing genetic research and breeding in sorghum. Over time, advancements in marker technology and map resolution have significantly improved the understanding and utilization of genetic information in sorghum breeding. Early linkage maps in sorghum utilized RFLPs, which, while useful for enhancing map coverage and marker density, could not fully represent the entire genome (Tao et al., 1998). A landmark study employed 38 sorghum and 33 maize genomic DNA probes, resulting in a linkage map across 15 groups covering 633 cM of genetic distance (Ragab et al., 1994). The subsequent merging of five key RFLP-based maps resulted in 1,036 markers being aligned into ten well-defined linkage groups, greatly improving QTL mapping (Subudhi & Nguyen, 2000).

Initially, RAPD markers were also explored, but their poor reproducibility led to the adoption of more powerful markers like AFLPs, which significantly improved map resolution and enabled saturation of the sorghum linkage maps (Shiringani & Friedt, 2011). Similarly, although SSR markers had limitations in availability, they were extensively used. The development of high-density linkage maps marked a significant leap in molecular mapping. One such map, constructed with 2,926 markers, including 2,454 AFLPs, 136 SSRs, and 336 RFLPs, achieved an average marker spacing of 0.5 cM, providing comprehensive genome coverage (Menz et al., 2002). Another map, with even tighter spacing of 0.4 cM and 2,512 RFLP loci, also offered extensive genome coverage (Bowers et al., 2003).

The advent of high-throughput marker technologies, such as DArT and SNPs has revolutionized linkage map construction, enabling the creation of ultra-high-density maps. DArT-based maps have now encompassed all ten sorghum chromosomes, with one ultra-high-density map featuring 3,418 bin markers, produced by resequencing 244 RILs from a cross between the 654 and LTR108 lines (Zou et al., 2012). SNPs, being abundant and highly specific, have become the gold standard for linkage mapping. For instance, a study of 1,171 Ethiopian sorghum landraces using 25,634 SNPs identified genomic regions associated with both abiotic and biotic stress tolerances, providing valuable genetic resources for breeding programs (Bantte et al., 2023). In another study, 3,710 SNP markers were generated from RADseq of 213 RILs from a cross between BTx623 and a landrace (Kajiya-Kanegae et al., 2020).

High-density SNP maps in sorghum have continued to evolve, providing valuable resources for genetic studies. For instance, Jin et al. (2021) constructed a map with 1,065 markers, achieving the highest density on chromosome 2 and an average marker interval of 0.88 cM. Similarly, Guden et al. (2023) used ddRAD-seq to develop a linkage map comprising 1,438 SNPs across ten linkage groups, spanning 1,349.72 cM with an average marker interval of 0.94 cM. A more comprehensive high-density map by Satrio et al. (2022) involved 33,421 SNP markers from 205 RILs derived from a cross between SC1103 and RTx430, covering 1,571.68 cM with 3,449 unique polymorphic markers and an average interval of 0.46 cM. These ultra-dense linkage maps are transforming the ability to dissect complex traits, fine-map genes, and accelerate breeding programs in sorghum. They are invaluable for: (i) QTL mapping of traits such as yield, drought tolerance, and disease resistance, (ii) comparative genomics, which aligns genetic data with related model crops, and (iii) MAS, which facilitates the transfer of desirable traits into elite cultivars. The integration of diverse marker systems and high-density mapping techniques has propelled sorghum genetic analysis to new heights, paving the way for the development of stress-resilient, high-yielding cultivars that can address the challenges of climate change and global food security.

4.4 ADVANCEMENTS IN TRAIT MAPPING IN SORGHUM

Trait mapping in sorghum has advanced significantly, driven by innovative molecular tools and genomic resources that have enhanced our understanding of the genetic architecture underlying key agronomic traits. Two key approaches, QTL mapping (linkage-based mapping) and GWAS (LD-based mapping), have been pivotal in uncovering the genetic basis of these traits. Despite this progress, the genetic mechanisms controlling many quantitative traits remain poorly understood and underutilized. Over the past two decades, approximately 150 QTL and GWAS studies in sorghum have generated a wealth of information, revealing the genomic locations, allelic effects, and epistatic interactions associated with numerous quantitative traits (https://aussorgm.org.au/sorghum-qtl-atlas/).

4.4.1 Advancements in Linkage-based Mapping in Sorghum

QTL mapping is a statistical technique used to analyze complex traits by associating them with molecular markers, providing a foundation for crop research and breeding. It helps locate genomic regions linked to traits such as stress tolerance, grain yield, and quality, enabling the development of improved crop varieties through MAS (Jin et al., 2021; Mace et al., 2019). Techniques such as Simple Interval Mapping (SIM), Composite Interval Mapping (CIM), Inclusive Composite Interval Mapping (ICIM), and Multiple Interval Mapping (MIM) have significantly improved the precision and reliability of QTL identification (Singh & Singh, 2015). SIM provides a basic framework for detecting QTL by testing marker intervals one at a time, while CIM refines this approach by incorporating background markers to account for genetic noise, enhancing detection power and accuracy. ICIM builds on CIM by using stepwise regression to select significant markers, further improving mapping efficiency and reducing false positives. MIM takes this a step further by simultaneously modelling multiple QTL and their interactions, providing a comprehensive understanding of the genetic architecture underlying complex traits (Singh & Singh, 2015). These advancements have been instrumental in dissecting the genetic basis of quantitative traits in crops like sorghum. Most studies in sorghum have utilized CIM and ICIM as the primary methods for QTL mapping, followed by MIM, with minimal research using SIM (Mace et al., 2019).

Numerous mapping populations have been utilized in genetic studies, with RILs being the most commonly employed. Among these, the BTx623 × IS3620C RIL population was developed from a cross between the sorghum reference genome line BTx623 and the genetically distinct Nigerian guinea margaritiferum line IS3620C (Camjin), which is believed to have undergone an independent domestication process (Burow et al., 2011). Due to the substantial genetic divergence between the parental lines, the large population size, and the presence of high-density GBS markers, this population has been a crucial tool for dissecting various agronomic traits (Kong et al., 2018). Additionally, three RIL populations specifically designed for cold tolerance research were developed from crosses between cold-sensitive U.S. sorghum lines and cold-tolerant Chinese parental lines. These populations, including RTx430 × Gai Gaoliang, BTx623 × Hong Ke Zi, and BTx623 × Niu Sheng Zi, are available through USDA-GRIN, along with their corresponding GBS markers, providing valuable resources for cold tolerance studies (Burow et al., 2011; Marla et al., 2019). Another RIL population, developed from the cross between BTx623 and Hongyingzi, has been genotyped using super-GBS technology and utilized for mapping grain traits (Cao et al., 2024). Similarly, a RIL population derived from Sorcoll 163/07 and Gambella has been genotyped with GBS and used to map yield-related traits (Takele et al., 2024). While these families were primarily developed for studies of cold tolerance and yield traits, they also segregate for other important traits, making them suitable for further trait mapping.

Beyond traditional RIL populations, several RIL families have been incorporated into Nested Association Mapping (NAM) populations, a strategy originally developed for maize but now widely applied across various crops. The NAM approach utilizes multiple RIL families derived from diverse founder lines crossed with a common parent, enabling high-resolution genetic mapping (Gage et al., 2020). In sorghum, a global NAM resource has been established using RTx430 as the shared founder, crossed with ten genetically diverse sorghum accessions, resulting in over 2,200 RILs across ten families (Perumal et al., 2021). RTx430 was selected due to its historical role as a key public pollinator line in sorghum breeding (Perumal et al., 2021). Notably, all NAM parental lines are also part of the sorghum association panel (SAP), facilitating comparative analyses and validation between NAM and SAP studies (Xin et al., 2021; Olatoye et al., 2020). SNP marker development for this sorghum NAM population initially relied on the v.2 BTx623 reference genome (Bouchet et al., 2017), but more recent datasets utilize the v.3 BTx623 genome assembly (Hu et al., 2019). Another significant NAM population, the carbon-partitioning NAM (CP_NAM), was developed to study biomass allocation in sorghum. This population consists of 2,484 F_6 individuals derived from 11 genetically diverse male founder lines crossed with the recurrent female parent "Grassl", with individual family sizes ranging from 193 to 287 (Kumar et al., 2022). Beyond NAM populations, multi-parental breeding strategies have been explored to capture broader genetic diversity. One such effort led to the development of a structured multi-parent advanced generation intercross (MAGIC) population, designed to enhance genetic variation within the sorghum seed parent (B-line) gene pool (Kumar et al., 2023). This structured MAGIC population was derived from four genetically distinct founder lines, including, SC630 (kafir), SC605 (guinea), BTx642 (durra), and BTxARG-1 (caudatum), representing different sorghum races and contributing unique allelic diversity to breeding programs (Kumar et al., 2023).

4.4.2 Genomic Selection in Sorghum

Traditional breeding methods initially relied on phenotypic selection and later advanced with the incorporation of genetic markers linked to QTLs through MAS. Although MAS significantly improved sorghum breeding, its reliance on both genotypic and phenotypic data from specific populations made it labor-intensive and time-consuming (Baloch et al., 2023). Genomic selection (GS) has since emerged as a superior approach, particularly for improving complex traits with low heritability, offering faster genetic gains, reduced costs, and more efficient breeding cycles (Crossa et al., 2017). By predicting the performance of individuals based on genome-wide markers, GS has proven highly effective for quantitative traits like yield (Velazco et al., 2019) and is gaining widespread recognition among plant breeders for traits that are challenging to evaluate (Crossa et al., 2017; Xu et al., 2020). Sorghum, with its well-characterized reference genome and extensive genomic resources, stands out as an ideal crop for applying genomic selection.

Genomic selection enables the estimation of an individual's breeding potential, referred to as genomic estimated breeding values (GEBVs), without direct phenotyping. This approach leverages both genetic and phenotypic data from a reference population to construct predictive models. The reference population, comprising individuals with known genotype and phenotype data, serves as the foundation for training the model. Once established, the model is applied to a target population that has only been genotyped, allowing for the prediction of GEBVs without the need for extensive field evaluations (Crossa et al., 2017). The accuracy of these predictions depends on the statistical models used and the relationship between families and progeny size. For example, in an experiment involving 791 sorghum hybrids at four locations in Australia, the performance of 544 genotyped hybrids, analyzed with 581 DArT markers, was used to predict the performance of other hybrids. By combining pedigree and marker-based information, the prediction accuracy was enhanced, which was further cross-validated (Hunt et al., 2018).

The Chibas sorghum breeding program in Haiti adopted GS utilizing the Practical Haplotype Graph (PHG) framework. This approach was applied to a training population comprising 207 genotypes, each characterized by phenotypic data for various essential traits (Jensen et al., 2020). The accuracies for PHG-based predictions ranged from 0.57 to 0.73 using 3,849 SNPs, demonstrating the potential for cost-effective genotyping in sorghum breeding. GS has proven to be especially beneficial for improving complex traits with low heritability. In a simulation using 400 genotypes, GS outperformed phenotypic selection in terms of genetic gain, prediction accuracy, and cost-efficiency per unit gain (Muleta et al., 2019).

For traits like antioxidants (e.g., anthocyanins and tannins), a study conducted in Italy involving 114 sorghum genotypes showed genomic prediction accuracies ranging from 0.49 to 0.58 based on 61,976 SNPs (Habyarimana & Lopez-Cruz, 2019). Biomass characteristics, which are crucial for biofuel production, have also been optimized using both single-trait and multi-trait GS approaches. Predicted accuracies for biomass-related traits, including plant height and moisture content, ranged from 0.33 to 0.65 using the GBLUP model. Additionally, incorporating correlated traits, such as plant height, led to improvements in prediction accuracy of up to 50% (Fernandes et al., 2018). Sorghum's resilience in drought-prone environments makes it an ideal candidate for GS studies. Multi-environment GS models for traits like grain yield, stay-green (delayed senescence), and flowering time showed 16–19% higher accuracy in predictions compared to single-trait models. Integration of genomic and pedigree data resulted in a 43–72% improvement in grain yield predictions across diverse environments (Velazco et al., 2020). In conclusion, GS is transforming sorghum breeding by accelerating improvements in complex traits like drought tolerance, biomass production, and grain quality. Advances in GS tools and genetic diversity are enhancing sorghum's adaptability and productivity, supporting sustainable agriculture and biofuel innovation.

4.5 CONCLUDING REMARKS AND FUTURE DIRECTIONS

Advancements in molecular mapping, including high-density genetic linkage maps, molecular marker utilization, and genomic technologies, have established a strong foundation for improving sorghum's productivity and resilience. Fine mapping of QTLs and integrating multi-environment trials will further refine our understanding of sorghum genetics and its responses to diverse agro-climatic conditions. Functional genomics research will be crucial for translating these findings into robust, high-yielding varieties. Looking ahead, rapid progress in genetic mapping technologies is expected to uncover novel genetic resources, enhancing sorghum's genetic diversity and adaptability. Integrating GWAS with advanced phenomics and environmental data can enable precise identification of loci associated with key traits, facilitating the development of varieties with superior yield, stress tolerance, and nutritional value.

Emerging multi-omics approaches, encompassing transcriptomics, proteomics, and metabolomics, promise to unravel the intricate networks governing trait expression and identify functional variants linked to key traits. Moreover, integrating artificial intelligence (AI) and machine learning (ML) into genomic research has the potential to transform large-scale data analysis. These advanced computational approaches can uncover complex relationships between genetic variations and phenotypic traits, enhance the efficiency of marker-assisted selection, and expedite the discovery of beneficial alleles in the initial breeding stages. Equally critical is the need to bridge the gap between technology-rich and resource-constrained regions to ensure equitable access to genomic advancements. Empowering smallholder farmers in Africa and Asia with these tools will be vital for realizing the full potential of molecular breeding innovations. By addressing these challenges and embracing these forward-looking strategies, molecular sorghum breeding stands poised to play a pivotal role in improving global food security and fostering resilience against the impacts of climate change.

REFERENCES

Assefa, Y., Holman, J. D., Obour, A. K., O'Brien, D., & Prasad, P. V. V. (2024). Historic grain sorghum production, value, yield gap, and weather relation trends. *Agronomy*, *14*(11), 2582.

Baloch, F. S., Altaf, M. T., Liaqat, W., Bedir, M., Nadeem, M. A., Cömertpay, G., . . . Sun, H. J. (2023). Recent advancements in the breeding of sorghum crop: Current status and future strategies for marker-assisted breeding. *Frontiers in Genetics*, *14*, Art. 1150616.

Bantte, K., Menamo, T. M., Borrell, A. K., Mace, E., Jordan, D. R., Tao, Y., & Hunt, C. (2023). Genetic dissection of root architecture in Ethiopian sorghum landraces. Theoretical and Applied Genetics, *136*(10), 209.

Bekele, W. A., Wieckhorst, S., Friedt, W., & Snowdon, R. J. (2013). High-throughput genomics in sorghum: From whole-genome resequencing to a SNP screening array. *Plant Biotechnology Journal*, *11*(9), 1112–1125.

Boatwright, J. L., Sapkota, S., Jin, H., Schnable, J. C., Brenton, Z., Boyles, R., & Kresovich, S. (2022). Sorghum Association Panel whole-genome sequencing establishes cornerstone resource for dissecting genomic diversity. *The Plant Journal*, *111*(3), 888–904.

Bouchet, S., Olatoye, M. O., Marla, S. R., Perumal, R., Tesso, T., Yu, J., . . . Morris, G. P. (2017). Increased power to dissect adaptive traits in global sorghum diversity using a nested association mapping population. *Genetics*, *206*(2), 573–585.

Bowers, J. E., Abbey, C., Anderson, S., Chang, C., Draye, X., Hoppe, A. H., . . . Paterson, A. H. (2003). A high-density genetic recombination map of sequence-tagged sites for sorghum, as a framework for comparative structural and evolutionary genomics of tropical grains and grasses. *Genetics*, *165*(1), 367–386.

Burow, G., Burke, J. J., Xin, Z., & Franks, C. D. (2011). Genetic dissection of early-season cold tolerance in sorghum (*Sorghum bicolor* (L.) Moench). *Molecular Breeding*, *28*, 391–402.

Cao, N., Ding, Y., Xu, J., Cheng, B., Gao, X., Li, W., . . . Zhang, L. (2024). QTL analysis of sorghum grain traits based on high-density genetic map. *Journal of Applied Genetics*, 1–11.

Choudhary, M., Kumar, P., Prakash, N. R., Gupta, M., Sheoran, S., Kumar, P., . . . Saini, D. K. (2024). QTLomics-based genetic dissection of key traits in major cereals: Challenges and prospects. In Saini, D.K. and Kole, C. (eds.), *Omics and system biology approaches for delivering better cereals* (pp. 71–110). Boca Raton, Florida: CRC Press.

Crossa, J., Pérez-Rodríguez, P., Cuevas, J., Montesinos-López, O., Jarquín, D., De Los Campos, G., . . . Varshney, R. K. (2017). Genomic selection in plant breeding: Methods, models, and perspectives. *Trends in plant science*, *22*(11), 961–975.

Fernandes, S. B., Dias, K. O., Ferreira, D. F., & Brown, P. J. (2018). Efficiency of multi-trait, indirect, and trait-assisted genomic selection for improvement of biomass sorghum. *Theoretical and Applied Genetics*, *131*, 747–755.

Gage, J. L., Monier, B., Giri, A., & Buckler, E. S. (2020). Ten years of the maize nested association mapping population: Impact, limitations, and future directions. *The Plant Cell*, *32*(7), 2083–2093.

Guden, B., Yol, E., Erdurmus, C., Lucas, S. J., & Uzun, B. (2023). Construction of a high-density genetic linkage map and QTL mapping for bioenergy-related traits in sweet sorghum [*Sorghum bicolor* (L.) Moench]. *Frontiers in Plant Science*, *14*, Art. 1081931.

Habyarimana, E., & Lopez-Cruz, M. (2019). Genomic selection for antioxidant production in a panel of *Sorghum bicolor* and *S. bicolor*× *S. halepense* Lines. *Genes*, *10*(11), 841.

Haussmann, B., Hess, D., Seetharama, N., Welz, H., & Geiger, H. (2002). Construction of a combined sorghum linkage map from two recombinant inbred populations using AFLP, SSR, RFLP, and RAPD markers, and comparison with other sorghum maps. *Theoretical and Applied Genetics*, *105*, 629–637.

Hu, Z., Olatoye, M. O., Marla, S., & Morris, G. P. (2019). An integrated genotyping-by-sequencing polymorphism map for over 10,000 sorghum genotypes. *The Plant Genome*, *12*, 1–15.

Hunt, C. H., van Eeuwijk, F. A., Mace, E. S., Hayes, B. J., & Jordan, D. R. (2018). Development of genomic prediction in sorghum. *Crop Science*, *58*(2), 690–700.

Jensen, S. E., Charles, J. R., Muleta, K., Bradbury, P. J., Casstevens, T., Deshpande, S. P., . . . Ramstein, G. P. (2020). A sorghum practical haplotype graph facilitates genome-wide imputation and cost-effective genomic prediction. *The Plant Genome*, *13*(1), Art. e20009.

Jin, P., Wang, L., Zhao, W., Zheng, J., Wang, Y. H., Liu, Y., . . . Li, J. (2021). Construction of high density genetic map and QTL mapping in sorghum× sudangrass. *Euphytica*, *217*, 1–11.

Kajiya-Kanegae, H., Takanashi, H., Fujimoto, M., Ishimori, M., Ohnishi, N., Wacera, W. F., . . . Sakamoto, W. (2020). RAD-seq-based high-density linkage map construction and QTL mapping of biomass-related traits in sorghum using the Japanese landrace Takakibi NOG. *Plant and Cell Physiology*, *61*(7), 1262–1272.

Kong, W., Kim, C., Zhang, D., Guo, H., Tan, X., Jin, H., . . . Paterson, A. H. (2018). Genotyping by sequencing of 393 *Sorghum bicolor* BTx623× IS3620C recombinant inbred lines improves sensitivity and resolution of QTL detection. *G3: Genes, Genomes, Genetics*, *8*(8), 2563–2572.

Kumar, N., Boatwright, J. L., Brenton, Z. W., Sapkota, S., Ballén-Taborda, C., Myers, M. T., . . . Boyles, R. E. (2023). Development and characterization of a sorghum multi-parent advanced generation intercross (MAGIC) population for capturing diversity among seed parent gene pool. *G3: Genes, Genomes, Genetics*, *13*(4), Art. jkad037.

Kumar, N., Brenton, Z., Myers, M. T., Boyles, R. E., Sapkota, S., Boatwright, J. L., . . . Kresovich, S. (2022). Registration of the sorghum carbon-partitioning nested association mapping (CP-NAM) population. *Journal of Plant Registrations*, *16*(3), 656–663.

Mace, E., Innes, D., Hunt, C., Wang, X., Tao, Y., Baxter, J., . . . Jordan, D. (2019). The Sorghum QTL Atlas: A powerful tool for trait dissection, comparative genomics and crop improvement. *Theoretical and Applied Genetics*, *132*, 751–766.

Mace, E. S., Xia, L., Jordan, D. R., Halloran, K., Parh, D. K., Huttner, E., . . . Kilian, A. (2008). DArT markers: Diversity analyses and mapping in *Sorghum bicolor*. *BMC genomics*, *9*, 1–11.

Marla, S. R., Burow, G., Chopra, R., Hayes, C., Olatoye, M. O., Felderhoff, T., . . . Morris, G. P. (2019). Genetic architecture of chilling tolerance in sorghum dissected with a nested association mapping population. *G3: Genes, Genomes, Genetics*, *9*(12), 4045–4057.

McCormick, R. F., Truong, S. K., Sreedasyam, A., Jenkins, J., Shu, S., Sims, D., . . . McKinley, B. (2018). The *Sorghum bicolor* reference genome: Improved assembly, gene annotations, a transcriptome atlas, and signatures of genome organization. *The Plant Journal*, *93*(2), 338–354.

Meena, R. P., Joshi, D., Bisht, J. K., & Kant, L. (2021). Global scenario of millets cultivation. In Kumar, A., Tripathi, M.K., Joshi, D., & Kumar, V. (eds.), *Millets and millet technology* (pp. 33–50). Singapore: Springer.

Menz, M. A., Klein, R. R., Mullet, J. E., Obert, J. A., Unruh, N. C., & Klein, P. E. (2002). A high-density genetic map of *Sorghum bicolor* (L.) Moench based on 2926 AFLP®, RFLP and SSR markers. *Plant Molecular Biology*, *48*, 483–499.

Muleta, K. T., Pressoir, G., & Morris, G. P. (2019). Optimizing genomic selection for a sorghum breeding program in Haiti: A simulation study. *G3: Genes, Genomes, Genetics*, *9*(2), 391–401.

Muluneh, M. G. (2021). Impact of climate change on biodiversity and food security: A global perspective—a review article. *Agriculture & Food Security*, *10*(1), 1–25.

Mwamahonje, A., Eleblu, J. S. Y., Ofori, K., Feyissa, T., Deshpande, S., Garcia-Oliveira, A. L., . . . Tongoona, P. (2021). Introgression of QTLs for drought tolerance into farmers' preferred sorghum varieties. *Agriculture*, *11*(9), 883.

Nelson, J. C., Wang, S., Wu, Y., Li, X., Antony, G., White, F. F., & Yu, J. (2011). Single-nucleotide polymorphism discovery by high-throughput sequencing in sorghum. *BMC Genomics*, *12*, 1–15.

Olatoye, M. O., Marla, S. R., Hu, Z., Bouchet, S., Perumal, R., & Morris, G. P. (2020). Dissecting adaptive traits with nested association mapping: Genetic architecture of inflorescence morphology in sorghum. *G3: Genes, Genomes, Genetics*, *10*(5), 1785–1796.

Paterson, A. H., Bowers, J. E., Bruggmann, R., Dubchak, I., Grimwood, J., Gundlach, H., . . . Rokhsar, D. S. (2009). The *Sorghum bicolor* genome and the diversification of grasses. *Nature*, *457*(7229), 551–556.

Perumal, R., Tesso, T. T., Morris, G. P., Jagadish, S. V. K., Little, C. R., Bean, S. R., . . . Tuinstra, M. R. (2021). Registration of the sorghum nested association mapping (NAM) population in RTx430 background. *Journal of Plant Registrations*, *15*(2), 395–402.

Ragab, R. A., Dronavalli, S., Maroof, M. S., & Yu, Y. G. (1994). Construction of a sorghum RFLP linkage map using sorghum and maize DNA probes. *Genome*, *37*(4), 590–594.

Satrio, R. D., Nikmah, I. A., Fendiyanto, M. H., Pratami, M. P., Awwanah, M., Sari, N. I. P., . . . Nurhadiyanta, N. (2022). Construction of an ultra-high-density consensus genetic map and analysis of recombination rate variation in *Sorghum bicolor*. *Asian Journal of Agriculture*, *6*(1).

Shiringani, A. L., & Friedt, W. (2011). QTL for fibre-related traits in grain× sweet sorghum as a tool for the enhancement of sorghum as a biomass crop. *Theoretical and Applied Genetics*, *123*(6), 999–1011.

Singh, B. D., & Singh, A. K. (2015). Mapping of quantitative trait loci. In Singh B.D. & Singh A.K. (eds.), *Marker-assisted plant breeding: Principles and practices* (pp. 185–216). New Delhi: Springer.

Singhal, D., Gupta, P., Sharma, P., Kashyap, N., Anand, S., & Sharma, H. (2011). In-silico single nucleotide polymorphisms (SNP) mining of *Sorghum bicolor* genome. *African Journal of Biotechnology*, *10*(5), 580–583.

Subudhi, P. K., & Nguyen, H. T. (2000). Linkage group alignment of sorghum RFLP maps using a RIL mapping population. *Genome*, *43*(2), 240–249.

Takele, A., Feyissa, T., & Disasa, T. (2024). High-density linkage map and QTL map for yield and yield components of sorghum recombinant inbreed lines developed from Gambella× Sorcoll 163/07 using GBS. *Cereal Research Communications*, *52*(2), 383–396.

Tao, Y., Luo, H., Xu, J., Cruickshank, A., Zhao, X., Teng, F., . . . Mace, E. (2021). Extensive variation within the pan-genome of cultivated and wild sorghum. *Nature Plants*, *7*(6), 766–773.

Tao, Y. Z., Jordan, D. R., Henzell, R. G., & McIntyre, C. L. (1998). Construction of a genetic map in a sorghum recombinant inbred line using probes from different sources and its comparison with other sorghum maps. *Australian Journal of Agricultural Research*, *49*(4), 729–736.

Velazco, J. G., Jordan, D. R., Hunt, C. H., Mace, E. S., & van Eeuwijk, F. A. (2020). Genomic prediction for broad and specific adaptation in sorghum accommodating differential variances of SNP effects. *Crop Science*, *60*(5), 2328–2342.

Velazco, J. G., Jordan, D. R., Mace, E. S., Hunt, C. H., Malosetti, M., & Van Eeuwijk, F. A. (2019). Genomic prediction of grain yield and drought-adaptation capacity in sorghum is enhanced by multi-trait analysis. *Frontiers in Plant Science*, *10*, 997.

Wondimu, Z., Dong, H., Paterson, A. H., Worku, W., & Bantte, K. (2021). Genetic diversity, population structure, and selection signature in Ethiopian sorghum [*Sorghum bicolor* L. (Moench)] germplasm. *G3*, *11*(6), Art. jkab087.

Wu, Y. Q., Huang, Y., Tauer, C. G., & Porter, D. R. (2006). Genetic diversity of sorghum accessions resistant to greenbugs as assessed with AFLP markers. *Genome*, *49*(2), 143–149.

Xin, Z., Wang, M., Cuevas, H. E., Chen, J., Harrison, M., Pugh, N. A., & Morris, G. (2021). Sorghum genetic, genomic, and breeding resources. *Planta*, *254*(6), 114.

Xu, Y., Liu, X., Fu, J., Wang, H., Wang, J., Huang, C., . . . Zhang, A. (2020). Enhancing genetic gain through genomic selection: From livestock to plants. *Plant Communications*, *1*(1), 100005.

Zhang, Z., Viana, J. P. G., Zhang, B., Walden, K. K., Paul, H. M., Moose, S. P., . . . Hudson, M. E. (2024). Major impacts of widespread structural variation on sorghum. *Genome Research*, *34*(2), 286–299.

Zou, G., Zhai, G., Feng, Q., Yan, S., Wang, A., Zhao, Q., . . . Tao, Y. (2012). Identification of QTLs for eight agronomically important traits using an ultra-high-density map based on SNPs generated from high-throughput sequencing in sorghum under contrasting photoperiods. *Journal of Experimental Botany*, *63*(15), 5451–5462.

5 Sorghum Genomics and Pan-Genomics in the Next-Generation Sequencing Era

Rintu Jha, Prashant Singh, Uday Chand Jha, Akanksha Singh, and Bandana Devi

5.1 INTRODUCTION

With the growing global population, food security remains a critical challenge worldwide (Jha et al., 2022a). Despite increased agricultural production, over 820 million people still face food insecurity, while at least 2 billion experience nutritional insecurity (Jha et al., 2024). Addressing this dual crisis requires a sustainable and innovative approach to global food production. In addition to staple food crops, the use of nutrient-dense, climate-resilient, and locally accessible crops, often referred to as "smart foods", is essential for future production, marketing, and consumption.

Sorghum (*Sorghum bicolor* (L.) Moench), a climate-smart crop, plays a pivotal role as a multipurpose cereal, particularly in South Asia and sub-Saharan Africa, where it serves as a major food source (Ruperao et al., 2021). Sorghum, a diploid species (2n = 20) with an estimated genome size of approximately 730 Mb (Price et al., 2005; Paterson et al., 2009), is valued for its versatility. It provides primary and secondary products, including food, fuel, starch, fiber, alcohol, dextrose syrup, and biofuels (Ruperao et al., 2021). With its rich genetic diversity, sorghum has been domesticated and bred for various uses, spanning food, fodder, and bioenergy, across diverse agroclimatic settings (Li et al., 2010; Kong et al., 2000). Significant strides have been made in sorghum genomics (Kaur et al., 2023; Saini et al., 2024). A basic genome assembly of 730 Mb was first reported by Paterson et al. (2009), followed by an improved assembly of 732.2 Mb, covering 91.5% of the genome (McCormick et al., 2018). Additionally, a 729 Mb reference genome assembly for the "Rio" variety was reported by Cooper et al. (2019). However, these genome assemblies represent only individual accessions and fail to capture the full genetic diversity of the species.

Recent genomic studies, including QTL mapping and genome-wide association studies (GWAS) have identified several candidate genes with crucial roles in sorghum improvement (Zhang et al., 2023; Wang et al., 2022; Mengistu et al., 2021). Genomic variation in sorghum includes single nucleotide polymorphisms (SNPs) and small insertions/deletions (InDels), which serve as valuable markers for developing robust varieties through genomic-assisted breeding (GAB). However, SNPs and InDels alone do not fully capture the genetic diversity of a species (Springer et al., 2009; Khan et al., 2020). Structural variations (SVs), such as copy number variations (CNVs), presence/absence variations (PAVs), inversions, transversions, and chromosomal translocations, represent another critical source of genetic variation. These variations have been shown to play essential roles in plant genetics (Khan et al., 2020; Feuk et al., 2006). To comprehensively capture SVs, including repetitive or transposable elements (TEs), pan-genomes, species-representative genomes encompassing

 DOI: 10.1201/9781003623717-5

the genetic diversity across all accessions, are essential (Jha et al., 2022b). Pan-genomes provide an in-depth characterization of gene content diversity, offering valuable insights for breeding programs.

Modern breeding concepts and genomic technologies are urgently needed to accelerate the development of sorghum varieties with high yield potential and stress resilience, catering to diverse end uses. This chapter discusses the latest advancements in sorghum genetics and genomics, highlighting key genes and allelic variants associated with agronomically significant traits. It also explores how emerging genomic tools and technologies can shape the future of sorghum molecular breeding. Additionally, the concept of the pan-genome, methods for constructing pan-genome assemblies, and the importance of SVs, PAVs, and CNVs in the dispensable genome are examined.

5.2 GENOMIC RESOURCES AVAILABLE FOR SORGHUM BREEDING

5.2.1 Sorghum Germplasm Resources

According to Hao et al. (2021), more than 41,000 accessions, including historic accessions, wild relatives, landraces, and improved breeding lines, are held by the two largest sorghum germplasm banks: the International Crops Research Institute for the Semi-Arid Tropics (ICRISAT, www.icrisat.org) and the United States Department of Agriculture's National Plant Germplasm System (USDA-NPGS, www.ars-grin.gov). The National Bureau of Plant Genetic Resources in India, with approximately 20,000 collections (https://nbpgr.org.in), and the Institute of Crop Science, Chinese Academy of Agricultural Sciences with approximately 18,263 collections.

5.2.2 NAM and MAGIC Populations

Limited published data exist on nested association mapping (NAM) and multi-parent advanced generation intercross (MAGIC) populations in sorghum. A NAM population developed using the common parent "RTx430", widely used in breeding programs, consists of 2,214 lines from ten RIL families (Perumal et al., 2021). The sorghum carbon-partitioning CP_NAM population includes 11 diverse male founder accessions crossed with the recurrent female parent "Grassl", resulting in 2,484 F_6 individuals across 11 RIL families (Kumar et al., 2022). The backcross NAM (BCNAM) approach, incorporating backcrossing and selection, offers enhanced breeding relevance. Garin et al. (2024) developed a BCNAM resource with 24 parents, including two populations with 2,189 BC_1F_4 lines (Patil et al., 2024). MAGIC populations, used for next-generation trait mapping, have also been developed. Ongom and Ejeta (2018) created a MAGIC population by intercrossing 19 diverse founder lines, followed by ten generations of random mating, resulting in 1,000 immortal MAGIC inbred lines. Kumar et al. (2023) developed a MAGIC B-Line population using four diverse grain sorghum founders, selected for their unique traits. These MAGIC populations, rich in allelic diversity, are ideal for gene mapping and marker-assisted breeding, providing valuable resources for characterizing complex traits and enhancing gene pools.

5.2.3 Mutagenized Populations

A well-categorized mutant library from an inbred line with a uniform genetic background and detailed phenotype annotations is an invaluable resource for isolating alleles linked to key traits. To create such a resource, Xin et al. (2008) developed a pedigreed mutant library using BTx623, the sorghum line employed for the first reference genome. M_1 seeds were mutagenized with ethyl methanesulfonate (EMS), and the resulting M_3 plants were propagated by single-seed descent, producing M_4 seed pools derived from independently mutagenized M_1 seeds. This library, which contains approximately 6,400 independent seed pools, preserves all mutations, including recessive lethal ones, and can be replenished for screening under various conditions. A similar approach at Purdue University generated around 10,000 pedigreed seed pools (Addo-Quaye et al., 2018).

In addition, Jiao et al. (2024) reported a sorghum mutant population containing 9.5 million EMS-induced mutations, covering 98% of annotated genes in the BTx623 inbred line. This population includes 610,320 mutations in the promoter and enhancer regions of 18,000 and 11,790 genes, respectively, enabling novel research on cis-regulatory elements. All mutation data are searchable through SorbMutDB (www.depts.ttu.edu/igcast/SorbMutDB.php) and SorghumBase (www.sorghumbase.org/). Expanding on these efforts, a recent initiative in Australia generated a library of 150,000 mutant seeds from BTx623 using EMS (Mason et al., 2024). Integrating these valuable variants into elite commercial lines offers a simpler and more cost-effective alternative to genome editing, providing a crucial breeding resource for enhancing sorghum adaptation to climate change and improving yield potential in marginal environments.

5.2.4 Genetic Diversity and Population Structure

Population structure is essential for understanding historical population dynamics, such as population growth, geographic shifts, and genetic differentiation, which are crucial for comprehending genetic diversity in sorghum breeding. Genetic variation plays a pivotal role in enhancing and maintaining the genetic diversity necessary for sorghum improvement. To explore the history of sorghum's domestication and improvement, Mace et al. (2013) analyzed the genetic diversity and population structure of 44 distinct sorghum lines, including 18 landraces, 17 improved inbreds, 7 wild and weedy sorghums, 2 guinea-margaritiferums, and 2 sorghum species. Similarly, Morris et al. (2013) examined the population structure of 971 accessions from global germplasm collections to investigate the diversity across various sorghum races. The population analysis revealed the presence of five distinct morphological types, including kafir, durra, bicolor, caudatum, and guinea, within the panel. Eleven populations, with pairwise F_{ST} values ranging from 0.11 to 0.47, were separated from a core selection of 374 USDA-NPGS Ethiopian sorghum germplasm (Cuevas et al., 2017).

Moreover, Girma et al. (2019) further dissected a complex collection of 1,425 Ethiopian landrace accessions into 11 subgroups, each exhibiting rich genetic diversity that had developed in response to the regional environment. Similarly, Cuevas and Prom (2020) analyzed five ancestral populations within the NPGS Sudan core set of 318 accessions, revealing F_{ST} values ranging from 0.17 to 0.39. Their IBS analysis further showed that the pairwise genetic distance in these populations ranged from 0.70 to 0.95, indicating higher genetic diversity compared to the NPGS Ethiopian germplasm. Additionally, genetic diversity analysis of a varied panel of 607 Nigerian accessions and 1,785 georeferenced global accessions, with a pairwise F_{ST} of 0.007, revealed that the Nigerian accessions were distinct from the global pool and primarily clustered with West African accessions. Olatoye et al. (2018) reported that LD in the Nigerian germplasm decreased to $R^2 = 0.1$ at 50 kb, while in West African germplasm, it decreased to 100 kb. Furthermore, a recent population genomic study on the *Dry* gene by Zhang et al. (2018) highlighted the multiple origins of the stem juiciness trait, noting that the selection of loss-of-function dry mutations played a crucial role in the evolution of sweet sorghum.

5.2.5 Sorghum Genome Sequencing

The sorghum genome is relatively small and non-duplicated, making it an ideal model for genomic studies. The initial assembly of the sorghum reference genome, BTx623, was achieved using Sanger sequencing (Paterson et al., 2009), and was later updated by McCormick et al. (2018). This updated assembly revealed approximately 34,129 identified genes, with 58.8% of the genome composed of retrotransposons and 8.7% of DNA transposons. Leveraging the genomic information of BTx623, numerous sequence variations, including SNPs, InDels, large genomic CNVs, and PAVs, have been identified (Zheng et al., 2011; Mace et al., 2019). Additionally, several sorghum SNP databases and microarray datasets are now available, offering valuable resources for genomic research (Shakoor et al., 2014; Luo et al., 2016). Advancements continued with the chromosome-scale *de novo* assembly of the Tx430 grain sorghum genome by Deschamps et al. (2018), utilizing Oxford Nanopore sequencing technology and Bionano Genomics Direct Label and Stain (DLS) optical mapping. In comparison to BTx623, Tx430 exhibited a lower median length of protein-coding genes and an increased gene count of 39,510 predicted genes. To better understand sugar metabolism, the genome of sweet sorghum Rio was assembled at the chromosome level, with its 729.4 Mb genome containing 35,476 predicted genes, including 54 genes unique to Rio, reflecting its distinct metabolic pathways (Cooper et al., 2019).

Despite these advances, all available sorghum genomes remain incomplete, particularly in unresolved centromeres and telomeres, limiting a comprehensive understanding of the genomic landscape in the sorghum gene pool. To address these gaps, Wei et al. (2024) assembled the complete genomes of BTx623 and Ji2055, using a combination of ONT ultra-long reads, PacBio HiFi long reads, Hi-C, and Illumina short reads. The genomes of BTx623, the reference genome, and Ji2055, an inbred line with red seeds, were sequenced with >150× ONT, >65× PacBio, >50× Hi-C, and >50× Illumina coverage. Initial assemblies with HiFi reads resulted in 246 contigs for BTx623 and 581 for Ji2055, which were then anchored into ten pseudomolecules using Hi-C data. Gaps were subsequently filled using ONT and HiFi reads, leading to gap-free assemblies. The final telomere-to-telomere assemblies yielded genome sizes of 719.90 Mb for BTx623-T2T and 722.96 Mb for Ji2055-T2T, offering a more complete genomic view of sorghum (Wei et al., 2024).

5.2.6 Comparative Genomics in Sorghum

To investigate the genetic variation in sorghum and related species, comparative genomics analysis has been employed. Previous resequencing of two sweet sorghum genomes and one grain sorghum genome revealed a significant number of SNPs, InDels, PAVs, and CNVs (Zheng et al., 2011). Additionally, nearly 1,500 genes were identified that distinguish grain-type sorghums from sweet sorghums, with some linked to the production of lignin, coumarin, and starch and sucrose (Cooper et al., 2019). Cooper et al. also discovered a notable frequency of nonsynonymous and potentially loss-of-function mutations in sugar metabolism genes by comparing the genomes of Rio (sweet sorghum) and BTx623 (grain sorghum), offering insights into the molecular mechanisms underlying sweet sorghum traits. However, comparative genomic studies of broom and forage sorghum varieties remain unexplored, indicating a need for further research into the genetic mechanisms behind the phenotypic divergence of different sorghum varieties for future breeding.

In a broader context, Lozano et al. (2021) analyzed ~13 million variants from 499 sorghum lines and 25 million variants (SNPs and InDels) from 1,218 maize lines to compare genomic variation between these species. They found that deleterious mutations were prevalent in pericentromeric regions, enriched in non-syntenic genes, and present at low allele frequencies. Unlike maize, sorghum deviated from the domestication-cost hypothesis, showing a lower deleterious burden in domesticated lines. This work, coupled with population genetic data, enabled the prediction of a gene deleterious index with >0.5 accuracy, contributing to the understanding of genetic variation in sorghum. Furthermore, Sathyamurthy et al. (2024) analyzed chloroplast genome data from multiple genera, including sorghum, to investigate their phylogeny and genome properties. While these genera showed conserved genome structures, codon bias, gene content, and GC content, variations were observed in genome size, SSRs, and SC/IR region boundaries. Phylogenetic analyses strengthened the species phylogeny but also revealed that each genus, including *Sorghum* is non-monophyletic, suggesting a need for re-evaluating millet genus taxonomy and phylogeny. This finding emphasizes that current classifications may not accurately represent the evolutionary history of these genera.

5.3 PAN-GENOMICS CONCEPT

Advancements in large-scale genomic studies have highlighted that genetic diversity at the species level cannot be adequately represented by a single reference genome. In contrast, the pan-genome encompasses the entire DNA sequences of a species, including both shared sequences (the core genome) and unique sequences specific to individual accessions (Gong et al., 2023). Pan-genomes provide a more comprehensive view of genetic variation within a species compared to single-reference genomes. By integrating genomic data from multiple accessions, pan-genomes facilitate the identification and annotation of complex DNA polymorphisms, such as SVs, which are key contributors to genetic diversity (Zanini et al., 2022). This approach is particularly valuable for studying genomic evolution, species origin and domestication, and plant improvement, as it captures the majority of genetic variation within a species (Li et al., 2022).

5.3.1 Strategies for Constructing Pan-genomes

De novo assembly, iterative assembly, and graph-based assembly are key techniques used to construct pan-genomes (Li et al., 2022). The simplest method for *de novo* assembly involves assembling the genomes of multiple samples and performing a comparative analysis to identify variant types, classifying genes as either essential or optional (Mahmoud et al., 2018). Iterative assembly begins by building a single reference genome, then sequentially mapping reads from additional samples. Unmapped reads are assembled and incorporated into the reference genome to create a pan-genome of non-redundant sequences (Golicz et al., 2016). This approach is more cost-effective than *de novo* assembly, as it allows pooling of hundreds of samples, each sequenced with a modest depth. In graph-based assembly, a graph is used to represent variations relative to the reference genome, offering significant advantages over traditional linear genomes by reducing reference bias (Garrison et al., 2018).

5.4 IMPORTANCE OF SVs, PAVs, TEs, AND CNVs IN THE DISPENSABLE GENOME

Genetic differences used as molecular markers have become a cornerstone in plant breeding. Over the past few decades, various molecular markers such as intergenic SSRs, random amplified polymorphic DNA, amplified fragment length polymorphism, and restriction fragment length polymorphism have been developed and widely utilized. Among these, microsatellites, or SSRs, gained popularity due to their codominance, multiallelic nature, high polymorphism, and ease of genotyping (Varshney et al., 2007). However, advancements in sequencing technologies enabled the identification of SNPs, leading to the development of highly reliable SNP markers. SNPs, being cost-effective and compatible with automation, are now the markers of choice.

In parallel, the focus has shifted to understanding genetic variation through SVs with the growing availability of sequenced genomes (Khan et al., 2020). Plant genomes differ significantly even within the same species due to variations in gene content, such as CNVs, PAVs, inversions, and the activity of TEs (Gaut et al., 2018; Jha et al., 2022b). Transposable elements, which can replicate and insert themselves into coding regions, disrupt gene functions and contribute to genome structural changes (Jha et al., 2022b). For instance, a pan-genome analysis of peas revealed that LTR Copia and LTR Gypsy sequences underwent 85.5% deletions and 77.4% duplications, significantly altering phenotypes through insertions and deletions (Yang et al., 2022). Similarly, SVs can arise from mechanisms such as unequal crossing over, recombination, chromosomal translocations, inversions, and transversions (Yandeau-Nelson et al., 2006; Jha et al., 2022b). These genomic changes highlight the limitations of relying on a single reference genome, as it cannot capture the complete gene repertoire of a species. In contrast, pan-genomes integrate both core (conserved) and dispensable (variable) genes, encompassing SVs in coding and non-coding regions across all members of a species, providing a comprehensive view of genomic diversity (Jha et al., 2022b).

5.5 MAJOR ACHIEVEMENTS IN SORGHUM PAN-GENOMICS

Sorghum improvement will greatly benefit from the development and characterization of its pan-genome, which provides insights into genome functionality and diversity. For instance, Ruperao et al. (2021) utilized reference genomes and 354 genetically diverse sorghum accessions from multiple races to construct a sorghum pan-genome. This pan-genome comprises 35,719 genes, including a core genome of 16,821 genes, with an average of 32,795 genes per cultivar. They further demonstrated that PAV occurs in 53% of the genes, with some linked to traits associated with drought adaptation. The pan-genome size was determined to be 954.8 Mb, which is 30% larger than the published reference genome of BTx623 (732.2 Mb). Within this pan-genome, 62% of the sequences belong to the core genome, while 38% consist of variable gene sequences (Tao et al., 2019; Sun et al., 2022). A GWAS study of seed color in sorghum revealed a SNP in the *Yellow seed1* gene, along with a 3,216 kb PAV when the pan-genome was integrated. These PAVs were shown to influence gene organization and agronomic traits functionally (Sun et al., 2022).

Recent advancements in pan-genomics have highlighted the enrichment of SVs and nonsynonymous SNPs in genes associated with bioenergy-related traits. Long-read sequencing has proven essential for detecting novel variations, as demonstrated in studies exploring genetic differences between sweet and cellulosic sorghum genotypes. These studies identified SVs and SNPs in genes and pathways relevant to bioenergy traits (Voelker et al., 2023). PAVs are a critical source of genome structural variation, significantly impacting phenotypic and genomic diversity across species, including humans and animals. However, PAVs in plant genomes remain relatively underexplored. In an earlier study, Zhang et al. (2014) resequenced three sorghum genomes with 12× coverage, identifying 5,364 PAVs. Among these, 51 large PAVs (>30 kb) spanning 2.92 Mb of the sorghum genome were thoroughly analyzed, uncovering 202 genes, including 38 involved in stress response and cell death. Advances in pan-genomic studies will continue to enrich genomic resources and provide deeper insights into the genetic mechanisms driving sorghum improvement.

5.6 RECENT ADVANCES IN MAPPING AND IDENTIFYING QTLs IN SORGHUM

Several studies have focused on uncovering the potential genomic regions responsible for different agronomic traits in sorghum through QTL mapping (https://aussorgm.org.au/sorghum-qtl-atlas/) (Table 5.1). One such study used two recombinant inbred line (RIL) populations (C1 and C2), created by crossing cold-tolerant (CT19, ICSV700) and cold-sensitive (TX430, M81E) sorghum parents, and identified a total of 16 and 39 QTLs for chilling tolerance at the seedling stage in the C1 and C2 populations, respectively (La Borde et al., 2023). Similarly, another study mapped seven QTLs for fertility restoration across three F_2 populations, offering insights into the genetics behind fertility restoration (Kante et al., 2018). Target leaf spot (TLS) resistance was evaluated in replicated trials across two sorghum RIL populations, developed from crosses of BTx623/BTx642 and BTx623/SC155-14E. The study identified four TLS resistance QTLs, two in each population (Kimball et al., 2019). In another study, comprehensive details on sorghum lectins and their role in conferring tolerance against abiotic stresses were provided, highlighting preferentially overrepresented lectin sequences in certain QTLs associated with cold, drought, salinity, panicle/grain composition, and leaf shape (Osman et al., 2022).

Several QTLs associated with important agronomic traits were also identified. For instance, seven QTLs were found to affect the length of the growing season, stomatal conductance, and root architecture, with specific QTLs identified for crown root angle (*qRA3*), days to maturity (*qDM6*, *qDM8*,

TABLE 5.1
A summary of genetic mapping studies conducted in sorghum for different traits

Traits	Material used	QTLs	Chromosomes	Marker type used	Phenotypic variation (%)	References
Crown root angle	Bi-parental cross between cultivars Early Hegari-Sart (EH) and Bk7	1 QTL (*qRA3*)	3	SNPs	2.60%	Lopez et al. (2017)
Chilling stress tolerance	RILs, (CT19, ICSV700) and (TX430, M81E)	16 QTLs for population C1 and 39 for population C2	1, 2, 4, 5, 6, and 9	SNPs	C1 (5.38–22.21) % and C2 (6.26–30.34) %	La Borde et al. (2023)
Six agronomic traits (tiller number, height, stem diameter, biomass, flowering time, aerial branching)	RILs, (*Sorghum propinquum* and *Sorghum bicolor* inbred Tx7000)	30 QTLs	1, 2, 3, 5, 6, 7, 8, and 9	SNPs	7.38–30.55%	Govindarajulu et al. (2021)
Rf (fertility restoration)	Three F_2 populations	5 QTLs	2, 3, 4, 5, and 10	SNPs	2.5–31.0%	Kante et al. (2018)
Days to maturity	Bi-parental cross between cultivars Early Hegari-Sart (EH) and Bk7	*qDM6*, *qDM8* and *qDM10*	6, 8, and 10	SNPs	25.69%	Lopez et al. (2017)
Leaf number	Bi-parental cross between cultivars Early Hegari-Sart (EH) and Bk7	*qLN6*	6	SNPs	52.74%	Lopez et al. (2017)
Stomatal conductance	Bi-parental cross between cultivars Early Hegari-Sart (EH) and Bk7	*qSC7* and *qSC10*	7 and 10	SNPs	1.25–4.32%	Lopez et al. (2017)
Days to flowering (FLW)	223 RILs (Brandes x Wray)	5 QTLs	1, 6, and 10	SNPs	3.31–45.89%	Souza et al. (2021)
Plant height (HGH, in cm),	223 RILs (Brandes x Wray)	5 QTLs	1, 3, 6, 9, and 10	SNPs	4.38–7.77%	Souza et al. (2021)
Total soluble solids (BRX, in °Brix)	223 RILs (Brandes x Wray)	5 QTLs	1, 3, 6, and 9	SNPs	3.31–6.24%	Souza et al. (2021)
Sucrose (SUC, in %)	223 RILs (Brandes x Wray)	5 QTLs	1, 3, 6, and 9	SNPs	4.09–7.09%	Souza et al. (2021)

TABLE 5.1 (Continued)
A summary of genetic mapping studies conducted in sorghum for different traits

Traits	Material used	QTLs	Chromosomes	Marker type used	Phenotypic variation (%)	References
Fibers (FIB, in %)	223 RILs (Brandes x Wray)	4 QTLs	3, 6, 8, and 9	SNPs	4.82–5.54%	Souza et al. (2021)
Fresh biomass yield (FBY, in $t.ha^{-1}$),	223 RILs (Brandes x Wray)	3 QTLs	1, 6, and 10	SNPs	6.14–8.87%	Souza et al. (2021)
Juice extraction yield (JUC, in %)	223 RILs (Brandes x Wray)	3 QTLs	1, 6, and 9	SNPs	3.07–6.59%	Souza et al. (2021)
Reducing sugar content (RSU, in %)	223 RILs (Brandes x Wray)	3 QTLs	1, 3, and 4	SNPs	4.20–8.63%	Souza et al. (2021)
Spikelet-related traits	213 RILs (BTx623 and the Japanese landrace NOG)	36 QTLs	1, 2, 3, 6, 7, and 9	SNPs	6.84–68.83%	Takanashi et al. (2021)
Grain-related traits	213 RILs (BTx623 and the Japanese landrace NOG)	7 QTLs	4 and 8	SNPs	5.71–28.84%	Takanashi et al. (2021)
Effective quantum yield of photosystem II and chlorophyll index, sink (grain yield per panicle), and panicle neck diameter	210 RILs ((RTx430) x SC35)	3 major QTL hotspots	1, 2, and 3	SNPs	10.15–25.35%	Chiluwal et al. (2022)
Ambemohor's coconut juice-like fragrance	F_2 population (Ambemohor x KU630 (nonfragrant))	*qAro6.1*	6	SNPs	8.07–11.47%	Monkhan et al. (2021)
Phosphorus use efficiency related traits	396 RILs (BR007xSC283)	The grain yield QTL (*Gy-3*), surface area of fine roots QTL (*SA2-3*), and a root diameter/grain yield QTL	1, 2, 3, 6, 7, and 9	SNPs	2.16–12.92%	Bernardino et al. (2019)
Dead hearts	243 RILs (ICSV 745xPB 15520-1)	4 QTLs	4, 5, 8, and 10	SNPs	7.6–9%	Muturi et al. (2021)

(continued)

TABLE 5.1 (Continued)
A summary of genetic mapping studies conducted in sorghum for different traits

Traits	Material used	QTLs	Chromosomes	Marker type used	Phenotypic variation (%)	References
Damage caused by feeding on leaves	243 RILs (ICSV 745xPB 15520-1)	6 QTLs	1, 3, 6, 7, and 8	SNPs	7.6–17.4%	Muturi et al. (2021)
Exit holes and stem tunneling	243 RILs (ICSV 745xPB 15520-1)	5 QTLs	2, 4, 6, and 9	SNPs	7–11.7%	Muturi et al. (2021)
Bloom waxiness, leaf glossiness, and seedling vigor	243 RILs (ICSV 745xPB 15520-1)	2 QTLs	3 and 9	SNPs	8.2–19.1	Muturi et al. (2021)
Trichome density	243 RILs (ICSV 745xPB 15520-1)	4 QTLs	1, 2, 3, and 6	SNPs	7.3–15.0%	Muturi et al. (2021)
Leaf toughness (African Stem Borer (*Busseola fusca*) and Spotted Stem Borer (*Chilo partellus*))	243 RILs (ICSV 745xPB 15520-1)	6 QTLs	2, 4, 6, 9, and 10	SNPs	7.1–10.6%	Muturi et al. (2021)
Target leaf spot (TLS) resistance	Two RIL populations (BTx623/ BTx642 and BTx623/ SC155-14E)	4 QTLs (two in each population)	3, 4, 5, and 9	SNPs	8.87–71.12%	Kimball et al. (2019)
Disease (charcoal rot)	107 sorghum accessions	4 novel major QTLs	2, 4, 5, and 9	SSR	9.47–18.87%	Mahmoud et al. (2018)
Anthracnose resistance	366 sorghum populations (GWAS)	8 major marker-trait associations (MTAs)	1, 4, 6, 8, 9, and 10	SNPs	4.86–15.9%	Mengistu et al. (2021)
Pericarp color, testa pigment, and tannin content	242 Chinese sorghum accession (GWAS)	73 QTLs	1, 2, 3, 4, 6, 7, 8, 9, and 10	SNPs	-	Zhang et al. (2023)
Grain size	(GWAS) was carried out with a BC-NAM population	81 QTLs	1, 2, 3, 4, 5, 6, 7, 8, 9, and 10	SNPs	-	Tao et al. (2020)
Leaf angle	729 accessions (GWAS)	33 QTLs	1, 5, 6, 7, and 9	SNPs	-	Zhi et al. (2022)
Fresh weight per plant (FW), stem diameter (SD), plant height (PH), and tiller number (TN)	245 sorghum accessions (GWAS)	338 Quantitative trait nucleotides (QTNs)/ SNPs	1, 3, 4, 5, 7, 8, 9, and 10	SNPs	4.83–35.83%	Wang et al. (2022)

and *qDM10*), leaf number (*qLN6*), and stomatal conductance (*qSC7* and *qSC10*) (Lopez et al., 2017). Additionally, a RIL population derived from an interspecific cross between *Sorghum propinquum* and *Sorghum bicolor* inbred Tx7000 identified 30 QTLs for six key agronomic traits: plant height, tiller number, stem diameter, biomass, flowering time, and aerial branching (Govindarajulu et al., 2021). A study on sweet sorghum, using 223 RILs derived from a cross between two sweet sorghum lines (Brandes X Wray), identified 33 QTLs for flowering time, plant height, total soluble solids, sucrose, fibers, fresh biomass yield, juice extraction yield, and reducing sugars (Souza et al., 2021). Similarly, another study identified 36 QTLs for spikelet-related traits and seven for grain-related traits (Takanashi et al., 2021).

Research on post-flowering drought tolerance in sorghum, using 210 RILs developed from a cross between the drought-susceptible cultivar RTx430 and the drought-tolerant cultivar SC35, identified three major QTL hotspots on chromosomes 1, 2, and 3, which control multiple traits including effective quantum yield of photosystem II, chlorophyll index, grain yield per panicle, and panicle neck diameter (Chiluwal et al., 2022). Additionally, a major QTL (*qAro6.1*) for coconut juice-like fragrance was detected on chromosome 6 in a population segregating for the "Aro" gene (Monkhan et al., 2021). Studies have also mapped QTLs for stay-green and high yield, including *Stg1*, *Stg2*, *Stg3*, *Stg4*, *Stg3A*, and *Stg3B* (Mwamahonje et al., 2021). Furthermore, QTLs for root morphology, grain yield, and phosphorus uptake were identified in a study under low phosphorus (P) availability. These QTLs included grain yield (*Gy-3*), root surface area (*SA2-3*), and root diameter/grain yield QTLs (Bernardino et al., 2019). Lastly, research on insect resistance in sorghum identified QTLs regulating various traits, such as dead hearts, leaf damage, stem tunneling, trichome density, and leaf toughness in response to African stem borer and spotted stem borer (Muturi et al., 2021). In total, 64 significant or very significant QTLs, along with 21 suggestive rust resistance QTLs, were found across 55 distinct genomic regions (Wang et al., 2022). These studies highlight the genetic complexity and potential of QTL mapping in sorghum improvement for a range of agronomic and stress-resilient traits.

Relying on a single reference genome for QTL mapping in sorghum is limiting because it fails to capture the full extent of genetic diversity, including SVs, PAVs, and rare alleles, all of which are crucial for understanding complex traits. This approach may result in biased mapping, incomplete representation of trait-associated loci, and the exclusion of genes unique to specific accessions (Della Coletta et al., 2021; Wang et al., 2023). Pan-genomics addresses these limitations by integrating multiple genomes into a comprehensive framework, enabling the accurate identification of QTLs across diverse accessions. By incorporating both the core and dispensable genomes, pan-genomics enhances mapping resolution, identifies novel candidate genes, and links traits to SVs and rare alleles. For sorghum, this methodology holds significant potential for dissecting complex traits such as root architecture, yield, and stress tolerance, thereby providing valuable insights for genomics-assisted breeding and crop improvement. However, to date, no study has utilized pan-genomic variation for QTL mapping in sorghum.

5.7 RECENT ADVANCES IN GENOME-WIDE ASSOCIATION STUDIES IN SORGHUM

Numerous GWAS studies have been conducted in sorghum to investigate the genetic control of traits related to stress tolerance, yield, and biofortification (https://aussorgm.org.au/sorghum-qtl-atlas/). Among these, more than 40 studies have exclusively utilized the Sorghum Association Panel (SAP), a widely used genetic resource (Mural et al., 2021). Employing GWAS and population structure analysis, a panel of 242 resequenced Chinese sorghum accessions was analyzed to identify QTLs influencing pericarp color, testa pigment, and tannin content (Zhang et al., 2023). Using 2,760,264 SNPs, GWAS was conducted across four contexts to explore the genetic basis of natural variation in sorghum grain color. This analysis identified 73 QTLs associated with the color of the exocarp, mesocarp, testa, and tannin concentration across all chromosomes except chromosome 5. Notably,

47 of these QTLs are potentially novel (Zhang et al., 2023). Building on earlier research, Morris et al. (2013) genotyped a widely used sorghum association panel using genotyping-by-sequencing (GBS), which yielded 265,487 SNPs. This panel has facilitated investigations into traits such as anthracnose resistance (Cuevas et al., 2017), stalk rot resistance (Adeyanju et al., 2015), grain mold resistance (Ackerman et al., 2021), and stress tolerances, including resistance to cold and heat (Ortiz et al., 2017; Chopra et al., 2017).

Environmental adaptation has also been explored in sorghum through GWAS. Lasky et al. (2015) analyzed allelic associations in a panel of 1,943 georeferenced landraces using 104,627 SNPs. Their findings revealed that environmental factors, such as drought and aluminum toxicity, significantly influenced SNP variation, with genic SNPs enriched for these associations. Similarly, Faye et al. (2019) examined genomic diversity and climate adaptation in 421 Senegalese sorghum landraces using 213,916 SNPs. Furthermore, Wang et al. (2020) identified adaptive variations in seed mass related to precipitation gradients in a set of 1,901 georeferenced landraces. In more focused studies, Tao et al. (2020) integrated GWAS with transcriptome analysis to dissect the genetic potential of grain size in a BC-NAM population (n = 1,421) and a diversity panel (n = 837). Their analyses identified 81 QTLs associated with grain size, revealing significant and highly heritable variation. Xu et al. (2023) highlighted genetic diversity in mucilage secretion capacity across 146 sorghum accessions, while Zhang et al. (2023) analyzed 214 global germplasms for six yield-related traits, identifying 2,811,016 SNPs and pinpointing candidate gene *SORBI_3008G116500*, a homolog of the *Arabidopsis thaliana VIP5* gene.

GWAS has also uncovered novel resistance mechanisms, including striga resistance (Ferguson et al., 2021) and anthracnose resistance in a panel of 366 sorghum populations evaluated over three seasons (Mengistu et al., 2021). Studies on sorghum aphid resistance have linked candidate genes to leucine-rich repeats (LRR), LOX, CAM kinases, WRKY transcription factors, and flavonoid biosynthesis (Punnuri et al., 2022). Other studies have focused on traits such as leaf senescence, where Wang et al. (2022) identified 31 senescence-associated genomic regions, and carotenoid content, with GWAS pinpointing zeaxanthin epoxidase as a key gene influencing lutein, β-carotene, and zeaxanthin levels (Cruet-Burgos et al., 2023). In another study, Wondimu et al. (2023) identified 338 QTNs associated with nine agronomic traits using a multi-locus GWAS approach in a panel of 304 accessions. Additionally, Ahn et al. (2023) combined phenotypic data from over 16,000 seeds with SNP data to reveal genes linked to seed morphology in 162 Senegalese germplasms. Together, these studies demonstrate the power of GWAS in discovering key genetic loci and candidate genes that govern complex traits in sorghum. This knowledge provides valuable resources for molecular marker-assisted breeding, driving progress toward the development of resilient sorghum varieties.

However, all previous GWAS and post-GWAS analyses in sorghum have relied on a single reference genome, with no studies to date utilizing a pan-genome approach. However, a pan-genome offers significant advantages over a single reference genome by capturing the full genetic diversity of a species, including SVs and population-specific alleles that are often missed in single-genome analyses (Della Coletta et al., 2021; Gupta, 2021). By reducing reference bias, a pan-genome improves variant detection and mapping accuracy, enabling the identification of novel genes and variants critical for understanding complex traits (Della Coletta et al., 2021; Gupta, 2021). This approach provides a more comprehensive representation of genomic variation underlying phenotypic diversity and is particularly beneficial for crop studies, where it enhances targeted breeding by incorporating diverse genotypes. For instance, Wang et al. (2023) developed a novel strategy for pan-genome construction and a downstream analysis pipeline in rice that captures genetic variants' positional information while maintaining a linearized layout. Using pan-genome-based GWAS, they identified a previously undetected locus (*qPH8-1*) associated with plant height on chromosome 8, which was missed by SNP-based GWAS. This highlights the power of pan-genomes to uncover hidden genetic variation and improve the resolution of association studies.

5.8 LEVERAGING PAN-GENOMICS AND TRANSCRIPTOMICS FOR TRAIT DISSECTION IN SORGHUM

The full genome sequencing and resequencing of several sorghum genotypes have facilitated numerous transcriptome studies on various stress responses and other traits in sorghum. For instance, in mucilage-secreting roots, transcriptome analysis revealed 4,461 differentially expressed genes (DEGs), with 1,844 upregulated and 2,617 downregulated. Notably, 82 of these DEGs were associated with glucuronidation and glycosyltransferase pathways. Both GWAS and transcriptome data converged to identify *Sobic.010G120200* as a candidate gene encoding UDP-glycosyltransferase, which may negatively regulate mucilage secretion in sorghum (Xu et al., 2023). Further insights into sorghum carbohydrate metabolism and its regulatory mechanisms were provided by transcriptomic studies on sweet sorghum stems. For instance, twofold upregulation of the mannose-binding lectin gene *Sb10g022730.1* and the LysM lectin gene *Sb01g049890.1* was observed, suggesting their role in cell-wall degradation (Calviño et al., 2008). Similarly, *Sobic.006G044200.1*, a CRA putative gene linked to traits such as brace roots, panicle length, and leaf morphology, was implicated in cell wall loosening, potentially affecting overall plant architecture (Rai et al., 2016; Osman et al., 2022). For root architecture traits, such as root angle and days to maturity, candidate genes *Sobic.003G052700* and *Sobic.006G057900* were identified through a combination of GWAS and transcriptomics (Lopez et al., 2017).

Transcriptome analysis has also shed light on the regulation of tiller bud elongation, identifying 6,189 DEGs and highlighting transcription factors such as Dormancy Associated Protein 1 *DRM1*. These findings suggest that genes located within tillering QTLs are critical regulators of tiller elongation in sorghum (Govindarajulu et al., 2021). In a broader context, transcriptome profiling of sorghum stems during key developmental stages, from floral initiation to grain maturity, identified approximately 200 DEGs linked to biomass accumulation and stem composition (McKinley et al., 2016). Building on these foundational studies, a comprehensive transcriptome analysis involving 47 samples from roots, leaves, stems, seeds, and panicles was conducted to annotate the v3.1 reference genome. This effort identified 3,396 senescence-associated genes (SAGs) enriched in 13 KEGG pathways, with 507 SAGs consistently expressed across developmental, salt, and dark-induced senescence conditions (Wu et al., 2016; McCormick et al., 2018).

Transcriptomic studies have also provided insights into sorghum's response to abiotic stresses. For example, a time-series analysis of RTx430 (pre-flowering drought-resistant but post-flowering drought-susceptible) and BTx642 (a stay-green genotype tolerant to post-flowering drought) revealed distinct transcriptional responses, underscoring genotype-specific stress adaptations (Varoquaux et al., 2019). Similarly, under low-phosphorus (low-P) stress, transcriptome analysis of 29 sorghum accessions identified 2,089 potential genes, including transcription factors and genes involved in plant hormone signal transduction, highlighting mechanisms underlying sorghum's adaptation to nutrient stress (Zhang et al., 2019).

Despite the advantages of the pan-genome approach, none of the transcriptomic studies conducted in sorghum have yet utilized pan-genome variation. Incorporating a pan-genome could improve gene expression analysis, RNA-seq read mapping, and overall gene function understanding. For instance, Wang et al. (2024) used a pan-genome-based GWAS on 414 rice accessions, identifying 74 QTLs for blast resistance, including the broad-spectrum *qPBR1* locus. Their transcriptomic analysis revealed 3,311 DEGs, and integration identified 6 candidate genes within *qPBR1* with no significant yield loss. Additionally, pan-transcriptomics, which analyzes transcriptomes from multiple individuals across populations, can uncover novel transcripts, enhance functional annotations, and reveal population-specific expression patterns, offering a more comprehensive view of transcriptomic variability (Zhong et al., 2024).

5.9 CONCLUSION AND FUTURE DIRECTIONS

To ensure food security, crop production must overcome significant challenges posed by a growing human population, shrinking arable land, and abrupt global climate changes. Addressing these challenges necessitates the adoption of novel breeding and genomics techniques. In this context, pan-genome resources of major crops offer valuable tools for understanding crop domestication, development, and the genetic basis of agronomic traits. These resources can be leveraged in crop breeding programs for genomic prediction, GWAS, QTL mapping, mining superior haplotypes for diverse traits, and the development of novel markers systems. As advancements in computational biology progress and sequencing costs decline, pan-genome studies hold the promise of constructing a "super pan-genome" encompassing all species within a genus. This comprehensive genomic repository would provide extensive information at both the genus and family levels, facilitating deeper insights into genetic diversity and evolutionary processes (Jha et al., 2022a, b).

The success of pan-genome construction, however, hinges on the choice of an appropriate strategy tailored to the study's data. *De novo* assembly offers distinct advantages over reference-based methods, particularly in mitigating potential biases caused by significant genome size and structural variations. Moreover, it effectively captures crucial sequences that are poorly aligned or unmatched to a reference genome, thereby reducing the likelihood of misalignment. Despite these

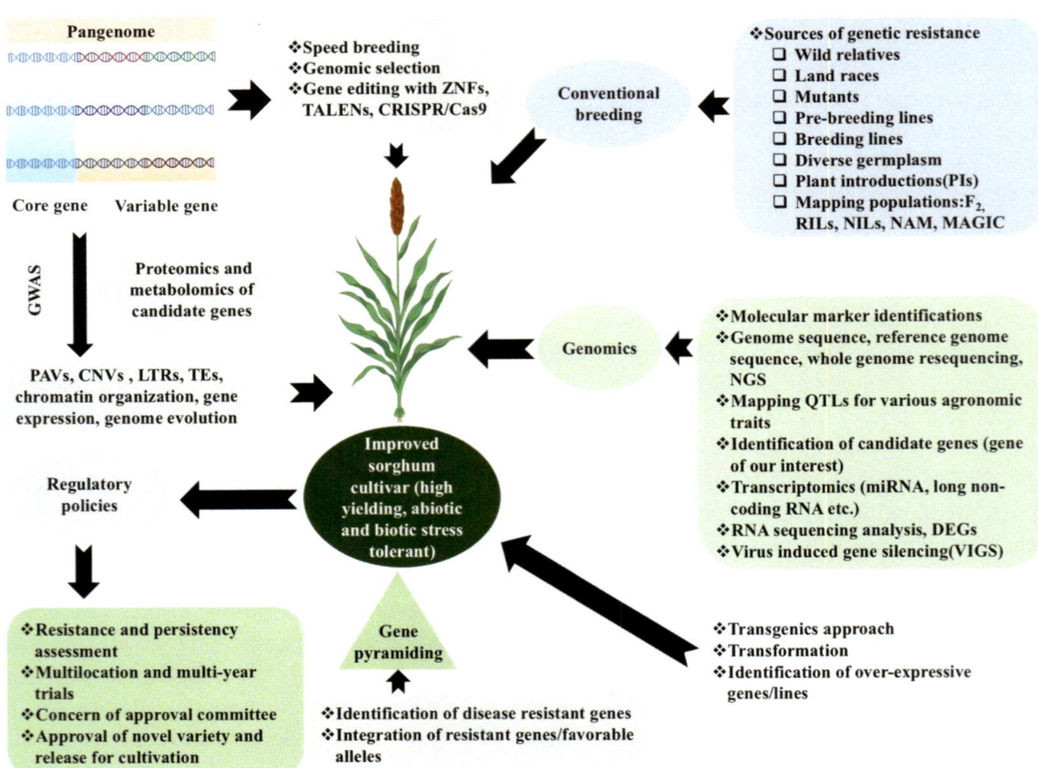

FIGURE 5.1 Illustrating genomics and pan-genomics approaches, the figure highlights genome sequencing, structural variation analysis, and core-accessory gene studies to enhance sorghum breeding for stress resilience and enhanced grain yield.

benefits, *de novo* assembly is resource-intensive, requiring substantial computational infrastructure and processing power. The quality of the pan-genome is also profoundly influenced by the accuracy of genome assemblies, gene model annotations, and the precise SVs detection. High-quality pan-genomes not only enhance the reliability of downstream analyses but also broaden the scope of genomic applications in crop improvement (Khan et al., 2020). Figure 5.1 illustrates how genomics and pan-genomics approaches, including genome sequencing, structural variation analysis, and core-accessory gene studies, contribute to advancing sorghum breeding efforts aimed at improving stress resilience and grain yield.

REFERENCES

Ackerman, A., Wenndt, A., & Boyles, R. (2021). The Sorghum grain mold disease complex: Pathogens, host responses, and the bioactive metabolites at play. *Frontiers in Plant Science*, *12*, Art. 660171.

Addo-Quaye, C., Tuinstra, M., Carraro, N., Weil, C., & Dilkes, B. P. (2018). Whole-genome sequence accuracy is improved by replication in a population of mutagenized sorghum. *G3 (Bethesda, Md.)*, *8*(3), 1079–1094.

Adeyanju, A., Little, C., Yu, J., & Tesso, T. (2015). Genome-wide association study on resistance to stalk rot diseases in grain sorghum. *G3 (Bethesda, Md.)*, *5*(6), 1165–1175.

Ahn, E., Botkin, J., Ellur, V., Lee, Y., Poudel, K., Prom, L. K., & Magill, C. (2023). Genome-wide association study of seed morphology traits in Senegalese sorghum cultivars. *Plants (Basel, Switzerland)*, *12*(12), 2344.

Bernardino, K. C., Pastina, M. M., Menezes, C. B., De Sousa, S. M., Maciel, L. S., Jr, G. C., . . . Magalhaes, J. V. (2019). The genetic architecture of phosphorus efficiency in sorghum involves pleiotropic QTL for root morphology and grain yield under low phosphorus availability in the soil. *BMC Plant Biology*, *19*, 1–15.

Calviño, M., Bruggmann, R. & Messing, J. (2008). Screen of genes linked to high-sugar content in stems by comparative genomics. *Rice*, *1*, 166–176.

Chiluwal, A., Perumal, R., Poudel, H. P., Muleta, K., Ostmeyer, T., Fedenia, L., . . . Jagadish, S. V. K. (2022). Genetic control of source-sink relationships in grain sorghum. *Planta*, *255*(2), 40.

Chopra, R., Burow, G., Burke, J. J., Gladman, N., & Xin, Z. (2017). Genome-wide association analysis of seedling traits in diverse Sorghum germplasm under thermal stress. *BMC Plant Biology*, *17*(1), 12.

Cooper, E. A., Brenton, Z. W., Flinn, B. S., Jenkins, J., Shu, S., Flowers, D., . . . Kresovich, S. (2019). A new reference genome for *Sorghum bicolor* reveals high levels of sequence similarity between sweet and grain genotypes: Implications for the genetics of sugar metabolism. *BMC Genomics*, *20*(1), 420.

Cruet-Burgos, C., Morris, G. P., & Rhodes, D. H. (2023). Characterization of grain carotenoids in global sorghum germplasm to guide genomics-assisted breeding strategies. *BMC Plant Biology*, *23*(1), 165.

Cuevas, H. E., & Prom, L. K. (2020). Evaluation of genetic diversity, agronomic traits, and anthracnose resistance in the NPGS Sudan Sorghum Core collection. *BMC genomics*, *21*(1), 88.

Cuevas, H. E., Rosa-Valentin, G., Hayes, C. M., Rooney, W. L., & Hoffmann, L. (2017). Genomic characterization of a core set of the USDA-NPGS Ethiopian sorghum germplasm collection: Implications for germplasm conservation, evaluation, and utilization in crop improvement. *BMC Genomics*, *18*(1), 108.

Della Coletta, R., Qiu, Y., Ou, S., Hufford, M. B., & Hirsch, C. N. (2021). How the pan-genome is changing crop genomics and improvement. *Genome Biology*, *22*, 1–19.

Deschamps, S., Zhang, Y., Llaca, V., Ye, L., Sanyal, A., King, M., . . . Lin, H. (2018). A chromosome-scale assembly of the sorghum genome using nanopore sequencing and optical mapping. *Nature Communications*, *9*(1), 4844.

Faye, J. M., Maina, F., Hu, Z., Fonceka, D., Cisse, N., & Morris, G. P. (2019). Genomic signatures of adaptation to Sahelian and Soudanian climates in sorghum landraces of Senegal. *Ecology and Evolution*, *9*(10), 6038–6051.

Ferguson, J. N., Fernandes, S. B., Monier, B., Miller, N. D., Allen, D., Dmitrieva, A., . . . Leakey, A. D. B. (2021). Machine learning-enabled phenotyping for GWAS and TWAS of WUE traits in 869 field-grown sorghum accessions. *Plant Physiology*, *187*(3), 1481–1500.

Feuk, L., Marshall, C. R., Wintle, R. F., & Scherer, S. W. (2006). Structural variants: Changing the landscape of chromosomes and design of disease studies. *Human Molecular Genetics*, *15*(Spec No. 1), R57–R66.

Garin, V., Diallo, C., Tékété, M. L., Théra, K., Guitton, B., Dagno, K., . . . Rami, J. F. (2024). Characterization of adaptation mechanisms in sorghum using a multireference back-cross nested association mapping design and envirotyping. *Genetics*, *226*(4), Art. iyae003.

Garrison, E., Sirén, J., Novak, A. M., Hickey, G., Eizenga, J. M., Dawson, E. T., . . . Durbin, R. (2018). Variation graph toolkit improves read mapping by representing genetic variation in the reference. *Nature Biotechnology*, *36*(9), 875–879.

Gaut, B. S., Seymour, D. K., Liu, Q., & Zhou, Y. (2018). Demography and its effects on genomic variation in crop domestication. *Nature Plants*, *4*(8), 512–520.

Girma, G., Nida, H., Seyoum, A., Mekonen, M., Nega, A., Lule, D., . . . Mengiste, T. (2019). A Large-scale genome-wide association analyses of Ethiopian sorghum landrace collection reveal loci associated with important traits. *Frontiers in Plant Science*, *10*, 691.

Golicz, A. A., Bayer, P. E., Barker, G. C., Edger, P. P., Kim, H., Martinez, P. A., . . . Edwards, D. (2016). The pangenome of an agronomically important crop plant *Brassica oleracea*. *Nature Communications*, *7*, Art. 13390.

Gong, Y., Li, Y., Liu, X., Ma, Y., & Jiang, L. (2023). A review of the pangenome: How it affects our understanding of genomic variation, selection and breeding in domestic animals?. *Journal of Animal Science and Biotechnology*, *14*(1), 73.

Govindarajulu, R., Hostetler, A. N., Xiao, Y., Chaluvadi, S. R., Mauro-Herrera, M., Siddoway, M. L., . . . Hawkins, J. S. (2021). Integration of high-density genetic mapping with transcriptome analysis uncovers numerous agronomic QTL and reveals candidate genes for the control of tillering in sorghum. *G3 (Bethesda, Md.)*, *11*(2), Art. jkab024.

Gupta, P. K. (2021). GWAS for genetics of complex quantitative traits: Genome to pangenome and SNPs to SVs and k-mers. *BioEssays*, *43*(11), Art. 2100109.

Hao, H., Li, Z., Leng, C., Lu, C., Luo, H., Liu, Y., . . . Jing, H. C. (2021). Sorghum breeding in the genomic era: Opportunities and challenges. *Theoretical and Applied Genetics*, *134*(7), 1899–1924.

Jha, R., Yadav, H. K., Raiya, R., Singh, R. K., Jha, U. C., Sathee, L., . . . Tripathi, S. (2022a). Integrated breeding approaches to enhance the nutritional quality of food legumes. *Frontiers in Plant Science*, *13*, Art. 984700.

Jha, R., Zhang, K., He, Y., Mendler-Drienyovszki, N., Magyar-Tábori, K., Quinet, M., . . . Zhou, M. (2024). Global nutritional challenges and opportunities: Buckwheat, a potential bridge between nutrient deficiency and food security. *Trends in Food Science & Technology*, *145,* p.104365.

Jha, U. C., Nayyar, H., von Wettberg, E. J. B., Naik, Y. D., Thudi, M., & Siddique, K. H. M. (2022b). Legume pangenome: Status and scope for crop improvement. *Plants (Basel, Switzerland)*, *11*(22), 3041.

Jiao, Y., Nigam, D., Barry, K., Daum, C., Yoshinaga, Y., Lipzen, A., . . . Xin, Z. (2024). A large sequenced mutant library–valuable reverse genetic resource that covers 98% of sorghum genes. *The Plant Journal*, *117*(5), 1543–1557.

Kante, M., Rattunde, H. F. W., Nébié, B., Weltzien, E., Haussmann, B. I. G., & Leiser, W. L. (2018). QTL mapping and validation of fertility restoration in West African sorghum A_1 cytoplasm and identification of a potential causative mutation for Rf_2. *Theoretical and Applied Genetics*, *131*(11), 2397–2412.

Kaur, I., Relan, A., Saini, D. K., Kaur, G., Biswas, A., Singh, L., . . . Sandhu, K. S. (2023). Revisiting the genomic approaches in the cereals and the path forward. In *Smart plant breeding for field crops in post-genomics era* (pp. 1–40). Singapore: Springer Nature.

Khan, A. W., Garg, V., Roorkiwal, M., Golicz, A. A., Edwards, D., & Varshney, R. K. (2020). Super-pangenome by integrating the wild side of a species for accelerated crop improvement. *Trends in Plant Science*, *25*(2), 148–158.

Kimball, J., Cui, Y., Chen, D., Brown, P., Rooney, W. L., Stacey, G., & Balint-Kurti, P. J. (2019). Identification of QTL for target leaf spot resistance in *Sorghum bicolor* and investigation of relationships between disease resistance and variation in the MAMP response. *Scientific Reports*, *9*(1), Art. 18285.

Kong, L., Dong, J., & Hart, G. (2000). Characteristics, linkage-map positions, and allelic differentiation of *Sorghum bicolor* (L.) Moench DNA simple-sequence repeats (SSRs). *Theoretical and Applied Genetics*, *101*, 438–448.

Kumar, N., Boatwright, J. L., Brenton, Z. W., Sapkota, S., Ballén-Taborda, C., Myers, M. T., . . . Boyles, R. E. (2023). Development and characterization of a sorghum multi-parent advanced generation intercross

(MAGIC) population for capturing diversity among seed parent gene pool. *G3: Genes, Genomes, Genetics*, *13*(4), Art. jkad037.

Kumar, N., Brenton, Z., Myers, M. T., Boyles, R. E., Sapkota, S., Boatwright, J. L., . . . Kresovich, S. (2022). Registration of the sorghum carbon-partitioning nested association mapping (CP-NAM) population. *Journal of Plant Registrations*, *16*(3), 656–663.

La Borde, N., Rajewski, J., & Dweikat, I. (2023). Novel QTL for chilling tolerance at germination and early seedling stages in sorghum. *Frontiers in Genetics*, *14*, Art. 1129460.

Lasky, J. R., Upadhyaya, H. D., Ramu, P., Deshpande, S., Hash, C. T., Bonnette, J., . . . Morris, G. P. (2015). Genome-environment associations in sorghum landraces predict adaptive traits. *Science Advances*, *1*(6), Art. e1400218.

Li, R., Zhang, H., Zhou, X., Guan, Y., Yao, F., Song, G., . . . Zhang, C. (2010). Genetic diversity in Chinese sorghum landraces revealed by chloroplast simple sequence repeats. *Genetic Resources and Crop Evolution*, *57*, 1–15.

Li, W., Liu, J., Zhang, H., Liu, Z., Wang, Y., Xing, L., He, Q., & Du, H. (2022). Plant pan-genomics: Recent advances, new challenges, and roads ahead. *Journal of Genetics and Genomics*, *49*(9), 833–846.

Lopez, J. R., Erickson, J. E., Munoz, P., Saballos, A., Felderhoff, T. J., & Vermerris, W. (2017). QTLs associated with crown root angle, stomatal conductance, and maturity in sorghum. *The Plant Genome*, *10*(2), 1–12.

Lozano, R., Gazave, E., Dos Santos, J. P., Stetter, M. G., Valluru, R., Bandillo, N., . . . Gore, M. A. (2021). Comparative evolutionary genetics of deleterious load in sorghum and maize. *Nature Plants*, *7*(1), 17–24.

Luo, H., Zhao, W., Wang, Y., Xia, Y., Wu, X., Zhang, L., . . . Luo, J. (2016). SorGSD: A sorghum genome SNP database. *Biotechnology for Biofuels*, *9*, 6.

Mace, E., Innes, D., Hunt, C., Wang, X., Tao, Y., Baxter, J., . . . Jordan, D. (2019). The Sorghum QTL Atlas: A powerful tool for trait dissection, comparative genomics and crop improvement. *Theoretical and Applied Genetics*, *132*(3), 751–766.

Mace, E. S., Tai, S., Gilding, E. K., Li, Y., Prentis, P. J., Bian, L., . . . Wang, J. (2013). Whole-genome sequencing reveals untapped genetic potential in Africa's indigenous cereal crop sorghum. *Nature Communications*, *4*, 2320.

Mahmoud, A. F., Abou-Elwafa, S. F., & Shehzad, T. (2018). Identification of charcoal rot resistance QTLs in sorghum using association and in silico analyses. *Journal of Applied Genetics*, *59*(3), 243–251.

Mason, P. J., Blaakmeer, A., Furtado, A., Stuart, P. N., Nomula, R., Bjarnholt, N., . . . Henry, R. J. (2024). Harnessing the power of an extensive EMS-induced sorghum population for rapid crop improvement. *Physiologia Plantarum*, *176*(4), Art. e14449.

McCormick, R. F., Truong, S. K., Sreedasyam, A., Jenkins, J., Shu, S., Sims, D., . . . Mullet, J. E. (2018). The *Sorghum bicolor* reference genome: Improved assembly, gene annotations, a transcriptome atlas, and signatures of genome organization. *The Plant Journal: For Cell and Molecular Biology*, *93*(2), 338–354.

McKinley, B., Rooney, W., Wilkerson, C., & Mullet, J. (2016). Dynamics of biomass partitioning, stem gene expression, cell wall biosynthesis, and sucrose accumulation during development of *Sorghum bicolor*. *The Plant Journal: For Cell and Molecular Biology*, *88*(4), 662–680.

Mengistu, G., Shimelis, H., Assefa, E., & Lule, D. (2021). Genome-wide association analysis of anthracnose resistance in sorghum [*Sorghum bicolor* (L.) Moench]. *PloS One*, *16*(12), Art. e0261461.

Monkhan, T., Chen, X., & Somta, P. (2021). *BADH1* is associated with fragrance in sorghum (*Sorghum bicolor* (L.) Moench) cultivar 'Ambemohor'. *Journal of Genetics*, *100*, 3.

Morris, G. P., Ramu, P., Deshpande, S. P., Hash, C. T., Shah, T., Upadhyaya, H. D., . . . Kresovich, S. (2013). Population genomic and genome-wide association studies of agroclimatic traits in sorghum. *Proceedings of the National Academy of Sciences of the United States of America*, *110*(2), 453–458.

Mural, R. V., Grzybowski, M., Miao, C., Damke, A., Sapkota, S., Boyles, R. E., . . . Schnable, J. C. (2021). Meta-analysis identifies pleiotropic loci controlling phenotypic trade-offs in sorghum. *Genetics*, *218*(3), Art. iyab087.

Muturi, P. W., Mgonja, M., Rubaihayo, P., & Mwololo, J. K. (2021). QTL mapping of traits associated with dual resistance to the African stem borer (*Busseola fusca*) and spotted stem borer (*Chilo partellus*) in sorghum (*Sorghum bicolor*). *International Journal of Genomics*, *2021*, Art. 7016712.

Mwamahonje, A., Eleblu, J. S. Y., Ofori, K., Deshpande, S., Feyissa, T., & Tongoona, P. (2021). Drought tolerance and application of marker-assisted selection in sorghum. *Biology*, *10*(12), 1249.

Olatoye, M. O., Hu, Z., Maina, F., & Morris, G. P. (2018). Genomic signatures of adaptation to a precipitation gradient in Nigerian sorghum. *G3 (Bethesda, Md.)*, *8*(10), 3269–3281.

Ongom, P. O., & Ejeta, G. (2018). Mating design and genetic structure of a multi-parent advanced generation intercross (MAGIC) population of sorghum (*Sorghum bicolor* (L.) Moench). *G3 (Bethesda, Md.)*, *8*(1), 331–341.

Ortiz, D., Hu, J., & Salas Fernandez, M. G. (2017). Genetic architecture of photosynthesis in *Sorghum bicolor* under non-stress and cold stress conditions. *Journal of Experimental Botany*, *68*(16), 4545–4557.

Osman, M. E. M., Dirar, A. I., & Konozy, E. H. E. (2022). Genome-wide screening of lectin putative genes from *Sorghum bicolor* L., distribution in QTLs and a probable implications of lectins in abiotic stress tolerance. *BMC Plant Biology*, *22*(1), 397.

Paterson, A. H., Bowers, J. E., Bruggmann, R., Dubchak, I., Grimwood, J., Gundlach, H., . . . Rokhsar, D. S. (2009). The *Sorghum bicolor* genome and the diversification of grasses. *Nature*, *457*(7229), 551–556.

Patil, N. Y., Hoffmann Jr, L., Winans, N., Perumal, R., Hayes, C., Emendack, Y., . . . Rooney, W. L. (2024). Registration of sorghum backcross-nested association mapping (BC-NAM) families in a BTx623 or RTx436 background. *Journal of Plant Registrations*, *18*(1), 204–219.

Perumal, R., Tesso, T. T., Morris, G. P., Jagadish, S. V. K., Little, C. R., Bean, S. R., . . . Tuinstra, M. R. (2021). Registration of the sorghum nested association mapping (NAM) population in RTx430 background. *Journal of Plant Registrations*, *15*(2), 395–402.

Price, H. J., Dillon, S. L., Hodnett, G., Rooney, W. L., Ross, L., & Johnston, J. S. (2005). Genome evolution in the genus *Sorghum* (Poaceae). *Annals of Botany*, *95*(1), 219–227.

Punnuri, S. M., Ayele, A. G., Harris-Shultz, K. R., Knoll, J. E., Coffin, A. W., Tadesse, H. K., . . . Wallace, J. G. (2022). Genome-wide association mapping of resistance to the sorghum aphid in *Sorghum bicolor*. *Genomics*, *114*(4), Art. 110408.

Rai, K. M., Thu, S. W., Balasubramanian, V. K., Cobos, C. J., Disasa, T., & Mendu, V. (2016). Identification, characterization, and expression analysis of cell wall related genes in *Sorghum bicolor* (L.) Moench, a food, fodder, and biofuel crop. *Frontiers in Plant Science*, *7*, 1287.

Ruperao, P., Thirunavukkarasu, N., Gandham, P., Selvanayagam, S., Govindaraj, M., Nebie, B., . . . Rathore, A. (2021). Sorghum pan-genome explores the functional utility for genomic-assisted breeding to accelerate the genetic gain. *Frontiers in Plant Science*, *12*, Art. 666342.

Saini, D. K., Kumar, S., & Kaur, R. (2024). Applying genomics resources to accelerate the development of climate resilient crops. In *Adapting to climate change in agriculture—theories and practices: Approaches for adapting to climate change in agriculture in India* (pp. 43–120). Cham, Switzerland: Springer Nature.

Sathyamurthy, D., Mannu, J., Natesan, S., Nathan, B., Nallusamy, S., Narayanan, M. B., & Duraisamy, K. (2024). Comparative chloroplast genome analysis of six millet species along with related Poaceae family members. *The Nucleus*, *68*, 13–24.

Shakoor, N., Nair, R., Crasta, O., Morris, G., Feltus, A., & Kresovich, S. (2014). A *Sorghum bicolor* expression atlas reveals dynamic genotype-specific expression profiles for vegetative tissues of grain, sweet and bioenergy sorghums. *BMC Plant Biology*, *14*, 35.

Souza, V. F., Pereira, G. D. S., Pastina, M. M., Parrella, R. A. D. C., Simeone, M. L. F., Barros, B. A., . . . Damasceno, C. M. B. (2021). QTL mapping for bioenergy traits in sweet sorghum recombinant inbred lines. *G3 (Bethesda, Md.)*, *11*(11), Art. jkab314.

Springer, N. M., Ying, K., Fu, Y., Ji, T., Yeh, C. T., Jia, Y., . . . Schnable, P. S. (2009). Maize inbreds exhibit high levels of copy number variation (CNV) and presence/absence variation (PAV) in genome content. *PLoS Genetics*, *5*(11), Art. e1000734.

Sun, Y., Wang, J., Li, Y., Jiang, B., Wang, X., Xu, W. H., . . . Kong, X. D. (2022). Pan-genome analysis reveals the abundant gene presence/absence variations among different varieties of melon and their influence on traits. *Frontiers in Plant Science*, *13*, Art. 835496.

Takanashi, H., Shichijo, M., Sakamoto, L., Kajiya-Kanegae, H., Iwata, H., Sakamoto, W., & Tsutsumi, N. (2021). Genetic dissection of QTLs associated with spikelet-related traits and grain size in sorghum. *Scientific Reports*, *11*(1), 9398.

Tao, Y., Zhao, X., Mace, E., Henry, R., & Jordan, D. (2019). Exploring and exploiting pan-genomics for crop improvement. *Molecular Plant*, *12*(2), 156–169.

Tao, Y., Zhao, X., Wang, X., Hathorn, A., Hunt, C., Cruickshank, A. W., . . . Jordan, D. R. (2020). Large-scale GWAS in sorghum reveals common genetic control of grain size among cereals. *Plant Biotechnology Journal*, *18*(4), 1093–1105.

Varoquaux, N., Cole, B., Gao, C., Pierroz, G., Baker, C. R., Patel, D., . . . Purdom, E. (2019). Transcriptomic analysis of field-droughted sorghum from seedling to maturity reveals biotic and metabolic responses. *Proceedings of the National Academy of Sciences of the United States of America*, *116*(52), 27124–27132.

Varshney, R. K., Marcel, T. C., Ramsay, L., Russell, J., Röder, M. S., Stein, N., . . . Graner, A. (2007). A high density barley microsatellite consensus map with 775 SSR loci. *Theoretical and Applied Genetics, 114*(6), 1091–1103.

Voelker, W. G., Krishnan, K., Chougule, K., Alexander, L. C., Jr, Lu, Z., Olson, A., . . . Cooper, E. A. (2023). Ten new high-quality genome assemblies for diverse bioenergy sorghum genotypes. *Frontiers in Plant Science, 13*, Art. 1040909.

Wang, J., Hu, H., Jiang, X., Zhang, S., Yang, W., Dong, J., . . . Zhao, J. (2024). Pangenome-wide association study and transcriptome analysis reveal a novel qtl and candidate genes controlling both panicle and leaf blast resistance in rice. *Rice, 17*(1), 27.

Wang, J., Hu, Z., Upadhyaya, H. D., & Morris, G. P. (2020). Genomic signatures of seed mass adaptation to global precipitation gradients in sorghum. *Heredity, 124*(1), 108–121.

Wang, J., Yang, W., Zhang, S., Hu, H., Yuan, Y., Dong, J., . . . Zhao, J. (2023). A pangenome analysis pipeline provides insights into functional gene identification in rice. *Genome Biology, 24*(1), 19.

Wang, L., Liu, Y., Gao, L., Yang, X., Zhang, X., Xie, S., . . . Shen, Y. (2022). Identification of candidate forage yield genes in sorghum (*Sorghum bicolor* L.) using integrated genome-wide association studies and RNA-Seq. *Frontiers in Plant Science, 12*, Art. 788433.

Wei, C., Gao, L., Xiao, R., Wang, Y., Chen, B., Zou, W., . . . Tao, Y. (2024). Complete telomere-to-telomere assemblies of two sorghum genomes to guide biological discovery. *iMeta, 3*(2).

Wondimu, Z., Dong, H., Paterson, A. H., Worku, W., & Bantte, K. (2023). Genome-wide association study reveals genomic loci influencing agronomic traits in Ethiopian sorghum (*Sorghum bicolor* (L.) Moench) landraces. *Molecular Breeding: New Strategies in Plant Improvement, 43*(5), 32.

Wu, X. Y., Hu, W. J., Luo, H., Xia, Y., Zhao, Y., Wang, L. D., . . . Jing, H. C. (2016). Transcriptome profiling of developmental leaf senescence in sorghum (*Sorghum bicolor*). *Plant Molecular Biology, 92*(4–5), 555–580.

Xin, Z., Wang, M. L., Barkley, N. A., Burow, G., Franks, C., Pederson, G., & Burke, J. (2008). Applying genotyping (TILLING) and phenotyping analyses to elucidate gene function in a chemically induced sorghum mutant population. *BMC Plant Biology, 8*, 103.

Xu, S., Li, X. Q., Guo, H., Wu, X. Y., Wang, N., Liu, Z. Q., . . . Jing, H. C. (2023). Mucilage secretion by aerial roots in sorghum (*Sorghum bicolor*): Sugar profile, genetic diversity, GWAS and transcriptomic analysis. *Plant Molecular Biology, 112*(6), 309–323.

Yandeau-Nelson, M. D., Xia, Y., Li, J., Neuffer, M. G., & Schnable, P. S. (2006). Unequal sister chromatid and homolog recombination at a tandem duplication of the A1 locus in maize. *Genetics, 173*(4), 2211–2226.

Yang, T., Liu, R., Luo, Y., Hu, S., Wang, D., Wang, C., . . . Zong, X. (2022). Improved pea reference genome and pan-genome highlight genomic features and evolutionary characteristics. *Nature Genetics, 54*(10), 1553–1563.

Zanini, S. F., Bayer, P. E., Wells, R., Snowdon, R. J., Batley, J., Varshney, R. K., . . . Golicz, A. A. (2022). Pangenomics in crop improvement-from coding structural variations to finding regulatory variants with pangenome graphs. *The Plant Genome, 15*(1), Art. e20177.

Zhang, J., Jiang, F., Shen, Y., Zhan, Q., Bai, B., Chen, W., & Chi, Y. (2019). Transcriptome analysis reveals candidate genes related to phosphorus starvation tolerance in sorghum. *BMC Plant Biology, 19*(1), 306.

Zhang, L., Xu, J., Ding, Y., Cao, N., Gao, X., Feng, Z., . . . Zou, G. (2023). GWAS of grain color and tannin content in Chinese sorghum based on whole-genome sequencing. *Theoretical and Applied Genetics, 136*(4), 77.

Zhang, L. M., Leng, C. Y., Luo, H., Wu, X. Y., Liu, Z. Q., Zhang, Y. M., . . . Jing, H. C. (2018). Sweet sorghum originated through selection of *Dry*, a plant-specific NAC transcription factor gene. *The Plant cell, 30*(10), 2286–2307.

Zhang, L. M., Luo, H., Liu, Z. Q., Zhao, Y., Luo, J. C., Hao, D. Y., & Jing, H. C. (2014). Genome-wide patterns of large-size presence/absence variants in sorghum. *Journal of Integrative Plant Biology, 56*(1), 24–37.

Zheng, L. Y., Guo, X. S., He, B., Sun, L. J., Peng, Y., Dong, S. S., . . . Jing, H. C. (2011). Genome-wide patterns of genetic variation in sweet and grain sorghum (*Sorghum bicolor*). *Genome Biology, 12*(11), R114.

Zhi, X., Tao, Y., Jordan, D., Borrell, A., Hunt, C., Cruickshank, A., . . . Mace, E. (2022). Genetic control of leaf angle in sorghum and its effect on light interception. *Journal of Experimental Botany, 73*(3), 801–816.

Zhong, Y., Luo, Y., Sun, J., Qin, X., Gan, P., Zhou, Z., . . . Song, J. M. (2024). Pan-transcriptomic analysis reveals alternative splicing control of cold tolerance in rice. *The Plant Cell, 36*(6), 2117–2139.

6 Mutagenesis, Transgenesis, and Targeted Gene Editing in Sorghum Improvement

Pallavi Dhiman[#]*, Zhiyuan Liu*[#]*, Yinping Jiao, and Gunvant B. Patil*

6.1 INTRODUCTION

As highlighted in earlier chapters, sorghum, with its approximately 730 megabase (Mb) genome and absence of genome duplication events, stands out as a valuable resource for genetic engineering and functional genomics studies (Paterson et al., 2009). The availability of genomic resources and the extensive genetic diversity in sorghum have greatly accelerated advancements in molecular breeding, genomic selection, and functional genomics research (Boyles et al., 2019; Xin et al., 2021). These advancements, leveraging natural and mutant sorghum populations alongside genomic resources, have been pivotal in understanding gene functions and driving sorghum improvement (Jiao et al., 2016; Xin et al., 2009).

Furthermore, sorghum is susceptible to many biotic and abiotic stresses, which cause limitation to its productivity, and there is also a need to improve its nutritional quality (Liu et al., 2016; Mofokeng et al., 2017; Tari et al., 2013). However, traditional breeding methods for introducing traits into diverse genetic backgrounds require multiple generations of backcrossing, which can be time-consuming and may introduce undesirable traits, negatively impacting agronomic performance and yield (Hao et al., 2021; Rooney, 2004). Genetic engineering has been instrumental in deepening our knowledge of plant biology, revolutionizing agriculture, and offering immense potential for accelerating trait improvement in sorghum (Pandey et al., 2010). The sorghum improvement using genetic engineering is hampered due to its recalcitrant nature (Azhakanandam & Zhang, 2015). However, genetic engineering with genomics could offer a more efficient approach to trait discovery and improvement. For instance, combining genome engineering with pan-genome studies allows researchers to pinpoint and modify variations linked to key agronomic traits specific to genotypes (Della Coletta et al., 2021; Tay Fernandez et al., 2022). Additionally, utilizing genomic resources alongside gene-editing technologies accelerates the functional validation of candidate genes, simplifying their identification and utilization (Thomson et al., 2022).

Among the achievements in sorghum genetic engineering, many efficient protocols for genetic modifications using *Agrobacterium*-mediated transformation or particle bombardment have been established (Liu & Godwin, 2012). Furthermore, recent advancements in genotype-independent transformation techniques using developmental regulator genes mark a pivotal milestone for further accelerating sorghum genetic engineering and tissue culture process (Aregawi et al., 2022; Che et al., 2022; Lowe et al., 2016; Massel et al., 2023). These innovations have enabled more efficient and flexible gene editing across diverse genotypes, addressing a longstanding challenge in the field. To date, there is no Genetically Modified (GM) sorghum commercially available, but it can be possible in the future due to this very efficient transformation system. This chapter focuses on the gene

[#] These authors contributed equally to this work.

DOI: 10.1201/9781003623717-6

function discovery using sorghum mutant population, and efforts towards sorghum genetic engineering, offering insights into overcoming existing barriers and driving further advancements in the discipline.

6.2 MUTAGENS AND MUTAGENESIS

With climate change and growing population, the breeding of new crop varieties has become crucial to addressing the global food challenge. Among various breeding methods, mutation breeding stands out due to its low cost, significant trait variation, plentiful variation types, and short breeding time (Bado et al., 2015; Garcia-Hernandez et al., 2002). Mutation is defined as a heritable change to the genetic make-up of an organism, and it has been the most important factor in evolution (Kimura & Kimura, 2020). Natural or spontaneous mutations occur at an extreme low frequency of only 10^{-7}–10^{-8} (Kovalchuk, 2021) which is insufficient to meet current breeding and genetic studies. Fortunately, starting in the early 20th century, scientists began exploring ways to increase the frequency of mutations. In 1926 and 1927, Muller observed that exposing *Drosophila* to X-rays induced heritable mutations, concluding that X-rays could be used to induce genetic changes (Muller, 1930). Shortly thereafter, in 1928, Stadler demonstrated this phenomenon in plants, showing that X-rays could significantly increase the mutation rate in maize, wheat, and barley (Stadler, 1944; Stadler, 1929). Such agents, like X-rays, that can permanently and effectively change genetics of organisms, are known as mutagens (Oladosu et al., 2016). Colchicine was found to induce the formation of polyploid *Datura stramonium*, demonstrating the role of chemical agents for the first time in inducing mutations (Bergner et al., 1940; Blakeslee, 1939, 1941). Later, mustard gas was found to induce mutations in *Drosophila*, further expanding the range of known chemical mutagens (Auerbach, 1945). Over time, hundreds of mutagens have been discovered, leading to the development of two major systems: physical mutagenesis and chemical mutagenesis (Oladosu et al., 2016). These systems continue to progress and develop, supporting various plant species breeding and genetic research (Mba et al., 2010). Among physical mutagens X-rays, gamma, fast neutrons, protons, and ion beams belonging to ionizing radiation, are widely used for mutation induction in higher plants (Çelik & Atak, 2017; Ehrenberg et al., 1961; Kodym & Afza, 2003).

Compared to physical mutagenesis, chemical mutagenesis offers advantages such as the ability to form stable point mutations, preserve desirable traits, ease of operation, and lower costs (Mba et al., 2010). Due to these benefits, numerous mutant libraries have been established by different chemical mutagens in crops (Bado et al., 2015; Roychowdhury & Tah, 2013; Sikora et al., 2011). Based on mechanisms of chemical mutagenesis, mutagens can be divided into four main categories including base modifiers, such as ethyl methane sulfonate (EMS) and nitrous acid, which induce mutations by altering DNA components, thus leading to base mispairing (Chen et al., 2023; Serrat et al., 2014). EMS is a widely used alkylating agent for generating mutagenized populations including major field, fruit and ornamental plant species (Nafisi et al., 2007). These mechanisms highlight the diverse ways chemical mutagens induce genetic alterations, which are valuable in understanding mutagenesis and its applications in genetic research. The forward genetics approach, particularly through mutant analysis in model plants like Arabidopsis, has led to the discovery of numerous important genes (Garcia-Hernandez et al., 2002). This process involves inducing mutations in a pure genetic line, selecting individuals based on observable phenotypes, and mapping the gene(s) responsible for those traits (Chopra, 2005). Mutations caused by chemical agents and ionizing radiation are distributed across the genome but occur more frequently than spontaneous mutations (Graham et al., 2020). As highlighted by Graham et al. (2020), over 3,200 mutant varieties have been released for commercial use across more than 210 plant species in over 70 countries. These include cereals (1,584; 49.5%), flowers and ornamentals (700; 21.9%), legumes (480; 15%), and other crops (435; 13.6%).

6.3 AVAILABLE SORGHUM MUTANT POPULATIONS AND REGISTERED SORGHUM MUTANTS

The development of high-throughput next-generation sequencing (NGS) has renewed the interest in gene function discovery using forward genetics (Kumawat et al., 2022; Ray & Satya, 2014). It starts with the identification of mutants developed using different methods including physical or chemical mutagenic agents as discussed above. Then the causal gene can be mapped using segregation analysis and linkage analysis (Jander et al., 2002).

In sorghum, chemical-induced mutation using EMS (ethyl methane sulfonate) is the most common approach for generating large-scale plant mutations. Leveraging extensive genomic resources and analyses of small and large structural variations through sequencing, researchers are increasingly focusing on functional genomics to uncover gene functions and their roles in phenotypic traits across various crops (Chen, Zou, et al., 2019). A well-characterized mutant library serves as an invaluable resource for elucidating the functions of sorghum genes. Genomic studies can identify genes lacking functional annotations (Abdel-Ghany et al., 2020), and a comprehensive mutant library, where each gene is represented by a corresponding mutant, enables systematic functional characterization. This approach facilitates the discovery of gene functions and their roles in biological processes, contributing significantly to advancements in sorghum genetics and crop improvement research. In sorghum, mutant populations have been developed for gene function studies using EMS treatment on the reference genome line BTx623. One study utilized 600 BTx623 mutants, identifying 3,274,606 single nucleotide polymorphisms (SNPs) across 30,285 sorghum genes, including medium- and high-impact mutations induced by EMS (Simons et al., 2022). Jiao et al. (2016) reported whole-genome sequencing of 256 phenotyped mutants and detected more than 1.8 million mutations induced by EMS. These mutations affect over 95% of sorghum genes and mutations of high impact are contained in more than half (57%) annotated genes (Jiao et al., 2016). Seeds of BTx623 line were selected, treated with EMS, and proceeded to M_3 through single seed descend (SSD) followed by bulking M_3 panicles to produce M_4 population. 256 mutant lines with phenotype were randomly selected from 6,400 M_4 pools developed (Xin et al., 2008) and sequenced for further analysis, such as mutations distribution and annotation. More recently, sequencing 1,256 mutants uncovered 9.5 million EMS-induced mutations, covering 98% of sorghum genes and providing a unique genetic diversity distinct from natural variations (Jiao et al., 2016; Jiao et al., 2024).

Likewise, around 12,000 BTx623 seeds were treated with EMS, and M_4 seeds were produced subsequently (Addo-Quaye et al., 2017, 2018). Mutations of 586 independently mutated lines were characterized covering 94% (31,163) genes with reliable mutations across the sorghum genome. Through whole-genome sequencing of the 586 lines, a database describing mutated genes as well as their functions based on corresponding homologs in maize and Arabidopsis was established (www.purdue.edu/sorghumgenomics/). Additional resources to facilitate access to sorghum EMS mutant resources can be explored through SorghumBase (Gladman et al., 2022) at www.sorghumbase.org/ and SorbMutDB at www.depts.ttu.edu/igcast/SorbMutDB.php. SorghumBase is a dynamic genomic resource for the sorghum supporting the integration of mutants' genomic contexts, gene and regulatory effects, and genotype frequencies. The SorbMutDB search interface enables users to locate genes of interest efficiently, identifying mutations and associated SIFT (Sorting Intolerant from Tolerant) scores (Jiao et al., 2024). Lately, many other EMS, fast neutron, and gamma-induced sorghum mutant libraries have been created (Ding et al., 2020; Ordonio, 2014; Yuan et al., 2024). In short, sorghum mutant populations and their annotated database have been established, leading to a new epoch in identifying sorghum genes and therefore paving the way for a deeper understanding of important gene function through employing high-technology like CRISPR-Cas9 systems.

6.4 APPLICATIONS OF MUTANTS FOR GENE FUNCTION ANALYSIS AND BREEDING

As mentioned earlier, mutant resources promote sorghum functional genetics research extensively. Numerous genes have been identified involved in disease resistance, development, nutrition, plant architecture, pollen, and seed formation through employing mutant resources in sorghum. For instance, a point mutation in *dhurrinase2* gene causing an acyanogenic phenotype was identified in the EMS population using the NGS. It revealed the deterrent role of cyanide release in insect herbivory, highlighting a genome-guided approach to identify causal mutations and gene functions in crops (Krothapalli et al., 2013). Tillering is an important part of plant architecture and critical in terms of forage sorghum breeders. To analyze the molecular mechanism of shoot branch initiation, a mutant with multiple tillers was screened from an EMS-induced mutant library. Afterward, *NAB1* (*Non-dormant Axillary Bud 1*) gene, with the homolog in rice being *HIGH-TILLERING DWARF1* (*HTD1*), was identified in sorghum. Functional experiments showed *NAB1* negatively regulates the excessive growth of auxiliary buds through taking part in strigolactone synthesis and auxin transportation (Chen et al., 2018).

Researchers identified a multi-seeded (multiseeded1, *msd1*) EMS mutant line to reveal the grain number regulation in sorghum. The two fertile pedicellate spikelets, which are sterile in the wild type, contributed to the production of two additional grains (Burow et al., 2014). Genetics analysis, bulk segregant analysis, and next-generation sequencing revealed that the *msd1* gene encodes a Teosinte-like transcription factor. Expression profiling revealed the phenotype was caused by reduced Jasmonic acid (JA) (Jiao et al., 2018b). Furthermore, Chen et al. (2020) characterized *BIOMASS YIELD 1* (*BY1*), which modulates grain yield. The *by1* mutant exhibits severely defective plant architecture and abnormal reproductive organ development. Map-based cloning identified *BY1* as encoding a 3-deoxy-D-arabino-heptulosonate-7-phosphate synthase (DAHPS), the enzyme responsible for catalyzing the initial step of the shikimate pathway (Chen et al., 2020). The male-sterile mutant *Ms9* was identified as the first nuclear male-sterile gene in sorghum, encoding a PHD-finger transcription factor essential for pollen development (Chen et al., 2019).

Several genes involved in the specific steps of monolignol biosynthesis which influences cell wall structure, and lignin have been identified and characterized through various brown midrib (*bmr*) mutants causing improved digestion in animals and ethanol production (Porter et al., 1978; Sattler et al., 2014). Sorghum *Bmr12* gene encodes caffeic acid O-methyltransferase (COMT), an enzyme that catalyzes the second-to-last step in monolignol biosynthesis (Bout & Vermerris, 2003; Porter et al., 1978). *Bmr6* derived from a sorghum population subjected to diethyl sulfate mutagenesis, exhibited a deficiency in cinnamyl alcohol dehydrogenase (CAD) activity which catalyzes the final step in monolignol synthesis (Pillonel et al., 1991). *Bmr2*, encoding 4-coumarate: CoA ligase (4CL), an enzyme producing monolignol precursors, was identified through F_2 population mapping (Saballos et al., 2012). More recently, *Bmr19* and *Bmr30* were characterized; *Bmr19* participates in one-carbon metabolism, and *Bmr30* encodes a chalcone isomerase, contributing to lignin biosynthesis and deposition (Adeyanju et al., 2021; Tetreault et al., 2021). Identifying all BMR genes in the entire lignin biosynthesis pathway has revealed the genetic resources to improve both agricultural productivity and industrial applications.

Another important trait is reduced plant height, which offers lodging resistance and allows more energy to be used for nourishing grains, resulting in a higher yield. In sorghum, three dwarf loci, *Dw1*, *Dw2*, and *Dw3* have been identified, while the *Dw4* locus remains uncharacterized (Quinby & Karper, 1953). Another dwarf mutant, BTx623dw5 (dw5), was registered by the USDA-ARS, which introduces a novel genetic approach to control plant height in sorghum. Unlike the known dwarf loci used in breeding, *dw5* is genetically distinct and reduces plant height significantly in different environments (Chen et al., 2019). Another lodging resistance gene, *SbBC1*, was identified through the brittle culm 1 (*bc1*) mutant, characterized by stems and leaves that bend easily. Further analysis revealed that the COBRA-like protein encoded by *SbBC1* plays a pivotal role in cellulose biosynthesis (Li et al., 2019). Plant vascular patterning has long intrigued biologists for its role in engineering C_4 traits into C_3 plants. A key difference between C_4 and C_3 species is the high vein density and Kranz anatomy in C_4 plants. Using mutagenized sorghum populations, forward genetics identified two recessive mutations in the *CYP90D2* gene, which encodes a protein involved in brassinosteroid biosynthesis. This pathway was found to promote high vein density in C_4 plants, contributing to anatomical differences between C_4 and C_3 species (Hong et al., 2003; Rizal et al., 2015). This novel gene sharpens the understanding of the denser vein of C_4 plants compared to C_3 plants.

Epicuticular wax on the aerial surface is a major trait contributing to the heat and drought resistance of sorghum compared to other crops. EMS mutants, namely *bloomless40-1* (*bm40-1*) and *bm40-2*, which exhibit little to no wax on their surfaces, were investigated. Subsequently, the gene *Sobic.001G228100*, encoding a GDSL-like lipase involved in cutin biosynthesis, was identified (Jiao et al., 2018a). Further, a new mutant, *bloomless 41* (*bm41*), was identified and causal gene was a leucine-rich repeat transmembrane protein kinase (Tian et al., 2024). These two studies show an extensive network of wax biosynthesis in sorghum with many genes involved. This will extensively help us understand the mechanisms of sorghum's resistance to various abiotic stresses. On all accounts, sorghum mutant resources have been well-established and plenty of genes were identified and functionally characterized. A list of sorghum mutant studies is provided in Table 6.1.

TABLE 6.1
Overview of studies on sorghum mutants

Type of Mutagen	Phenotype	Mapped causal gene / pathway	Method	Associated trait	References
EMS	lobed-leaf	*Sblob1/Sobic.003G010300* /plant serotonin/ melatonin biosynthesis pathway	MutMap and whole-genome sequencing	biomass yield	Li et al. (2024b)
EMS	late booting stage	*Sbld1/Sobic.001G242900*	Bulk segregant analysis (BSA)	flowering time	De Riseis et al. (2023)
EMS	bloomless	*BM41/Sobic.001G061400/* lipid biosynthesis and steroid metabolism	BSA	heat and drought tolerance	Tian et al. (2024)
EMS	stunted and yellow ripple leaf	*SbYR1/Sobic.007G014500/* NLR protein	BSA	head smut disease tolerance	Li et al. (2025)
EMS	less sensitive to osmotic stress	*SbBES1*/Brassinosteroid signaling	DNA Affinity Purification sequencing	osmotic and drought stress tolerance	Fontanet-Manzaneque, Laibach et al. (2024)
EMS	leaf necrosis and short roots	*SbPHO2/ Sobic.009G228100*/Pi signaling	BSA	absorption of multiple nutrients	Zhu et al. (2023)
EMS	thick leaf	*SbKOR1/Sb01g036480*/cell wall biosynthesis	BSA	biofuel production	Mendu et al. (2022)

TABLE 6.1 (Continued)
Overview of studies on sorghum mutants

Type of Mutagen	Phenotype	Mapped causal gene / pathway	Method	Associated trait	References
EMS	increased tillering and reduced plant height	*NAB1/Sobic.006G170300/* auxin-related gene	BSA	tillering	Chen et al. (2018)
EMS	all spikelets are fertile and set grain	*MSD1/Sb07g021140/* Jasmonic acid pathway	BSA	grain number per panicle	Jiao et al. (2018b)
EMS	reduced plant height, narrow stems, erect and narrow leaves, and abnormal floral organs	*BY1/Sobic.002G379600/* shikimate pathway	BSA	biomass production and grain yield	Chen et al. (2020)
EMS	small pale-colored anthers that contained no pollen grains	*Ms9/Sobic.002G221000/* anther and pollen development	BSA	hybrid crop production	Chen et al. (2023)
diethyl sulfate mutagenesis	brown midrib	caffeic acid O-methyltransferase, COMT/lignin monomer biosynthesis	candidate-gene approach	biofuel production	Bout and Vermerris (2003)
diethyl sulfate mutagenesis	brown midrib	cinnamyl alcohol dehydrogenase, CAD/ *Sb04g005950*/lignin biosynthesis	candidate gene approach	biofuel production	Pillonel et al. (1991); Sattler et al. (2014)
diethyl sulfate mutagenesis	brown midrib	*FPGS*/folate (C1) metabolism	BSA	biofuel production	Adeyanju et al. (2021)
EMS	brittle culm	*SbBC1/Sobic.001G336700/* cellulose biosynthesis	map-based cloning	lodging resistance	Li et al. (2019)
γ and EMS	reduced vein density	*CYP90D2*/brassinosteroid biosynthesis pathway	BSA	vein density	Rizal et al. (2015)
EMS	no epi-cuticular wax on sorghum's aerial surfaces	GDSL-like lipase/wax biosynthesis	BSA	drought-tolerant	Jiao et al. (2018a)

Undoubtedly, mutant populations, a cornerstone of forward genetics, are valuable resources. However, these resources have notable limitations, including the restricted availability of mutants in only a few genotypes, which limits their broader application. Additionally, traditional backcrossing requires several generations to introduce mutations into other genotypes. In contrast, an efficient transformation system can significantly accelerate the breeding process. Genetic transformation plays a vital role by allowing precise manipulation of specific genes identified through population studies, enabling the assessment of their contributions to key traits. But the lack of efficient transformation systems poses a significant challenge, hindering the critical functional validation of these genes. The below sections focus on transgenic approaches to overcome these challenges and enable the functional validation of key genes contributing to important traits in sorghum.

6.5 GENETIC ENGINEERING

Genetic engineering introduces new traits, significantly enhancing plant productivity and usefulness by incorporating genes (transgenes) from both related and unrelated species (Kohli et al., 2003; Meyer, 1995). Transgenic plants also provide a platform to study gene functions and understand gene regulation by facilitating gene overexpression or gene knockout using genetic engineering tools (Gasser & Fraley, 1989; Tenea et al., 2009). These genetic modifications enable improved traits such as pest and disease resistance, increased yield, and stress tolerance (Sharma et al., 2002). Advances in plasmid biology have facilitated the use of *Agrobacterium tumefaciens* plasmids as vectors for DNA transfer into plant genomes, enabling precise studies of gene activity, regulation, and control through enhancer and silencer sequences (Bevan, 1984; Wang et al., 1984). The commercial production of genetically modified crops is widespread in some countries bringing benefits like pest resistance, herbicide tolerance, and virus resistance (Aragão & Faria, 2009; Clark & Lehman, 2001). Examples include golden rice, enriched with beta-carotene to address vitamin A deficiency (Potrykus, 2001). GM crops support agricultural productivity by increasing yields, reducing pesticide reliance, enhancing nutritional value, and improving food quality. This agricultural biotechnology contributes to food security and, in some cases, provides medical benefits, addressing the needs of an expanding global population.

6.5.1 GENETIC TRANSFORMATION IN SORGHUM

Since the first transgenic plants in 1983 (Fraley et al., 1983; Herrera-Estrella et al., 1983), scientists have consistently worked toward advancing the genetic engineering of plants, including sorghum. As previously mentioned, sorghum, a valuable crop, has been enhanced through traditional breeding methods. In the last decade, significant progress has been made to achieve transformation in cereal crops successfully using biolistic methods or *Agrobacterium* methods. Sorghum is considered a recalcitrant crop in terms of transformation and regeneration efficiency and lags behind other major crops like rice and maize.

Sorghum transformation using particle bombardment was first reported in 1993 (Casas et al., 1993), and seven years later, *Agrobacterium*-mediated transformation was achieved in 2000 (Zhao et al., 2000). Theoretically, any differentiated cell, tissue, or organ holds the potential to serve as an explant and can be induced to de-differentiate under appropriate *in vitro* conditions (Su et al., 2021). In sorghum, various explants, including immature and mature embryos (Tadesse et al., 2003), immature inflorescences (Casas et al., 1993), leaf whorls (Tadesse et al., 2003), shoot meristems (Yellisetty et al., 2015), and pollens (Wang et al., 2007) have been utilized. But immature embryos (IEs) are the most preferred explants due to high regeneration capabilities. As mentioned earlier, IEs are preferred explants for transformation but they need a long period since the plant must reach the reproductive stage within specific seasons, with only a narrow 10–15-day window to collect immature seeds. To address this, leaf whorls were used as explants from 4-week-old sorghum seedlings to obtain embryogenic callus formation and regeneration in Tx430 and P898012 (Silva et al., 2020). Sorghum pollen has been identified as a promising explant for genetic transformation, providing an alternative approach for sorghum transformation (Wang et al., 2007). Despite the development of a tissue culture system for sorghum nearly 40 years ago, advancements in sorghum transformation have lagged behind those in other cereals. Sorghum transformation is faced with challenges like release of phenolics and low regeneration. The below sections discuss the challenges in sorghum transformation. The

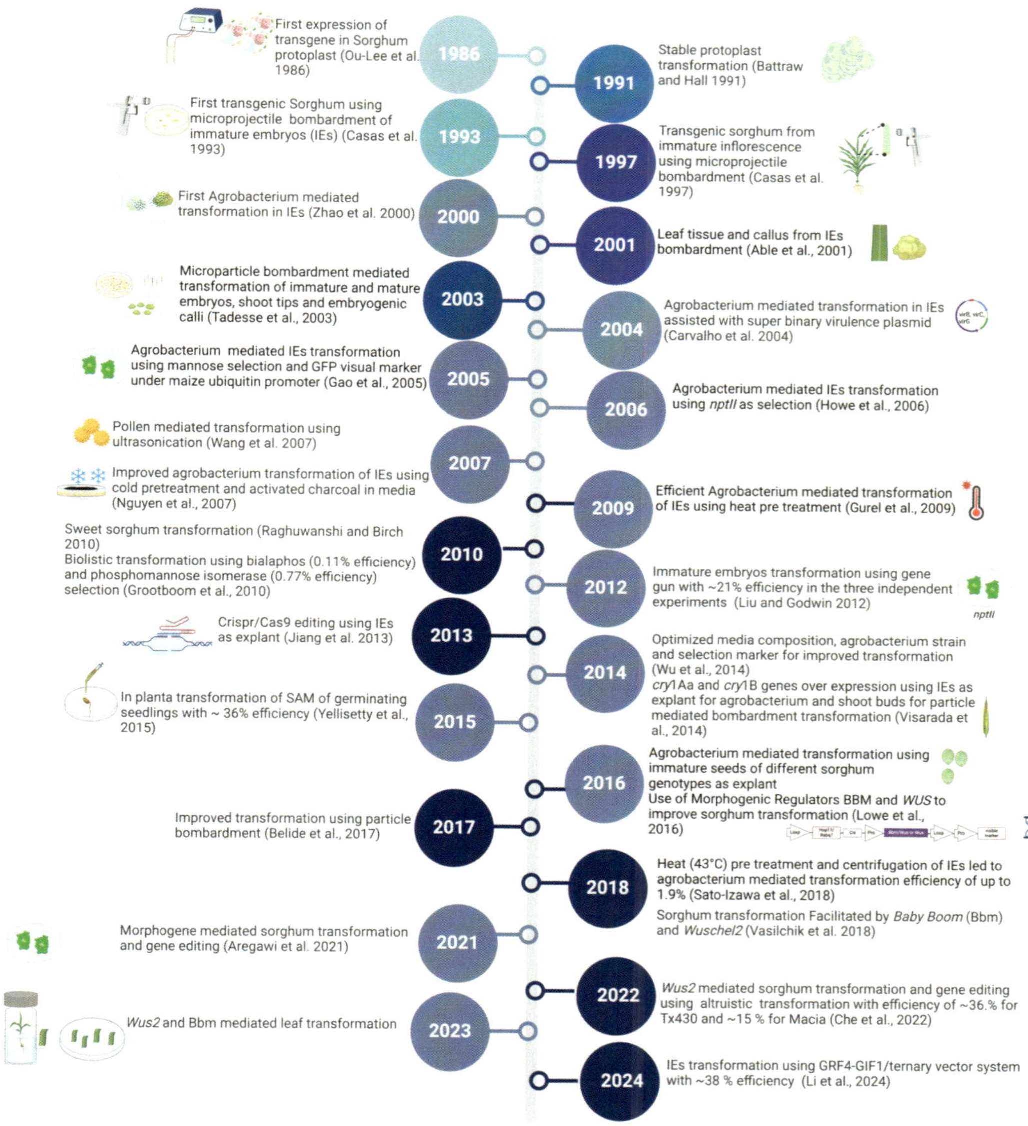

FIGURE 6.1 Overview of milestone studies in sorghum genetic transformation.

timeline of sorghum transformation studies using different transformation methods and explants is shown in Figure 6.1.

6.5.2 TARGETED GENE EDITING IN SORGHUM

Genome editing with programmable nucleases has revolutionized functional genomics and the generation of genetic diversity for breeding. The use of advanced genome-editing tools, particularly CRISPR/Cas9 (Doudna & Charpentier, 2014), has accelerated sorghum genetic research, offering promising potential for improved crop resilience, yield, and nutritional value (Jiang et al., 2013; Liu et al., 2019). This technology functions by creating a targeted double-strand

DNA break (DSB) *in vivo*, which is subsequently repaired by cellular DNA repair mechanisms (Schmidt et al., 2019, 2020). The origin of editing tools such as the CRISPR/Cas9 system is adapted from a prokaryotic immune mechanism against foreign nucleic acids (Jinek et al., 2012). CRISPR relies on RNA-guided DNA cleavage to induce site-specific DSBs (Jinek et al., 2012). These are subsequently repaired via one of three pathways: homology-directed repair (HDR), non-homologous end joining (NHEJ), or microhomology-mediated end joining (MMEJ) (Kadam et al., 2023; Samanta et al., 2016). CRISPR/Cas9 is the most widely used genome editing tool because of its straightforward design, high accuracy, and capacity for multiplex editing (Prajapati & Tyagi, 2024). Sorghum, a staple crop known for its climate adaptability, benefits from CRISPR/Cas9 applications that can enhance resistance to biotic and abiotic stresses and introduce valuable genetic variations. Recent advancements in CRISPR techniques have made precise modifications in sorghum more feasible, fostering gene functionality studies and facilitating crop improvement efforts. The first gene editing in sorghum through *Agrobacterium*-mediated transformation of immature embryos was achieved in 2013 (Jiang et al., 2013). They used a binary vector with a synthetic Cas9, guide RNA, and *GFP-NptII* fusion as a selectable marker. The system targeted a frameshift in *DsRED2*, with *GFP* confirming T-DNA integration and red fluorescence indicating successful mutagenesis. Furthermore, U6 promoter was used to enhance CRISPR/Cas9 editing efficiency in sorghum (Massel et al., 2022).

In sorghum, CRISPR has enabled nutrient fortification and climate resilience by targeting specific genes, providing a promising tool for modern sorghum breeding programs aimed at addressing food security and environmental challenges. Kafirins, the main storage proteins in sorghum, hinder grain digestibility due to their proteolytic-resistant γ- and β- forms encapsulating α-kafirins. Using CRISPR/Cas9 gene-editing technology, the *K2G* gene was targeted to develop low-kafirin sorghum lines (Li et al., 2024c). Similarly, CRISPR/Cas9 mediated mutagenesis of α- and γ-kafirin genes *k1C5* and *gKAF1* respectively, and increased endosperm protein digestibility (Elkonin et al., 2023). Another targeted genome editing using CRISPR/Cas9 induced *LIGULELESS-1* (*LG1*) gene mutations in sorghum, leading to change in leaf angle phenotype potential for improved yield (Brant et al., 2021). CRISPR/Cas9 knockout of carotenoid cleavage dioxygenase genes *SbCCD8a* and *SbCCD8b* reduced orobanchol, lowered Striga germination, altered plant structure, and impacted root microbial diversity, offering sustainable Striga control strategies (Hao et al., 2023). An efficient protoplast transfection assay to screen single guide RNAs (sgRNAs) for CRISPR-based genome editing was developed in sorghum. Testing plasmid and Ribonucleoprotein (RNP) based CRISPR delivery achieved mutation rates of up to 77.8% and 18.5%, respectively, providing a fast and efficient method to assess sgRNA effectiveness for advancing sorghum genome editing (Lee et al., 2023). CRISPR/Cas9 genome editing in sorghum using biolistic bombardment targeting *cinnamyl alcohol dehydrogenase*, *COD* and *phytoene desaturase, PDS* genes resulted into edits inherited in the T_1 generation (Liu et al., 2019). An efficient transformation and regeneration system was established in two Sudangrass cultivars using CRISPR/Cas9 by editing the *Caffeic acid O-methyl transferase COMT* gene to decrease lignin level (Assem et al., 2023). Multiplex gene editing in sorghum was achieved by targeting *SbFT* (*Sb10G045100*) and *SbGA2ox5* (*Sb09G230800*) via *Agrobacterium*-mediated transformation. Editing efficiency reached 33.3% for *SbFT* and 83.3% for *SbGA2ox5*. Mutations were heritable, with *SbFT* mutants showing delayed flowering, while *SbGA2ox5* biallelic mutants were absent, suggesting lethality (Char et al., 2020). A list of sorghum transformation studies for gene function validation or trait improvement is provided in Table 6.2. In summary, CRISPR/Cas9 has significantly advanced sorghum genetic research, enabling precise modifications that can enhance trait improvement and trait discovery. These developments not only bolster sorghum breeding programs but also provide a framework for addressing global food security and environmental sustainability challenges. The technology's versatility, from improving protein

TABLE 6.2
Studies for trait improvement using sorghum genetic engineering

Genotype	Explant	Technique/ Approach	Trait of improvement	Targeted gene(s)	Vector /promoters used	Transformation method	Efficiency	References
SDW5	Pollen	CRISPR/ Cas9	Protein digestibility	*SbK2G γ-kafirin*	OsU3:gRNA 2x35s:Cas9	Sonication	34% targeted mutation rate	Li et al. (2024)
Avans	Immature embryos	CRISPR/ Cas9	Protein digestibility	*k1C5*	OsU3:gRNA ZmUbi:Cas9	Agrobacterium tumefaciens, strain AGL0	~9% independent transgenic events	Elkonin et al. (2023)
P898012	Immature embryos	CRISPR/ Cas9	Delayed flowering, NA	*SbFT, SbGA2ox5*	OsU6:gRNA ZmUbi:Cas9	Agrobacterium tumefaciens	33.3%, 83.3% editing efficiency	Char et al. (2020)
Tx430	Immature embryos	CRISPR/ Cas9	Leaf angle	*SbLG1*	OsU6:gRNA CaMV35S:Cas9	Biolistic	33.3% editing efficiency	Brant et al. (2024)
Zheltozyornoe 10 (Zh10) (Sorghum bicolor (L.) Moench)	Immature embryos	RNAi silencing	Improved digestibility	*γ-kafirin*	Binary silencing vector pNRKAFSIL	Agrobacterium tumefaciens, strain GV3101	NA	Elkonin et al. (2016)
BT×623	Shoot apices	Over expression	Resistance against Stem borer	synthetic *cry1Ac* gene over expression	Wound-inducible promoter from the maize protease inhibitor gene (mpiC1)	Particle bombardment	1.5% (5 t0 plants harbouring the cry1Ac transgene)	Girijashankar et al. (2005)
Sorghum bicolor variety SPV-462	Shoot-tip derived callus cultures	Over expression	Salt stress resistance	*pyrroline-5-carboxylate synthetase (P5CSF129A)*	CaMV35S:P5CSF129A	A. tumefaciens strain LBA 4404 and bombardment methods	NA	Reddy et al. (2015)
Tx430	Immature embryos	Over expression	Delayed flowering	*Sobic.006G061100 (SbSNF4-2)*	ZmUbi	Biolistic	NA	Upadhyaya et al. (2022)

(continued)

TABLE 6.2 (Continued)
Studies for trait improvement using sorghum genetic engineering

Genotype	Explant	Technique/ Approach	Trait of improvement	Targeted gene(s)	Vector /promoters used	Transformation method	Efficiency	References
P898012	Immature embryos	Over expression	Growth and development	*SbGln1*	ZmUbi or the CaMV 35S	Agrobacterium-mediated transformation	NA	Urriola and Rathore (2015)
Macia	NA	CRISPR/ Cas9	Parasitic weed control	*SbLGS1*	NA	NA	NA	Bellis et al. (2020)
Wheatland	Immature embryos	CRISPR/ Cas9	Aromatic sorghum	*SbBADH2*	pYLCRISPR/Cas9Pubi-B	Agrobacterium-mediated	NA	Zhang et al. (2022)
Sudangrass (Sorghum sudanense) Giza-1 and Giza-2	Callus derived from immature inflorescence	CRISPR/ Cas9	Reduced lignin	*Caffeic acid O-methyl transferase (COMT)*	U6:gRNA 35S:Sas9	particle bombardment	30.6% transformation efficiency (no. of Cas9 transgenic events/no. of Geneticin-resistant events*100).	Assem et al. (2023)

digestibility to controlling parasitic plants like Striga, underscores its promise for revolutionizing sorghum and related crop species.

6.6 CHALLENGES IN SORGHUM TRANSFORMATION

The successful production of transgenic plants requires an efficient tissue culture system, but sorghum is regarded as the most challenging cereal crop due to its poor *in vitro* response. The challenges in sorghum tissue culture and transformation are listed below and mentioned in Figure 6.2.

6.6.1 Genotype Dependence: The inbred lines Tx430 and P898012 are currently the primary genotypes used for transformation. However, the exact cause of genotype dependence is not fully understood. These genotypes are known for their superior embryogenic capacity and reduced phenolic release during the tissue culture process (Do et al., 2016).

This poses a limitation, especially for commercial applications where engineering elite cultivars would offer significant benefits. The use of developmental regulators, such as *BABY BOOM* (*ZmBbm*) and *Wuschel2* (*ZmWus2*) from maize, has been effective in reducing genotype dependency in sorghum (Lowe et al., 2016), and these are discussed in detail in the sections below.

6.6.2 Low Transformation Efficiency: Currently, sorghum transformation relies on indirect somatic embryogenesis via a callus phase, which is labor-intensive and prone to somaclonal variations (Azhakanandam & Zhang, 2015). The bombardment method initially showed low transformation efficiency (0.08–1%) but was improved to 1.3% by optimizing parameters like acceleration pressure, target distance, gap width, and using various explants (Tadesse et al., 2003). Through further optimization of tissue culture and the use of the biolistic method, Liu and Godwin (2012) achieved a

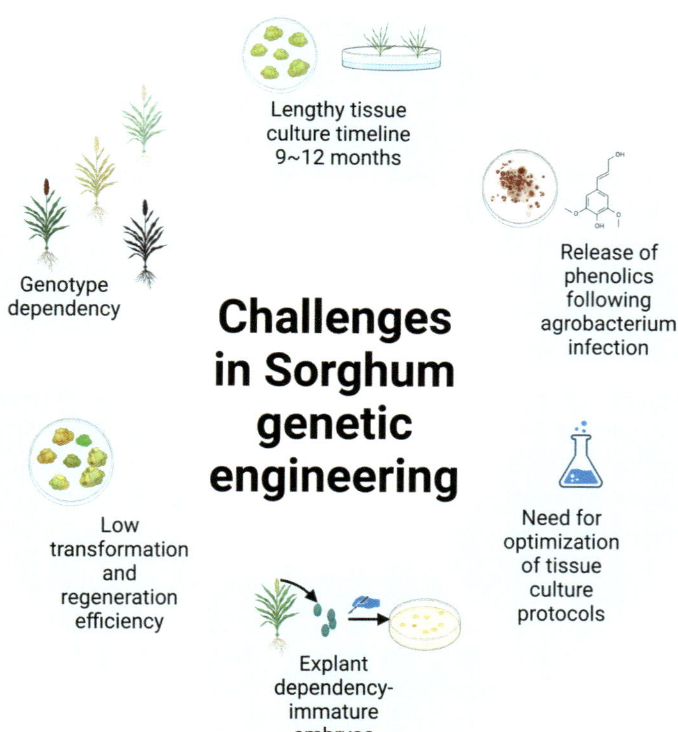

FIGURE 6.2 Major technical challenges in sorghum tissue culture and genetic transformation.

20% transformation frequency. However, the biolistic method causes unintended DNA breaks and multiple insertions (Ozyigit & Yucebilgili Kurtoglu, 2020).

Agrobacterium-mediated transformation with different *Agrobacterium* strains including LBA4404, EHA105, AGL1 has been established in sorghum (Carvalho et al., 2004; Char et al., 2020; Gao et al., 2005) and further improved with the use of "Super-binary" vectors containing additional virulence genes to enhance the transformation efficiency (Zhao et al., 2000). Direct somatic embryogenesis, successfully applied in maize, could be adapted for sorghum to improve efficiency by bypassing the callus stage. This approach has been made possible using developmental regulators. Additionally, embryogenic cell suspension cultures, used effectively in switchgrass (Ondzighi-Assoume et al., 2019) and cotton (Rajasekaran et al., 2000), offer potential for sorghum, as they can improve transformation efficiency by minimizing somaclonal variation and enhancing transgenic survival.

6.6.3 Release of Phenolics: The release of phenolics in culture hinders callus growth and somatic embryogenesis. This issue is particularly pronounced in tannin-rich sorghum varieties but also affects low-tannin types to a lesser extent (Liang et al., 2023). To counter phenolic effects, culture media have been optimized with additives like polyvinylpropyl (PVP) (Liu et al., 2015), activated charcoal (Nguyen et al., 2007), ascorbic acid (Zhao et al., 2000), and L-cysteine (Kumar et al., 2011) which improve transformation efficiency. Lipoic acid, effective in other crops, has similarly improved callus induction in Tx430 from 65% to 85% (Belide et al., 2017; Kumar et al., 2011). However, Liu et al. (2015) used the sorghum inbred line SA281, known for its strong embryogenic callus regeneration yet high phenolic production, to optimize the culture medium and extended trials to Tx430 and 91419R. L-Asparagine, L-proline, $CuSO_4{\cdot}5H_2O$ and KH_2PO_4 were utilized to decrease phenolic compounds in the callus induction medium and enhance the callus induction rate (Liu et al., 2015).

6.7 SORGHUM TRANSFORMATION METHODS

Efforts to enhance sorghum tissue culture and transformation have involved testing various genotypes, methods, and explant sources, yielding improvements in transformation efficiency from 0.3% to as high as 46.6% (Liu et al., 2020). However, these advances are limited to specific genotypes and are affected by the seasonal availability of explants. Three primary methods for stable and transient gene expression in sorghum have been developed: electroporation, particle bombardment, and *Agrobacterium*-mediated transformation, with the latter two widely studied. This section focuses on the commonly used biolistic transformation and *Agrobacterium*-mediated approaches.

6.7.1 Biolistic Transformation

Biolistic transformation, also known as particle bombardment, involves propelling high-speed microparticles (gold or tungsten) coated with foreign genes into plant cells, facilitating gene integration into the host genome (Sanford, 1990). This technique, which bypasses biological constraints, is versatile and applicable across diverse plant species, proving particularly effective for organelle transformations, including chloroplasts and mitochondria. Sorghum, historically a recalcitrant crop, saw its first successful transgenic transformation via particle bombardment PDS1000/HE system, albeit with less than 1% efficiency (Casas et al., 1993). Subsequent optimizations have significantly improved efficiency, with Belide et al. (2017) achieving up to 27%. More recent studies demonstrated stable transformation using gold particles on BTx430 leaf whorls, though with limited efficiency (Miller et al., 2023). An optimized tissue culture system combined with particle bombardment raised sorghum transformation efficiency to 20.7% (Liu et al., 2014). In another instance, transforming embryogenic callus derived from mature seeds using *nptII* and *gus* genes achieved a 13.33% efficiency (Wang et al., 2021). An overexpression vector containing the *chitinase* gene under the Actin promoter and the *bar* gene under the *CaMV 35S* promoter was introduced into sorghum via particle bombardment on shoot clumps yielding five transgenic out of eleven regenerated T_0 plants (Devi & Sticklen, 2003). Particle bombardment of single explant with multiple plasmids enables the introduction of multiple genes using a single selectable marker. This approach is particularly important for improving agronomic traits that rely on the interaction of multiple genes (Ozyigit & Yucebilgili Kurtoglu, 2020; Zorrilla López et al., 2013). Nevertheless, particle bombardment has inherent challenges, including random transgene integration, potential gene rearrangements, size DNA fragments limitation, and possible silencing. Additionally, it can induce significant genome damage (Klein et al., 1992; Taylor & Fauquet, 2002).

6.7.2 *Agrobacterium*-Mediated Transformation

Agrobacterium tumefaciens is a soil-borne plant pathogen known to cause crown gall disease in dicots and a few monocots. This bacterium is a natural tool for biotechnology, where a fragment called Transfer (T)-DNA from Tumor-inducing, Ti plasmid can transfer to the host genome. *Agrobacterium*-mediated transformation has been effectively utilized in monocots and is generally preferred due to its propensity for single or low-copy gene insertions (Cheng et al., 2004). Zhao et al. (2000) demonstrated *Agrobacterium*-mediated transformation in sorghum. Subsequently, despite modification of transformation protocols including addition of coconut water and reducing bacterial load, increased embryo survival following *Agrobacterium* infection in P898012, transformation efficiency was still low (Carvalho et al., 2004). An optimized protocol using immature embryos of Tx430 achieved transformation frequencies as high as 33% with the AGL1 strain, which doubled high-quality transgenic event rates compared to strain LBA4404 (Wu et al., 2014). Transformation of the sorghum genotype P898012 with the AGL1 strain and *bar* gene as a selectable marker reached frequencies above 14%, with up to 50% single-copy T-DNA integrations (Do et al., 2016). Further, the introduction of a ternary vector system using the virulence plasmid pVIR achieved efficient *Agrobacterium*-mediated transformation in Tx430 with regeneration frequencies between 6% and 29%, producing 45%–66% single-copy, backbone-free events (Anand et al., 2019). This ternary vector system also facilitated successful transformation in difficult-to-transform African varieties such as Macia, Malisor 84-7, and Tegemeo, and enabled the first stable CRISPR/Cas9 gene knockouts in Tx430 (Anand et al., 2019; Che et al., 2018).

6.8 OPTIMIZATION IN SORGHUM TISSUE CULTURE

In plant transformation and tissue culture each genotype and explant can respond differently to the protocols used. Many factors including genetic variability, hormone requirements, and stress and wounding responses influence the efficiency of transformation and regeneration, including genetic variability, physiological state of the tissue, and environmental conditions during culture.

6.8.1 Selection of Genotype and Explant

Sorghum transformation is highly genotype-dependent, with Tx430 and P898012 remaining the most favorable. It remains unclear if this dependency stems from T-DNA delivery or *in vitro* tissue culture response. One reason is that they are non-tannin genotypes and cause low release of phenolics during tissue culture process (Liang et al., 2023). Studies using immature embryos from various sorghum genotypes, including cultivars and hybrids, have reported varied transformation rates (Mishra & Khurana, 2003; Silva et al., 2020; Silva et al., 2022). Using optimized tissue culture for high regeneration and a selection system for hygromycin resistance could achieve 0.09% transformation efficiency in the sweet sorghum cultivar Ramada (Raghuwanshi & Birch, 2010). Another group attained a transformation frequency of 27% in tannin-rich sorghum Hongyingzi by using optimizing geneticin (G418) concentration and culture medium (Liang et al., 2023). Initially, immature embryos and mature seeds were identified as optimal for *Agrobacterium*-mediated transformation in rice, due to active cell division (Hiei & Komari, 2008). This approach was also applied to other monocots, including sorghum. Freshly harvested immature embryos, particularly those isolated 12–16 days post-pollination, have proven effective for callus induction and are frequently used in sorghum transformation studies (Pola et al., 2008). Although immature embryos are ideal for callus induction, their availability is limited and time-dependent, posing challenges. Alternative explants like shoot apices (Syamala & Devi, 2004; Zhong et al., 1998), and immature inflorescences (Brandão et al., 2012; Chou et al., 2020; Huang, 2005) have been tested, although with generally lower transformation rates.

6.8.2 Optimization of Media Composition and Selection Agents

As discussed earlier, sorghum transformation protocols have been modified to improve transformation and regeneration efficiency. The process of sorghum tissue culture involves the sequential use of callus induction, regeneration, and rooting media, each formulated with distinct combinations and concentrations of hormones and plant growth regulators to optimize specific developmental stages. For instance, plant growth regulators, including 2,4-D, BA, IAA, IBA, and NAA, in combination with additives such as KH_2PO_4, $CuSO_4 \cdot 5H_2O$, L-asparagine, L-proline, and PVP, was found to significantly enhance the efficiency of callus regeneration and overall tissue culture performance in sorghum (Elkonin & Pakhomova, 2000; Liu et al., 2015). Optimized geneticin (G418) concentrations at 250 mg/L yielded a transformation efficiency of up to 26% in Tx430, with around 45% of these being quality events. Incorporation of copper sulfate and 6-benzylaminopurine in the resting and selection media, transformation frequencies over 10% with *Agrobacterium* strain LBA4404 and up to 33% with strain AGL1 (Che et al., 2018). A study developed an *Agrobacterium*-mediated method to produce marker-free transgenic sorghum line P898012 using binary vectors with the *bar* gene as a selectable marker. Among eight tested co-cultivation conditions, transformation efficiencies of 0.4% and 0.7% were achieved, with the inclusion of PVP in the co-cultivation media yielding the highest efficiency (Lu et al., 2009). Research has highlighted that both cold pretreatment (Nguyen et al., 2007) and heat pretreatment (Gurel et al., 2009) can significantly enhance transformation frequency in sorghum. A dual-marker plasmid carrying the *manA* gene for mannose selection and *sgfp* as a reporter was used to transform immature sorghum embryos via *Agrobacterium*-mediated transformation, driven by the *ubi1* promoter. The *pmi* gene from *E. coli* enabled mannose-based selection, allowing the identification of transformed cells with GFP fluorescence in genotypes C401 and Pioneer 8505, achieving transformation efficiencies of 2.88% and 3.30%, respectively (Gao et al., 2005).

6.8.3 Importance of Promoter Selection

A promoter sequence capable of expression in target cells is essential for effective foreign gene introduction and is crucial for optimizing DNA delivery systems (Porto et al., 2014). Numerous promoters have been explored for plant transformation, though their effectiveness varies, leading to different gene expression levels in target tissues (Able et al., 2001; Gao et al., 2005; Tadesse et al., 2003). To identify suitable promoters, the transient expression of the GUS gene under dual 35S cauliflower mosaic virus, rice actin-1, maize alcohol dehydrogenase-1, and maize ubiquitin-1 promoters was evaluated in sorghum but resulted in low β-glucuronidase activity in embryonic tissue (Hill-Ambroz & Troy Weeks, 2001). Other studies have shown that the *ubi1* promoter results in significantly higher GUS foci compared to *actin1* and CaMV 35S (Able et al., 2001; Gao et al., 2005; Tadesse et al., 2003). Further analysis using GUS histochemical staining and enzymatic assays ranked promoter strength as follows: ubi1 > act1D > adh1 > CaMV 35S (Tadesse et al., 2003). CaMV 35S activity was found to be significantly low in T_0 sorghum transgenic plants (Kumar et al.,

2011), making the use of *ubi1* promoter the most effective in sorghum transformation (Grootboom et al., 2010).

6.9 SORGHUM TRANSFORMATION STUDIES

6.9.1 Improved Nutrition and Quality

Sorghum, like other cereals, has low protein quality due to a deficiency in essential amino acids such as lysine and tryptophan. Additionally, sorghum proteins are poorly digestible (Duodu et al., 2003). The main seed proteins in sorghum are prolamins, known as kafirins, which make up over 70% of total seed protein and act as a nitrogen reservoir for seedling development. Kafirins are categorized into four classes – α, β, γ, and δ – based on their molecular weight and sequence (El Nour et al., 1998). Enhanced seed digestibility in sorghum was achieved by knocking out specific gamma-kafirin genes, including *SbK2G* (Li et al., 2024c) and *SbK1C5* (Elkonin et al., 2023), as well as the alpha-kafirin *SbK1C* gene family (Li et al., 2018). In contrast, although *beta-kafirin* was also targeted for knockout using CRISPR/Cas9, this modification did not lead to significant improvements in grain quality (Massel et al., 2023). Additionally, aromatic sorghum was created by knocking out the *Betaine aldehyde dehydrogenase 2 (BADH2)* gene via CRISPR/Cas9 (Zhang et al., 2022).

6.9.2 Biotic and Abiotic Stress Resistance

Overexpression and CRISPR/Cas9 approaches have been instrumental in enhancing sorghum stress resistance traits, as demonstrated in different studies. Lignin, a complex polymer integral to plant cell walls, is known for its protective role against environmental stresses (Bucholtz et al., 1980; Pillonel et al., 1991). However, its role in defending sorghum against sugarcane aphids (Fontanet-Manzaneque, Laibach, et al., 2024) was not well understood. Using *brown midrib (bmr)* sorghum mutants with disrupted monolignol synthesis, Grover et al. (2024) found that *Bmr12* loss increased SCA resistance. Overexpression of the *Bmr12* gene in the Tx430 background resulted in increased susceptibility (Grover et al., 2024). Resistance to stem borer was achieved by overexpressing synthetic *cry1Ac* (Girijashankar et al., 2005). Similarly, in sorghum variety SPV-462, overexpression of the proline biosynthetic pathway gene *P5CSF129A* significantly improved tolerance to salt stress (Reddy et al., 2015).

6.10 RECENT ADVANCEMENTS IN SORGHUM TRANSFORMATION USING *AGROBACTERIUM* (USE OF DEVELOPMENTAL REGULATORS)

A thorough grasp of embryogenesis and regeneration based on cellular totipotency is necessary for successful plant genetic engineering (Vasil & Vasil, 1972). Hormone signaling networks, dynamic changes in gene expression, and epigenetic modifications all play a role in this intricate process, which coordinates cell differentiation and de-differentiation for efficient regeneration (Asghar et al., 2023; Fambrini et al., 2022; Sivanesan et al., 2022). Important regulatory genes that aid in plant transformation include *BABY BOOM* (*BBM*) and *WUSCHEL2* (*Wus2*). In monocots, such as sorghum, combinations such as *Wus2* and *BBM* have significantly improved transformation success, allowed direct somatic embryogenesis, and successfully avoided lengthy callus development stages (Jones et al., 2019; Lowe et al., 2016; Nelson-Vasilchik et al., 2018). The utilization of ternary vector systems and strategic promoter selection, which fine-tunes gene expression, have allowed for further optimization (Fontanet-Manzaneque, Haeghebaert, et al., 2024). Growth regulatory factors (*GRF*) and *GRF* interacting factors *GIF1* have been overexpressed in recent developments, such as the *GRF:GIF* chimera, which has greatly decreased genotype reliance and expanded transformation efficacy in sorghum (Li et al., 2024a). Furthermore, sorghum has successfully undergone a leaf-base transformation mediated by *BBM* and *Wus2* (Wang et al., 2023). These advances underscore the potential of morphogenic or developmental regulators and refined gene expression systems to enable efficient, genotype-independent sorghum transformation.

6.11 INTEGRATION OF MUTANTS WITH GENETIC ENGINEERING

Numerous genetic alterations within a genome are produced by induced mutations, which are more common and brought on by chemical agents, radiation, and other factors (Brock, 1971). Historically, breeding programs have used these mutations to create current cultivars, frequently through mutagenesis operations that started in the middle of the 20_{th} century. By introducing foreign genes, genetic engineering and transgenesis provide a more focused method of mutation that permits the modification of particular traits (Gasser & Fraley, 1989). This method, being faster and more precise than traditional mutagenesis, is instrumental in developing and studying desired traits. The information of transgenesis conferring the desired impacts is based on knowledge of gene function already known using genomics resources or studying any novel gene with an unknown function. By supplying useful allelic variants for deconstructing complicated traits and speeding up breeding efforts, mutant populations enhance natural diversity. The integration of genetic engineering with mutant populations allows researchers to manipulate and study specific genes precisely, complementing traditional breeding and accelerating the discovery of key traits (Figure 6.3).

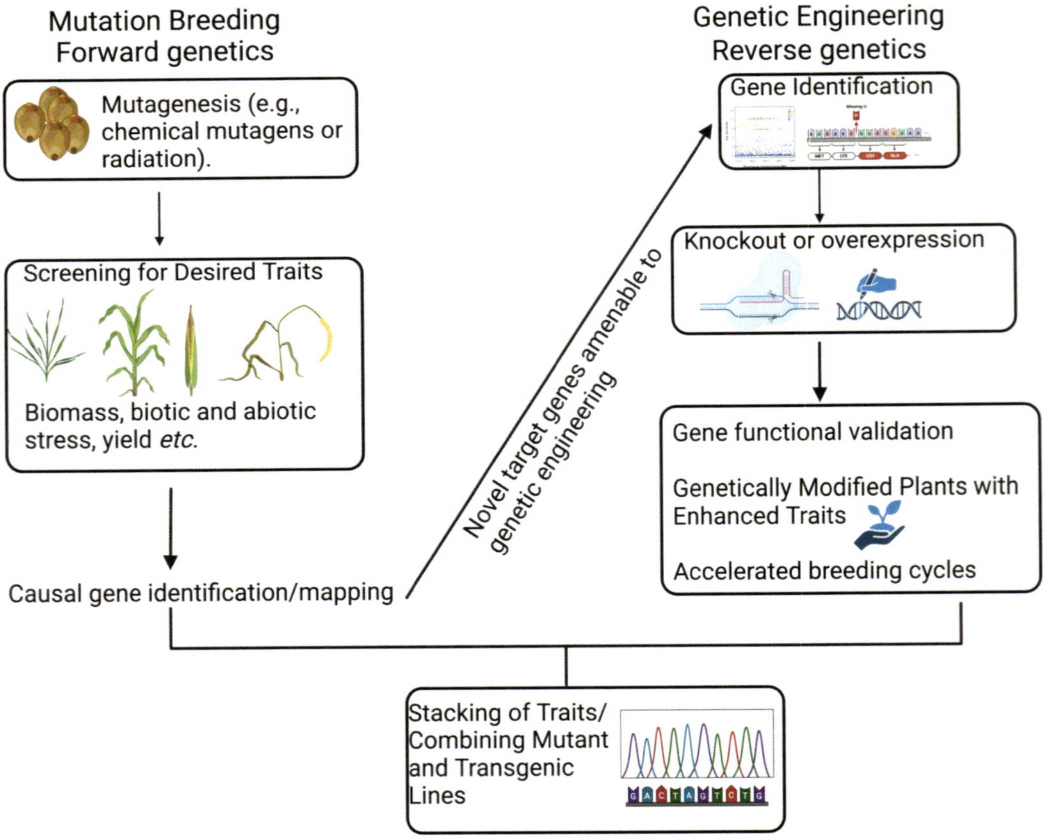

FIGURE 6.3 Integration of forward genetics through mutation breeding and reverse genetics via genetic engineering to accelerate functional genomics and crop improvement.

However, due to the concerns related to GMOs, e.g., transgene flow to wild relatives and herbicide resistance, to the best of our knowledge there is no transgenic sorghum that is commercialized. However, by combining the strengths of natural diversity, induced mutations, and cutting-edge genomic tools and getting marker-free transgenic plants, it becomes possible to address complex agronomic challenges more efficiently.

6.12 CONCLUDING REMARKS AND FUTURE PROSPECTS

Sorghum, a C_4 crop species, is well-adapted to hot, arid climates and marginal soils, making it a vital resource for food and bioenergy, particularly in regions affected by climate change. As the fifth most cultivated crop globally, sorghum's economic and nutritional importance drives the need for advanced genetic improvements to meet global food security demands. The recent surge in genomic resources, such as availability of high-quality reference genome and comprehensive germplasm collections, has unlocked new avenues for understanding and harnessing sorghum's genetic diversity for enhancing yield and quality. However, genetic improvement in sorghum faces hurdles, particularly the difficulty of genotype-dependent transformation, limiting the scope of genetic engineering. Advances in innovative genotype-independent transformation methods using morphogenic genes have begun to break these barriers, enabling efficient and targeted trait improvements. Because of the advent of developmental regulators in sorghum transformation and its recent highly efficient outcome, there will be more interest in studying sorghum gene function analysis using genetic engineering.

ACKNOWLEDGMENT

GP and YJ are grateful for the State of Texas Governor's University Research (GURI) funding. PD would like to thank Texas Tech University Graduate School for their Distinguished Graduate Student Assistantship.

REFERENCES

Abdel-Ghany, S. E., Ullah, F., Ben-Hur, A., & Reddy, A. S. (2020). Transcriptome analysis of drought-resistant and drought-sensitive sorghum (*Sorghum bicolor*) genotypes in response to PEG-induced drought stress. *International Journal of Molecular Sciences*, *21*(3), 772.

Able, J. A., Rathus, C., & Godwin, I. D. (2001). The investigation of optimal bombardment parameters for transient and stable transgene expression in sorghum. *In Vitro Cellular & Developmental Biology-Plant*, *37*, 341–348.

Addo-Quaye, C., Buescher, E., Best, N., Chaikam, V., Baxter, I., & Dilkes, B. P. (2017). Forward genetics by sequencing EMS variation-induced inbred lines. *G3: Genes, Genomes, Genetics*, *7*(2), 413–425.

Addo-Quaye, C., Tuinstra, M., Carraro, N., Weil, C., & Dilkes, B. P. (2018). Whole-genome sequence accuracy is improved by replication in a population of mutagenized sorghum. *G3: Genes, Genomes, Genetics*, *8*(3), 1079–1094.

Adeyanju, A., Sattler, S., Rich, P., Rivera-Burgos, L., Xu, X., & Ejeta, G. (2021). Sorghum brown midrib19 (*Bmr19*) gene links lignin biosynthesis to folate metabolism. *Genes*, *12*, 660.

Anand, A., Che, P., Wu, E., & Jones, T. J. (2019). Novel ternary vectors for efficient sorghum transformation. *Sorghum: Methods and Protocols*, *1931*, 185–196.

Aragão, F. J., & Faria, J. C. (2009). First transgenic geminivirus-resistant plant in the field. *Nature Biotechnology*, *27*(12), 1086–1088.

Aregawi, K., Shen, J., Pierroz, G., Sharma, M. K., Dahlberg, J., Owiti, J., & Lemaux, P. G. (2022). Morphogene-assisted transformation of *Sorghum bicolor* allows more efficient genome editing. *Plant Biotechnology Journal*, *20*(4), 748–760.

Asghar, S., Ghori, N., Hyat, F., Li, Y., & Chen, C. (2023). Use of auxin and cytokinin for somatic embryogenesis in plant: A story from competence towards completion. *Plant Growth Regulation*, *99*(3), 413–428.

Assem, S. K., Basry, M. A., Taha, T. A., Abd El-Aziz, M., Alwa, T., & Fouad, W. M. (2023). Development of an in vitro regeneration system from immature inflorescences and CRISPR/Cas9-mediated gene editing in sudangrass. *Journal of Genetic Engineering and Biotechnology*, *21*(1), 58.

Auerbach, C. (1945). The induction by mustard gas of chromosomal instabilities in Drosophila melanogaster. *Proceedings of the Royal Society of Edinburgh, Section B: Biological Sciences*, *62*(3), 307–320.

Azhakanandam, K., & Zhang, Z. J. (2015). Sorghum transformation: Achievements, challenges, and perspectives. In Azhakanandam, K., Silverstone, A., Daniell, H., Davey, M. (eds.), *Recent advancements in gene expression and enabling technologies in crop plants* (pp. 291–312). New York, NY: Springer.

Bado, S., Forster, B. P., Nielen, S., Ali, A. M., Lagoda, P. J., Till, B. J., & Laimer, M. (2015). Plant mutation breeding: Current progress and future assessment. *Plant Breeding Reviews*, *39*, 23–88.

Belide, S., Vanhercke, T., Petrie, J. R., & Singh, S. P. (2017). Robust genetic transformation of sorghum (*Sorghum bicolor* L.) using differentiating embryogenic callus induced from immature embryos. *Plant Methods*, *13*, 1–12.

Bellis, E. S., Kelly, E. A., Lorts, C. M., Gao, H., DeLeo, V. L., Rouhan, G., . . . Muscarella, R. (2020). Genomics of sorghum local adaptation to a parasitic plant. *Proceedings of the National Academy of Sciences*, *117*(8), 4243–4251.

Bergner, A. D., Avery, A. G., & Blakeslee, A. F. (1940). Chromosomal deficiencies in *Datura stramonium* induced by colchicine treatment. *American Journal of Botany*, 676–683.

Bevan, M. (1984). Binary *Agrobacterium* vectors for plant transformation. *Nucleic Acids Research*, *12*(22), 8711–8721.

Blakeslee, A. F. (1939). The present and potential service of chemistry to plant breeding. *American Journal of Botany*, *26*(3), 163–172.

Blakeslee, A. F. (1941). Effect of induced polyploidy in plants. *The American Naturalist*, *75*(757), 117–135.

Bout, S., & Vermerris, W. (2003). A candidate-gene approach to clone the sorghum Brown midrib gene encoding caffeic acid O-methyltransferase. *Molecular Genetics and Genomics*, *269*, 205–214.

Boyles, R. E., Brenton, Z. W., & Kresovich, S. (2019). Genetic and genomic resources of sorghum to connect genotype with phenotype in contrasting environments. *The Plant Journal*, *97*(1), 19–39.

Brandão, R. L., Carneiro, N. P., de Oliveira, A. C., Coelho, G. T., & Carneiro, A. A. (2012). Genetic transformation of immature sorghum inflorescence via microprojectile bombardment. In Çiftçi, Y. Ö. (ed.), *Transgenic Plants–Advances and Limitations* (pp.133–148). *InTech, Croatia*.

Brant, E. J., Baloglu, M. C., Parikh, A., & Altpeter, F. (2021). CRISPR/Cas9 mediated targeted mutagenesis of *LIGULELESS-1* in sorghum provides a rapidly scorable phenotype by altering leaf inclination angle. *Biotechnology Journal*, *16*(11), Art. 2100237.

Brant, E. J., Eid, A., Kannan, B., Baloglu, M. C., & Altpeter, F. (2024). The extent of multiallelic, co-editing of *LIGULELESS1* in highly polyploid sugarcane tunes leaf inclination angle and enables selection of the ideotype for biomass yield. *Plant Biotechnology Journal, 22*(10), 2660–2671.

Brock, R. (1971). The role of induced mutations in plant improvement. *Radiation Botany*, *11*(3), 181–196.

Bucholtz, D. L., Cantrell, R. P., Axtell, J. D., & Lechtenberg, V. L. (1980). Lignin biochemistry of normal and brown midrib mutant sorghum. *Journal of Agricultural and Food Chemistry*, *28*(6), 1239–1241.

Burow, G., Xin, Z., Hayes, C., & Burke, J. (2014). Characterization of a multiseeded (msd1) mutant of sorghum for increasing grain yield. *Crop Science*, *54*(5), 2030–2037.

Carvalho, C. H. S., Zehr, U. B., Gunaratna, N., Anderson, J., Kononowicz, H. H., Hodges, T. K., & Axtell, J. D. (2004). *Agrobacterium*-mediated transformation of sorghum: Factors that affect transformation efficiency. *Genetics and Molecular Biology*, *27*, 259–269.

Casas, A. M., Kononowicz, A. K., Zehr, U. B., Tomes, D. T., Axtell, J. D., Butler, L. G., . . . Hasegawa, P. M. (1993). Transgenic sorghum plants via microprojectile bombardment. *Proceedings of the National Academy of Sciences*, *90*(23), 11212–11216.

Çelik, Ö., & Atak, Ç. (2017). Applications of ionizing radiation in mutation breeding. *New Insights on Gamma Rays*, *6*, 111–132.

Char, S. N., Wei, J., Mu, Q., Li, X., Zhang, Z. J., Yu, J., & Yang, B. (2020). An *Agrobacterium*-delivered CRISPR/Cas9 system for targeted mutagenesis in sorghum. *Plant Biotechnology Journal*, *18*(2), 319.

Che, P., Anand, A., Wu, E., Sander, J. D., Simon, M. K., Zhu, W., . . . Liu, D. (2018). Developing a flexible, high-efficiency *Agrobacterium*-mediated sorghum transformation system with broad application. *Plant Biotechnology Journal*, *16*(7), 1388–1395.

Che, P., Wu, E., Simon, M. K., Anand, A., Lowe, K., Gao, H., . . . Gordon-Kamm, W. (2022). Wuschel2 enables highly efficient CRISPR/Cas-targeted genome editing during rapid de novo shoot regeneration in sorghum. *Communications Biology*, *5*(1), 344.

Chen, J., Jiao, Y., Laza, H., Payton, P., Ware, D., & Xin, Z. (2019). Identification of the first nuclear male sterility gene (Male-sterile 9) in sorghum. *The Plant Genome*, *12*(3), Art. 190020.

Chen, L., Duan, L., Sun, M., Yang, Z., Li, H., Hu, K., Yang, H., & Liu, L. (2023). Current trends and insights on EMS mutagenesis application to studies on plant abiotic stress tolerance and development. *Frontiers in Plant Science*, *13*, Art. 1052569.

Chen, J., Xin, Z., & Laza, H. (2019). Registration of BTx623dw5, a new sorghum dwarf mutant. *Journal of Plant Registrations*, 13(2), 254–257.

Chen, J., Zhang, L., Zhu, M., Han, L., Lv, Y., Liu, Y., . . . Cai, H. (2018). Non-dormant Axillary Bud 1 regulates axillary bud outgrowth in sorghum. *Journal of Integrative Plant Biology*, *60*(10), 938–955.

Chen, J., Zhu, M., Liu, R., Zhang, M., Lv, Y., Liu, Y., . . . Cai, H. (2020). BIOMASS YIELD 1 regulates sorghum biomass and grain yield via the shikimate pathway. *Journal of Experimental Botany*, *71*(18), 5506–5520.

Chen, J., Zou, G., & Xin, Z. (2019). Development of a pedigreed sorghum mutant library. In Zhao, Z.Y. & Dahlberg, J. (eds.), *Sorghum: Methods and protocols* (vol. 1931, pp. 61–73). New York: Humana Press.

Chen, L., Duan, L., Sun, M., Yang, Z., Li, H., Hu, K., Yang, H., & Liu, L. (2023). Current trends and insights on EMS mutagenesis application to studies on plant abiotic stress tolerance and development. *Frontiers in Plant Science*, *13*, Art. 1052569.

Cheng, M., Lowe, B. A., Spencer, T. M., Ye, X., & Armstrong, C. L. (2004). Factors influencing *Agrobacterium*-mediated transformation of monocotyledonous species. *In Vitro Cellular & Developmental Biology-Plant*, *40*, 31–45.

Chopra, V. (2005). Mutagenesis: Investigating the process and processing the outcome for crop improvement. *Current Science*, *89*, 353–359.

Chou, J., Huang, J., & Huang, Y. (2020). Simple and efficient genetic transformation of sorghum using immature inflorescences. *Acta Physiologiae Plantarum*, *42*(3), 41.

Clark, E. A., & Lehman, H. (2001). Assessment of GM crops in commercial agriculture. *Journal of Agricultural and Environmental Ethics*, *14*, 3–28.

De Riseis, S., Chen, J., Xin, Z., & Harmon, F. G. (2023). *Sorghum bicolor INDETERMINATE1* is a conserved primary regulator of flowering. *Frontiers in Plant Science*, *14*, Art. 1304822.

Della Coletta, R., Qiu, Y., Ou, S., Hufford, M. B., & Hirsch, C. N. (2021). How the pan-genome is changing crop genomics and improvement. *Genome Biology*, *22*, 1–19.

Devi, P. B., & Sticklen, M. B. (2003). In vitro culture and genetic transformation of sorghum by microprojectile bombardment. *Plant Biosystems—An International Journal Dealing with all Aspects of Plant Biology*, *137*(3), 249–254.

Ding, Y.-Q., Cao, N., Zhou, L.-B., Cheng, B., Gao, X., Wang, C., . . . Zhang, L.-Y. (2020). Construction of EMS-induced mutant library and mutant screening in liquor-making waxy sorghum. *Journal of Southern Agriculture*, *51*(12), 2884–2891.

Do, P. T., Lee, H., Mookkan, M., Folk, W. R., & Zhang, Z. J. (2016). Rapid and efficient Agrobacterium-mediated transformation of sorghum (*Sorghum bicolo*r) employing standard binary vectors and bar gene as a selectable marker. *Plant Cell Reports*, *35*, 2065–2076.

Doudna, J. A., & Charpentier, E. (2014). The new frontier of genome engineering with CRISPR-Cas9. *Science*, *346*(6213), Art. 1258096.

Duodu, K., Taylor, J., Belton, P., & Hamaker, B. (2003). Factors affecting sorghum protein digestibility. *Journal of Cereal Science*, *38*(2), 117–131.

Ehrenberg, L., Gustafsson, Å., & Lundqvist, U. (1961). Viable mutants induced in barley by ionizing radiations and chemical mutagens. *Hereditas*, *47*(2), 243–282.

El Nour, I., Peruffo, A., & Curioni, A. (1998). Characterisation of sorghum kafirins in relation to their cross-linking behaviour. *Journal of Cereal Science*, *28*(2), 197–207.

Elkonin, L., Italianskaya, J., Domanina, I., Selivanov, N. Y., Rakitin, A., & Ravin, N. (2016). Transgenic sorghum with improved digestibility of storage proteins obtained by *Agrobacterium*-mediated transformation. *Russian Journal of Plant Physiology*, *63*, 678–689.

Elkonin, L. A., Gerashchenkov, G. A., Borisenko, N. V., Kenzhegulov, O. A., Sarsenova, S. K., Rozhnova, N. A., & Panin, V. M. (2023). Development of sorghum mutants with improved in vitro protein digestibility by CRISPR/Cas9 editing of kafirin genes. *The Crop Journal*, *11*(5), 1411–1418.

Elkonin, L. A., & Pakhomova, N. V. (2000). Influence of nitrogen and phosphorus on induction embryogenic callus of sorghum. *Plant Cell, Tissue and Organ Culture*, *61*, 115–123.

Fambrini, M., Usai, G., & Pugliesi, C. (2022). Induction of somatic embryogenesis in plants: Different players and focus on WUSCHEL and WUS-RELATED HOMEOBOX (WOX) transcription factors. *International Journal of Molecular Sciences*, *23*(24), Art. 15950.

Fontanet-Manzaneque, J. B., Haeghebaert, J., Aesaert, S., Coussens, G., Pauwels, L., & Caño-Delgado, A. I. (2024). Efficient sorghum and maize transformation using a ternary vector system combined with morphogenic regulators. *The Plant Journal, 120*(5), 2076–2088.

Fontanet-Manzaneque, J. B., Laibach, N., Herrero-García, I., Coleto-Alcudia, V., Blasco-Escámez, D., Zhang, C., . . . Bjarnholt, N. (2024). Untargeted mutagenesis of brassinosteroid receptor SbBRI1 confers drought tolerance by altering phenylpropanoid metabolism in *Sorghum bicolor. Plant Biotechnology Journal*, *22*(12), 3406–3423.

Fraley, R. T., Rogers, S. G., Horsch, R. B., Sanders, P. R., Flick, J. S., Adams, S. P., . . . Fry, J. S. (1983). Expression of bacterial genes in plant cells. *Proceedings of the National Academy of Sciences*, *80*(15), 4803–4807.

Gao, Z., Xie, X., Ling, Y., Muthukrishnan, S., & Liang, G. H. (2005). Agrobacterium tumefaciens-mediated sorghum transformation using a mannose selection system. *Plant Biotechnology Journal*, *3*(6), 591–599.

Garcia-Hernandez, M., Berardini, T., Chen, G., Crist, D., Doyle, A., Huala, E., . . . Mueller, L. A. (2002). TAIR: A resource for integrated Arabidopsis data. *Functional & Integrative Genomics*, *2*, 239–253.

Gasser, C. S., & Fraley, R. T. (1989). Genetically engineering plants for crop improvement. *Science*, *244*(4910), 1293–1299.

Girijashankar, V., Sharma, H. C., Sharma, K. K., Swathisree, V., Prasad, L. S., Bhat, B., . . . Altosaar, I. (2005). Development of transgenic sorghum for insect resistance against the spotted stem borer (*Chilo partellus*). *Plant Cell Reports*, *24*, 513–522.

Gladman, N., Olson, A., Wei, S., Chougule, K., Lu, Z., Tello-Ruiz, M., . . . Wang, B. (2022). SorghumBase: A web-based portal for sorghum genetic information and community advancement. *Planta*, *255*(2), 35.

Graham, N., Patil, G. B., Bubeck, D. M., Dobert, R. C., Glenn, K. C., Gutsche, A. T., . . . May, G. D. (2020). Plant genome editing and the relevance of off-target changes. *Plant Physiology*, *183*(4), 1453–1471.

Grootboom, A. W., Mkhonza, N., O'Kennedy, M. M., Chakauya, E., Kunert, K. J., & Chikwamba, R. K. (2010). Biolistic mediated sorghum (*Sorghum bicolor* L. Moench) transformation via mannose and bialaphos based selection systems. *International Journal of Botany*, *6*(2), 89–94.

Grover, S., Mou, D. F., Shrestha, K., Puri, H., Pingault, L., Sattler, S. E., & Louis, J. (2024). Impaired Brown midrib12 function orchestrates sorghum resistance to aphids via an auxin conjugate indole-3-acetic acid–aspartic acid. *New Phytologist, 244*(4), 1597–1615.

Gurel, S., Gurel, E., Kaur, R., Wong, J., Meng, L., Tan, H.-Q., & Lemaux, P. G. (2009). Efficient, reproducible Agrobacterium-mediated transformation of sorghum using heat treatment of immature embryos. *Plant Cell Reports*, *28*, 429–444.

Hao, H., Li, Z., Leng, C., Lu, C., Luo, H., Liu, Y., . . . Jing, H.-C. (2021). Sorghum breeding in the genomic era: Opportunities and challenges. *Theoretical and Applied Genetics*, *134*, 1899–1924.

Hao, J., Yang, Y., Futrell, S., Kelly, E. A., Lorts, C. M., Nebie, B., . . . Lasky, J. R. (2023). CRISPR/Cas9-mediated mutagenesis of carotenoid cleavage dioxygenase (CCD) genes in sorghum alters strigolactone biosynthesis and plant biotic Interactions. *Phytobiomes Journal*, *7*(3), 339–351.

Herrera-Estrella, L., Depicker, A., Van Montagu, M., & Schell, J. (1983). Expression of chimaeric genes transferred into plant cells using a Ti-plasmid-derived vector. *Nature*, *303*(5914), 209–213.

Hiei, Y., & Komari, T. (2008). Agrobacterium-mediated transformation of rice using immature embryos or calli induced from mature seed. *Nature Protocols*, *3*(5), 824–834.

Hill-Ambroz, K. L., & Troy Weeks, J. (2001). Comparison of constitutive promoters for sorghum [*Sorghum bicolor* (L.) Moench] transformation. *Cereal Research Communications*, *29*(1), 17–24.

Hong, Z., Ueguchi-Tanaka, M., Umemura, K., Uozu, S., Fujioka, S., Takatsuto, S., . . . Matsuoka, M. (2003). A rice brassinosteroid-deficient mutant, ebisu dwarf (d2), is caused by a loss of function of a new member of cytochrome P450. *The Plant Cell*, *15*(12), 2900–2910.

Huang, Y. (2005). Development of a new genetic transformation system for sorghum using *Agrobacterium* and immature inflorescences. Sorghum Improvement Conference of North America, Lubbock, USA; International Crops Research Institute for the Semi-Arid Tropics (ICRISAT).

Jander, G., Norris, S. R., Rounsley, S. D., Bush, D. F., Levin, I. M., & Last, R. L. (2002). Arabidopsis map-based cloning in the post-genome era. *Plant Physiology*, *129*(2), 440–450.

Jiang, W., Zhou, H., Bi, H., Fromm, M., Yang, B., & Weeks, D. P. (2013). Demonstration of CRISPR/Cas9/sgRNA-mediated targeted gene modification in *Arabidopsis*, tobacco, sorghum and rice. *Nucleic Acids Research*, *41*(20), e188–e188.

Jiao, Y., Burke, J., Chopra, R., Burow, G., Chen, J., Wang, B., . . . Xin, Z. (2016). A sorghum mutant resource as an efficient platform for gene discovery in grasses. *The Plant Cell*, *28*(7), 1551–1562.

Jiao, Y., Burow, G., Gladman, N., Acosta-Martinez, V., Chen, J., Burke, J., . . . Xin, Z. (2018a). Efficient identification of causal mutations through sequencing of bulked F 2 from two allelic bloomless mutants of *Sorghum bicolor*. *Frontiers in Plant Science*, *8*, 2267.

Jiao, Y., Lee, Y. K., Gladman, N., Chopra, R., Christensen, S. A., Regulski, M., . . . Ware, D. (2018b). MSD1 regulates pedicellate spikelet fertility in sorghum through the jasmonic acid pathway. *Nature Communications*, *9*(1), 822.

Jiao, Y., Nigam, D., Barry, K., Daum, C., Yoshinaga, Y., Lipzen, A., . . . Lu, Z. (2024). A large sequenced mutant library–valuable reverse genetic resource that covers 98% of sorghum genes. *The Plant Journal*, *117*(5), 1543–1557.

Jinek, M., Chylinski, K., Fonfara, I., Hauer, M., Doudna, J. A., & Charpentier, E. (2012). A programmable dual-RNA–guided DNA endonuclease in adaptive bacterial immunity. *Science*, *337*(6096), 816–821.

Jones, T., Lowe, K., Hoerster, G., Anand, A., Wu, E., Wang, N., . . . Gordon-Kamm, W. (2019). Maize transformation using the morphogenic genes *Baby Boom* and *Wuschel2*. In Kumar, S., Barone, P., & Smith, M. (eds.), *Transgenic Plants: Methods and Protocols* (vol. 1864, pp. 81–93). Singapore: Springer.

Kadam, S., Lee, D., & Dhiman, P. (2023). Genome-editing approaches for biofortification of cereal crops. In *Biofortification in cereals: Progress and prospects* (pp. 93–126).

Kimura, M., & Kimura, M. (2020). Mutation as an evolutionary factor. In Kimura, M. (ed.), *My thoughts on biological evolution* (pp. 49–63). Singapore: Springer.

Klein, T. M., Arentzen, R., Lewis, P. A., & Fitzpatrick-McElligott, S. (1992). Transformation of microbes, plants and animals by particle bombardment. *Biotechnology*, *10*(3), 286–291.

Kodym, A., & Afza, R. (2003). Physical and chemical mutagenesis. In Grotewold, E. (ed.), *Plant functional genomics* (vol. 236, pp. 189–203). Totowa, NJ: Humana Press.

Kohli, A., Twyman, R. M., Abranches, R., Wegel, E., Stoger, E., & Christou, P. (2003). Transgene integration, organization and interaction in plants. *Plant Molecular Biology*, *52*, 247–258.

Kovalchuk, I. (2021). Genome stability: An evolutionary perspective. In Kovalchuk. I. & Kovalchuk, O. (eds.), *Genome Stability* (pp. 1–19). Cambridge, MA: Academic Press..

Krothapalli, K., Buescher, E. M., Li, X., Brown, E., Chapple, C., Dilkes, B. P., & Tuinstra, M. R. (2013). Forward genetics by genome sequencing reveals that rapid cyanide release deters insect herbivory of *Sorghum bicolor*. *Genetics*, *195*(2), 309–318.

Kumar, V., Campbell, L. M., & Rathore, K. S. (2011). Rapid recovery-and characterization of transformants following Agrobacterium-mediated T-DNA transfer to sorghum. *Plant Cell, Tissue and Organ Culture (PCTOC)*, *104*, 137–146.

Kumawat, S., Raturi, G., Dhiman, P., Sudhakarn, S., Rajora, N., Thakral, V., . . . Rachappanavar, V. (2022). Opportunity and challenges for whole-genome resequencing-based genotyping in plants. In Sonah, H., Goyal, V., Shivaraj, S.M. & Deshmukh, R. K. (eds.), *Genotyping by sequencing for crop improvement* (pp. 38–51). Hoboken, NJ: Wiley.

Lee, J. S., Bae, S.-J., Kim, J.-S., Kim, C., & Kang, B.-C. (2023). A streamlined guide RNA screening system for genome editing in *Sorghum bicolor*. *Plant Methods*, *19*(1), 90.

Li, A., Jia, S., Yobi, A., Ge, Z., Sato, S. J., Zhang, C., Angelovici, R., . . . Holding, D. R. (2018). Editing of an alpha-kafirin gene family increases, digestibility and protein quality in sorghum. *Plant Physiology*, *177*(4), 1425–1438.

Li, D., Zhu, Z., Qu, K., Li, J., Ma, D., & Lu, X. (2025). A coiled-coil domain mutation in the NLR receptor SbYR1 coordinates plant growth and stress tolerance in sorghum. *Plant Science*, *350*, Art. 112246.

Li, J., Pan, W., Zhang, S., Ma, G., Li, A., Zhang, H., & Liu, L. (2024a). A rapid and highly efficient sorghum transformation strategy using GRF4-GIF1/ternary vector system. *The Plant Journal*, *117*(5), 1604–1613.

Li, M., Cai, Q., Liang, Y., Zhao, Y., Hao, Y., Qin, Y., . . . Li, H. (2024b). Mapping and screening of candidate gene regulating the biomass yield of sorghum (*Sorghum bicolor* L.). *International Journal of Molecular Sciences*, *25*(2), 796.

Li, P., Liu, Y., Tan, W., Chen, J., Zhu, M., Lv, Y., . . . Cai, H. (2019). *Brittle Culm 1* encodes a COBRA-like protein involved in secondary cell wall cellulose biosynthesis in sorghum. *Plant and Cell Physiology*, *60*(4), 788–801.

Li, X., Liu, W., Wang, G., Sun, S. S. M., Yuan, L., & Wang, J. (2024c). Improving digestibility of sorghum proteins by CRISPR/Cas9-based genome editing. *Food and Energy Security*, *13*(1), Art. e506.

Liang, Y., Yan, X., Xu, J., Liu, Y., Xie, K., Li, J., & Zhan, Q. (2023). An efficient transformation method for tannin-containing sorghum. *PeerJ*, *11*, Art. e15066.

Liu, G., Gilding, E. K., & Godwin, I. D. (2014). Genetic transformation and breeding. In Wang, Y.H., Upadhyaya, H.D. & Kole, C. (eds.), *Genetics, genomics and breeding of sorghum* (p. 251). Boca Raton, FL: CRC Press.

Liu, G., Gilding, E. K., & Godwin, I. D. (2015). A robust tissue culture system for sorghum [*Sorghum bicolor* (L.) Moench]. *South African Journal of Botany*, *98*, 157–160.

Liu, G., & Godwin, I. D. (2012). Highly efficient sorghum transformation. *Plant Cell Reports*, *31*, 999–1007.

Liu, G., Han, Y., Jiang, Y., Wang, Y., Lv, P., & Li, H. (2016). Genomics approaches to biotic stress resistance. In Rakshit, S. & Wang, Y.H. (eds.), *The sorghum genome* (pp. 149–167). Cham: Springer.

Liu, G., Li, J., & Godwin, I. D. (2019). Genome editing by CRISPR/Cas9 in sorghum through biolistic bombardment. In Zhao, Z.Y. & Dahlberg, J. (eds.), *Sorghum: Methods and Protocols*, (vol. 1931, pp. 169–183). New York, NY: Humana Press.

Liu, G., Massel, K., Tabet, B., & Godwin, I. D. (2020). Biolistic DNA delivery and its applications in *Sorghum bicolor*. In Rustgi, S. & Luo, H. (eds.), *Biolistic DNA delivery in plants: Methods and protocols* (pp. 197–215). New York, NY: Humana Press.

Lowe, K., Wu, E., Wang, N., Hoerster, G., Hastings, C., Cho, M.-J., . . . Cushatt, J. (2016). Morphogenic regulators Baby boom and Wuschel improve monocot transformation. *The Plant Cell*, *28*(9), 1998–2015.

Lu, L., Wu, X., Yin, X., Morrand, J., Chen, X., Folk, W. R., & Zhang, Z. J. (2009). Development of marker-free transgenic sorghum [*Sorghum bicolor* (L.) Moench] using standard binary vectors with bar as a selectable marker. *Plant Cell, Tissue and Organ Culture (PCTOC)*, *99*, 97–108.

Massel, K., Hintzsche, J., Restall, J., Kerr, E. D., Schulz, B. L., & Godwin, I. D. (2023). CRISPR-knockout of β-kafirin in sorghum does not recapitulate the grain quality of natural mutants. *Planta*, *257*(1), 8.

Massel, K., Lam, Y., Hintzsche, J., Lester, N., Botella, J. R., & Godwin, I. D. (2022). Endogenous U6 promoters improve CRISPR/Cas9 editing efficiencies in *Sorghum bicolor* and show potential for applications in other cereals. *Plant Cell Reports*, *41*(2), 489–492.

Mba, C., Afza, R., Bado, S., & Jain, S. M. (2010). Induced mutagenesis in plants using physical and chemical agents. *Plant Cell Culture: Essential Methods*, *20*, 111–130.

Mendu, L., Jalathge, G., Dhillon, K. K., Singh, N. P., Balasubramanian, V. K., Fewou, R., . . . Mendu, V. (2022). Mutation in the Endo-β-1, 4-glucanase (*KORRIGAN*) is responsible for thick leaf phenotype in sorghum. *Plants*, *11*(24), 3531.

Meyer, P. (1995). Understanding and controlling transgene expression. *Trends in Biotechnology*, *13*(9), 332–337.

Miller, S., Rønager, A., Holm, R., Fontanet-Manzaneque, J. B., Caño-Delgado, A. I., & Bjarnholt, N. (2023). New methods for sorghum transformation in temperate climates. *AoB Plants*, *15*(3), Art. plad030.

Mishra, A., & Khurana, P. (2003). Genotype dependent somatic embryogenesis and regeneration from leaf base cultures of *Sorghum bicolor*. *Journal of Plant Biochemistry and Biotechnology*, *12*, 53–56.

Mofokeng, A. M., Shimelis, H., & Laing, M. (2017). Breeding strategies to improve sorghum quality. *Australian Journal of Crop Science*, *11*(2), 142–148.

Muller, H. J. (1930). Radiation and genetics. *The American Naturalist*, *64*(692), 220–251.

Nafisi, S., Saboury, A. A., Keramat, N., Neault, J.-F., & Tajmir-Riahi, H.-A. (2007). Stability and structural features of DNA intercalation with ethidium bromide, acridine orange and methylene blue. *Journal of Molecular Structure*, *827*(1–3), 35–43.

Nelson-Vasilchik, K., Hague, J., Mookkan, M., Zhang, Z. J., & Kausch, A. (2018). Transformation of recalcitrant sorghum varieties facilitated by baby boom and Wuschel2. *Current Protocols in Plant Biology*, *3*(4), Art. e20076.

Nguyen, T.-V., Thanh Thu, T., Claeys, M., & Angenon, G. (2007). Agrobacterium-mediated transformation of sorghum (*Sorghum bicolor* (L.) Moench) using an improved in vitro regeneration system. *Plant Cell, Tissue and Organ Culture*, *91*, 155–164.

Oladosu, Y., Rafii, M. Y., Abdullah, N., Hussin, G., Ramli, A., Rahim, H. A., . . . Usman, M. (2016). Principle and application of plant mutagenesis in crop improvement: A review. *Biotechnology & Biotechnological Equipment*, *30*(1), 1–16.

Ondzighi-Assoume, C. A., Willis, J. D., Ouma, W. K., Allen, S. M., King, Z., Parrott, W. A. L., . . . Stewart, C. N. (2019). Embryogenic cell suspensions for high-capacity genetic transformation and regeneration of switchgrass (*Panicum virgatum* L.). *Biotechnology for Biofuels*, *12*, 1–14.

Ordonio, R. L. (2014). *Analysis of gamma ray-induced sweet sorghum mutants*. 名古屋大学. NAGOYA University, Nagoya, Japan.

Ozyigit, I. I., & Yucebilgili Kurtoglu, K. (2020). Particle bombardment technology and its applications in plants. *Molecular Biology Reports*, *47*(12), 9831–9847.

Pandey, A. K., Bhat, B. V., Balakrishna, D., & Seetharama, N. (2010). Genetic transformation of sorghum (*Sorghum bicolor* (L.) Moench.). *Int J Biotechnol Biochem*, *6*(1), 45–53.

Paterson, A. H., Bowers, J. E., Bruggmann, R., Dubchak, I., Grimwood, J., Gundlach, H., . . . Poliakov, A. (2009). The *Sorghum bicolor* genome and the diversification of grasses. *Nature*, *457*(7229), 551–556.

Pillonel, C., Mulder, M. M., Boon, J. J., Forster, B., & Binder, A. (1991). Involvement of cinnamyl-alcohol dehydrogenase in the control of lignin formation in *Sorghum bicolor* L. Moench. *Planta*, *185*, 538–544.

Pola, S., Mani, N. S., & Ramana, T. (2008). Plant tissue culture studies in *Sorghum bicolor:* Immature embryo explants as the source material. *International Journal of Plant Production*, *2*(1), 1–14.

Porter, K., Axtell, J., Lechtenberg, V., & Colenbrander, V. (1978). Phenotype, fiber composition, and in vitro dry matter disappearance of chemically induced brown midrib (bmr) mutants of sorghum 1. *Crop Science*, *18*(2), 205–208.

Porto, M. S., Pinheiro, M. P. N., Batista, V. G. L., dos Santos, R. C., de Albuquerque Melo Filho, P., & de Lima, L. M. (2014). Plant promoters: An approach of structure and function. *Molecular Biotechnology*, *56*, 38–49.

Potrykus, I. (2001). Golden rice and beyond. *Plant Physiology*, *125*(3), 1157–1161.

Prajapati, R., & Tyagi, K. (2024). CRISPR/Cas-mediated multiplex genome editing in plants and applications. In Upadhyay, S.K. (ed.), *Applications of genome engineering in plants* (pp. 20–39). Hoboken, NJ: Wiley.

Quinby, J., & Karper, R. (1953). Inheritance of height in sorghum. *Agronomy Journal*, *46*(5), 211–216.

Raghuwanshi, A., & Birch, R. G. (2010). Genetic transformation of sweet sorghum. *Plant Cell Reports*, *29*, 997–1005.

Rajasekaran, K., Hudspeth, R., Cary, J., Anderson, D., & Cleveland, T. (2000). High-frequency stable transformation of cotton (*Gossypium hirsutum* L.) by particle bombardment of embryogenic cell suspension cultures. *Plant Cell Reports*, *19*, 539–545.

Ray, S., & Satya, P. (2014). Ray, S., & Satya, P. (2014). Next generation sequencing technologies for next generation plant breeding. *Frontiers in Plant Science*, *5*, 367.

Reddy, P. S., Jogeswar, G., Rasineni, G. K., Maheswari, M., Reddy, A. R., Varshney, R. K., & Kishor, P. K. (2015). Proline over-accumulation alleviates salt stress and protects photosynthetic and antioxidant enzyme activities in transgenic sorghum [*Sorghum bicolor* (L.) Moench]. *Plant Physiology and Biochemistry*, *94*, 104–113.

Rizal, G., Thakur, V., Dionora, J., Karki, S., Wanchana, S., Acebron, K., . . . Montecillo, F. (2015). Two forward genetic screens for vein density mutants in sorghum converge on a *cytochrome P450* gene in the brassinosteroid pathway. *The Plant Journal*, *84*(2), 257–266.

Rooney, W. L. (2004). Sorghum improvement-integrating traditional and new technology to produce improved genotypes. *Advances in Agronomy*, *83*(10.1016), S0065–2113.

Roychowdhury, R., & Tah, J. (2013). Mutagenesis—A potential approach for crop improvement. In Hakeem, K., Ahmad, P. & Ozturk, M. (eds.), *Crop improvement: New approaches and modern techniques* (pp. 149–187). Boston, MA: Springer.

Saballos, A., Sattler, S. E., Sanchez, E., Foster, T. P., Xin, Z., Kang, C., . . . Vermerris, W. (2012). Brown midrib2 (Bmr2) encodes the major 4-coumarate: Coenzyme A ligase involved in lignin biosynthesis in sorghum (*Sorghum bicolor* (L.) Moench). *The Plant Journal*, *70*(5), 818–830.

Samanta, M. K., Dey, A., & Gayen, S. (2016). CRISPR/Cas9: An advanced tool for editing plant genomes. *Transgenic Research*, *25*, 561–573.

Sanford, J. C. (1990). Biolistic plant transformation. *Physiologia Plantarum*, *79*(1), 206–209.

Sattler, S. E., Saballos, A., Xin, Z., Funnell-Harris, D. L., Vermerris, W., & Pedersen, J. F. (2014). Characterization of novel sorghum brown midrib mutants from an EMS-mutagenized population. *G3: Genes, Genomes, Genetics*, *4*(11), 2115–2124.

Schmidt, C., Pacher, M., & Puchta, H. (2019). DNA break repair in plants and its application for genome engineering. In Kumar, S., Barone, P. & Smith, M. (eds.), *Transgenic plants: Methods and protocols* (vol. 1864, pp. 237–266). New York, NY: Humana Press.

Schmidt, C., Schindele, P., & Puchta, H. (2020). From gene editing to genome engineering: Restructuring plant chromosomes via CRISPR/Cas. *Abiotech*, *1*(1), 21–31.

Serrat, X., Esteban, R., Guibourt, N., Moysset, L., Nogués, S., & Lalanne, E. (2014). EMS mutagenesis in mature seed-derived rice calli as a new method for rapidly obtaining TILLING mutant populations. *Plant Methods*, *10*, 1–14.

Sharma, H., Crouch, J., Sharma, K., Seetharama, N., & Hash, C. (2002). Applications of biotechnology for crop improvement: Prospects and constraints. *Plant Science*, *163*(3), 381–395.

Sikora, P., Chawade, A., Larsson, M., Olsson, J., & Olsson, O. (2011). Mutagenesis as a tool in plant genetics, functional genomics, and breeding. *International Journal of Plant Genomics*, *2011*(1), Art. 314829.

Silva, T. N., Kelly, M. E., & Vermerris, W. (2020). Use of *Sorghum bicolor* leaf whorl explants to expedite regeneration and increase transformation throughput. *Plant Cell, Tissue and Organ Culture (PCTOC)*, *141*(2), 243–255.

Silva, T. N., Thomas, J. B., Dahlberg, J., Rhee, S. Y., & Mortimer, J. C. (2022). Progress and challenges in sorghum biotechnology, a multipurpose feedstock for the bioeconomy. *Journal of Experimental Botany*, *73*(3), 646–664.

Simons, J. M., Herbert, T. C., Kauffman, C., Batete, M. Y., Simpson, A. T., Katsuki, Y., . . . Weil, C. (2022). Systematic prediction of EMS-induced mutations in a sorghum mutant population. *Plant Direct*, *6*(5), Art. e404.

Sivanesan, I., Nayeem, S., Venkidasamy, B., Kuppuraj, S. P., Rn, C., & Samynathan, R. (2022). Genetic and epigenetic modes of the regulation of somatic embryogenesis: A review. *Biologia Futura*, *73*(3), 259–277.

Stadler, L. (1944). The effect of X-rays upon dominant mutation in maize. *Proceedings of the National Academy of Sciences*, *30*(6), 123–128.

Stadler, L. J. (1929). Chromosome number and the mutation rate in Avena and Triticum. *Proceedings of the National Academy of Sciences*, *15*(12), 876–881.

Su, Y. H., Tang, L. P., Zhao, X. Y., & Zhang, X. S. (2021). Plant cell totipotency: Insights into cellular reprogramming. *Journal of Integrative Plant Biology*, *63*(1), 228–243.

Syamala, D., & Devi, P. (2004). Development of plant regeneration and Genetic Transformation system from shoot apices of *Sorghum bicolor* (L.) Moench. *Journal of Plant Biotechnology*, *6*(2), 77–85.

Tadesse, Y., Sagi, L., Swennen, R., & Jacobs, M. (2003). Optimisation of transformation conditions and production of transgenic sorghum (*Sorghum bicolor*) via microparticle bombardment. *Plant Cell, Tissue and Organ Culture*, *75*, 1–18.

Tari, I., Laskay, G., Takács, Z., & Poór, P. (2013). Response of sorghum to abiotic stresses: A review. *Journal of Agronomy and Crop Science*, *199*(4), 264–274.

Tay Fernandez, C. G., Nestor, B. J., Danilevicz, M. F., Marsh, J. I., Petereit, J., Bayer, P. E., . . . Edwards, D. (2022). Expanding gene-editing potential in crop improvement with pangenomes. *International Journal of Molecular Sciences*, *23*(4), 2276.

Taylor, N. J., & Fauquet, C. M. (2002). Microparticle bombardment as a tool in plant science and agricultural biotechnology. *DNA and Cell Biology*, *21*(12), 963–977.

Tenea, G. N., Spantzel, J., Lee, L.-Y., Zhu, Y., Lin, K., Johnson, S. J., & Gelvin, S. B. (2009). Overexpression of several Arabidopsis histone genes increases Agrobacterium-mediated transformation and transgene expression in plants. *The Plant Cell*, *21*(10), 3350–3367.

Tetreault, H. M., Gries, T., Liu, S., Toy, J., Xin, Z., Vermerris, W., . . . Sattler, S. E. (2021). The sorghum (*Sorghum bicolor*) *brown midrib 30* gene encodes a chalcone isomerase required for cell wall lignification. *Frontiers in Plant Science*, *12*, Art. 732307.

Thomson, M. J., Biswas, S., Tsakirpaloglou, N., & Septiningsih, E. M. (2022). Functional allele validation by gene editing to leverage the wealth of genetic resources for crop improvement. *International Journal of Molecular Sciences*, *23*(12), 6565.

Tian, R., Nájera-González, H.-R., Nigam, D., Khan, A., Chen, J., Xin, Z., . . . Jiao, Y. (2024). Leucine-rich repeat receptor kinase BM41 regulates cuticular wax deposition in sorghum. *Journal of Experimental Botany*, *75*(20), 6331–6345.

Upadhyaya, H. D., Wang, L., Prakash, C. S., Liu, Y., Gao, L., Meng, R., . . . Singh, S. K. (2022). Genome-wide association mapping identifies an SNF4 ortholog that impacts biomass and sugar yield in sorghum and sugarcane. *Journal of Experimental Botany*, *73*(11), 3584–3596.

Urriola, J., & Rathore, K. S. (2015). Overexpression of a glutamine synthetase gene affects growth and development in sorghum. *Transgenic Research*, *24*, 397–407.

Vasil, I. K., & Vasil, V. (1972). Totipotency and embryogenesis in plant cell and tissue cultures. *In Vitro*, *8*(3), 117–125.

Wang, K., Herrera-Estrella, L., Van Montagu, M., & Zambryski, P. (1984). Right 25 bp terminus sequence of the nopaline T-DNA is essential for and determines direction of DNA transfer from agrobacterium to the plant genome. *Cell*, *38*(2), 455–462.

Wang, L., Gao, L., Liu, G., Meng, R., Liu, Y., & Li, J. (2021). An efficient sorghum transformation system using embryogenic calli derived from mature seeds. *PeerJ*, *9*, Art. e11849.

Wang, N., Ryan, L., Sardesai, N., Wu, E., Lenderts, B., Lowe, K., . . . Van Dyk, D. (2023). Leaf transformation for efficient random integration and targeted genome modification in maize and sorghum. *Nature Plants*, *9*(2), 255–270.

Wang, W., Wang, J., Yang, C., Li, Y., Liu, L., & Xu, J. (2007). Pollen-mediated transformation of *Sorghum bicolor* plants. *Biotechnology and Applied Biochemistry*, *48*(2), 79–83.

Wu, E., Lenderts, B., Glassman, K., Berezowska-Kaniewska, M., Christensen, H., Asmus, T., . . . Zhao, Z.-Y. (2014). Optimized Agrobacterium-mediated sorghum transformation protocol and molecular data of transgenic sorghum plants. *In Vitro Cellular & Developmental Biology-Plant*, *50*, 9–18.

Xin, Z., Li Wang, M., Barkley, N. A., Burow, G., Franks, C., Pederson, G., & Burke, J. (2008). Applying genotyping (TILLING) and phenotyping analyses to elucidate gene function in a chemically induced sorghum mutant population. *BMC Plant Biology*, *8*, 1–14.

Xin, Z., Wang, M., Cuevas, H. E., Chen, J., Harrison, M., Pugh, N. A., & Morris, G. (2021). Sorghum genetic, genomic, and breeding resources. *Planta*, *254*(6), 114.

Xin, Z., Wang, M. L., Burow, G., & Burke, J. (2009). An induced sorghum mutant population suitable for bioenergy research. *BioEnergy Research*, *2*, 10–16.

Yellisetty, V., Reddy, L., & Mandapaka, M. (2015). In planta transformation of sorghum (*Sorghum bicolor* (L.) Moench) using TPS1 gene for enhancing tolerance to abiotic stresses. *Journal of Genetics*, *94*, 425–434.

Yuan, N., Liang, S., Zhou, L., Yuan, X., Li, C., Chen, X., & Zhao, H. (2024). Comparison of Mutations Induced by Different Doses of Fast-Neutron Irradiation in the M1 Generation of Sorghum (*Sorghum bicolor*). *Genes*, *15*(8), 976.

Zhang, D., Tang, S., Xie, P., Yang, D., Wu, Y., Cheng, S., . . . Yu, F. (2022). Creation of fragrant sorghum by CRISPR/Cas9. *Journal of Integrative Plant Biology*, *64*(5), 961–964.

Zhao, Z.-y., Cai, T., Tagliani, L., Miller, M., Wang, N., Pang, H., . . . Seltzer, J. (2000). Agrobacterium-mediated sorghum transformation. *Plant Molecular Biology*, *44*, 789–798.

Zhong, H., Wang, W., & Sticklen, M. (1998). In vitro morphogenesis of *Sorghum bicolor* (L.) Moench: Efficient plant regeneration from shoot apices. *Journal of Plant Physiology*, *153*(5-6), 719–726.

Zhu, Z., Qu, K., Li, D., Zhang, L., Wang, C., Cong, L., Bai, C., & Lu, X. (2023). SbPHO2, a conserved Pi starvation signalling gene, is involved in the regulation of the uptake of multiple nutrients in sorghum. *Plant Science*, *327*, Art. 111556.

Zorrilla López, U., Masip Vilà, G., Arjó Pont, G., Bai, C., Banakar, R., Bassie Rene, L., . . . Pérez Massot, E. (2013). Engineering metabolic pathways in plants by multigene transformation. *International Journal of Developmental Biology*, *57*, 565–576.

7 Genomic Selection as a Tool to Accelerate Genetic Gain in Sorghum Improvement

Current Status, Integration with Phenomics and Future Prospects

Pradeep Kumar and Harsimardeep Singh Gill

7.1 INTRODUCTION

Sorghum (*Sorghum bicolor* (L.) Moench) is an important crop for global agriculture, ranking as the fifth most important cereal in terms of production and acreage (Khalifa & Eltahir, 2023). Known for its versatility, sorghum is cultivated for a wide range of uses, such as food, feed, biofuels, and industrial applications. As a C_4 crop, its efficient use of nutrients and water, combined with its ability to thrive in drought-prone environments, makes it a superior crop to be grown on marginal land without competing with other food crops (Rooney & Serna-Saldivar, 2000). Furthermore, it contains biologically active compounds that offer health benefits and help prevent chronic diseases, further increasing its demand and importance in both the agricultural and health sectors. Considering these factors, sorghum plays a crucial role in global food security and agricultural industries. Nevertheless, enhancing the productivity of this crop remains crucial to addressing the anticipated demand for a twofold increase in global food production by 2050, driven by population expansion and rising economic prosperity (Falcon et al., 2022).

Sorghum improvement is being carried out across the globe by many national and international programs, including the International Crops Research Institute for the Semi-Arid Tropics (ICRISAT), FAO/UNDP (Food and Agricultural Organization and the United Nations Development Program), and many national research organizations in different countries. Long-term sorghum improvement has been characterized by increases in grain yields but at a notably slower rate compared to other major crops (Pfeiffer et al., 2019). Despite this slower progress, sorghum continues to be cultivated in diverse agroecosystems worldwide, primarily due to its adaptability to marginal lands and lower input requirements. Over the past few decades, conventional breeding methods have yielded progress in sorghum improvement, albeit at a more modest pace than observed in other crops. The development of improved sorghum varieties and hybrids that integrate beneficial genetic traits suited to diverse environments and end-use applications remains a lengthy and intricate endeavor. Conventional breeding approaches typically require around six years to advance male parental lines from initial crosses to preliminary yield trials, while the process extends to nearly eight years for female parental lines. Additionally, an extra two to four years may be necessary before these lines are selected as parents for subsequent breeding cycles, further prolonging the development timeline (Rizal et al., 2014). Conventional breeding also depends heavily on multi-environment trials, which are both costly and time-consuming. This lengthy process slows the rate of genetic gain and highlights the need for modern breeding technologies that can accelerate sorghum improvement, particularly by developing varieties with higher genetic yield potential and resistance to

 DOI: 10.1201/9781003623717-7

environmental stresses for various end uses. The urgency to improve sorghum varieties is further magnified by the deteriorating soil conditions and the reduction in available water resources, both of which are exacerbated by climate change. Under these conditions, achieving the desired rate of genetic gain through traditional breeding techniques becomes increasingly difficult.

In response to these challenges, advanced strategies such as genomic selection (GS) and high-throughput phenotyping (HTP) may provide a potential advantage in crop breeding by accelerating the genetic improvement of crops per unit of time through the reduction in cost per breeding cycle and shortening the generation interval. These methods are particularly valuable for traits with low heritability, where traditional phenotypic selection may be less effective. However, the accurate selection of complex quantitative traits such as yield, disease resistance, and quality traits across different growing environments and breeding lines is a critical challenge for plant breeders. These traits are influenced by numerous genetic and environmental factors, making selection a formidable challenge. Improving the selection accuracy of these traits can greatly enhance the efficiency of breeding programs, as it allows for better selection of high-performing genotypes at earlier stages of the breeding cycle. Thus, genomic and phenomic prediction together provide a powerful tool for plant breeders, enabling the prediction and selection of superior genotypes earlier and more accurately than ever before.

Genomic prediction (GP) involves using genome-wide markers to estimate the genomic breeding value of individuals, allowing breeders to make informed decisions based on their genomic data (Meuwissen et al., 2001). This technology leverages high-density genomic data to predict the performance of untested genotypes, thus accelerating the breeding cycle by allowing early selection based on genetic merit. Phenomic selection, on the other hand, utilizes HTP platforms to collect large-scale phenotypic data, such as spectral imaging or sensor-based measurements, which can then be used to predict traits like yield, biomass, and drought tolerance at a larger scale (Montes et al., 2007). In sorghum, genomic prediction models are commonly used to predict hybrid performance as the hybrid development process is resource-intensive, involving hundreds of testcross hybrids evaluated across environments to identify the best-performing hybrids. GS further enhances breeding efficiency by predicting the performance of gene bank accessions, facilitating pre-breeding efforts, and enabling the targeted selection of individuals for inclusion in breeding programs. By employing GS, breeders can improve key traits such as yield, and performance under biotic and abiotic stress conditions, accelerating the development of superior hybrids and cultivars. Furthermore, genomic and phenomic prediction models can help breeders anticipate how different sorghum lines will respond to these changing climatic conditions, allowing for more targeted breeding strategies. These models act as a valuable supplement to multi-environment trials, enhancing selection accuracy while reducing the overall dependency on extensive field evaluations. This chapter discusses the role of genomic selection, its methods, prediction models, and integration with high-throughput phenotyping in sorghum breeding, emphasizing their potential to accelerate genetic gain and improve breeding efficiency. Additionally, the integration of these prediction tools into breeding programs is explored, highlighting their ability to enhance selection accuracy, reduce costs, and expedite the development of new sorghum cultivars.

7.2 GENETIC GAIN: A KEY INDICATOR FOR MONITORING PROGRESS IN BREEDING INITIATIVES

The fundamental objective of any breeding program is to enhance germplasm by introducing superior genetic traits. The pace at which these genetic improvements accumulate over time is referred to as genetic gain (Heffner et al., 2010). Increasing the rate of genetic gain is essential for ensuring sustainable advancements in crop productivity, particularly in response to the growing global demand for plant-derived commodities. Within breeding programs, genetic gain is quantified as:

$$\Delta G = (i \times r \times \sigma A) / t \text{ (Xu et al., 2017)}$$

Where "ΔG" represents the response to selection, "i" represents the selection intensity, measured as the mean deviation of selected individuals in phenotypic standard deviation units, "r" denotes the selection accuracy, while "σA" is the genetic standard deviation, calculated as the square root of the additive genetic variance (Falconer & Mackay, 1996). The variable "t" represents the breeding cycle duration. In genomic selection (GS), "r" signifies the correlation between true breeding values (TBVs) and genomic-estimated breeding values (GEBVs). For phenotypic selection, "r" equals the square root of narrow-sense heritability (h_2) (Heffner et al., 2010; Bassi et al., 2016).

The breeder's equation serves as a crucial analytical tool for pinpointing constraints within breeding programs that hinder genetic progress. It has been suggested that designing new breeding strategies based on comparative analyses of genetic gain across various breeding methodologies could enhance efficiency (Rutkoski, 2019). Conventional phenotype-based selection methods have yielded only modest genetic gains, prompting a shift toward marker-assisted selection (MAS), which targets a limited number of major QTLs associated with key traits (Xu & Crouch, 2008). In contrast, GS evaluates the cumulative effects of all markers to predict GEBVs. Several studies have demonstrated that incorporating genomic data leads to higher genetic gains than phenotypic selection in simulation-based analyses (Tessema et al., 2020). Additionally, the real-world application of GS has driven substantial genetic advancements in dairy cattle and numerous crop species (Schaeffer, 2006; Chen et al., 2009; Bassi et al., 2016; Bernardo & Yu, 2007). One of the most significant advantages of GS is its ability to accelerate selection by shortening breeding cycles relative to conventional phenotypic selection, ultimately boosting the annual rate of genetic gain in breeding programs (Lorenzana & Bernardo, 2009). This highlights the transformative potential of GS in expediting genetic improvements in sorghum breeding.

Genetic gain in sorghum yield has been relatively slow compared to other major crops, as observed in both public and private breeding programs over several decades. This rate of improvement, while contributing significantly to total yield gains in US sorghum production, lags behind the genetic gains observed in crops like corn, cotton, and soybeans (Pfeiffer et al., 2019). Various factors have contributed to this slower progress, including the need for multiple trait breeding, lack of well-defined heterotic groups, breeding process limitations, reduced research investments, shifting production environments, and grower-imposed phenotypic constraints. However, the implementation of GS in sorghum breeding programs presents a promising avenue for accelerating genetic gains. By leveraging high-density marker data and statistical models to predict breeding values, GS could enhance the efficiency and speed of sorghum improvement. When combined with other emerging technologies such as speed breeding and high-throughput phenotyping, it has the potential to significantly impact sorghum breeding, potentially leading to faster and more substantial yield improvements in the future.

7.3 HISTORY OF GENOMIC SELECTION AND BENEFITS OVER CONVENTIONAL BREEDING METHODS

A successful crop improvement program depends on the availability of genetic variation and sufficient resources to conduct large-scale evaluations, which makes it both cost- and time-intensive. The development of new crop varieties can take 10–12 years, depending on the species, and requires evaluating candidate genotypes with high precision across multiple locations and over several years. This lengthy process demands significant resources and limits the number of candidates that can be advanced in the varietal development pipeline. However, in the quest for crop improvement, various plant breeding systems have evolved beyond traditional phenotypic selection methods by incorporating innovative approaches such as cytological, biochemical, and molecular markers (Hasan et al., 2021; Jiang et al., 2015), which help to accelerate the breeding process. Molecular marker technologies have provided tools like MAS, marker-assisted backcrossing, and marker-assisted recurrent selection (Crossa et al., 2010), which are widely used to address limitations in commercial varieties through gene pyramiding and the introgression of new alleles to popular cultivars. However, MAS is limited in its ability to enhance complex quantitative traits due to its reliance on low- or medium-density markers. MAS relies on markers associated with major QTLs, excluding those not significantly linked to traits. However, it has limitations in improving complex traits due to challenges in QTL detection, such as environmental interactions and genetic background differences. This inconsistency, combined with limited QTL mapping power in biparental populations, particularly in low-recombination regions, has hindered its effectiveness (Bernardo, 2016). To overcome these challenges, association mapping emerged in the early 2000s, enabling better detection of marker-trait associations in diverse populations and fine-mapping in high-recombination regions (Pritchard et al., 2000). However, it struggles with identifying rare variants and QTLs with small effects on traits influenced by environmental factors. Advancements in genotyping technologies and the reduced cost of SNP assays have dramatically expanded the availability of DNA markers, allowing breeders to select plants based on genetic marker composition rather than phenotypic performance (Crossa et al., 2017). This paved the way for GS, a concept first introduced by Lande and Thompson (1990) and later popularized by Meuwissen et al. (2001). Unlike MAS, which focuses on a limited number of QTLs, GS leverages genome-wide markers to predict genetic values without the need for prior QTL discovery and integrates markers for both major and minor effects, enabling robust phenotype predictions for untested individuals (Meuwissen et al., 2001). This genome-wide prediction greatly improves the precision of breeding decisions, especially for complex traits like yield and drought tolerance, which are time- and resource-intensive to enhance using traditional methods. Furthermore, GS reduces the time and expenses involved in breeding programs by reducing the number of generations. This streamlined process enables breeders to shorten breeding cycles and make more efficient use of resources. Empirical studies have demonstrated superiority of GS over traditional methods in accelerating genetic gains, making it a revolutionary tool in crop breeding (Crossa et al., 2010, 2017). Figure 7.1 illustrates the key steps in the GS process.

7.4 GENOMIC PREDICTION MODELS

Genomic prediction (GP) refers to the utilization of statistical models for estimating genomic breeding values based on genotypic data, usually derived from single nucleotide polymorphism (SNP) markers. GS utilizes GP models to compute GEBVs, which reflect an individual's total genetic potential for a specific heritable trait. These values serve as a dependable resource for breeders to make effective and informed selection choices without the necessity of extensive phenotypic assessments. A key aspect of GS is prediction accuracy, which measures the correlation between GEBVs and true breeding values (TBVs). Although TBVs cannot be directly observed, prediction accuracy is generally assessed through the correlation between GEBVs and empirically estimated breeding values (EBVs). The prediction accuracy calculated by a genomic prediction model should reflect its ability to reliably predict genetic merit. Additionally, predictive ability generally refers to how well a model can predict phenotypes or breeding values of individuals that were not used in model training.

GP models integrate pedigree data, genotypic data, phenotypic data, and/or environmental covariates to estimate GEBVs and improve prediction accuracy (Meuwissen et al., 2001; Heslot et al., 2012; Malosetti et al., 2016). These models leverage a multidisciplinary framework that combines computer science, genetics, quantitative genetics, statistics, bioinformatics, mathematics, physics, and machine learning to calculate breeding values (Crossa et al., 2017).

A substantial body of research in GP has been dedicated to refining both parametric and non-parametric statistical as well as computational methodologies to enhance the accuracy of predicting

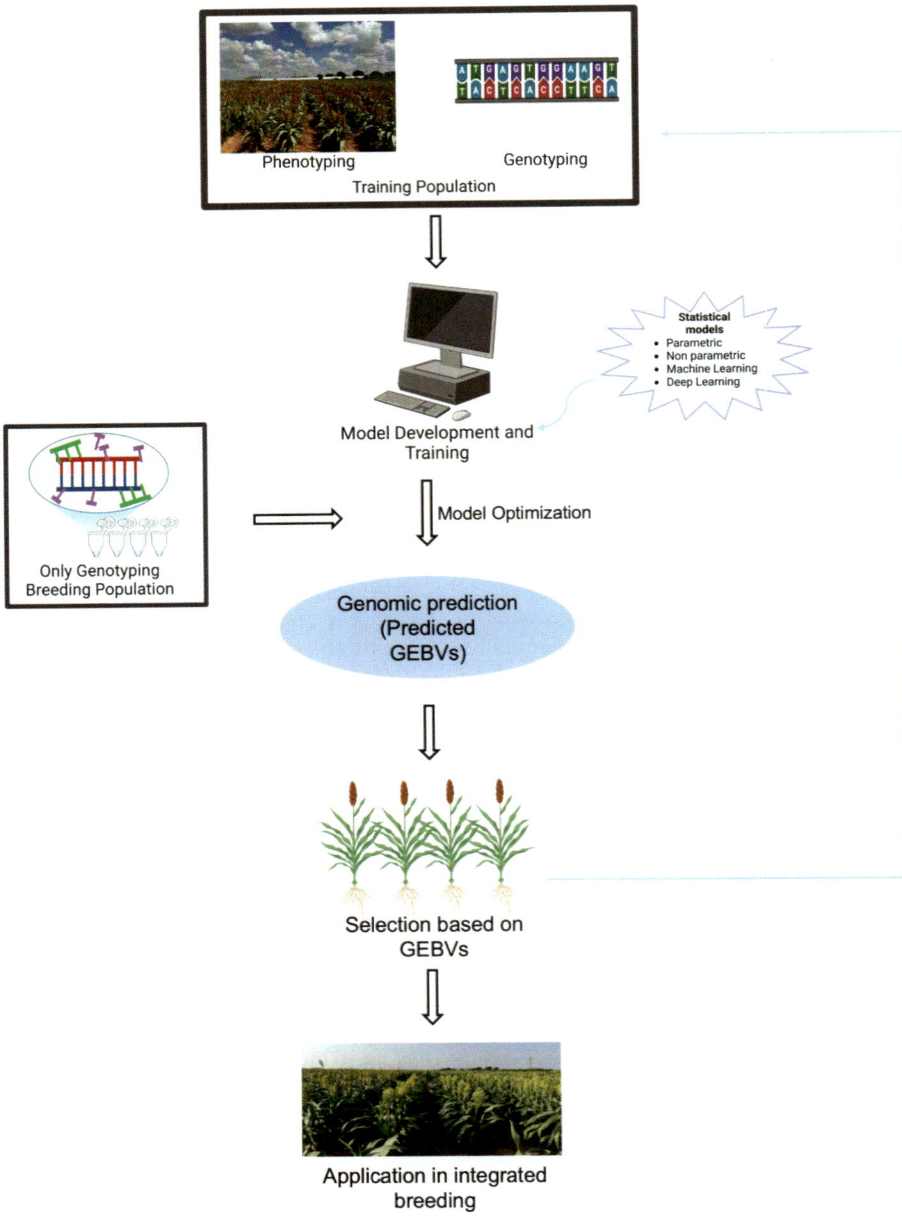

FIGURE 7.1 A schematic overview of the key steps involved in the genomic selection process.

breeding values for individuals that have not been phenotyped. Broadly, GP models are categorized into three principal groups: parametric, semi-parametric, and non-parametric approaches (Alemu et al., 2024). Parametric models encompass classical statistical techniques such as modified least squares regression, ridge regression, best linear unbiased prediction (BLUP), penalized regression models, and Bayesian models, which have been widely applied in GS (Meuwissen et al., 2001; Montesinos López et al., 2022; Solberg et al., 2009). Among parametric models, Bayesian and shrinkage-based methods have become widely adopted due to their effectiveness in handling high-dimensional genomic data. These include ridge regression best linear unbiased prediction (RR-BLUP) and genomic

BLUP (G-BLUP) (VanRaden, 2008), as well as a suite of Bayesian models such as BayesA, BayesB, and BayesC, which have been extensively utilized to improve prediction accuracy by incorporating prior distributions (Meuwissen et al., 2001; George & McCulloch, 1993; Habier et al., 2011; Erbe et al., 2012; Pong-Wong & Woolliams, 2014). Additionally, various penalized regression methods have been widely adopted in GP to improve prediction accuracy. Some of the common penalized regression methods include the least absolute shrinkage and selection operator (LASSO) (Usai et al., 2009), adaptive LASSO (Zou, 2006), and their Bayesian counterparts (Perez et al., 2010; Zou & Hastei, 2005).

Semi-parametric models, such as the Reproducing Kernel Hilbert Space (RKHS) model, combine features of traditional parametric and fully non-parametric models, and are known for their ability to model complex trait architectures. These RKHS models are often implemented within mixed-model or Bayesian modeling frameworks to improve predictive accuracy for traits influenced by intricate genetic architectures (Gianola et al., 2006). Non-parametric genomic prediction models leverage machine learning algorithms to capture complex non-linear relationships between genotypic and phenotypic data. Common approaches include gradient boosting machines (GBM) (Li et al., 2018), extreme gradient boosting (XGBoost) (Chen & He, 2015), and support vector machines (SVM) (Maenhout et al., 2007). Additionally, ensemble-based techniques such as random forests (RF) (Chen & Ishwaran, 2012) and Bayesian additive regression trees (BART) (Waldmann, 2016) have been widely applied to improve prediction accuracy by aggregating multiple weak learners. Neural network-based approaches have gained prominence in genomic prediction due to their ability to capture intricate interactions and hierarchical feature representations. These include multilayer perceptrons (MLP) (Gianola et al., 2011), probabilistic neural networks (PNN) (González-Camacho et al., 2016), and radial basis function (RBF) networks (Chen & Ishwaran, 2012). Furthermore, Bayesian-regularized neural networks and advanced deep learning architectures, such as convolutional neural networks (CNNs) and other deep neural networks, have emerged as powerful tools for predicting complex agronomic traits by extracting deep feature representations from high-dimensional genomic datasets (Alemu et al., 2024).

In genomic studies focusing on crop improvement, two of the most widely implemented GP models are genomic best linear unbiased prediction (G-BLUP) and ridge regression best linear unbiased prediction (RR-BLUP). RR-BLUP operates under the assumption that all genetic markers contribute to the trait of interest with uniform variance while maintaining small but nonzero effects. However, this assumption does not necessarily translate to identical effects for each marker (Bernardo & Yu, 2007; Buckler et al., 2009). On the other hand, G-BLUP utilizes genome-wide marker information to construct a genomic relationship matrix, which is then used to estimate genetic values for individuals, thereby improving selection accuracy. Both models share the assumption that marker effects share a common variance; however, this is often not the case in practice since most markers exhibit minimal or no effects while a few have substantial effects. Recent advancements have led to the development of GBLUP extensions that incorporate genotype-by-environment (G × E) interactions to enhance predictive accuracy for unobserved genotypes across different environments (Islam et al., 2021; Vos-Fels et al., 2021; Ornella et al., 2012). Various models addressing GP accuracy for discrete response variables such as ordinal disease data have emerged alongside Bayesian models that analyze multiple traits across various environments (Gonzalez Camacho et al., 2012; Perez-Rodriguez et al., 2012; Sun et al., 2012; Heffner et al., 2011; Isidro et al., 2015).

Bayesian approaches generally yield superior prediction accuracies due to their flexibility in modeling varying marker effects and variances. Bayesian models like BayesA, BayesB, BayesC, and Bayesian-LASSO are well-suited for realistic conditions, assuming different prior distributions on marker effects. BayesA model assigns the same prior distribution to the variances of all marker effects, and is particularly effective for traits influenced by a moderate number of loci due to its weaker shrinkage. BayesB and BayesC can be viewed as direct extensions of the BayesA model, through the inclusion of a parameter π, which specifies the prior proportion of covariates expected to have non-zero effects (Munkvold et al., 2009; Pérez and de los Campos, 2014). Generally, Bayesian models achieve superior prediction accuracies by capturing large-effect QTLs. Bayesian LASSO combines subset selection with Bayesian regression shrinkage. Parametric Bayesian models can also incorporate population structure using kernel methods, representing genetic similarities via a genomic covariance matrix (VanRaden, 2008; de los Campos et al., 2009). These kernels integrate dense whole-genome marker information through a genomic relationship matrix (GRM), effectively reducing the impact of data dimensionality (Gianola & de los Campos, 2008). Non-linear kernels, used in semi-parametric models, have shown improved accuracy in genomic predictions across various studies (Morota & Gianola, 2014; Costa-Neto et al., 2021). The application of kernel methods has significantly expanded opportunities for plant breeders. Kernels can flexibly integrate diverse data types, including covariates, strings, images, and graphs. Their flexibility allows for the integration of multiple information sources within a single genomic prediction model, enhancing the potential for more comprehensive and accurate predictions in plant breeding programs (de los Campos et al., 2009; Crossa et al., 2010).

Interdisciplinary data science researchers are focused on enhancing predictive accuracy through the application of novel machine learning models. Neural networks, a prominent tool in machine learning, consist of interconnected layers of neurons. Each neuron's output is calculated as the weighted sum of inputs from a specific network layer plus a bias, which is then processed through an activation function. In GP, neural networks are structured with markers as input neurons, followed by multiple hidden layers, and culminating in an output layer for phenotype prediction. Recent advancements in neural network architectures and computational power have enabled the widespread adoption of deep learning models (LeCun, 2015). In GP, these models can capture complex interactions among genetic markers that underlie phenotypic traits. Multi-layer perceptrons (MLPs), a type of artificial neural network, have become a common deep learning architecture in machine learning-based genomic prediction. MLPs excel at learning complex relationships between phenotypes and genotypes, making them particularly effective for highly quantitative traits (Patterson & Gibson, 2017; Montesinos-López et al., 2018; González-Camacho et al., 2018; Montesinos-López et al., 2020; Sandhu et al., 2021a). MLP networks often outperform conventional machine learning approaches, including convolutional neural networks, in improving genomic prediction accuracy for complex traits, such as grain yield and end-use quality (Sandhu et al., 2021a, 2021b).

7.5 FACTORS AFFECTING GP ACCURACY

The application of GP in breeding is inherently complex and influenced by multiple factors at various levels. The accuracy of GP is primarily determined by several genetic and statistical factors, including the size, structure, and genetic diversity of the training population, the genetic complexity of the target trait, the choice of cross-validation schemes, marker density, statistical methods and G × E interactions. The accuracy of GP is significantly influenced by the size and structure of the training population. Larger training populations enhance prediction accuracy by providing a more comprehensive representation of genetic diversity and trait variation within a breeding population. For instance, a study showed that increasing the training population size from 41 to 163 individuals significantly improved prediction accuracy in sorghum, with the accuracy for grain yield rising from 0.46 to 0.59 (Maulana et al., 2023). Population structure, characterized by distinct allele frequencies in specific populations, arises from founder effects and selection processes (Isidro et al., 2015; Norman et al., 2018). Fluctuations in allele frequencies can create misleading correlations between genetic markers and phenotypic traits, even when these markers are not physically linked to the actual QTLs responsible for trait variation. If not accounted for using appropriate statistical corrections, this can introduce bias into genomic predictions. Another key determinant of GP accuracy is the genetic diversity present within the training population (Habier et al., 2007; Lorenzana & Bernardo, 2009).

A diverse training set is crucial for capturing a comprehensive range of genetic variants influencing target traits, ensuring that predictive models can accurately estimate genetic effects across different genetic backgrounds. To enhance the detection of contributing QTLs, the training population should be designed to include a broad representation of allelic diversity associated with the traits under selection (Norman et al., 2018). This comprehensive approach to training set composition enhances the robustness and reliability of genomic predictions across diverse genetic landscapes. Further, GP accuracy is significantly influenced by the genetic relatedness between the training and validation populations. When these sets are closely related, the LD patterns are better captured, resulting in higher prediction accuracy. However, as the generational gap between the reference and validation populations widens, the LD between markers and causal variants diminishes, leading to a decline in prediction accuracy (Jarquín et al., 2021; Clark et al., 2012). To mitigate this effect, incorporating pedigree information and modeling relationships between individuals can help maintain prediction accuracy across generations (Clark et al., 2012). Integration of environmental data, such as climate information, can enhance genomic prediction accuracy, particularly when the training and validation sets originate from different environments. Modeling genotype-by-environment interactions is crucial for accurate predictions in novel environmental conditions (Jarquín et al., 2021). Additionally, accounting for spatial and temporal variations in environmental factors can further improve prediction accuracy. These strategies collectively contribute to more robust and accurate genomic predictions, especially in the context of changing environments and across multiple generations.

7.6 GENOMIC SELECTION IN SORGHUM: METHODOLOGY, NEED AND IMPORTANCE

GS accelerates plant breeding by using genome-wide markers to estimate GEBVs and identify superior genotypes. It typically involves a two-step process. The first step entails developing a prediction equation based on a training population. The training population consists of a representative set of individuals that have been both genotyped (i.e., their DNA has been analyzed to identify genetic markers) and phenotyped (i.e., their traits of interest have been measured). For GS to be effective, quantitative trait loci (QTL) affecting the target trait must be in linkage disequilibrium (LD) with the DNA markers used for genotyping. If this condition is met, trait effects for DNA markers can be estimated and incorporated into the prediction equation (Spindle et al., 2015; Heslot et al., 2013). In the second step, these marker effects are used to calculate GEBVs for selecting candidates that have only genomic data available (breeding population). The breeding population consists of individuals that have been genotyped but not phenotyped. Genotypes can then be ranked based on their GEBVs to inform selection decisions in breeding programs. The success of GS relies on the relatedness between the training population and breeding population in terms of LD between marker loci and trait loci. Unlike traditional marker-assisted breeding, GS considers the effects of minor QTLs, potentially capturing a greater portion of genotypic variation. Marker effects are estimated throughout the genome of the breeding population based on the predictive model trained using the training population. Ideally, the breeding population should comprise training population descendants or closely related elite breeding lines. The GEBV represents the estimated breeding value of breeding population individuals, calculated using the training population data through statistical modeling. Superior lines selected based on high GEBVs in the breeding population can serve as potential parents in breeding programs, without the need for immediate phenotypic evaluation under field conditions.

The need for GS in sorghum breeding is well-established due to the crop's high genetic diversity and its potential for improving various agronomic traits. Sorghum exhibits diverse end-uses such as food, feed, fuel, and forage (Habyarimana & Lopez-Cruz, 2019; Baloch et al., 2023). Implementing GS across these different sorghum classes holds immense significance for improving productivity, adaptability, and quality, thereby strengthening its role as a key global food and industrial crop. GS plays a crucial role in predicting hybrid performance based on parental genotypes, which is essential for hybrid breeding programs. This approach significantly reduces the time and resources required to develop A and B lines for hybrid development. Studies have demonstrated that GP models can effectively predict grain yield, plant height, and days to anthesis in sorghum hybrids (Maulana et al., 2023). Additionally, for hybrid breeding, GS facilitates the prediction of general and specific combining abilities (GCA and SCA) in sorghum, as well as the understanding of G × E interactions. This approach allows breeders to increase selection intensity and, consequently, the rates of genetic gain. Additionally, GS facilitates the prediction of performance for gene bank accessions, enabling efficient pre-breeding and targeted selection of individuals for inclusion in breeding programs. This approach allows breeders to harness the extensive genetic diversity available in sorghum germplasm collections, unlocking their potential for improving key traits and enhancing breeding efficiency (Yu et al., 2016). Speed breeding is a transformative approach that significantly accelerates genetic gain by allowing for multiple generations of crops to be produced within a single year, thereby reducing the overall breeding cycle duration (Gudi et al., 2022). Standardizing the protocol for speed breeding and its integration with GS enables breeders to rapidly advance desirable traits and improve selection accuracy, ultimately leading to faster development of improved cultivars. Recent advancements in sequencing technologies have made genotyping more cost-effective, enabling the generation of high-density marker data for sorghum. As sequencing costs continue to decrease and prediction models become more sophisticated, the integration of GS in sorghum breeding programs is likely to become increasingly prevalent and effective.

7.7 APPLICATIONS IN SORGHUM

Genomic selection presents a significant opportunity for sorghum improvement, particularly due to the availability of its well-characterized reference genome and a wealth of genomic resources. While the application of GS in sorghum lags behind that of other cereal crops, such as wheat and maize, several recent studies have revealed the immense potential of genomic selection to achieve significant genetic gains in sorghum breeding programs (de Oliveira et al., 2018; Hunt et al., 2018; Habyarimana & Lopez-Cruz, 2019; Velazco et al., 2019). Nevertheless, additional research integrating phenotypic and genotypic data from natural and breeding populations is essential to unravel the potential of GS in sorghum for improving important agronomic traits. Subsequent research has expanded the scope of GS applications in sorghum, encompassing a diverse array of agronomically important traits, including grain yield, biomass production, drought adaptation, and phenological characteristics. The rich genetic diversity of sorghum, preserved in the form of germplasm, offers an invaluable resource for identifying new alleles and genes associated with desirable traits (Cowan et al., 2020). Thus, comprehensive phenotypic and genotypic characterization of the sorghum germplasm collection is crucial for maximizing and supporting conservation efforts. This enables better parent selection, and identification of key genes and molecular markers and ensures targeted trait expression through GS. The implementation of GS offers significant advantages for sorghum breeders across various market classes. For instance, in forage sorghum breeding, high-density genotyping enables the evaluation of genetic potential in untested individuals, facilitating genomic prediction-based selection of high yielding biomass genotypes. This approach has the potential to reduce breeding cycle time, enhance selection accuracy, and accelerate genetic gain in forage varieties. Recent studies have successfully implemented genomic prediction models for various traits in sorghum, including biomass yield, plant architecture, phenology, and abiotic stress tolerance. This multi-trait GS approach enhances the overall efficiency and effectiveness of sorghum breeding programs, potentially leading to more rapid development of improved cultivars tailored to specific end-uses and environmental conditions.

7.7.1 GS for Biomass, Yield, and Yield-related Traits in Sorghum

Yield traits in sorghum are a complex quantitative trait influenced by multiple genetic loci, each contributing small to moderate effects. GS offers a promising approach to predicting agronomic traits, such as yield, biomass, plant height, etc., which can be employed at different levels depending on the study objectives. In sorghum breeding, where hybrids are the predominant cultivar type, GS can assist in identifying high-yielding parental lines in segregating populations or inbred lines, thereby facilitating the prediction of hybrid performance for these agronomic traits and narrowing down the candidates for advanced varietal field trials.

7.7.1.1 Genomic Selection in Inbred Lines

Inbred lines, which are homogenous and homozygous populations used as parents in hybrid breeding, serve as the foundation for GS in sorghum. The primary goal of GS in these populations is to support parental selection and assess its feasibility for specific traits. Charles et al. (2024) evaluated GP models for sweet sorghum under varying water regimes in Haiti, using 250 breeding lines derived from three subpopulations developed at CHIBAS, Haiti. The study predicted agronomic and drought-related traits, emphasizing the need for resource-efficient strategies to develop drought-tolerant sorghum cultivars. Bayesian whole-genome regression models, specifically Bayesian ridge regression (BRR) and Bayes B, estimated genomic breeding values. Within environments, prediction accuracies ranged from 0.30 to 0.71, with total soluble solids showing the highest accuracy (0.71) and grain yield the lowest (0.37–0.55). However, across the environments, predictive accuracy significantly declined, particularly in severe drought conditions, highlighting the need for improved multi-trait models for better prediction accuracy, especially for highly correlated traits. Another study by Dos Santos et al. (2020) focused on multi-trait genomic prediction (MTGP) models for developmental traits in biomass sorghum. Using a panel of 869 sorghum lines and 100,435 SNP markers, five different genomic prediction models were assessed, with prediction accuracies ranging from 0.46 to 0.75 for dry biomass yield and plant height. These models showed significant improvement when leveraging plant height measurements over time, suggesting that early selection for plant height could indirectly predict dry biomass. Dynamic Bayesian Networks and multi-time GBLUP models, which leverage repeated measures over time, significantly improved prediction accuracy compared to models that assume trait independence over time.

While multi-trait GS did not always significantly improve predictions over single-trait GS, integrating correlated traits in validation populations boosted performance. Fernandes et al. (2018) demonstrated that incorporating correlated traits like plant height into a trait-assisted GS approach improved accuracy by up to 50% (0.60) for biomass yield. Similarly, Habyarimana et al. (2020) assessed GS in biomass sorghum using a panel of 369 genotypes and implemented genomic selection indices, which combined plant height, dry biomass yield, and dry mass fraction traits. The study found that integrating multiple correlated traits using genomic selection indices can improve predictions, particularly for traits like dry mass fraction, which shows lower individual accuracy when using single-trait GS.

Maintaining a genetically diverse training population that includes mixed or intermediate races can enhance prediction accuracy, particularly when the training population size is limited. This approach is beneficial for small or early-stage breeding programs with limited resources for constructing separate training populations for different breeding lines (Muleta et al., 2019). Furthermore, incorporating new phenotypic data from diverse lines helps maintain beneficial minor alleles within the gene pool, improving the genetic diversity and performance of sorghum breeding programs. Sapkota et al. (2020a) explored the influence of sorghum's racial structure on genomic prediction accuracy using the GBLUP model, revealing that prediction accuracy was higher for grain weight and plant height compared to days to anthesis, a trait more strongly influenced by racial structure. This study highlighted the importance of interracial diversity in improving genetic prediction models, leading to increased resistance to diseases and enhanced yield potential (see Table 7.1).

7.7.1.2 Genomic Selection in Hybrid Performance

GS plays a critical role in predicting hybrid performance among available inbred lines within the genetic space from which hybrid combinations must be selected. For hybrid crops with well-established heterotic patterns, such as maize, rye, and sorghum, the number of potential hybrids can be determined by multiplying the number of inbred lines from different heterotic groups. Hybrids are typically generated through simple crosses between individuals, which may or may not belong to distinct genetic groups like heterotic groups (Labroo et al., 2021), and parents are evaluated based on the hybrid performance resulting from crosses with known elite inbred testers. High-performing hybrids advance to further evaluations conducted across a growing number of representative environments or target population environments, often with additional testers for comparison. Thus, it is a resource-intensive process requiring the synthesis of hundreds of testcross hybrids and their evaluation across multiple environments to identify only a small subset of highly promising hybrids. This process demands considerable time, labor, and financial resources. To optimize this process, methods that predict hybrid performance prior to field testing offer a more efficient breeding pipeline (Beyene et al., 2015; Fonseca et al., 2021; Heffner et al., 2009). In sorghum hybrid breeding programs, genomic selection can be valuable for identifying optimal crosses, estimating hybrid performance in untested environments, predicting the genetic potential of untested lines, and determining specific line combinations likely to produce superior hybrids.

Hunt et al. (2018) studied the potential of GS in hybrid sorghum breeding by comparing pedigree-based and marker-based prediction models. The single trait marker-based models outperformed pedigree-based models, with prediction accuracy improving further when both sources of relationship information were combined. The study focused on predicting grain yield in sorghum hybrids, with prediction accuracies ranging from 0.41 to 0.70 across different models and environments and the prediction accuracy varied across environments, with marker-based additive variance contributing up to 50% of the total genetic variation. Cross-validation results showed that prediction accuracy decreased for families less related to the training population, emphasizing the importance of genetic relatedness for accurate predictions.

Unlike single-trait models, MTGP leverages genetic correlations between traits to improve prediction accuracy for hybrid performance, particularly for complex characteristics like grain yield. These models exploit shared genetic information across traits, enabling more precise estimations even for low-heritability traits or those challenging to phenotype. Velazco et al. (2019) assessed the predictive improvement of MTGP over single-trait G-BLUP, incorporating a blended kinship matrix of pedigree and genomic data. Multi-trait analysis increased grain yield predictive ability by up to 16% when plant height data was leveraged in the training set. For stay-green, a 19% improvement was achieved only when grain yield data was available for the predicted lines. Predictive ability, unbiasedness, and accuracy were further optimized using a combined pedigree-genomic relationship matrix. These findings underscore the importance of trait correlations in hybrid prediction and suggest that multi-trait genomic evaluation is a powerful tool for predicting complex traits in grain sorghum hybrids.

Standard GBLUP models assume uniform variance across all loci and a homogeneous population structure, but hybrid crops inherently exhibit population structure, such as heterotic groups, which necessitates the use of less restrictive genomic prediction models (de los Campos et al., 2015). Most GP studies have focused on estimating breeding values based solely on general combining ability (GCA), which does not fully account for the heterotic groups used in hybrid production. To overcome these limitations, hierarchical Bayesian framework, have been introduced to model genetic variation more effectively. Fonseca et al. (2021) integrated GCA-SCA models with G × E interactions and population structure within a unified genomic framework to predict hybrid

TABLE 7.1
A summary of genomic selection studies conducted in sorghum

Objective	Traits	Population	Statistical Method	Results	References
Genomic prediction of grain sorghum hybrids across diverse environments	Days to anthesis, plant height, grain yield	100 grain sorghum hybrids developed from 20 grain sorghum inbred lines	Different combinations of gBLUP kernels models	Incorporating envirotype data and dominance effects into genomic prediction models can significantly improve prediction accuracy for grain sorghum traits	Winans et al. (2024)
Prediction of hybrid grain yield and days to mid-anthesis in sorghum B-lines using TFMSA	Grain yield, days to mid-anthesis	434 hybrids	GBLUP	Prediction accuracies ranging from 0.31 to 0.66 for grain yield and 0.61 to 0.84 for days to mid-anthesis	Kent et al. (2023)
GP using phenotypic information from inbred parents, dominance effects, G × E interactions	Days to mid-anthesis, grain yield, plant height	395 hybrids	Kernel based (GBLUP)	G × E effects and inbred phenotypic data improved GP accuracy	Crozier et al. (2023)
Grain sorghum hybrid performance	Plant height, days to flowering, panicle length & weight, thousand kernel weight	204 hybrids	GBLUP	PA of 0.67 for grain yield using both additive and dominance effect model	Maulana et al. (2023)
To compute accuracies of GP of hybrid genetic value and male GCA	Plant height, days to flowering, thousand kernel weight, grain yield	602 hybrids	GBLUP, Genomic GCA+SCA	Prediction of GCA ranged from 0.33 to 0.62 for all traits	Sapkota et al. (2023)
To investigate using a genomic selection index for biomass yield	Aboveground dry biomass, plant height, dry mass fraction fresh material	180 Sorghum bicolor landraces and lines, 189 *S. bicolor* × *S. halepense* RILs	Bayesian approach through the Multiple-Trait Model	Three-trait genomic selection index showed higher accuracy to single-trait models	Habyarimana et al. (2020)
Facilitation of genome-wide imputation and GP	Grain yield, plant height, brix content, juice weight, stem weight, leaf weight, precocity	973 diverse sorghum lines	Imputation based on haplotype graph	Developed a haplotype graph to enable imputation and GP	Jensen et al. (2020)

(continued)

TABLE 7.1 (Continued)
A summary of genomic selection studies conducted in sorghum

Objective	Traits	Population	Statistical Method	Results	References
Influence of sorghum racial composition and genetic diversity on the GP of grain yield traits	Grain yield and component traits (grain number & weight, panicle length)	Diverse sorghum lines	Bayesian Lasso and RR-BLUP	Accuracies varied significantly depending on the statistical method used and trait	Sapkota et al. (2020a)
GS for antioxidant production	Antioxidant content	Diverse lines of *S.bicolor*and *S. bicolor* x *S. halepense*	RR-BLUP	Accuracies varied from 0.20 to 0.56 for various traits	Habyarimana et al. (2019)
Comparison of GS methods	Plant height, biomass yield, stem diameter, sugar content	Diverse F_3 lines	BayesCπ, GBLUP, BayesR, Bayesian Lasso	GBLUP provided the highest accuracy for all the traits	Fernandes et al. (2018)
Optimization of GS for a sorghum breeding program	Flowering time, plant height, grain yield, stay green	2000 diverse individuals (simulated population)	Bayesian LASSO, GBLUP	Optimized GS methods to improve genetic gain in the breeding program	Muleta et al. (2019)
Development of GS models	Flowering time, plant height, grain yield, stay green	544 genotyped lines	Pedigree and marker-based relationship models	Implementing genomic selection in highlighting factors such as genetic relatedness and G X E	Hunt et al. (2018)

performance. The inclusion of G × E interaction effects in GP models consistently led to higher prediction accuracies across all traits studied. Their work showed that including G × E interactions consistently improved prediction accuracy across traits, with plant height and days to anthesis being more predictable than grain yield. The study also demonstrated the benefits of population-specific predictions, where hybrids bred for a particular environment exhibited improved prediction accuracy, especially when G × E interactions were incorporated. The GB-MMGE (genomic Bayesian model with multi-environment and G × E effects) model, which accounts for G × E effects, was particularly effective, achieving high coincidence indices even at low selection intensities.

Furthermore, the inclusion of inbred phenotypic data and G × E effects in GP models can further enhance breeding efficiency, particularly for untested lines. Crozier et al. (2023) found that incorporating inbred phenotypic data significantly improved prediction accuracy by up to 33% for traits like plant height, particularly when no testcross data for hybrids were available. While dominance effects were evaluated, their contribution to predictive ability was inconsistent and did not show clear benefits in the validation set, suggesting limited utility under the current population structure. Importantly, modeling G × E interactions consistently improved the accuracy of predictions across traits, emphasizing its value in optimizing genomic prediction for hybrid sorghum.

Increasing training population sizes has also been shown to improve prediction accuracy, with both additive and dominance marker effects contributing to better predictions for grain yield. Maulana et al. (2023) emphasize that incorporating both additive and dominance effects in prediction models further improves accuracy, with grain yield achieving the highest prediction accuracy when using a larger training population hybrid. The importance of population size and genomic relatedness was further demonstrated by Sapkota et al. (2023), who optimized training population design to achieve up to 0.93 prediction accuracy when both parental lines were included in the training set.

Traditional methods of developing A/B-line pairs in sorghum are time-consuming and have limited genetic diversity in elite germplasm. Conventionally, new B-lines must first be developed and subsequently converted to A-lines through backcrossing to introduce the cytoplasmic male sterility (CMS) system. This two-phase process creates significant limitations in sorghum breeding, including reduced genetic diversity among elite A/B-line germplasm and the inability to evaluate B-line hybrid performance early in the breeding pipeline. However, the use of trifluoromethanesulfonamide (TFMSA), a chemical gametocide, offers a solution by inducing temporary male sterility in B-lines, enabling breeders to test hybrid combinations earlier in the breeding pipeline without requiring the labor-intensive conversion to A-lines. Kent et al. (2023) showed that combining genomic prediction with TFMSA-induced male sterility improved hybrid performance screening for grain yield and days to mid-anthesis. Despite variations in prediction accuracy, the models effectively ranked hybrids, accurately identifying the bottom-performing hybrids for grain yield, even when prediction accuracies were low. This combination of GP and TFMSA provides a cost-effective approach to evaluating hybrid performance in early breeding stages.

7.7.2 Genomic Selection for Quality Traits in Sorghum

Sorghum serves as a vital dietary staple for more than 500 million people across the semi-arid tropics of Africa and Asia, playing a crucial role in global food security (Lindsay, 2010). Efforts to enhance key quality attributes such as antioxidant content, grain composition, and overall nutritional value are central breeding objectives aimed at improving the crop's economic viability and adaptability. A comprehensive understanding of the intricate interactions among genetic factors, environmental conditions, and agronomic management practices can facilitate the identification of superior genotypes tailored for specific applications. While grain sorghum is predominantly cultivated for livestock feed in regions such as the United States and Australia, its utilization has diversified significantly in recent years, now encompassing applications in baking, malting, brewing, and biofortification (Taylor et al., 2006; Taylor et al., 2012; Zhu, 2014). Consequently, refining the genetic makeup of sorghum is essential to combat global malnutrition, optimize feed efficiency for livestock production, and capitalize on the growing demand for gluten-free grains in niche markets. Numerous grain quality traits exhibit high heritability; however, their precise phenotyping remains resource-intensive. Despite being controlled by multiple genes, these traits are well-suited for GS.

While GS has been widely employed to enhance grain quality in major crops such as wheat (Heffner et al., 2011; Battenfield et al., 2016; Haile et al., 2018), rye (Schulthess et al., 2016), maize (Guo et al., 2014), and soybean (Duhnen et al., 2017), its application in sorghum remains relatively unexplored. One notable study by Sapkota et al. (2020b) investigated the feasibility of GP for improving grain composition traits using multi-trait, multi-environment models. Their findings revealed that the Single-Trait Single-Environment (STSE) model produced average prediction accuracies ranging between 0.4 and 0.6 for most traits, with the exception of amylose, which exhibited a notably lower accuracy of 0.25. Interestingly, the Bayesian Multi-Output Regressor Stacking (BMORS) model significantly outperformed STSE, enhancing prediction accuracy by an average of 41% and 32%, respectively. Although the Bayesian multi-environment (BME) model provided only marginal improvements, BMORS demonstrated a remarkable ability to predict trait performance in unobserved environments, achieving accuracies of 0.5 or higher for the majority of trait-environment interactions. These findings underscore the potential of multi-trait models such as BMORS, which leverage trait-environment relationships to enhance genomic predictions, ultimately reducing the reliance on labor-intensive phenotyping while improving sorghum breeding efficiency.

Another study by Habyarimana and Lopez-Cruz (2019) evaluated the performance of four GS models, GBLUP, BRR, Bayesian LASSO, and BayesB, for predicting antioxidant traits in sorghum, including polyphenols, flavonoids, total antioxidant capacity, and condensed tannins. The GS models achieved prediction accuracies ranging from 0.49 to 0.58, with polyphenols and flavonoids achieving accuracies of 0.53–0.56, total antioxidant capacity at 0.49–0.52, and condensed tannins at 0.55–0.58. The study also found that *S. bicolor* × *S. halepense* recombinant inbred lines outperformed pure *S. bicolor* lines for all antioxidant traits, showing higher total antioxidant capacity values (59.16 mmol TE (Trolox equivalents) kg^{-1}) compared to 40.17 mmol TE kg^{-1} in *S. bicolor*. This research highlights the potential of GS to accelerate breeding for antioxidant traits, which are traditionally costly and labor-intensive to measure, while incorporating genetic diversity from wild relatives like *S. halepense* to enhance antioxidant production in sorghum.

7.7.3 Genomic Selection for Abiotic Stress Tolerance in Sorghum

Sorghum exhibits remarkable drought tolerance, making it a valuable crop in water-limited regions. The stay-green trait, which allows plants to maintain green leaves and photosynthetic capacity after anthesis, is a crucial contributor to drought tolerance in sorghum (Thomas & Ougham, 2014; Abebe et al., 2021). The *Stg1-Stg4* QTLs are key to post-flowering drought tolerance and have been linked to reduced tillering, smaller upper leaves, and lower leaf numbers, affecting water use efficiency and grain yield (Borrell et al., 2014; Harris et al., 2007; Tuinstra et al., 1997). Recent advancements in genomic technologies have facilitated the identification of numerous genomic regions associated with natural variation in photosynthetic traits under drought conditions, revealing promising candidate genes involved in stress signaling, oxidative stress protection, hormonal response to stress, and dehydration protection (Ortiz & Salas-Fernandez, 2022). The availability of genome-wide markers presents an opportunity to leverage genomic prediction models for improving drought tolerance in sorghum breeding programs. However, despite the potential of GS to enhance genetic gain rates in sorghum, studies on the application of genomic models in this crop for drought tolerance are limited compared to other cereals such as maize, wheat, and rice (Kulwal, 2016). Velazco et al. (2019) highlighted the predictive ability for stay-green increased by 19% when grain yield data was available for both training and validation lines, showcasing the benefit of incorporating correlated traits in multi-trait models for predicting drought tolerance trait. While stay-green exhibited moderate heritability, its strong positive genetic correlation with grain yield allowed additional predictive gains when the joint analysis was performed. The study utilized G-BLUP models and further enhanced performance by combining pedigree and genomic information in a method termed K-BLUP. This combined approach produced the highest increase in the predictive ability of about 30% for prediction of stay-green.

7.7.4 Genomic Selection for Pre-breeding in Sorghum

GS has emerged as a powerful tool for pre-breeding, offering significant potential to accelerate genetic gains and optimize the utilization of gene bank resources. With approximately 7.4 million plant accessions stored in global gene banks (Yu et al., 2016), GP models can enhance the management and evaluation of these resources by predicting traits and improving breeding efficiency. Despite the vast genetic diversity available, gene banks are underutilized due to limited phenotypic data and the large volume of material. However, advancements in genomic technologies now enable more effective mining of this genetic pool. GS can help establish a coordinated international effort to better exploit gene bank resources by predicting the phenotypic performance of untested accessions, making breeding programs more efficient. For sorghum, a study demonstrated the effectiveness of RR-BLUP in estimating biomass yield (Yu et al., 2016). A reference set of 962 photoperiod-sensitive sorghum accessions was selected from 34,844 accessions, representing five sorghum races. These accessions underwent genotyping-by-sequencing, identifying 340,496 high-quality SNPs, and a training set of 299 accessions was phenotypically evaluated for biomass yield. The study found that multiple prediction models, including Bayes LASSO and Gaussian kernel, yielded similar accuracies, indicating that breeders can select models based on specific needs without compromising performance. The careful design of the training set, incorporating accessions with extreme predicted values, further enhanced prediction reliability. This approach not only facilitates the efficient evaluation of germplasm but also accelerates the identification of superior traits, making GS a vital tool in pre-breeding efforts aimed at improving sorghum's resilience and productivity. As genomic prediction continues to evolve, its integration into pre-breeding strategies promises to bridge the gap between untapped genetic resources and the development of elite cultivars, ultimately contributing to food security and sustainable agricultural practices (Crossa et al., 2017).

7.8 INTEGRATION OF GS WITH HIGH-THROUGHPUT PHENOTYPING IN SORGHUM IMPROVEMENT

Integrating GS with high-throughput phenotyping offers a powerful advancement in sorghum improvement by enabling the rapid and efficient evaluation of large populations across diverse environments for different traits. HTP has revolutionized the way plant traits are analyzed and characterized, enabling the rapid, precise, and non-destructive collection of large amount of phenotypic data. This technological advancement significantly reduces the reliance on traditional manual phenotyping, which is often labor-intensive, error-prone, and time-consuming (Gill et al., 2022; Kundu et al., 2024). Utilizing advanced tools such as spectral imaging, drones, and remote sensing, facilitates the efficient monitoring of complex traits and crop growth dynamics under varying environmental conditions (Singh et al., 2016). GS models are highly dependent on the accuracy of phenotypic data, making precision phenotyping a critical component for reliable prediction. HTP bridges this gap by providing large-scale, high-resolution phenotypic data that enhances the accuracy and robustness of GS models. GS seeks to minimize the expenses associated with phenotyping while expediting genetic improvement by leveraging molecular markers, whereas HTP is centered on capturing extensive phenotypic data across large breeding populations over time and diverse environments through remote or proximal sensing technologies. The synergistic integration of these two methodologies can significantly enhance selection accuracy and intensity, ultimately boosting genetic gains while simultaneously reducing the costs of phenotyping. Moreover, early-generation prediction plays a pivotal role in maximizing genetic improvement during the initial breeding stages. However, applying GS at these early stages is economically unfeasible due to the vast number of field-grown plants requiring evaluation. Although pedigree-based relationships provide a cost-free means of assessing relatedness among individuals, they fail to capture Mendelian sampling variance, which GS can exploit. Therefore, adopting multivariate pedigree-based prediction models that integrate predictor traits derived from HTP platforms presents a more cost-effective strategy for predicting genetic potential within and among breeding families during early-generation testing (Rutkoski et al., 2016). In addition to facilitating early-stage selection, HTP platforms provide valuable data on secondary traits that exhibit genetic correlations with grain yield. These secondary traits can be incorporated into multivariate pedigree-based and genomic prediction frameworks to enhance indirect selection for grain yield. Empirical research indicates that utilizing statistical models that estimate the breeding values of secondary traits, in conjunction with multivariate pedigree and genomic models, can result in an approximate 70% increase in grain yield selection accuracy when secondary traits are included in both training and validation populations (Sun et al., 2017). This underscores the potential of integrating HTP-derived trait information into genomic and pedigree-based selection models to optimize breeding efficiency and accelerate genetic progress.

The integration of HTP data with GS approaches remains relatively limited in sorghum. Watanabe et al. (2017) evaluated the potential of UAV remote sensing for phenotyping in genomic prediction.

The NIR-GB camera and the 50th percentile of the digital surface model effectively estimated plant height. UAV-measured height (PHUAV) correlated with ruler-measured height (PHR) at 0.523, improving to 0.678 by selecting the lower PHUAV replication. Under low-fertilization conditions, PHUAV overestimation was reduced, enhancing genomic prediction. PHUAV and PHR predictions were highly correlated ($r = 0.842$), indicating comparable performance. UAV remote sensing shows strong potential for high-throughput, cost-effective phenotyping in genomics-assisted breeding. Future studies could focus on integrating advanced machine learning algorithms to enhance the accuracy of image analysis for high-throughput phenotyping. Additionally, the deployment of multispectral or hyperspectral sensors could enable the capture of a broader range of plant traits, such as canopy temperature, chlorophyll content, and stress indices. These traits, when incorporated as predictors in genomic prediction models, have the potential to improve the precision of breeding value estimates, thereby accelerating genetic gain and enhancing the overall efficiency of sorghum breeding programs.

7.9 CONCLUSIONS AND FUTURE PROSPECTS

To address the growing global food demand, accelerating genetic gain in sorghum is crucial. GS is a transformative tool in this process, and as DNA sequencing costs decline, its adoption in sorghum breeding is expected to rise. GS can reduce breeding cycle durations, optimize population sizes, and aid in selecting superior parents for variety development. However, compared to crops like wheat, maize, and rice, the application of GS in sorghum remains limited. Future research should enhance GS model predictions by incorporating multi-trait, multi-environment models that account for genetic covariance and G × E interactions, improving hybrid performance prediction. Optimizing training population sizes and refining prediction models will be key to advancing GS in sorghum. While linear and Bayesian models have been used, machine learning and deep learning models offer substantial opportunities for hybrid prediction.

Research should prioritize generating hybrid combinations through diallel or factorial mating designs, leveraging genetically diverse individuals from distinct heterotic groups. This approach would create robust training sets for accurately predicting the genetic value of untested single crosses, thereby enhancing the efficiency of hybrid development programs. The use of TFMSA male gametocide could expedite hybrid breeding by enabling the production of hybrids from B lines without conversion to A lines, thus reducing costs and time associated with cytoplasmic male sterility line development. This approach facilitates the rapid assessment of hybrid performance using GP models. Integration of GS with complementary technologies such as HTP, gene editing, and speed breeding will accelerate breeding cycles and further enhance genetic gains. Developing standardized speed breeding protocols in sorghum and their integration with GS would allow fast-track breeding. Furthermore, leveraging HTP technologies for precise large-scale phenotyping and incorporating these data into GS pipelines can enhance the accuracy and efficiency of genomic predictions. Continuous efforts to dissect the genetic architecture of traits, refine optimization strategies, and integrate advanced computational tools will be pivotal for maximizing the impact of GS in sorghum breeding. Ultimately, these advancements will not only accelerate genetic gain but also enhance the adaptability of sorghum to the challenges posed by climate change and global food demand.

REFERENCES

Abebe, T., Belay, G., Tadesse, T., & Keneni, G. (2021). Stay-green genes contributed for drought adaptation and performance under post-flowering moisture stress on sorghum (*Sorghum bicolor* L. Moench). *Journal of Plant Breeding and Crop Science*, *13*(4), 190–202.

Alemu, A., Åstrand, J., Montesinos-Lopez, O. A., y Sanchez, J. I., Fernandez-Gonzalez, J., Tadesse, W., . . . Chawade, A. (2024). Genomic selection in plant breeding: Key factors shaping two decades of progress. *Molecular Plant, 17*(4), 552–578.

Baloch, F. S., Altaf, M. T., Liaqat, W., Bedir, M., Nadeem, M. A., Cömertpay, G., . . . Sun, H. J. (2023). Recent advancements in the breeding of sorghum crop: Current status and future strategies for marker-assisted breeding. *Frontiers in Genetics*, *14*, Art. 1150616.

Bassi, F. M., Bentley, A. R., Charmet, G., Ortiz, R., & Crossa, J. (2016). Breeding schemes for the implementation of genomic selection in wheat (*Triticum* spp.). *Plant Science*, *242*, 23–36.

Battenfield, S. D., Guzmán, C., Gaynor, R. C., Singh, R. P., Peña, R. J., Dreisigacker, S., . . . Poland, J. A. (2016). Genomic selection for processing and end-use quality traits in the CIMMYT spring bread wheat breeding program. *The Plant Genome*, *9*(2), 1–12.

Bernardo, R. (2016). Bandwagons I, too, have known. *Theoretical and Applied Genetics*, *129*, 2323–2332.

Bernardo, R., & Yu, J. (2007). Prospects for genome wide selection for quantitative traits in maize. *Crop Science*, *47*(3), 1082–1090.

Beyene, Y., Semagn, K., Mugo, S., Tarekegne, A., Babu, R., Meisel, B., . . . Crossa, J. (2015). Genetic gains in grain yield through genomic selection in eight bi-parental maize populations under drought stress. *Crop Science*, *55*(1), 154–163.

Borrell, A. K., van Oosterom, E. J., Mullet, J. E., George-Jaeggli, B., Jordan, D. R., Klein, P. E., & Hammer, G. L. (2014). Stay-green alleles individually enhance grain yield in sorghum under drought by modifying canopy development and water uptake patterns. *New Phytologist*, *203*(3), 817–830.

Buckler, E. S., Holland, J. B., Bradbury, P. J., Acharya, C. B., Brown, P. J., Browne, C., . . . McMullen, M. D. (2009). The genetic architecture of maize flowering time. *Science*, *325*(5941), 714–718.

Charles, J. R., Dorval, M. D., Durone, J. B., Ferrão, L. F. V., Amadeu, R. R., Munoz, P. R., . . . Pressoir, G. (2024). Genomic prediction of sweet sorghum agronomic performance under drought and irrigated environments in Haiti. *Crop Science, 64*(3), 1595–1607.

Chen, G. K., Marjoram, P., & Wall, J. D. (2009). Fast and flexible simulation of DNA sequence data. *Genome Research*, *19*(1), 136–142.

Chen, T., & He, T. (2015, August). Higgs boson discovery with boosted trees. In *NIPS 2014 workshop on high-energy physics and machine learning* (pp. 69–80). PMLR: Workshop and Conference Proceedings *42*, 69–80..

Chen, X., & Ishwaran, H. (2012). Random forests for genomic data analysis. *Genomics*, *99*(6), 323–329.

Clark, S. A., Hickey, J. M., Daetwyler, H. D., & van der Werf, J. H. (2012). The importance of information on relatives for the prediction of genomic breeding values and the implications for the makeup of reference data sets in livestock breeding schemes. *Genetics Selection Evolution*, *44*, 1–9.

Costa-Neto, G., Fritsche-Neto, R., & Crossa, J. (2021). Nonlinear kernels, dominance, and envirotyping data increase the accuracy of genome-based prediction in multi-environment trials. *Heredity*, *126*(1), 92–106.

Cowan, M. F., Blomstedt, C. K., Norton, S. L., Henry, R. J., Møller, B. L., & Gleadow, R. (2020). Crop wild relatives as a genetic resource for generating low-cyanide, drought-tolerant Sorghum. *Environmental and Experimental Botany*, *169*, 103884.

Crossa, J., Campos, G. D. L., Pérez, P., Gianola, D., Burgueno, J., Araus, J. L., . . . Braun, H. J. (2010). Prediction of genetic values of quantitative traits in plant breeding using pedigree and molecular markers. *Genetics*, *186*(2), 713–724.

Crossa, J., Pérez-Rodríguez, P., Cuevas, J., Montesinos-López, O., Jarquín, D., de los Campos, G., . . . Varshney, R. K. (2017). Genomic selection in plant breeding: Methods, models, and perspectives. *Trends in Plant Science*, *22*(11), 961–975.

Crozier, D., Leon, F., Fonseca, J. M., Klein, P. E., Klein, R. R., & Rooney, W. L. (2023). Inbred phenotypic data and non-additive effects can enhance genomic prediction models for hybrid grain sorghum. *Crop Science*, *63*(3), 1183–1196.

de los Campos, G., Gianola, D., & Rosa, G. J. (2009). Reproducing kernel Hilbert spaces regression: A general framework for genetic evaluation. *Journal of Animal Science*, *87*(6), 1883–1887.

de los Campos, G., Veturi, Y., Vazquez, A. I., Lehermeier, C., & Pérez-Rodríguez, P. (2015). Incorporating genetic heterogeneity in whole-genome regressions using interactions. *Journal of Agricultural, Biological, and Environmental Statistics*, *20*, 467–490.

de Oliveira, A. A., Pastina, M. M., de Souza, V. F., da Costa Parrella, R. A., Noda, R. W., Simeone, M. L. F., . . . Margarido, G. R. A. (2018). Genomic prediction applied to high-biomass sorghum for bioenergy production. *Molecular Breeding*, *38*, 1–16.

Dos Santos, J. P., Fernandes, S. B., McCoy, S., Lozano, R., Brown, P. J., Leakey, A. D., . . . Gore, M. A. (2020). Novel bayesian networks for genomic prediction of developmental traits in biomass sorghum. *G3: Genes, Genomes, Genetics*, *10*(2), 769–781.

Duhnen, A., Gras, A., Teyssèdre, S., Romestant, M., Claustres, B., Daydé, J., & Mangin, B. (2017). Genomic selection for yield and seed protein content in soybean: A study of breeding program data and assessment of prediction accuracy. *Crop Science*, *57*(3), 1325–1337.

Erbe, M., Hayes, B. J., Matukumalli, L. K., Goswami, S., Bowman, P. J., Reich, C. M., . . . Goddard, M. E. (2012). Improving accuracy of genomic predictions within and between dairy cattle breeds with imputed high-density single nucleotide polymorphism panels. *Journal of Dairy Science*, *95*(7), 4114–4129.

Falcon, W. P., Naylor, R. L., & Shankar, N. D. (2022). Rethinking global food demand for 2050. *Population and Development Review*, *48*(4), 921–957.

Falconer, D. S. (1996). *Introduction to quantitative genetics*. India: Pearson Education.

Fernandes, S. B., Dias, K. O., Ferreira, D. F., & Brown, P. J. (2018). Efficiency of multi-trait, indirect, and trait-assisted genomic selection for improvement of biomass sorghum. *Theoretical and Applied Genetics*, *131*, 747–755.

Fonseca, J. M., Klein, P. E., Crossa, J., Pacheco, A., Perez-Rodriguez, P., Ramasamy, P., . . . Rooney, W. L. (2021). Assessing combining abilities, genomic data, and genotype× environment interactions to predict hybrid grain sorghum performance. *The Plant Genome*, *14*(3), Art. e20127.

George, E. I., & McCulloch, R. E. (1993). Variable selection via Gibbs sampling. *Journal of the American Statistical Association*, *88*(423), 881–889.

Gianola, D., & de los Campos, G. (2008). Inferring genetic values for quantitative traits non-parametrically. *Genetics Research*, *90*(6), 525–540.

Gianola, D., Fernando, R. L., & Stella, A. (2006). Genomic-assisted prediction of genetic value with semiparametric procedures. *Genetics*, *173*(3), 1761–1776.

Gianola, D., Okut, H., Weigel, K. A., & Rosa, G. J. (2011). Predicting complex quantitative traits with Bayesian neural networks: A case study with Jersey cows and wheat. *BMC Genetics*, *12*, 1–14.

Gill, T., Gill, S. K., Saini, D. K., Chopra, Y., de Koff, J. P., & Sandhu, K. S. (2022). A comprehensive review of high throughput phenotyping and machine learning for plant stress phenotyping. *Phenomics*, *2*(3), 156–183.

González-Camacho, J. M., Crossa, J., Pérez-Rodríguez, P., Ornella, L., & Gianola, D. (2016). Genome-enabled prediction using probabilistic neural network classifiers. *BMC Genomics*, *17*, 1–16.

González-Camacho, J. M., de los Campos, G., Pérez, P., Gianola, D., Cairns, J. E., Mahuku, G., . . . Crossa, J. (2012). Genome-enabled prediction of genetic values using radial basis function neural networks. *Theoretical and Applied Genetics*, *125*, 759–771.

González-Camacho, J. M., Ornella, L., Pérez-Rodríguez, P., Gianola, D., Dreisigacker, S., & Crossa, J. (2018). Applications of machine learning methods to genomic selection in breeding wheat for rust resistance. *The Plant Genome*, *11*(2), Art. 170104.

Gudi, S., Kumar, P., Singh, S., Tanin, M. J., & Sharma, A. (2022). Strategies for accelerating genetic gains in crop plants: Special focus on speed breeding. *Physiology and Molecular Biology of Plants*, *28*(10), 1921–1938.

Guo, Z., Tucker, D. M., Basten, C. J., Gandhi, H., Ersoz, E., Guo, B., . . . Gay, G. (2014). The impact of population structure on genomic prediction in stratified populations. *Theoretical and Applied Genetics*, *127*, 749–762.

Habier, D., Fernando, R. L., & Dekkers, J. (2007). The impact of genetic relationship information on genome-assisted breeding values. *Genetics*, *177*(4), 2389–2397.

Habier, D., Fernando, R. L., Kizilkaya, K., & Garrick, D. J. (2011). Extension of the Bayesian alphabet for genomic selection. *BMC Bioinformatics*, *12*, 1–12.

Habyarimana, E., & Lopez-Cruz, M. (2019). Genomic selection for antioxidant production in a panel of *Sorghum bicolor* and *S. bicolor* × *S. halepense* lines. *Genes*, *10*(11), 841.

Habyarimana, E., Lopez-Cruz, M., & Baloch, F. S. (2020). Genomic selection for optimum index with dry biomass yield, dry mass fraction of fresh material, and plant height in biomass sorghum. *Genes*, *11*(1), 61.

Haile, J. K., N'Diaye, A., Clarke, F., Clarke, J., Knox, R., Rutkoski, J., . . . Pozniak, C. J. (2018). Genomic selection for grain yield and quality traits in durum wheat. *Molecular Breeding*, *38*, 1–18.

Harris, K., Subudhi, P. K., Borrell, A., Jordan, D., Rosenow, D., Nguyen, H., . . . Mullet, J. (2007). Sorghum stay-green QTL individually reduce post-flowering drought-induced leaf senescence. *Journal of Experimental Botany*, *58*(2), 327–338.

Hasan, N., Choudhary, S., Naaz, N., Sharma, N., & Laskar, R. A. (2021). Recent advancements in molecular marker-assisted selection and applications in plant breeding programmes. *Journal of Genetic Engineering and Biotechnology*, *19*(1), 128.

Heffner, E. L., Jannink, J. L., Iwata, H., Souza, E., & Sorrells, M. E. (2011). Genomic selection accuracy for grain quality traits in biparental wheat populations. *Crop Science*, *51*(6), 2597–2606.

Heffner, E. L., Lorenz, A. J., Jannink, J. L., & Sorrells, M. E. (2010). Plant breeding with genomic selection: Gain per unit time and cost. *Crop Science*, *50*(5), 1681–1690.

Heffner, E. L., Sorrells, M. E., & Jannink, J. L. (2009). Genomic selection for crop improvement. *Crop Science*, *49*(1), 1–12.

Heslot, N., Rutkoski, J., Poland, J., Jannink, J. L., & Sorrells, M. E. (2013). Impact of marker ascertainment bias on genomic selection accuracy and estimates of genetic diversity. *PloS One*, *8*(9), Art. e74612.

Heslot, N., Yang, H. P., Sorrells, M. E., & Jannink, J. L. (2012). Genomic selection in plant breeding: A comparison of models. *Crop Science*, *52*(1), 146–160.

Hunt, C. H., van Eeuwijk, F. A., Mace, E. S., Hayes, B. J., & Jordan, D. R. (2018). Development of genomic prediction in sorghum. *Crop Science*, *58*(2), 690–700.

Isidro, J., Jannink, J. L., Akdemir, D., Poland, J., Heslot, N., & Sorrells, M. E. (2015). Training set optimization under population structure in genomic selection. *Theoretical and Applied Genetics*, *128*, 145–158.

Islam, M. S., McCord, P. H., Olatoye, M. O., Qin, L., Sood, S., Lipka, A. E., & Todd, J. R. (2021). Experimental evaluation of genomic selection prediction for rust resistance in sugarcane. *The Plant Genome*, *14*(3), Art. e20148.

Jarquin, D., De Leon, N., Romay, C., Bohn, M., Buckler, E. S., Ciampitti, I., . . . Lorenz, A. (2021). Utility of climatic information via combining ability models to improve genomic prediction for yield within the genomes to fields maize project. *Frontiers in Genetics*, *11*, Art. 592769.

Jensen, S. E., Charles, J. R., Muleta, K., Bradbury, P. J., Casstevens, T., Deshpande, S. P., . . . Ramstein, G. P. (2020). A sorghum practical haplotype graph facilitates genome-wide imputation and cost-effective genomic prediction. *The Plant Genome*, *13*(1), Art. e20009.

Jiang, G. L. (2015). Molecular marker-assisted breeding: A plant breeder's review. In Al-Khayri, J., Jain, S., Johnson, D. (Eds.) *Advances in plant breeding strategies: Breeding, biotechnology and molecular tools* (pp. 431–472), Springer, Cham.

Kent, M. A., Fonseca, J. M., Klein, P. E., Klein, R. R., Hayes, C. M., & Rooney, W. L. (2023). Use of genomic prediction to screen sorghum B-lines in hybrid testcrosses. *The Plant Genome*, *16*(3), Art. e20369.

Khalifa, M., & Eltahir, E. A. (2023). Assessment of global sorghum production, tolerance, and climate risk. *Frontiers in Sustainable Food Systems*, *7*, Art. 1184373.

Kulwal, P. L. (2016). Association Mapping and Genomic Selection—Where Does Sorghum Stand?. In: Rakshit, S., Wang, YH. (eds) The Sorghum Genome. Compendium of Plant Genomes. Springer, Cham.

Kundu, S., Saini, D. K., Meena, R. K., Bahuguna, R. N., & Jagadish, S. V. K. (2024). High-throughput phenotyping and AI technologies for deciphering crop resilience to heat stress. *Plant Physiology Reports*, 29, 699–715.

Labroo, M. R., Studer, A. J., & Rutkoski, J. E. (2021). Heterosis and hybrid crop breeding: A multidisciplinary review. *Frontiers in Genetics*, *12*, Art. 643761.

LeCun, Y., Bengio, Y., & Hinton, G. (2015). Deep learning. *Nature*, *521*(7553), 436–444.

Li, B., Zhang, N., Wang, Y. G., George, A. W., Reverter, A., & Li, Y. (2018). Genomic prediction of breeding values using a subset of SNPs identified by three machine learning methods. *Frontiers in Genetics*, *9*, 237.

Lindsay, J. (2010). Sorghum: An ancient, healthy and nutritious old world cereal. *NTSORMIL Scientific Publications*, *7*, 1–28.

Lorenzana, R. E., & Bernardo, R. (2009). Accuracy of genotypic value predictions for marker-based selection in biparental plant populations. *Theoretical and Applied Genetics*, *120*, 151–161.

Maenhout, S., De Baets, B., Haesaert, G., & Van Bockstaele, E. (2007). Support vector machine regression for the prediction of maize hybrid performance. *Theoretical and Applied Genetics*, *115*, 1003–1013.

Malosetti, M., Bustos-Korts, D., Boer, M. P., & van Eeuwijk, F. A. (2016). Predicting responses in multiple environments: Issues in relation to genotype× environment interactions. *Crop Science*, *56*(5), 2210–2222.
Maulana, F., Perumal, R., Serba, D. D., & Tesso, T. (2023). Genomic prediction of hybrid performance in grain sorghum (*Sorghum bicolor* L.). *Frontiers in Plant Science*, *14*, Art. 1139896.
Meuwissen, T. H., Hayes, B. J., & Goddard, M. (2001). Prediction of total genetic value using genome-wide dense marker maps. *Genetics*, *157*(4), 1819–1829.
Montes, J. M., Melchinger, A. E., & Reif, J. C. (2007). Novel throughput phenotyping platforms in plant genetic studies. *Trends in Plant Science*, *12*(10), 433–436.
Montesinos-López, A., Montesinos-López, O. A., Gianola, D., Crossa, J., & Hernández-Suárez, C. M. (2018). Multi-environment genomic prediction of plant traits using deep learners with dense architecture. *G3: Genes, Genomes, Genetics*, *8*(12), 3813–3828.
Montesinos-López, O. A., Montesinos-López, A., Kismiantini, Roman-Gallardo, A., Gardner, K., Lillemo, M., . . . Crossa, J. (2022). Partial least squares enhances genomic prediction of new environments. *Frontiers in Genetics*, *13*, Art. 920689.
Montesinos-López, O. A., Montesinos-López, J. C., Singh, P., Lozano-Ramirez, N., Barrón-López, A., Montesinos-López, A., & Crossa, J. (2020). A multivariate Poisson deep learning model for genomic prediction of count data. *G3: Genes, Genomes, Genetics*, *10*(11), 4177–4190.
Morota, G., & Gianola, D. (2014). Kernel-based whole-genome prediction of complex traits: A review. *Frontiers in Genetics*, *5*, 363.
Muleta, K. T., Pressoir, G., & Morris, G. P. (2019). Optimizing genomic selection for a sorghum breeding program in Haiti: A simulation study. *G3: Genes, Genomes, Genetics*, *9*(2), 391–401.
Munkvold, J. D., Tanaka, J., Benscher, D., & Sorrells, M. E. (2009). Mapping quantitative trait loci for preharvest sprouting resistance in white wheat. *Theoretical and Applied Genetics*, *119*, 1223–1235.
Norman, A., Taylor, J., Edwards, J., & Kuchel, H. (2018). Optimising genomic selection in wheat: Effect of marker density, population size and population structure on prediction accuracy. *G3: Genes, Genomes, Genetics*, *8*(9), 2889–2899.
Ornella, L., Singh, S., Perez, P., Burgueño, J., Singh, R., Tapia, E., . . . Crossa, J. (2012). Genomic prediction of genetic values for resistance to wheat rusts. *The Plant Genome*, *5*(3), 136–148.
Ortiz, D., & Salas-Fernandez, M. G. (2022). Dissecting the genetic control of natural variation in sorghum photosynthetic response to drought stress. *Journal of Experimental Botany*, *73*(10), 3251–3267.
Patterson, J., & Gibson, A. (2017). *Deep learning: A practitioner's approach*. O'Reilly Media, Inc.: Newton, MA, USA.
Pérez, P., de los Campos, G., Crossa, J., & Gianola, D. (2010). Genomic-enabled prediction based on molecular markers and pedigree using the Bayesian linear regression package in R. *The Plant Genome*, *3*(2), 106–116.
Pérez-Rodríguez, P., Gianola, D., González-Camacho, J. M., Crossa, J., Manès, Y., & Dreisigacker, S. (2012). Comparison between linear and non-parametric regression models for genome-enabled prediction in wheat. *G3: Genes| Genomes| Genetics*, *2*(12), 1595–1605.
Pfeiffer, B. K., Pietsch, D., Schnell, R. W., & Rooney, W. L. (2019). Long-term selection in hybrid sorghum breeding programs. *Crop Science*, *59*(1), 150–164.
Pong-Wong, R., & Woolliams, J. (2014, August). Bayes U: A genomic prediction method based on the horseshoe prior. In Proceedings, *10th world congress on genetics applied to livestock Production (WCGALP)* (p. 679), 10th World Congress on Genetics Applied to Livestock production (WCGALP), Vancouver, Canada, 17/08/14.
Pritchard, J. K., Stephens, M., Rosenberg, N. A., & Donnelly, P. (2000). Association mapping in structured populations. *The American Journal of Human Genetics*, *67*(1), 170–181.
Rizal, G., Karki, S., Alcasid, M., Montecillo, F., Acebron, K., Larazo, N., . . . Quick, W. P. (2014). Shortening the breeding cycle of sorghum, a model crop for research. *Crop Science*, *54*(2), 520–529.
Rooney, L. W., & Serna-Saldivar, S. O. (2000). Sorghum. In *Handbook of cereal science and technology, revised and expanded* (pp. 149–175). CRC Press: Boca Raton, Florida.
Rutkoski, J., Poland, J., Mondal, S., Autrique, E., Pérez, L. G., Crossa, J., . . . Singh, R. (2016). Canopy temperature and vegetation indices from high-throughput phenotyping improve accuracy of pedigree and genomic selection for grain yield in wheat. *G3: Genes, Genomes, Genetics*, *6*(9), 2799–2808.
Rutkoski, J. E. (2019). A practical guide to genetic gain. *Advances in Agronomy*, *157*, 217–249.

Sandhu, K. S., Lozada, D. N., Zhang, Z., Pumphrey, M. O., & Carter, A. H. (2021a). Deep learning for predicting complex traits in spring wheat breeding program. *Frontiers in Plant Science*, *11*, Art. 613325.

Sandhu, K. S., Patil, S. S., Pumphrey, M., & Carter, A. (2021b). Multitrait machine-and deep-learning models for genomic selection using spectral information in a wheat breeding program. *The Plant Genome*, *14*(3), Art. e20119.

Sapkota, S., Boatwright, J. L., Jordan, K., Boyles, R., & Kresovich, S. (2020b). Multi-trait regressor stacking increased genomic prediction accuracy of sorghum grain composition. *Agronomy*, *10*(9), 1221.

Sapkota, S., Boatwright, J. L., Kumar, N., Myers, M., Cox, A., Ackerman, A., . . . Kresovich, S. (2023). Genomic prediction of hybrid performance for agronomic traits in sorghum. *G3: Genes, Genomes, Genetics*, *13*(4), Art. jkac311.

Sapkota, S., Boyles, R., Cooper, E., Brenton, Z., Myers, M., & Kresovich, S. (2020a). Impact of sorghum racial structure and diversity on genomic prediction of grain yield components. *Crop Science*, *60*(1), 132–148.

Schaeffer, L. R. (2006). Strategy for applying genome-wide selection in dairy cattle. *Journal of animal Breeding and genetics*, *123*(4), 218–223.

Schulthess, A. W., Wang, Y., Miedaner, T., Wilde, P., Reif, J. C., & Zhao, Y. (2016). Multiple-trait-and selection indices-genomic predictions for grain yield and protein content in rye for feeding purposes. *Theoretical and Applied Genetics*, *129*, 273–287.

Singh, A., Ganapathysubramanian, B., Singh, A. K., & Sarkar, S. (2016). Machine learning for high-throughput stress phenotyping in plants. *Trends in Plant Science*, *21*(2), 110–124.

Solberg, T. R., Sonesson, A. K., Woolliams, J. A., & Meuwissen, T. H. (2009). Reducing dimensionality for prediction of genome-wide breeding values. *Genetics Selection Evolution*, *41*, 1–8.

Spindel, J., Begum, H., Akdemir, D., Virk, P., Collard, B., Redona, E., . . . McCouch, S. R. (2015). Genomic selection and association mapping in rice (*Oryza sativa*): Effect of trait genetic architecture, training population composition, marker number and statistical model on accuracy of rice genomic selection in elite, tropical rice breeding lines. *PLoS Genetics*, *11*(2), Art. e1004982.

Sun, J., Rutkoski, J. E., Poland, J. A., Crossa, J., Jannink, J. L., & Sorrells, M. E. (2017). Multitrait, random regression, or simple repeatability model in high-throughput phenotyping data improve genomic prediction for wheat grain yield. *The Plant Genome*, *10*(2), 1–12.

Sun, X., Ma, P., & Mumm, R. H. (2012). Nonparametric method for genomics-based prediction of performance of quantitative traits involving epistasis in plant breeding. *PLoS One*, *7*(11), Art. e50604.

Taylor, J. R. N. (2012). Food product development using sorghum and millets: Opportunities and challenges. *Quality Assurance and Safety of Crops & Foods*, *4*(3), 151–151.

Taylor, J. R. N., Schober, T. J., & Bean, S. R. (2006). Novel food and non-food uses for sorghum and millets. *Journal of Cereal Science*, *44*(3), 252–271.

Tessema, B. B., Liu, H., Sørensen, A. C., Andersen, J. R., & Jensen, J. (2020). Strategies using genomic selection to increase genetic gain in breeding programs for wheat. *Frontiers in Genetics*, *11*, Art. 578123.

Thomas, H., & Ougham, H. (2014). The stay-green trait. *Journal of Experimental Botany*, *65*(14), 3889–3900.

Tuinstra, M. R., Grote, E. M., Goldsbrough, P. B., & Ejeta, G. (1997). Genetic analysis of post-flowering drought tolerance and components of grain development in *Sorghum bicolor* (L.) Moench. *Molecular Breeding*, *3*, 439–448.

Usai, M. G., Goddard, M. E., & Hayes, B. J. (2009). LASSO with cross-validation for genomic selection. *Genetics Research*, *91*(6), 427–436.

VanRaden, P. M. (2008). Efficient methods to compute genomic predictions. *Journal of Dairy Science*, *91*(11), 4414–4423.

Velazco, J. G., Jordan, D. R., Mace, E. S., Hunt, C. H., Malosetti, M., & Van Eeuwijk, F. A. (2019). Genomic prediction of grain yield and drought-adaptation capacity in sorghum is enhanced by multi-trait analysis. *Frontiers in Plant Science*, *10*, 997.

Voss-Fels, K. P., Wei, X., Ross, E. M., Frisch, M., Aitken, K. S., Cooper, M., & Hayes, B. J. (2021). Strategies and considerations for implementing genomic selection to improve traits with additive and non-additive genetic architectures in sugarcane breeding. *Theoretical and Applied Genetics*, *134*, 1493–1511.

Waldmann, P. (2016). Genome-wide prediction using Bayesian additive regression trees. *Genetics Selection Evolution*, *48*, 1–12.

Watanabe, K., Guo, W., Arai, K., Takanashi, H., Kajiya-Kanegae, H., Kobayashi, M., . . . Iwata, H. (2017). High-throughput phenotyping of sorghum plant height using an unmanned aerial vehicle and its application to genomic prediction modeling. *Frontiers in Plant Science*, *8*, 421.

Winans, N. D., Fonseca, J. M., Perumal, R., Klein, P. E., Klein, R. R., & Rooney, W. L. (2024). Envirotyping can increase genomic prediction accuracy of new environments in grain sorghum trials depending on mega-environment. *Crop Science, 64*(5), 2519–2533.

Xu, Y., & Crouch, J. H. (2008). Marker-assisted selection in plant breeding: From publications to practice. *Crop Science*, *48*(2), 391–407.

Xu, Y., Li, P., Zou, C., Lu, Y., Xie, C., Zhang, X., . . . Olsen, M. S. (2017). Enhancing genetic gain in the era of molecular breeding. *Journal of Experimental Botany*, *68*(11), 2641–2666.

Yu, X., Li, X., Guo, T., Zhu, C., Wu, Y., Mitchell, S. E., . . . Yu, J. (2016). Genomic prediction contributing to a promising global strategy to turbocharge gene banks. *Nature Plants*, *2*(10), 1–7.

Zhu, F. (2014). Structure, physicochemical properties, modifications, and uses of sorghum starch. *Comprehensive Reviews in Food Science and Food Safety*, *13*(4), 597–610.

Zou, H. (2006). The adaptive lasso and its oracle properties. *Journal of the American Statistical Association*, *101*(476), 1418–1429.

Zou, H., & Hastie, T. (2005). Regularization and variable selection via the elastic net. *Journal of the Royal Statistical Society Series B: Statistical Methodology*, *67*(2), 301–320.

8 Progress in Hybrid Breeding for Enhanced Grain Yield, Stress Tolerance, and Quality Improvement in Grain Sorghum

Dinesh Kumar Saini

8.1 INTRODUCTION

Sorghum (*Sorghum bicolor* (L.] Moench) is the world's fifth most cultivated crop by acreage and grain yield, following wheat, maize, rice, and barley (Ngidi et al., 2024a, b). It is widely cultivated in the semi-arid regions of the world due to its high productivity under low-input and drought conditions. It is valued not only for its use in food and fodder but also as an industrial crop for biofuel production. Its C_4 photosynthetic pathway enhances its efficiency, leading to higher dry matter production and carbon sequestration, making it an eco-friendly option. Sorghum production globally averages nearly 58.7 million tons per year, with ten countries accounting for approximately 78.6% of the total production (USDA-FAS, 2024). The major sorghum-producing countries, including the United States, Nigeria, Brazil, and India, dominate the global market. The United States leads with 14% of production at 8.07 million metric tons in 2023/2024, followed by Nigeria at 11% (6.4 million metric tons), and both Brazil and India at 8%, with 4.91 and 4.4 million metric tons, respectively (USDA-FAS, 2024).

Despite advances in sorghum breeding, its average grain yield remains low. Global sorghum production averages 58.7 million tons annually, compared to 648.6 million tons for wheat, 1.16 billion tons for maize, and 778 million tons for paddy rice in 2023, with significantly larger harvested areas for these other cereals (KNOEMA, 2024). There are also significant regional disparities in global sorghum production, with yields varying widely due to differences in genotype and environmental conditions. Developed countries typically achieve yields about three times higher than the global average. To address these imbalances, there is an urgent need for modern breeding concepts and technologies to accelerate the development of sorghum varieties with higher genetic yield potential, improved stress resistance, and enhanced grain quality to meet diverse end-use demands.

Sorghum breeding primarily aims to enhance yield stability, improve stress resilience, and optimize grain quality. As a self-pollinated crop, sorghum is predominantly cultivated as hybrids in regions like the United States and Australia. Since the 1950s, hybrid breeding in the U.S. has relied on the cytoplasmic male sterility (CMS) system for cultivar development (Stephens & Holland, 1953). This system involves three distinct genetic lines: male sterile (A) lines, maintainers (B) that preserve sterility, and restorers (R) that restore fertility (Bohra et al., 2016). While these A and R lines function similarly to heterotic pools in maize, they do not always fall into well-defined heterotic groups (Niebur et al., 2004; Duvick, 2005). Though hybrid seed production is straightforward, developing inbred parental lines, particularly A/B lines, is labor-intensive due to

DOI: 10.1201/9781003623717-8

cytoplasm introgression and backcrossing requirements (Crozier et al., 2020). Hybrid breeding focuses on improving inter-population performance via reciprocal recurrent selection, where individuals are evaluated based on their parental contributions to hybrid vigor (Hallauer et al., 2010). The two primary goals of hybrid breeding programs are: (1) progressively enhancing combining ability within each population and (2) identifying superior hybrids with stable, high yield potential (Sapkota et al., 2023). Additionally, researchers are increasingly studying the genetic and molecular mechanisms underlying male sterility and fertility restoration, given their role in promoting heterosis (Kariyannanavar et al., 2024). Sorghum's diverse CMS sources, observed instability, and fertility restorer genes make it an ideal model for exploring male sterility-fertility dynamics.

This chapter reviews key advances in hybrid breeding systems to enhance sorghum productivity and adaptability. It covers male sterility mechanisms, fertility restoration strategies, and essential breeding objectives, including yield improvement, stress resilience, and grain quality. Topics such as heterotic grouping, combining ability, and hybrid performance evaluation are examined. Genetic advancements for dryland and irrigated systems are highlighted, emphasizing strategies to enhance yield stability under environmental stress. The discussion also addresses global challenges such as climate change and food security, underscoring the critical role of hybrid breeding in ensuring sorghum remains a resilient and sustainable crop for the future.

8.2 PRIMARY BREEDING OBJECTIVES IN GRAIN SORGHUM BREEDING

Defining clear objectives is critical for the success of any breeding program. In grain sorghum breeding, the focus is on improving yield and stability, enhancing stress resistance, and optimizing grain quality. Ideotype breeding targets developing ideal plants with traits that maximize yield and align with specific agricultural needs. However, this approach is time-consuming and rarely fully achieved in any crop (Carbajal-Friedrich & Burgess, 2024). Therefore, identifying key traits that directly or indirectly influence yield and adaptability is crucial for timely improvements, such as breeding disease or pest resistance, enhancing grain quality, or incorporating stress tolerance traits into high-yielding varieties.

8.2.1 Higher Grain Yield

Yield stagnation is a widespread issue that affects multiple crops and regions globally. Current yield trends in the top four crops are below the 2.4% annual increase required to meet global food demands by 2050 (Ray et al., 2012). Sorghum, like many other crops, has experienced stagnating yield trends in key regions, including the United States, China, Nigeria (Rakshit et al., 2014), and Argentina (Parra et al., 2020), underscoring the urgent need for focused improvement efforts.

Despite advancements in breeding programs that have produced superior sorghum varieties, achieving grain yield levels comparable to those of other cereal crops remains a significant challenge. Yield and adaptation have consistently been the primary objectives in sorghum improvement programs. Adaptation reflects a cultivar's ability to thrive and perform in its target environment, with successful adaptation relying on traits that enable consistent production even under stress. A high-yielding, widely adapted cultivar responds positively to varying environmental conditions, making it critical to address both yield and adaptation concurrently (Aruna & Cheruku, 2019). The performance of sorghum genotypes is further influenced by genotype × environment (G × E) interactions, which cause variability in performance across locations. These interactions significantly impact yield and its components, emphasizing the need for a thorough understanding of environmental influences in breeding programs (Chapman et al., 2000; Fonseca et al., 2021). Phenotypic screening has proven to be a vital tool for leveraging genetic variability and predicting genotype responses to diverse environments, thereby improving selection efficiency (Kumar et al., 2013).

Grain yield is a complex polygenic trait shaped by multiple component traits, such as grain size, grain number, thousand-grain weight, and harvest index. These traits, along with other morphophysiological characteristics, indirectly affect yield but are more amenable to improvement compared to grain yield itself, which is strongly influenced by environmental factors and G × E interactions. Due to its low heritability, direct selection for grain yield is less effective (Ngidi et al., 2024a, b). Consequently, breeders prioritize enhancing yield-contributing traits to maximize genetic gains. Ultimately, the primary goal of breeding programs is to achieve higher economic yields, as farmers prioritize maximum productivity. This goal may be attained by selecting transgressive segregants or developing hybrids that exhibit a high degree of heterosis. By addressing yield stagnation through a combination of improving yield-contributing traits and understanding G × E interactions, sorghum breeding programs can meet the challenges posed by yield stagnation and contribute significantly to global food security.

8.2.2 Stress Tolerance

Grain sorghum is susceptible to a wide range of diseases, which can lead to significant losses in both production and profitability (Das & Rajendrakumar, 2016). Seedling blights, often called "damping off diseases", affect sorghum seedlings due to soil-borne pathogens, weather conditions, or low seedling vigor. These blights can cause seeds to rot or seedlings to die, with fungi like *Pythium, Fusarium, Aspergillus,* and *Rhizoctonia* as common causal organisms. Leaf diseases, such as leaf blight caused by *Exserohilum turcicum*, and sooty stripe from *Ramulispora* species, impact sorghum leaves, reducing yield (Das & Rajendrakumar, 2016; Aslam et al., 2024). Similarly, head smut, downy mildew, and Fusarium stalk and root diseases also pose significant threats to sorghum (Little et al., 2018). Plant diseases can rapidly mutate, creating new pathotypes or races. Due to this variability and fluctuating environmental conditions, routine monitoring of sorghum for disease incidence is essential.

Further, insect pest damage significantly threatens sorghum productivity, with pests attacking the plant at various stages, from seeds to mature grain. Around 150 insect species across 29 families affect sorghum worldwide (Guo et al., 2011), with only a few being major pests of economic importance (Okosun et al., 2021). Key pests include shoot fly, midge, stem borer, and head bug during the rainy season, and sugarcane aphid, shoot fly, and shot bug in the post-rainy season (Okosun et al., 2021). Shoot flies, midges, stem borers, and head bugs cause significant sorghum damage in Africa, with losses exceeding 85% in some cases (Kahate et al., 2014; Okosun et al., 2021). Stem borer infestations alone reduce yields by 11–88% across various regions (Overholt et al., 2001), contributing to annual losses of $1.1 billion in Africa and Asia (Nwanze et al., 1995; Okosun et al., 2021). In China, sorghum yield losses due to insect pests range from 50–70% (Guo et al., 2011), while in India, losses are estimated at 35%, equating to $580 million (Reddy & Zehr, 2004; Okosun et al., 2021). In Australia, sorghum midge control and yield losses cost over $8 million annually (Henzell et al., 1994; Okosun et al., 2021), and in Brazil, losses from *S. sorgicola* and other pests range between $7 and $18 million (Oliveira et al., 2013). In the U.S., insect damage led to 90% losses in southern Florida in 2013 (Nuessly et al., 2013) and a $31.6 million economic impact in Texas from 2014 to 2015 (Zapata et al., 2016). Developing resistant cultivars via various breeding strategies is the most effective and cost-efficient way to reduce losses from diseases and pests. Resistance breeding relies on identifying donor genotypes within germplasm collections, with wild relatives serving as key sources of resilience traits (Ananda et al., 2020).

Sorghum has steeper shoot and root angles, along with more primary and secondary roots, allowing it to draw water from deeper soil layers and resist drought better than other cereals (Hadebe et al., 2017; Zheng et al., 2023). Unlike maize, which suffers from heat stress, sorghum remains largely unaffected by high temperatures (Chen et al., 2012, 2017). Certain sorghum accessions also exhibit greater tolerance to cold, salt, and heavy metals (Maulana et al., 2017; Sun et al., 2014;

Mishra et al., 2021). Despite inherent resilience of sorghum to different abiotic stresses, further breeding efforts are essential to enhance its tolerance. This is due to the increasing intensity and frequency of environmental stresses caused by climate change, which are pushing the limits of sorghum's natural tolerance (Eggen et al., 2019). Additionally, breeding for improved tolerance can ensure stable yields in diverse and unpredictable environments, address the need for better adaptation to specific local conditions, and improve performance of sorghum cultivars under extreme stress scenarios, ensuring food security in vulnerable regions.

8.2.3 Grain Quality

Sorghum is an essential staple crop in semi-arid and tropical regions, providing a rich source of carbohydrates, protein, ash, oil, and fiber (Queiroz et al., 2015). It also contributes more than 50% of the dietary micronutrients for millions of people across Asia, Africa, and Central America (Djanaguiraman et al., 2024). Compared to other cereals like rice and wheat, sorghum offers higher levels of key minerals such as iron and zinc (Chan et al., 2007). However, diets based primarily on sorghum are often insufficient in these vital nutrients (Garg et al., 2018). Additionally, sorghum contains anti-nutritional factors, including tannins, phytic acids, and oxalates, which limit protein absorption and digestibility (Abdelhalim et al., 2021; de Morais Cardoso et al., 2017). Enhancing sorghum's micronutrient content is crucial to combat hidden hunger in regions where it is a dietary staple. Biofortification through crop breeding is key to addressing malnutrition (Ali et al., 2024). Sorghum exhibits considerable variation in its macronutrient (starch, protein, fat) and micronutrient (minerals, antioxidants, vitamins) content across diverse germplasm (Shewayrga et al., 2012; Sukumaran et al., 2012; Djanaguiraman et al., 2024). This genetic diversity holds promise for breeding programs that target nutritional enhancement and specific end-use qualities.

Breeding objectives for grain quality in sorghum focus on improving its nutritional value, digestibility, and processing traits for both human consumption and industrial use. Key goals include increasing protein content, particularly essential amino acids like lysine, and reducing anti-nutritional factors such as tannins and phytates that hinder micronutrient absorption. Enhancing starch composition for better digestibility, improving grain texture for processing, and breeding for optimal grain size, shape, and color are also significant. Biofortification to boost iron and zinc levels remains crucial in addressing malnutrition in sorghum-dependent regions.

8.3 BREEDING SYSTEMS IN SORGHUM

Breeding methods in sorghum are guided by its self-pollinating nature, with self-pollination techniques used for variety development, while hybrids leverage heterosis through male sterility and partial cross-pollination. Breeders utilize various self-pollination techniques for improving sorghum, including mass selection, pure-line selection, pedigree, bulk, backcross, and single seed descent (Rakshit & Bellundagi, 2019). The choice of method depends on the breeding objectives and the type of population, with mass and pure-line selections suited for variable and homozygous populations, while bulk, pedigree, backcross, and SSD methods are used for segregating populations (Poehlman, 2013; Rakshit & Bellundagi, 2019). Selection procedures in sorghum breeding include the following two methods. (i) Mass selection involves selecting phenotypically similar individuals and mixing their seeds to form an improved variety. This method focuses on visual traits and natural resistance rather than progeny testing, which helps maintain genetic variability but lacks precision in determining zygosity (Foster et al., 1980). (ii) Pure-line Selection targets individual plants from genetically uniform populations, which are then evaluated for yield and agronomic traits. This approach ensures genetic purity within the selected lines but may introduce variability between different lines (Poehlman, 2013).

Methods based on hybridization in sorghum breeding include the following. (i) *Pedigree method*: this involves selecting plants from segregating generations (F_2 and beyond) while carefully tracking parent-progeny relationships. It aims to maintain genetic variation and achieve uniformity in advanced generations, resulting in many current sorghum varieties. (ii) *Bulk method*: in this approach, plants are bulked together in early generations, with individual selection based on progeny performance occurring later. The method relies on natural selection and reduces the need for intensive management. (iii) *Single Seed Descent method*: this is a variation of bulk selection where one seed per plant is advanced from early generations. This method efficiently captures genetic variation and is commonly used to develop recombinant inbred lines. (iv) *Backcross method*: this technique is used to introduce specific traits into improved varieties. Repeated backcrossing of a donor parent to a recurrent parent aims to transfer desirable traits while minimizing undesirable genetic material (Rakshit & Bellundagi, 2019). Further, population improvement focuses on isolating superior inbreds from a base population by enhancing desirable genes and minimizing inbreeding. Recurrent selection and random crossing, alongside male sterility systems, have significantly advanced grain yield, disease resistance, and other traits in sorghum (Chisi et al., 1996; Rattunde et al., 2009).

8.4 DEVELOPMENT OF HYBRIDS IN GRAIN SORGHUM

Heterosis, or hybrid vigor, is a key strategy for boosting grain yield in crops like sorghum. Successful hybrid production relies on generating pure male-sterile female parents, enabling large-scale hybrid seed production through cross-pollination with fertile males. A reasonable level of heterosis for the targeted traits is essential in hybrid breeding programs. The degree of heterosis depends on the genetic diversity in the germplasm and is measured through methods like mid-parent heterosis (MPH), better parent heterosis (BPH), and standard heterosis, which compare hybrid performance to their parents and standard checks. Heterosis, successfully applied in corn breeding, led researchers to explore male sterile lines in other crops. Conner and Karper (1927) were the first to identify heterosis in sorghum, while Stephens and Quinby (1952) demonstrated how hand emasculation enhanced yield and shortened maturity in sorghum hybrids.

The breakthrough came with Stephens and Holland (1953) developing CMS systems, making hybrid seed production economically feasible and enabling widespread adoption of sorghum hybrids. In sorghum, CMS operates through a three-line system involving A, B, and R lines (Kim & Zhang, 2018). Male sterile "A" lines, with sterility-inducing mitochondrial genes, are grown with male fertile "R" lines (in isolated crossing blocks) to produce F_1 individuals. "B" lines, which lack sterile cytoplasm, serve as female parent donors for converting into A lines via backcrossing. A and B lines share the same nuclear genome but differ in cytoplasm. "R" lines carry dominant restorer genes that restore fertility in hybrids by counteracting the sterility effects in A lines (Jordan et al., 2011; Kim & Zhang, 2018).

The extent of heterosis in grain sorghum can vary significantly based on genotype and environmental factors. Quinby (1963) reported heterosis ranging from 39% to 80% for grain yield. Similarly, Blum et al. (1990) observed 23.9–39.6% heterosis for grain yield and significant heterosis for biomass, grain yield per plant, and grain number per panicle. The Indian National Program Testing reported 18–31% heterosis for grain yield using A1 cytoplasm in 1999–2000 (Reddy et al., 2008). Wagaw and Tadesse (2020) found MPH for grain yield ranged from 162.1–57.3%. Wang et al. (2020) noted stronger MPH and BPH for grain yield, grain weight, and plant height in sweet sorghum, with MPH for thousand-grain weight being lowest at 17.24%. Zhang et al. (2024) reported high MPH and BPH for plant height (40.26% and 29.95%) and grain weight per panicle (18.67% and 9.36%).

8.4.1 Diversity of Male-Sterility-Inducing Cytoplasm and Stability of Diverse CMS Systems

The milo (A1) cytoplasm, the first male sterility-inducing cytoplasm in sorghum, was incorporated into a kafir nuclear background by Stephens and Holland (1953). Since then, the majority of commercial sorghum hybrids worldwide have utilized the milo (A1) cytoplasm. However, this widespread reliance on a single cytoplasmic source increases the risk of devastating pest and disease outbreaks, mirroring the 1970 maize leaf blight epidemic caused by genetic uniformity in Texas (T) cytoplasm (Tatum, 1971). In addition to cytoplasmic vulnerability, the use of a single CMS system restricts the nuclear genetic diversity of A and R lines in hybrid breeding. To mitigate these risks and enhance genetic diversity, incorporating alternative CMS sources into sorghum hybrid breeding is crucial (Schertz, 1994).

Several alternative CMS systems have been identified across various sorghum races and species worldwide (Mital et al., 1958; Hussaini & Rao, 1964; Appathurai, 1964; Nagur & Menon, 1974; Schertz, 1977; Schertz & Ritchey, 1978; Quinby, 1982; Schertz & Pring, 1982; Worstell et al., 1984). The milo cytoplasm was classified as A1 by Stephens and Holland (1953), while Schertz (1977) designated the Ethiopian variety IS12662C (TAM428) as A2. The A3 cytoplasm, derived from Nilwa and IS1112C, was introduced by Quinby (1982), whereas A4 cytoplasms originated from IS-7920C, Guntur, VZM, and Maldandi (Worstell et al., 1984; Quinby, 1982; Senthil et al., 1994; Kariyannanavar et al., 2024). Additionally, a male-sterile source from Ghana, identified within the 9E selection, was discovered (Webster & Singh, 1964; Elkonin et al., 2015). Male-sterile lines developed in Kansas, incorporating a kafir nucleus with cytoplasm from wild grassy sorghum, were designated KS34-KS39 (Ross, 1965; Ross & Hackerott, 1972). Detailed information on these CMS sources is available in Choudhary et al. (2022).

Male sterility in sorghum is defined by the absence of pollen shedding and the inability to self-pollinate, regardless of environmental conditions. The stability of CMS systems depends on restorer gene interactions and heterosis levels (Schertz & Pring, 1982; Senthil et al., 1994; Reddy et al., 2004a, b, 2010). Instability, leading to unintended pollen shedding, increases seed production costs and reduces breeding efficiency, necessitating the elimination of unstable lines. Most commercial sorghum hybrids rely on the A1 CMS system due to its high stability, with a low incidence of pollen shedders (<1%) and minimal environmental influence (Reddy et al., 2004,a, b 2010). Environmental factors influencing male sterility and fertility restoration have been extensively studied (Schertz et al., 1997; Reddy et al., 2004a, b; Jordan et al., 2024; Kariyannanavar et al., 2024). Research indicates that low night temperatures before flowering in the post-rainy season reduce fertility restoration, while daytime temperatures exceeding 42°C compromise male sterility in CMS lines (Reddy & Stenhouse, 1994; Jordan et al., 2024). High temperatures provide an effective means of screening for stable A lines (seed parents), whereas low night temperatures help identify consistent fertility restorers (Jordan et al., 2024). Among CMS sources, A1 exhibits the highest stability in maintaining sterility, followed by A3, A2, and A4 (Reddy & Stenhouse, 1994). Furthermore, Devi and Murthy (1993) reported that A2 cytoplasm is particularly unstable, with its tapetum remaining intact in sterile pollen during winter (<10°C) but degrading in summer (>42°C).

8.4.2 Fertility Restoration in CMS System

Cytoplasmic male sterility (CMS), inherited maternally, results in plants producing non-functional pollen due to disruptions in mitochondrial genome organization, reflecting complex nuclear-mitochondrial interactions (Hanson & Bentolila, 2004; Ingle et al., 2023). Sporophytic CMS affects the tapetum and meiocytes, leading to pollen abortion, while gametophytic CMS impacts microspore development (Guo & Liu, 2012; Chen & Liu, 2014). Male sterility occurs through the interaction of dominant alleles in the S-cytoplasm with nuclear recessive alleles, neither of which alone controls sterility (Ingle et al., 2023). Understanding fertility restoration is essential for hybrid seed production in sorghum. Fertility restoration involves nuclear-encoded Rf genes that counterbalance male sterility by restoring harmony between the mitochondrial and nuclear genomes (Bohra et al., 2016). The emergence of Rf genes reflects selection pressure driven by male sterility, affecting population sex ratios.

Mendelian inheritance principles, alongside mono- and digenic models and allelism tests, have been instrumental in identifying key Rf genes responsible for fertility restoration in sorghum. Investigations into the A1 CMS system have revealed the presence of two major non-allelic Rf genes, whose expression is modulated by additional genomic factors that contribute to restoring male fertility (Jordan et al., 2011; Ingle et al., 2023). Murty (1986) demonstrated that fertility restoration in A1 cytoplasm is controlled by a single gene, based on the segregation patterns observed in F_2 progeny. However, subsequent research by Murty and Gangadhar (1990) proposed a complementary digenic model for the A2 cytoplasm, identifying *Msc1* and *Msc2* as key loci responsible for fertility restoration, exhibiting a 9:7 segregation ratio. Further genetic mapping efforts have linked *Rf1* to LG8 and *Rf2* to SBI-02, both of which play crucial roles in fertility restoration within the A1 CMS system (Klein et al., 2005; Jordan et al., 2010). Jordan et al. (2011) later discovered *Rf5* on SBI-05, which is capable of fully restoring fertility, while another locus on SBI-04 was identified as contributing to partial fertility restoration in both A1 and A2 cytoplasms. Expanding on this research, Kiyosawa et al. (2022) conducted fine-mapping of *Rf5* in the SBI-05 region using a mapping population derived from the cross between "Nakei MS-3A" and "JN43", successfully narrowing its position to a 140-kb interval in the BTx623 reference genome. Meanwhile, Madugula et al. (2018) identified a novel *Rf6* gene on SBI-04, which facilitates fertility restoration in both A1 and A2 cytoplasms.

Comparative analyses of fertility restoration frequencies across different cytoplasmic backgrounds indicate that A3 cytoplasm exhibits the lowest restoration efficiency, followed in increasing order by A4, A2, and A1 (Reddy et al., 2010; Madugula et al., 2018). Restoration in A3 cytoplasm has been attributed to interactions between *Rf3* and *Rf4*, which operate in a gametophytic manner (Pring et al., 1999; Tang et al., 2007), although Tang and Pring (2003) also reported instances of single-gene inheritance. In contrast, sporophytic restoration mechanisms, involving one or two genes, have been observed in the 9E and A4 cytoplasms (El'konin et al., 1998). The genetic mechanisms underlying fertility restoration in sorghum have been extensively studied, with key findings summarized in Table 8.1.

8.4.3 Nuclear Male Sterility in Sorghum

Developing all three lines simultaneously is time-consuming, costly, and complex. Many sorghum germplasms, especially bioenergy types, do not fit the strict B-line or R-line requirements for the three-line hybrid system. This limits the number of accessions available for hybrid breeding, restricting the potential to exploit heterosis fully. Additionally, the reliance on A1 CMS in nearly all commercial grain sorghum hybrids raises the risk of vulnerability to diseases and stresses. A historical example is the 1970 southern corn leaf blight epidemic, where widespread use of T-cytoplasm maize led to over 80% yield loss (Ullstrup, 1972). To avoid similar losses in sorghum, it is crucial to develop hybrids using diverse sources of male sterility.

In recent years, two-line hybrid breeding systems using nuclear male sterility (NMS) have been explored in different crops (Chang et al., 2016). Since NMS lines segregate sterile and fertile plants in a 1:3 ratio, the challenge lies in selecting pure (100%) sterile plants from the population. Chang et al. (2016) addressed this by rescuing sterility in the *osnp1* mutant through a transgene containing the *OsNP1* gene, a pollen-deactivating element, and a red fluorescence marker. Transgene-free seeds, identified via fluorescence sorting, served as female parents in hybrid breeding. Fertility in the NMS system is restored by any non-mutant line, as the sterile mutation is recessive. This

TABLE 8.1
Genetic determinants of fertility restoration across sorghum cytoplasms.

Cytoplasm	Genetic factors influencing fertility restoration	References
A1	Four genes; although the presence of a fifth gene was hypothesized for complete restoration of the sterile lines	Tripathi et al. (1985)
	Single dominant gene	Murty and Gangadhar (1990)
	One to three genes	Lonkar and Borikar (1994)
	A single dominant gene	Arunkumar et al. (2004)
	One basic gene and two duplicate complimentary genes	Sanjana Reddy et al. (2010)
	Single dominant gene (*Rf6*)	Praveen et al. (2015), Madugula et al. (2018)
	Single dominant gene (*Rf5*)	Jordan et al. (2011); Kiyosawa et al. (2022)
A2	Two genes (*Msc1* and *Msc2*; duplicate recessive epistasis)	Murty and Gangadhar (1990)
	Two to four genes	Lonkar and Borikar (1994)
	Three dominant duplicate-complimentary genes	Sanjana Reddy et al. (2010)
	Single dominant gene (*Rf6*)	Praveen et al. (2015); Madugula et al. (2018)
	Single dominant gene (*Rf5*)	Jordan et al. (2011); Kiyosawa et al. (2022)
A3	Two-gene gametophytic mechanism	Pring et al. (1999); Tang et al. (2007)
	Three dominant duplicate-complimentary genes	Sanjana Reddy et al. (2010)
	Complementary interaction of two gametophytic genes *Rf3* and *Rf4*	Elkonin et al. (2019)
A4	One or two dominant genes based sporophytic regulation	Elkonin et al. (1998)
	Two dominant genes with duplicate epistasis	Arunkumar et al. (2004)
	Two to three genes	Sanjana Reddy et al. (2010)
	Single dominant gene (*Rf6*)	Praveen et al. (2015); Kumar et al. (2023)
	Duplicate dominant gene action	Kumar et al. (2023)
9E	One or two dominant genes based sporophytic regulation	Elkonin et al. (1998)
	Epigenetic regulation	Elkonin et al. (2009)

"two-line system" simplifies seed production and expands the breeding pool, achieving a 5–10% yield boost compared to the three-line system based on cytoplasmic sterility (Chang et al., 2016). In sorghum, there is growing interest in isolating male sterile mutants and identifying causal mutations that can be incorporated via MAS or gene editing to establish a similar two-line breeding system.

In sorghum, eight NMS lines, *ms1* through *ms8* and *msal* (antherless), have been reported (Andrews & Webster, 1971; Xin et al., 2018). While *ms4*, *ms5*, and *ms6* are no longer available, the remaining lines have been preserved and integrated into different genetic backgrounds (Pedersen & Toy, 2001). These NMS lines hold potential for studying male gametophyte development and developing new hybrid breeding systems. However, the genes responsible for male sterility in these lines remain unidentified, limiting their application in breeding. Chen et al. (2019) identified the first NMS gene in sorghum through a mutant from an ethyl methane sulfonate (EMS) population in the BTx623 inbred line (Jiao et al., 2016). This mutant, representing a new NMS locus, was named *male sterile 9 (ms9)*, with the causal gene designated as *Ms9*. Recently, two EMS-induced *ms9* mutants (BTx623ms9-1 (Reg. no. GS-791, PI 699654) and BTx623ms9-3 (Reg. no. GS-792, PI 699655)) were registered (Chen et al., 2023). These mutants, along with the identification of the

Ms9 gene, provide valuable resources for introducing male sterility into other sorghum accessions and breeding lines.

8.4.4 Selective Induction of Male Sterility Using Gametocides

Sorghum, though produced as a hybrid crop, is primarily self-pollinating with complete flowers, requiring specialized mechanisms to facilitate cross-pollination. CMS enables commercial hybrid seed production but is unsuitable for other breeding activities, such as developing new inbred lines. Different crossing methods are required for large-scale seed production versus breeding crosses, and these evolve with new technologies such as genomic selection. For instance, a doubled haploid (DH) system, successful in maize and barley, offers more efficient genotype development and integrates well with genomic selection models. However, adapting this system to sorghum requires a practical, temporary male sterility mechanism. While maize achieves this through tassel removal, the presence of complete flowers in sorghum makes it unfeasible. Cytoplasmic or genetic male sterility may not guarantee sterility due to allele interactions, and resulting haploid plants would remain sterile (an undesirable outcome). Other methods like hot water, hand, or plastic bag emasculation are impractical due to seed limitations and logistical challenges (Schertz, 1963). Thus, developing an effective temporal sterility system is essential for inbred development in sorghum.

Further, in sorghum breeding, fertile × fertile crosses are essential for developing new inbred lines. However, existing methods such as hand and plastic-bag emasculations, hot water treatment, and genetic male sterility have limitations (Yahaya et al., 2020). Hand emasculations are labor-intensive with low F_1 seed yield, while plastic-bag emasculations produce both self-pollinated and hybrid seeds, requiring phenotypic or genotypic sorting (Schertz & Clark, 1967). Genetic male sterility is useful in recurrent selection but depends on the availability of suitable lines (Doggett & Eberhart, 1968). Hot water treatment generates more hybrid seeds but is challenging to implement and has genotype-specific effects.

Given the limitations of current methods, an alternative approach that induces temporal male sterility is needed. Gametocides are easy to apply and, when effective, generate high frequencies of temporary male sterility (Saxena & Hingane, 2015). The use of chemical hybridizing agents (CHAs) for selective male sterility was first demonstrated in the mid-20th century (Naylor, 1950). Since then, many chemicals have been tested across various crops (Sharma & Sharma, 2005; Boerman et al., 2019), with crop-specific agents identified for maize and Brassicas (Li et al., 2015; Wei et al., 2012). However, most CHAs also impair female fertility and cause phytotoxic effects, such as stunted growth and reduced seed production (Sharma & Sharma, 2005). In sorghum, compounds like ethyl-4-fluorooxanilate (E4FO) and Ethrel have induced male sterility but with similar drawbacks, including female sterility and toxicity at higher doses (Amelework et al., 2016). Recently, trifluoromethanesulfonamide (TFMSA), originally tested in maize (Loussaert, 2004), showed promise in sorghum, inducing male sterility without significant phytotoxicity in both greenhouse (Hodnett & Rooney, 2018) and field conditions (Boerman et al., 2019). Beyond breeding crosses, TFMSA could enhance early-generation testcross evaluations of new seed parent lines (Kent et al., 2023). By temporarily sterilizing parents, breeders can rapidly assess hybrids and discard undesirable lines before permanent sterilization, streamlining hybrid seed production in sorghum.

8.4.5 Heterotic Grouping in Sorghum

A heterotic group is defined as a set of related or unrelated genotypes from the same or different populations that exhibit similar combining abilities and heterotic responses when crossed with genotypes from other genetically distinct germplasm groups (Melchinger & Gumber, 1998). While this concept is well-established in maize (Melchinger & Gumber, 1998), it is comparatively less developed in sorghum. Historically, sorghum hybrid parents came from contrasting races such as Kafir and Milo (durra), with Kafir typically serving as the female parent (House, 1985; Nguyen, 2024). The discovery of CMS played a pivotal role in hybrid vigor by enabling Kafir genes to induce male sterility in Milo cytoplasm, while Milo and Caudatum races contributed fertility restoration genes (Schertz & Pring, 1982). The principle of gene pool complementarity between male and female lines remains a cornerstone for increasing heterosis. Sorghum breeders generate new lines by crossing within or with unrelated germplasm and evaluate their combining ability using opposite pools to preserve and enhance heterotic potential. The discovery of alternative male sterile cytoplasms further enables the construction of novel heterotic pools beyond the traditional R and B classifications. For instance, Begna et al. (2024) identified two heterotic groups based on specific combining ability (SCA) and three groups based on general combining ability (GCA). Similarly, Wagaw et al. (2020) grouped parental lines into dualistic heterotic groups using SCA and GCA data.

Maintaining robust genetic variation within these heterotic groups is essential for continued genetic gains. Previous studies using genotypic analyses to delineate heterotic groups in sorghum have yielded mixed results. Some reports indicated separate clustering of B-lines and R-lines (Ahnert et al., 1996), while others suggested clustering was more aligned with ancestral racial origins (Brown et al., 2011; Menz et al., 2004; Mindaye et al., 2015; Ramu et al., 2013). Moreover, the necessity of using CMS to produce commercial hybrid seed has tied heterotic grouping to fertility reaction scores. However, certain genotypes do not cluster strictly by fertility reaction, with instances of R-lines clustering with B-lines, reflecting the inbreeding in foundational accessions before heterotic pools were fully established. In a subset of 24 elite grain inbreds, B-line and R-line classifications formed mutually exclusive clusters, representing genetically distinct heterotic pools (Crozier et al., 2020). Similarly in another study, B-line accessions, primarily *bicolor* types from wet highlands, clustered distinctly from R-line groups, with some overlap (Girma et al., 2024). This distinction highlights progress in sorghum breeding, which has led to the development of heterotic pools, albeit with less divergence compared to outcrossing species like maize (Niebur et al., 2004; Duvick, 2005). The increase in heterosis since hybrid development (Gizzi & Gambin, 2016; Pfeiffer et al., 2019) underscores the importance of these efforts, even though heterotic differentiation remains narrower than in maize.

8.4.6 Selection for Combining Ability

Information on combining ability is essential for identifying superior parents and hybrids while providing insights into the gene effects underlying the expression of quantitative traits (Kenga et al., 2004). GCA reflects the additive genetic effects and higher-order additive interactions, making it critical for developing superior genotypes, whereas SCA captures non-additive dominance and epistatic effects, offering insights into hybrid performance. By analyzing GCA and SCA, breeders can infer the nature of gene effects and devise strategies to select desirable parents and implement efficient breeding procedures for trait improvement (Fasahat et al., 2016).

Diallel mating designs are widely employed to evaluate progeny performance and discern the roles of GCA and SCA in yield and related traits in different crops including sorghum (Sprague & Tatum, 1942; Yan & Hunt, 2002). Similarly, line × tester mating designs are commonly used to measure GCA, SCA, and various gene actions, providing valuable information on the genetic architecture of quantitative traits in sorghum (Williams-Alanís et al., 2022). Combining ability studies are crucial for trait improvement programs in sorghum, as they reveal *per se* performance, combining ability effects of parents and hybrids, and the genetic basis of inheritance (Kenga et al., 2004). Many studies have demonstrated that both GCA and SCA significantly influence key traits, including those contributing to higher yield and stress tolerance (Schaffasz et al., 2019; Mengistu et al., 2020;

Fonseca et al., 2021). By understanding the genetic effects underlying trait expression, breeders can adopt targeted sorghum breeding schemes to develop genotypes with enhanced agronomic performance and adaptability.

8.4.7 Key Steps in the Production of Hybrids in Sorghum

This section outlines the fundamental steps involved in the production of hybrid sorghum, emphasizing the collaborative efforts within breeding programs. The development and commercialization of a new sorghum hybrid requires the involvement of multiple stakeholders. Breeders initiate the process by generating new parental lines, which undergo rigorous testing in breeder trials. Once superior hybrid combinations are identified, breeder seeds are provided to seedstock personnel to facilitate the initial hybrid seed production and increase parental lines. These seeds are subsequently assessed in large-scale field trials by agronomists. High-performing hybrids that demonstrate commercial potential are further considered for market release. Marketing teams analyze breeding and trial data to determine the most suitable regions for hybrid adoption, while sales teams introduce the final product to dealers and customers. This entire cycle generally spans 10–12 generations. The process can be expedited through the use of off-season nurseries and greenhouse facilities for early-stage development, while genetic marker technologies and approaches such as DH production can further accelerate hybrid deployment.

The foundation of sorghum hybrid development lies in the development of robust parental lines. Given that sorghum produces perfect flowers, a thorough understanding of the CMS system is essential for hybrid seed production. The CMS system relies on three distinct parental components: (i) Seed Parent (Female): This category includes sterile A-lines and fertile B-lines. (ii) Pollen Parent (Male): The R-line, which carries restorer genes necessary to recover fertility in F_1 hybrids. The propagation of A-lines occurs through controlled crosses with corresponding B-lines of the same genotype. B-lines, often termed maintainers, are fully fertile and capable of pollen production, but they lack restorer genes, ensuring that when crossed with A-lines, the resulting progeny remain sterile. In contrast, both B-lines and R-lines are maintained through self-pollination. The final step in hybrid production involves crossing A-lines with R-lines, resulting in F_1 hybrid progeny that are fertile due to the presence of restorer genes from the R-line.

8.4.7.1 Parental Line Development

The process of developing both B-lines and R-lines follows an identical pathway up to the F_3 generation. It begins with controlled crosses between R-lines and B-lines, strictly maintaining the principle that R-lines are only crossed with other R-lines, while B-lines are exclusively crossed with B-lines. This ensures the genetic integrity and stability of each parental type throughout the early generations of line development (Lambright, 2019). To avoid mixing restorer and maintainer lines, elite R- and B-lines are selected based on their strong hybrid performance, with the potential addition of specific traits. Since both B-lines and R-lines are fertile, emasculation is necessary for successful recombination. Emasculation is generally done through: (i) *Hand Emasculation*: manually removing the anthers from 40–50 delicate sorghum flowers, which is tedious and requires precision. (ii) *Plastic Bag Method*: covering part of the sorghum head with a plastic bag to destroy pollen through condensation (Begna, 2022). The hand emasculation method typically results in minimal self-pollination, while the plastic bag method may yield a higher rate.

In both methods, pollen from the other parent generates an F_1 breeding population. The F_1 seeds are selfed, avoiding suspected self-pollinations, and then planted in F_2 nurseries in representative target geographies. Genetic markers identify plants with specific traits, allowing breeders to select based on visual and agronomic characteristics. F_2 population sizes may vary, but a minimum of 1,000 plants is recommended. Selected plants are then grown in F_3 nurseries, with each selection planted in a row of about 100 plants. Genetic markers are confirmed in F_3 rows, leading to selections based on trait positivity, agronomic quality, and uniformity. Maturity comparisons are facilitated by including rows of elite parents in F_3 plantings.

8.4.7.2 R- and A-Line Development

Although R-line and B-line populations follow the same management approach from F_1 to F_3, their development paths diverge beginning at the F_4 generation. At this stage, F_4 selections of R-lines undergo test crosses with two or three elite seed parents to generate F_1 hybrids, enabling the evaluation of new male parental lines for their combinability (Reddy et al., 2008). In contrast, F_4 selections of B-lines are crossed with elite-related A-lines to initiate the CMS process. Employing a related female in the initial sterilization cross often minimizes the necessity for sterile backcrosses. In the subsequent generation, CMS_1 A-line crosses are combined with F_5 selections of the new B-line, facilitating further introgression into sterile cytoplasm ($CMS_{1\text{-}2}$). This process continues as CMS_2 A-line crosses are paired with F_6 selections of the new B-line, resulting in a $CMS_{2\text{-}3}$ version of the A-line. Simultaneously, test crosses involving two or three elite pollen parents are performed on the same row of CMS_2 plants to assess the combining ability of new A-lines (Reddy et al., 2008; Lambright, 2019). At each generation, sterility evaluations of candidate seed parents are conducted, with any lines exhibiting uncertain sterility being discarded.

8.4.7.3 Testcross and Hybrid Evaluation

The testcross generation marks the final phase in parental line development, requiring multiple generations. To accelerate progress, contra-season nurseries and greenhouse plantings facilitate two to three generations per year during early-stage development and B-line sterilization. Once new parental lines undergo testcrossing, hybrid evaluation begins. Each R-line is testcrossed with two or three elite A-lines, while each A-line is paired with two or three elite R-lines. These testcross hybrids are evaluated in single-rep trials at a minimum of three to four locations, alongside adapted commercial check hybrids. Only 5–10% of testcross entries advance based on yield performance, maturity, and agronomic traits (Lambright, 2019). Parental lines that perform well across multiple testers are identified as potential general combiners. Lines exhibiting strong GCA are crossed with additional elite lines to produce new F_1 hybrids, bypassing further testcross evaluations and proceeding directly to intensive trials. Additionally, promising GCA lines enter a recombination pool to generate new F_1 breeding populations, contributing to future parental line development.

8.4.7.4 Advanced Hybrid Testing

Following testcross evaluations, only 5–10% of the highest-performing entries advance to intensive multi-rep, multi-location hybrid trials conducted across five to six sites within targeted regions (Lambright, 2019; Leon et al., 2024). Each trial includes five to six well-adapted commercial check hybrids for comparison. Selection criteria encompass overall yield performance, maturity, stalk and root strength, and agronomic traits. Hybrids that outperform the checks progress to advanced testing, where they undergo multi-rep trials across 10–12 locations, with an expanded panel of eight to ten maturity-appropriate check hybrids (Lambright, 2019; Leon et al., 2024). Advanced testing prioritizes total yield performance, incorporating three key factors: yield stability, yield for maturity, and harvestable yield. Yield stability measures a hybrid's consistency relative to competing products across diverse environments and climatic conditions, reflecting broad adaptation. Yield for maturity identifies hybrids achieving the highest yield within a specific maturity class, expressed as total yield divided by days to mid-bloom (yield/day). Harvestable yield accounts for biotic and abiotic stresses, such as insect pressure, disease susceptibility, lodging, drought, and heat stress, that may limit a hybrid's genetic yield potential.

8.4.7.5 Cost-of-Goods Evaluation

Alongside advanced hybrid testing, Cost-of-Goods (COG) evaluations begin for parental lines used in these trials (Lambright, 2019; Begna, 2024). Assessments focus on A-line seed yield potential, R-line pollen production, and the synchronization of parental bloom timing. COG evaluations occur in replicated trials with parental lines used in advanced entries, planted in seed production areas aligned with standard schedules. Seed parent yield potential directly affects COG, while pollen volume and bloom synchronization influence green seed yields and genetic purity. Adequate pollen volume during female bloom is crucial for maintaining genetic purity and minimizing outcrossing; R-lines with low pollen volume are discarded. Bloom synchronization is managed through split planting or delay plantings; greater timing differences increase the risk of genetic purity issues. Split plantings aim to ensure sufficient pollen is available throughout the female bloom period, often involving three male plantings. However, female delay plantings for achieving desired splits carry higher risk than male delays due to potential weather impacts.

8.4.7.6 Breeder's Seed Increase

During the same generation as advanced testing and COG evaluations, breeder seed increases are initiated for all parental lines used in hybrids advancing to trial evaluations (Reddy et al., 2004a). Nursery seed increase targets typically range from five to ten pounds for each new parental line, including both A-line and B-line versions of seed parents. To support pilot hybrid seed production, an AxB increase is planted over one to two acres in preparation for potential large-scale hybrid production. Additionally, A-line and R-line seed increases are conducted to enable pilot seed production for hybrids with commercial potential. Seedstock personnel oversee these pilot production plots, which span one to three acres, aiming to generate 2,500–5,000 pounds of hybrid seed. This seed is then utilized by agronomists and sales managers for strip trial evaluations in the subsequent growing season.

8.4.7.7 Strip Trials

Field-scale strip trials with producers are conducted across the target release area, featuring experimental hybrids alongside commercially available hybrids for comparison (Lambright, 2019; Lofton et al., 2024). Yield, grain moisture, plant height, stalk quality, root strength, and pest or disease responses are recorded and integrated into a breeding database alongside small plot evaluation data. Experimental hybrids demonstrating at least a 5% yield advantage over comparable commercial hybrids, stable multi-location performance, strong agronomics, and acceptable Cost-of-Goods advance to commercial status (Lambright, 2019). These hybrids are typically intended to replace existing portfolio products. In the season before commercialization, seed production targets 500–1,500 units to meet initial market demand.

8.4.7.8 Seedstock Development

Seedstock for new parental lines is sourced from pollen parents harvested in pilot productions and seed parents from the previously mentioned one-to-two-acre AxB increases. Planting splits for initial seed production are based on COG evaluations and pilot production data. To ensure sufficient pollen coverage during the seed parent's bloom period, the pollen parent is typically planted on three different dates.

8.4.7.9 Hybrid Seed Production

Hybrid sorghum seed production must be isolated from grain sorghum, forage sorghum, sudangrass, and wild sorghum species like johnsongrass (*S. halepense*) and shattercane (*S. bicolor*) (Kannababu et al., 2023). Isolation is achieved through distance or staggered planting. After fall harvest, grow-out samples are planted in winter locations to assess genetic purity before commercial sales. Seed lots that meet genetic purity and germination standards are then conditioned and bagged.

8.4.8 Progress and Challenges in Sorghum Hybrid Breeding

Over the past six decades, advancements in plant breeding have significantly enhanced the genetic potential of cereals, including sorghum, leading to the release of numerous improved sorghum hybrids. Sorghum hybrids, as highlighted by Duvick (1999), accounted for approximately 48% of the grain sorghum planted globally at the time, delivering yields 19% higher than those from non-hybrid counterparts. This pivotal shift marked the onset of steadily increasing hybrid adoption, which, in turn, has resulted in substantial yield gains over the years. A study by Assefa and Staggenborg (2010) further elucidated these gains by analyzing yield improvements in Kansas from 1957 to 2008. Two key drivers were identified: (i) the expansion of dryland sorghum production facilitated by hybrid advancements and (ii) productivity enhancements derived from the transition from open-pollinated varieties to hybrids, alongside continuous improvements within the hybrid era. Over the 52-year analysis, dryland hybrid yields increased by nearly 0.05 t ha^{-1} annually, underscoring the synergistic impact of genetic innovations and agronomic advancements. Building on this foundation, Assefa et al. (2024) reported that sorghum productivity in Kansas averaged 4.3 Mg ha^{-1} between 2000 and 2022. However, alongside these gains, yield variability rose markedly, with fluctuations growing from ±0.5 Mg ha^{-1} (1929–1956) to ±2 Mg ha^{-1} (1957–2022), reflecting an increasing sensitivity of sorghum yields to environmental and management factors. This trend of steady genetic gains is further supported by other studies. Gizzi and Gambin (2016) observed annual yield increases of 0.0087 t ha^{-1} in sorghum hybrids from Advanta Seeds in Argentina between 1984 and 2014. Pfeiffer et al. (2019) documented similar improvements (0.008 t ha^{-1} per year) in the Texas A&M AgriLife Research Sorghum Breeding Program (1959–2016). More recently, Mayor et al. (2023) reported genetic gains of 0.0263 t ha^{-1} per year across maturity groups in the U.S. sorghum belt, while Leon et al. (2024) highlighted an annual yield increase of 0.03 t ha^{-1} from hybrids tested across 514 environments in Texas between 1970 and 2021. These consistent gains underscore the sustained progress achieved through hybrid development.

The introduction of the CMS system between 1957 and 1962 revolutionized U.S. grain sorghum production by facilitating the rapid adoption of hybrids and driving significant yield increases. However, the initial momentum waned over the subsequent decades due to disease and pest challenges, as noted by Miller and Kebede (1984). To address this, breeding programs began incorporating conversion germplasm in the 1970s, which led to the development of hybrids with improved resistance to pests like midge (Johnson et al., 1973) and greenbug (Johnson et al., 1974). Global breeding efforts have since diversified sorghum hybrids to suit varying agro-climatic conditions. The release of Coordinated Sorghum Hybrid 1 (CSH 1) in 1964 marked a significant milestone, initiating a transition toward the development of parental lines with enhanced traits such as yield, maturity, pest tolerance, and quality (Rao, 1982; House et al., 1996). Between 1970 and 2009, genetic gains in India's rainy-season hybrid trials averaged 0.0204 t ha^{-1} per year, though post-rainy season gains were less pronounced (Rakshit et al., 2014). Subsequently, the 1990s saw a focus on dual-purpose varieties with tolerance to shoot fly and grain mold, further boosting yield potential (Das et al., 2020). To date, India has developed approximately 80 stress-tolerant cultivars with good grain quality, providing resilience and profitability for dryland farmers (Hariprasanna et al., 2024).

Modern hybrids have demonstrated significant advantages under stress conditions. For instance, drought-tolerant hybrids maintain comparable yields to standard hybrids in low-water environments while exhibiting stability under fluctuating moisture conditions (Leon et al., 2024). Historical trends reveal the success of targeted breeding strategies in improving both yield potential and stress resilience. Hybrids such as 1G282, DKB540, and A9721R have shown remarkable resilience to post-flowering drought (Batista et al., 2017), while others, like A9904 and 1G282, have demonstrated resilience to both pre- and post-flowering drought, sustaining yields near optimal levels (Menezes et al., 2022). In conclusion, the consistent progress in hybrid sorghum breeding has significantly advanced global productivity by enhancing genetic potential and improving agronomic practices. However, the growing yield variability highlights the importance of addressing genotype × environment interactions and environmental challenges. Continued efforts in hybrid development are critical to meeting the rising global demand for food and feed while ensuring sustainability and resilience.

8.5 CONCLUSION

Hybrid breeding in grain sorghum has achieved remarkable success over the past decades, driven by innovations in CMS systems, fertility restoration, and combining ability selection. These advancements have enabled the development of high-yielding hybrids with enhanced grain yield and stress resilience, catering to diverse agro-climatic conditions and market demands. The steady genetic gains observed in global breeding programs underscore the effectiveness of targeted hybridization strategies and the integration of indigenous and exotic germplasm for trait improvement. Despite significant progress, challenges such as increasing yield variability, heightened environmental stresses, and evolving pest and disease threats necessitate further innovation. The application of precision breeding tools, marker-assisted selection, and genomic prediction offers promising pathways for accelerating hybrid development. Additionally, greater focus on sustainability, including cost-effective hybrid production and seedstock development, will ensure the broader adoption of improved hybrids. In conclusion, hybrid breeding remains a cornerstone of sorghum improvement, delivering solutions to enhance productivity, quality, and resilience under diverse and often challenging growing conditions. Continued investments in breeding research, coupled with adaptive agronomic practices, will be instrumental in realizing the full potential of sorghum to meet future needs.

ACKNOWLEDGEMENT

The author gratefully acknowledges the financial support provided by the United Sorghum Checkoff Program to Dr. Dinesh Kumar Saini through the TeCSIS-TAWC program, led by Dr. Jagadish at Texas Tech University.

REFERENCES

Abdelhalim, T. S., Abdelhalim, N. S., Kamal, N. M., Mohamed, E. E., & Hassan, A. B. (2021). Exploiting the potential of Sudanese sorghum landraces in biofortification: Physicochemical quality of the grain of sorghum (*Sorghum bicolor* L. Moench) landraces. *Food Chemistry*, *337*, Art. 127604.

Ahnert, D., Lee, M., Austin, D. F., Livini, C., Woodman, W. L., Openshaw, S. J., . . . Dalton, G. (1996). Genetic diversity among elite sorghum inbred lines assessed with DNA markers and pedigree information. *Crop Science*, *36*(5), 1385–1392.

Ali, H. M. W., Ahmad, M. Q., Saleem, M. A., Rehman, H. M., Qayyum, A., Malik, W., . . . Khan, S. H. (2024). Biofortified sorghum: A prospectus of combating malnutrition. In *Biofortification of grain and vegetable crops* (pp. 115–138). London EC2Y 5AS, United Kingdom: Academic Press.

Amelework, A., Laing, M., & Shimelis, H. (2016). Evaluation of effective gametocides for selective induction of male sterility in sorghum, *The Indian Journal of Genetics and Plant Breeding, 59*(4), 429-436.

Ananda, G. K., Myrans, H., Norton, S. L., Gleadow, R., Furtado, A., & Henry, R. J. (2020). Wild sorghum as a promising resource for crop improvement. *Frontiers in Plant Science*, *11*, 1108.

Andrews, D. J., & Webster, O. J. (1971). A new factor for genetic male sterility in *Sorghum bicolor* (L.) Moench 1. *Crop Science*, *11*(2), 308–309.

Appathurai, R. (1964). Diverse plasmons in male-sterile sorghum, *Madras Agricultural Journal*, 51, 276–278..

Aruna, C., & Cheruku, D. (2019). Genetic improvement of grain sorghum. In: Aruna, C., Visarada, K.B.R.S., Bhat, B.V., Tonapi, V.A. (Eds), *Breeding sorghum for diverse end uses* (pp. 157–173). Cambridge, UK : Woodhead Publishing.

Arunkumar, B., Biradar, B. D., & Salimath, P. M. (2004). Inheritance of fertility restoration on milo and maldandi sources of male sterility in rabi sorghum [*Sorghum bicolor* (L.) Moench], *The Indian Journal of Genetics and Plant Breeding*, 64, 325–326.

Aslam, H. M. U., Peters, N. T., Killiny, N., Riaz, H., Hameed, A., Aslam, S., & Shakeel, Q. (2024). In: Haq, I.U., Ijaz, S., Ali, H. (Eds), An insight into infectious and noninfectious diseases of sorghum species. *Sustainable Summer Fodder*, 58–69. Boca Raton, FL, USA: CRC Press.

Assefa, Y., Holman, J. D., Obour, A. K., O'Brien, D., & Prasad, P. V. V. (2024). Historic grain sorghum production, value, yield gap, and weather relation trends. *Agronomy*, *14*(11), 2582.

Assefa, Y., & Staggenborg, S. A. (2010). Grain sorghum yield with hybrid advancement and changes in agronomic practices from 1957 through 2008. *Agronomy Journal*, *102*(2), 703–706.

Batista, P. S. C., Menezes, C. B., Carvalho, A. J., Portugal, A. F., Bastos, E. A., Cardoso, M. J., . . . Julio, M. P. M. (2017). Performance of grain sorghum hybrids under drought stress using GGE biplot analyses, *Genetics and Molecular Research: GMR,* 16(10), 4238.

Begna, T. (2022). Hybrid sorghum development mechanisms to enhance production and productivity. *International Journal of Research*, *8*(2), 20–31.

Begna, T., Birhan, T., & Tadesse, T. (2024). Estimation of standard heterosis and heterotic grouping of elite sorghum [*Sorghum bicolor* (L) Moench] inbred lines under moisture stress conditions. *Ecological Genetics and Genomics*, *31*, Art. 100242.

Blum, A., Ramaiah, S., Kanemasu, E. T., & Paulsen, G. M. (1990). The physiology of heterosis in sorghum with respect to environmental stress. *Annals of Botany*, *65*(2), 149–158.

Boerman, N. A., Hlavinka, K. B., Zhu, W., Dabney, A. R., Hodnett, G. L., & Rooney, W. L. (2019). Efficacy of the chemical trifluoromethanesulfonamide as a male gametocide in field-grown sorghum [*Sorghum bicolor* (L.) Moench]. *Euphytica*, *215*, 1–10.

Bohra, A., Jha, U. C., Adhimoolam, P., Bisht, D., & Singh, N. P. (2016). Cytoplasmic male sterility (CMS) in hybrid breeding in field crops. *Plant Cell Reports*, *35*, 967–993.

Brown, P. J., Myles, S., & Kresovich, S. (2011). Genetic support for phenotype-based racial classification in sorghum. *Crop Science*, *51*(1), 224–230.

Carbajal-Friedrich, A. A., & Burgess, A. J. (2024). The role of the ideotype in future agricultural production. *Frontiers in Plant Physiology*, *2*, Art. 1341617.

Chan, S. S., Ferguson, E. L., Bailey, K., Fahmida, U., Harper, T. B., & Gibson, R. S. (2007). The concentrations of iron, calcium, zinc and phytate in cereals and legumes habitually consumed by infants living in East Lombok, Indonesia. *Journal of Food Composition and Analysis*, *20*(7), 609–617.

Chang, Z., Chen, Z., Wang, N., Xie, G., Lu, J., Yan, W., . . . Deng, X. W. (2016). Construction of a male sterility system for hybrid rice breeding and seed production using a nuclear male sterility gene. *Proceedings of the National Academy of Sciences*, *113*(49), 14145–14150.

Chapman, S. C., Cooper, M., Butler, D. G., & Henzell, R. G. (2000). Genotype by environment interactions affecting grain sorghum. I. Characteristics that confound interpretation of hybrid yield. *Australian Journal of Agricultural Research*, *51*(2), 197–208.

Chen, J., Chopra, R., Hayes, C., Morris, G., Marla, S., Burke, J., . . . Burow, G. (2017). Genome-wide association study of developing leaves' heat tolerance during vegetative growth stages in a sorghum association panel. *The Plant Genome*, *10*(2), 1–15.

Chen, J., Jiao, Y., Laza, H., Payton, P., Ware, D., & Xin, Z. (2019). Identification of the first nuclear male sterility gene (Male-sterile 9) in sorghum. *The Plant Genome*, *12*(3), Art. 190020.

Chen, J., Laza, H., Burow, G., Hayes, C., Burke, J. J., Emendack, Y., & Xin, Z. (2023). Registration of two novel grain sorghum nuclear male sterile mutants: BTx623ms9-1 and BTx623ms9-3. *Journal of Plant Registrations*, *17*(1), 209–213.

Chen, J., Xu, W., Velten, J., Xin, Z., & Stout, J. (2012). Characterization of maize inbred lines for drought and heat tolerance. *Journal of Soil and Water Conservation*, *67*(5), 354–364.

Chen, L., & Liu, Y. G. (2014). Male sterility and fertility restoration in crops. *Annual Review of Plant Biology*, *65*(1), 579–606.

Chisi, M., Bramel-Cox, P. J., Witt, M. D., Claassen, M. M., & Andrews, D. J. (1996). Breeding for grain yield stability using full-sib family recurrent selection in sorghum. *Crop Science*, *36*(5), 1083–1087.

Choudhary, K. B., Khandelwal, V., & Sharma, S. R. (2022). Sorghum improvement: Male sterility and hybrid breeding approaches. In *Plant male sterility systems for accelerating crop improvement* (pp. 73–89). Singapore: Springer Nature.

Conner, A. B. (1927). *Hybrid vigor in sorghum* (No. 359). Texas Agricultural and Mechanical College System. Texas Agricultural Experiment Station, Texas A&M University, College Station.

Crozier, D., Hoffmann Jr, L., Klein, P. E., Klein, R. R., & Rooney, W. L. (2020). Predicting heterosis in grain sorghum hybrids using sequence-based genetic similarity estimates. *Journal of Crop Improvement*, *34*(5), 600–617.

Das, I. K., Aruna, C., & Tonapi, V. A. (2020). Sorghum grain mold. *ICAR-Indian Institute of Millets Research, Hyderabad, India*, 1-86.

Das, I. K., & Rajendrakumar, P. (2016). Disease resistance in sorghum. In *Biotic Stress resistance in millets* (pp. 23–67). Cambridge, MA, United States: Academic Press.

de Morais Cardoso, L., Pinheiro, S. S., Martino, H. S. D., & Pinheiro-Sant'Ana, H. M. (2017). Sorghum (*Sorghum bicolor* L.): Nutrients, bioactive compounds, and potential impact on human health. *Critical Reviews in Food Science and Nutrition*, *57*(2), 372–390.

Devi, D. R., & Murthy, U. R. (1993). Genotype environment interaction in tapetal development A2 based sorghum genotypes, *Sorghum Newsletter*, 34, 41.

Djanaguiraman, M., Vimala, K., Sofi, P. A., Perumal, R., & Prasad, P. V. V. (2024). Genetic variation for grain iron and zinc concentration in the US sorghum [*Sorghum bicolor* (L.) Moench] association panel. *Crop Science*, *64*(5), 2652–2665.

Doggett, H., & Eberhart, S. A. (1968). Recurrent selection in sorghum 1. *Crop Science*, *8*(1), 119–121.

Duvick, D. N. (1999). Heterosis: Feeding people and protecting natural resources. In: Coors J.G., Pandey, S. (Eds), *Genetics and exploitation of heterosis in crops* (pp. 19–29). Madison, WI, American Society of Agronomy/Crop Science Society of America.

Duvick, D. N. (2005). The contribution of breeding to yield advances in maize (*Zea mays* L.). *Advances in Agronomy*, *86*, 83–145.

Eggen, M., Ozdogan, M., Zaitchik, B., Ademe, D., Foltz, J., & Simane, B. (2019). Vulnerability of sorghum production to extreme, sub-seasonal weather under climate change. *Environmental Research Letters*, *14*(4), Art. 045005.

Elkonin, L. A., Gerashchenkov, G. A., Domanina, I. V., & Rozhnova, N. A. (2015). Inheritance of reversions to male fertility in male-sterile sorghum hybrids with 9E male-sterile cytoplasm induced by environmental conditions. *Russian Journal of Genetics*, *51*, 251–261.

Elkonin, L. A., Kozhemyakin, V. V., & Ishin, A. G. (1998). Nuclear-cytoplasmic interactions in restoration of male fertility in the '9E'and A4 CMS-inducing cytoplasms of sorghum. *Theoretical and Applied Genetics*, *97*, 626–632.

Elkonin, L. A., Kozhemyakin, V. V., & Tsvetova, M. I. (2009). Epigenetic control of the expression of fertility-restoring genes for the'9E'CMS-inducing cytoplasm of sorghum. *Maydica*, *54*(2), 243.

Elkonin, L. A., Kozhemyakin, V. V., & Tsvetova, M. I. (2019). The sporophytic type of fertility restoration in the A3 CMS-inducing cytoplasm of sorghum and its modification by plant water availability conditions, *Vavilov Journal of Genetics and Breeding*, 23, 412–421.

Fasahat, P., Rajabi, A., Rad, J. M., & Derera, J. J. B. B. I. J. (2016). Principles and utilization of combining ability in plant breeding. *Biometrics & Biostatistics International Journal*, *4*(1), 1–24.

Fonseca, J. M., Klein, P. E., Crossa, J., Pacheco, A., Perez-Rodriguez, P., Ramasamy, P., . . . Rooney, W. L. (2021). Assessing combining abilities, genomic data, and genotype× environment interactions to predict hybrid grain sorghum performance. *The Plant Genome*, *14*(3), Art. e20127.

Foster, K. W., Jain, S. K., & Smeltzer, D. G. (1980). Responses to 10 cycles of mass selection in an inbred population of grain sorghum 1. *Crop Science*, *20*(1), 1–4.

Garg, M., Sharma, N., Sharma, S., Kapoor, P., Kumar, A., Chunduri, V., & Arora, P. (2018). Biofortified crops generated by breeding, agronomy, and transgenic approaches are improving lives of millions of people around the world. *Frontiers in nutrition*, *5*, 12.

Girma, G., Tirfessa, A., Bejiga, T., Seyoum, A., Mekonen, M., Nega, A., . . . Tesso, T. (2024). Assessing genetic, racial, and geographic diversity among Ethiopian sorghum landraces and implications for heterotic potential for hybrid sorghum breeding. *Molecular Breeding*, *44*(7), 46.

Gizzi, G., & Gambin, B. L. (2016). Eco-physiological changes in sorghum hybrids released in Argentina over the last 30 years. *Field Crops Research*, *188*, 41–49.

Guo, C., Cui, W., Feng, X., Zhao, J., & Lu, G. (2011). Sorghum insect problems and Managementf. *Journal of Integrative Plant Biology*, *53*(3), 178–192.

Guo, J. X., & Liu, Y. G. (2012). Molecular control of male reproductive development and pollen fertility in rice F. *Journal of Integrative Plant Biology*, *54*(12), 967–978.

Hadebe, S. T., Modi, A. T., & Mabhaudhi, T. (2017). Drought tolerance and water use of cereal crops: A focus on sorghum as a food security crop in sub-Saharan Africa. *Journal of Agronomy and Crop Science*, *203*(3), 177–191.

Hallauer AR, M. J. Carena, Filho JBM (2010) Selection: experimental results. In: Carena, M.J., Hallauer, A.R., Filho, J.B.M. (Eds), Quantitative genetics in maize breeding. Handbook of plant breeding, vol 6. Springer, New York, pp 291–383.

Hanson, M. R., & Bentolila, S. (2004). Interactions of mitochondrial and nuclear genes that affect male gametophyte development. *The Plant Cell*, *16*(suppl_1), S154–S169.

Hariprasanna, K., Aruna, C., & Madhusudhana, R. (2024). High-yielding grain sorghum cultivars released in India for enhanced productivity. *Indian Farming*, *73*(12), 13–18.

Henzell, R. G., Franzmann, B. A., & Brengman, R. L. (1994). Sorghum midge resistance research in Australia, *International Sorghum Millets Newsletter*, 35, 41–47.

Hodnett, G. L., & Rooney, W. L. (2018). Male sterility induction of sorghum using chemical hybridizing agent TFMSA, trifluoromethanesulfonamide. *Canadian Journal of Plant Science*, *98*(5), 1102–1108.

House, L. R. (1985). In: International Crops Res Institute for the Semi-Arid Tropics Patancheru, India, *A guide to sorghum breeding* (2nd ed.).

House, L. R., Verma, B. N., Ejeta, G., Rana, B. S., Kapran, I., Obilana, A. B., & Reddy, B. V. S. (1996, September). Developing countries breeding and potential of hybrid sorghum. In *Proceedings of the international conference on genetic improvement of sorghum and pearl millet, Lubbock, TX* (pp. 23–27). Madison: INTSORMIL, University of Wisconsin.

Hussaini, S. H., & Rao, P. V. (1964). A note on the spontaneous occurrence of cytoplasmic male sterility in Indian sorghums. *Sorghum Newsl*, *27*, 27–28.

Ingle, K. P., Moharil, M. P., Gahukar, S. J., Jadhav, P. V., Ghorade, R., Abdi, G., . . . Singh, A. (2023). Mendelian to genomics and bioinformatics approaches in cytoplasmic male sterility and fertility restoration in sorghum breeding. In *Advanced crop improvement, Volume 2: Case studies of economically important crops* (pp. 393–417). Springer, Cham, Switzerland.

Jiao, Y., Burke, J., Chopra, R., Burow, G., Chen, J., Wang, B., . . . Xin, Z. (2016). A sorghum mutant resource as an efficient platform for gene discovery in grasses. *The Plant Cell*, *28*(7), 1551–1562.

Johnson, J. W., Rosenow, D. T., & Teetes, G. L. (1973). Resistance to the sorghum midge in converted exotic sorghum cultivars 1. *Crop Science*, *13*(6), 754–755.

Johnson, J. W., Rosenow, D. T., & Teetes, G. L. (1974). Response of greenbug-resistant grain sorghum lines and hybrids to a natural infestation of greenbugs 1. *Crop Science*, *14*(3), 442–443.

Jordan, D., Klein, R., Melonek, J., Small, I., Cruickshank, A., Bradburn, L., . . . Mace, E. (2024). The genetic architecture of temperature-induced partial fertility restoration in A1 cytoplasm in sorghum (*Sorghum bicolor* (L.) Moench).

Jordan, D. R., Klein, R. R., Sakrewski, K. G., Henzell, R. G., Klein, P. E., & Mace, E. S. (2011). Mapping and characterization of Rf 5: A new gene conditioning pollen fertility restoration in A1 and A2 cytoplasm in sorghum (*Sorghum bicolor* (L.) Moench). *Theoretical and Applied Genetics*, *123*, 383–396.

Jordan, D. R., Mace, E. S., Henzell, R. G., Klein, P. E., & Klein, R. R. (2010). Molecular mapping and candidate gene identification of the Rf2 gene for pollen fertility restoration in sorghum [*Sorghum bicolor* (L.) Moench]. *Theoretical and Applied Genetics*, *120*, 1279–1287.

Kahate, N. S., Raut, S. M., Ulemale, P. H., & Bhogave, A. F. (2014). Management of sorghum shoot fly. *Popular Kheti*, *2*, 72–74.

Kannababu, N., Raghunath, K., & Bhat, B. V. (2023). Sorghum hybrid seed production and quality maintenance. *Indian Farming*, *73*(1), 79–83.

Kariyannanavar, P., Wali, M. C., Biradar, B. D., Chattannavar, S. N., Kubsad, V. S., Verma, L. K., . . . Revanasiddayya. (2024). Deciphering the inheritance of fertility restoration and influence of environmental factors on maldandi source of male sterility in rabi sorghum [*Sorghum bicolor* (L.) Moench]. *Plant Breeding*, *143*(2), 190–194.

Kenga, R., Alabi, S. O., & Gupta, S. C. (2004). Combining ability studies in tropical sorghum (*Sorghum bicolor* (L.) Moench). *Field Crops Research*, *88*(2–3), 251–260.

Kent, M. A., Fonseca, J. M., Klein, P. E., Klein, R. R., Hayes, C. M., & Rooney, W. L. (2023). Use of genomic prediction to screen sorghum B-lines in hybrid testcrosses. *The Plant Genome*, *16*(3), Art. e20369.

Kim, Y. J., & Zhang, D. (2018). Molecular control of male fertility for crop hybrid breeding. *Trends in Plant Science*, *23*(1), 53–65.

Kiyosawa, A., Yonemaru, J. I., Mizuno, H., Kanamori, H., Wu, J., Kawahigashi, H., & Goto, K. (2022). Fine mapping of Rf5 region for a sorghum fertility restorer gene and microsynteny analysis across grass species. *Breeding Science*, *72*(2), 141–149.

Klein, R. R., Klein, P. E., Mullet, J. E., Minx, P., Rooney, W. L., & Schertz, K. F. (2005). Fertility restorer locus Rf1 of sorghum (*Sorghum bicolor* L.) encodes a pentatricopeptide repeat protein not present in the colinear region of rice chromosome 12. *Theoretical and Applied Genetics*, *111*, 994–1012.

KNOEMA. (2024). https://knoema.com/atlas/topics/Agriculture

Kumar, A. A., Sharma, H. C., Sharma, R., Blummel, M., Reddy, P. S., & Reddy, B. V. (2013). Phenotyping in sorghum [*Sorghum bicolor* (L.) Moench]. In *Phenotyping for plant breeding: Applications of phenotyping methods for crop improvement* (pp. 73–109). Springer, New York, NY, USA.

Kumar, N. P., Biradar, B. D., Bagewadi, B., Hanamaratti, N. G., Bhat, S., Shekharappa, N. P., . . . Kavyashree, N. M. (2023). Identification of SSR markers linked to new fertility restoration trait in sorghum [*Sorghum bicolor* (L.) Moench] for A4 (maldandi) male sterile cytoplasm. *Plant Breed*, 143(2), 1–9.

Lambright, L. (2019). Hybrid sorghum product development and production. *Sorghum: Methods and Protocols*, 3–9.

Leon, F., Harvey, L., & Rooney, W. L. (2024). In: Zhao, ZY., Dahlberg, J. (Eds), Historical multi-environmental trials reveal trends in hybrid grain sorghum performance from 1970 to 2021 in Texas. *Crop Science*, 1931, 3014–3027. New York: Humana Press.

Li, Z., Cheng, Y., Cui, J., Zhang, P., Zhao, H., & Hu, S. (2015). Comparative transcriptome analysis reveals carbohydrate and lipid metabolism blocks in *Brassica napus* L. male sterility induced by the chemical hybridization agent monosulfuron ester sodium. *BMC Genomics*, *16*, 1–19.

Little, C. R., Bandara, A. Y., Todd, T. C., & Perumal, R. (2018). Sorghum diseases and their management in cultivation: Stalk, root and other diseases. In: Rooney, W. (Ed.), *Achieving sustainable cultivation of sorghum* (Vol. 1, pp. 447–484). Sawston Cambridge, UK: Burleigh Dodds Science Publishing.

Lofton, J., G. Strickland, C. Harris, J. Rice, M. Drendel, and S. Sharma. 2024. Grain sorghum performance trials in Oklahoma, 2023. Oklahoma Cooperative Extension Service, Stillwater, OK. CR-2162.

Lonkar, S. G., & Borikar, S. T. (1994). Inheritance of A1 and A2 cytoplasmic genetic male sterility in sorghum, *Journal of the Maharashtra Agricultural Universities*, *19*(3), 450.

Loussaert, D. (2004). Trihalogenated methylsulfonamides as specific male gametocides. *Sexual Plant Reproduction*, *16*, 299–307.

Madugula, P., Uttam, A. G., Tonapi, V. A., & Ragimasalawada, M. (2018). Fine mapping of Rf2, a major locus controlling pollen fertility restoration in sorghum A1 cytoplasm, encodes a PPR gene and its validation through expression analysis. *Plant Breeding*, *137*(2), 148–161.

Maulana, F., Weerasooriya, D., & Tesso, T. (2017). Sorghum landrace collections from cooler regions of the world exhibit magnificent genetic differentiation and early season cold tolerance. *Frontiers in Plant Science*, *8*, 756.

Mayor, L., Demarco, P., Lira, S., Fang, Y., Abadie, T., Gambin, B. L., . . . Messina, C. (2023). Retrospective study in US commercial sorghum breeding: I. Genetic gain in relation to relative maturity. *Crop Science*, *63*(2), 501–510.

Melchinger, A. E., & Gumber, R. K. (1998). Overview of heterosis and heterotic groups in agronomic crops. *Concepts and Breeding of Heterosis in Crop Plants*, *25*, 29–44.

Menezes, C. B. D., Silva, K. J. D., Teodoro, L. P. R., Santos, C. V. D., Julio, B. H. M., Portugal, A. F., . . . Teodoro, P. E. (2022). Grain sorghum hybrids under drought stress and full-irrigation conditions in the Brazilian Semiarid. *Journal of Agronomy and Crop Science*, *208*(6), 868–875.

Mengistu, G., Shimelis, H., Laing, M., Lule, D., & Mashilo, J. (2020). Combining ability and heterosis among sorghum (*Sorghum bicolor* [L.] Moench) lines for yield, yield-related traits, and anthracnose resistance in western Ethiopia. *Euphytica*, *216*(2), 33.

Menz, M. A., Klein, R. R., Unruh, N. C., Rooney, W. L., Klein, P. E., & Mullet, J. E. (2004). Genetic diversity of public inbreds of sorghum determined by mapped AFLP and SSR markers. *Crop Science*, *44*(4), 1236–1244.

Miller, F. R., & Kebede, Y. (1984). Genetic contributions to yield gains in sorghum, 1950 to 1980. *Genetic Contributions to Yield Gains of Five Major Crop Plants*, *7*, 1–14.

Mindaye, T. T., Mace, E. S., Godwin, I. D., & Jordan, D. R. (2015). Genetic differentiation analysis for the identification of complementary parental pools for sorghum hybrid breeding in Ethiopia. *Theoretical and Applied Genetics*, *128*, 1765–1775.

Mishra, D., Kumar, S., & Mishra, B. N. (2021). An overview of morpho-physiological, biochemical, and molecular responses of sorghum towards heavy metal stress. *Reviews of Environmental Contamination and Toxicology*, *256*, 155–177.

Mital, S. P., Vishnu Swarup, V. S., & Joshi, A. B. (1958). Cytoplasmic male sterility in jowar (*Sorghum vulgare* Pers.), *Current Science*, *27*, 314.

Murty, U. R. (1986). Effect of A2 cytoplasm on the inheritance of plant height in temperate × tropical sorghum crosses, *Sorghum Newsletter 29*, p. 77.

Murty, U. R., & Gangadhar, G. (1990). Milo and non-milo sources of cytoplasms in *Sorghum bicolor* (L.) Moench. III. Genetics of fertility restoration. *Cereal Research Communications*, 18(1), 111–116.

Nagur, T., & Menon, P. M. (1974). Characterization of different male-sterility inducing cytoplasms in sorghum. *Sorghum Newsletter*, *17*, 18.

Naylor, A. W. (1950). Observations on the effects of maleic hydrazide on flowering of tobacco, maize and cocklebur. *Proceedings of the National Academy of Sciences*, *36*(4), 230–232.

Ngidi, A., Shimelis, H., Abady, S., Chaplot, V., & Figlan, S. (2024a). Genetic variation and association of yield, yield components, and carbon storage in sorghum (*Sorghum bicolor* [L.] Moench) genotypes. *BMC Genomic Data*, *25*(1), 74.

Ngidi, A., Shimelis, H., Abady, S., Figlan, S., & Chaplot, V. (2024b). Response of *Sorghum bicolor* genotypes for yield and yield components and organic carbon storage in the shoot and root systems. *Scientific Reports*, *14*(1), 9499.

Nguyen, H. T. (2024). Sorghum improvement: Past achievements and future prospects. In: Kang, M.S.(Ed.), *Crop improvement: Challenges in the twenty-first century* (p. 109). Boca Raton, Florida: CRC Press.

Niebur, W. S., Rafalski, J. A., Smith, O. S., & Cooper, M. (2004, September). Applications of genomics technologies to enhance rate of genetic progress for yield of maize within a commercial breeding program. In *New directions for a diverse planet: Proceedings for the 4th international crop science congress*. Gosford, Australia: Regional Institute. www.cropscience.org.au/icsc2004.

Nuessly, G. S., Wang, Y., Sandhu, H., Larsen, N., & Cherry, R. H. (2013). Entomologic and agronomic evaluations of 18 sweet sorghum cultivars for biofuel in Florida. *Florida Entomologist*, 96, 512–528.

Nwanze, K. F., Seetharama, N., Sharma, H. C., & Stenhouse, J. W. (1995). Biotechnology in pest management: Improving resistance in sorghum to insect pests. *African Crop Science Journal*, *3*(2), 209–215.

Okosun, O. O., Allen, K. C., Glover, J. P., & Reddy, G. V. (2021). Biology, ecology, and management of key sorghum insect pests. *Journal of Integrated Pest Management*, *12*(1), 4.

Oliveira, C. M., Auad, A. M., Mendes, S. M., & Frizzas, M. R. (2013). Economic impact of exotic insect pests in Brazilian agriculture. *Journal of Applied Entomology*, *137*(1-2), 1–15.

Overholt, W. A., Maes, K. V. N., & Goebel, F. R. (2001). *Field guide to the stemborer larvae of maize, sorghum and sugarcane in Eastern and Southern Africa*. International Centre of Insect Physiology and Ecology (ICIPE), Nairobi, Kenya: ICIPE Science Press.

Parra, G., Borras, L., & Gambin, B. L. (2020). Maize long-term genetic progress explains current dominance over sorghum in Argentina. *European Journal of Agronomy*, *119*, 126122.

Pedersen, J. F., & Toy, J. J. (2001). Registration of N316-N320 sorghum nuclear male-sterility genetic stocks. *Crop Science*, *41*(2), 607–607.

Pfeiffer, B. K., Pietsch, D., Schnell, R. W., & Rooney, W. L. (2019). Long-term selection in hybrid sorghum breeding programs. *Crop Science*, *59*(1), 150–164.

Poehlman, J. M. (2013). *Breeding field crops*. Springer Science & Business Media, Springer, Netherlands.

Praveen, M., Suneetha, N., Av, U., Patil, J. V., & Madhusudhana, R. (2015). Inheritance and molecular mapping of *Rf6* locus with pollen fertility restoration ability on A1 and A2 cytoplasms in sorghum. *Plant Science*, *238*, 73–80.

Pring, D. R., Tang, H. V., Howad, W., & Kempken, F. (1999). A unique two-gene gametophytic male sterility system in sorghum involving a possible role of RNA editing in fertility restoration. *Journal of Heredity*, *90*(3), 386–393.

Queiroz, V. A. V., da Silva, C. S., de Menezes, C. B., Schaffert, R. E., Guimarães, F. F. M., Guimarães, L. J. M., . . . Tardin, F. D. (2015). Nutritional composition of sorghum [*Sorghum bicolor* (L.) Moench] genotypes cultivated without and with water stress. *Journal of Cereal Science*, *65*, 103–111.

Quinby, J. R. (1963). Manifestations of hybrid vigor in sorghum, *Crop Science*, *3*, 288–291.

Quinby, J. R. (1982). Interaction of genes and cytoplasms in sex expression in sorghum. *Sorghum in the Eighties*, *385*, pp 175–184. Hyderabad, India: International Crops Research Institute for the Semi Arid Tropics.

Rakshit, S., & Bellundagi, A. (2019). Conventional breeding techniques in sorghum. In: Aruna, C., Visarada, K.B.R.S., Bhat, B.V., Tonapi, V.A.(Eds), *Breeding sorghum for diverse end uses* (pp. 77–91). Cambridge, MA, United States: Woodhead Publishing.

Rakshit, S., Hariprasanna, K., Gomashe, S., Ganapathy, K. N., Das, I. K., Ramana, O. V., . . . Patil, J. V. (2014). Changes in area, yield gains, and yield stability of sorghum in major sorghum-producing countries, 1970 to 2009. *Crop Science*, *54*(4), 1571–1584.

Ramu, P., Billot, C., Rami, J. F., Senthilvel, S., Upadhyaya, H. D., Ananda Reddy, L., & Hash, C. T. (2013). Assessment of genetic diversity in the sorghum reference set using EST-SSR markers. *Theoretical and Applied Genetics*, *126*, 2051–2064.

Rao, N. G. P. (1982). Transforming traditional sorghums in India. In L. R. House et al. (Eds.), *Sorghum in the eighties* (pp. 175–184). International Crops Research Institute for the Semi Arid Tropics. Hyderabad, India..

Rattunde, F., Vom Brocke, K., Weltzien, E., & Haussmann, B. I. G. (2009). Selection methods 4. Developing open-pollinated varieties using recurrent selection methods. In S. Ceccarelli, E. P. Guimaraes, & E. Weltzien (Eds.), *Plant breeding and farmer participation* (pp. 259–273). Rome: FAO.

Ray, D. K., Ramankutty, N., Mueller, N. D., West, P. C., & Foley, J. A. (2012). Recent patterns of crop yield growth and stagnation. *Nature Communications*, *3*(1), 1293.

Reddy, B. V. S., Ashok Kumar, A., & Kaul, S. L. (2008). Alternative cytoplasmic male sterility systems in sorghum and their utilization. In: Reddy, B.V.S., Ramesh, S., Kumar, A.A., Gowda, C.L.L. (Eds), *Sorghum improvement in the new millennium* (pp. 132–144). International Crops Research Institute for the Semi-Arid Tropics, Patancheru, Andhra Pradesh.

Reddy, B. V. S., Ashok Kumar, A., Sanjana Reddy, P., & Ramaiah, B. (2008). *Development of male-sterile lines in sorghum* (pp. 72–78). Patancheru, Andhra Pradesh, India: International Crops Research Institute for the Semi-Arid Tropics.

Reddy, B. V. S., Rai, K. N., Sarma, N. P., Kumar, I., & Saxena, K. B. (2004b). Cytoplasmic-nuclear male sterility: Origin, evaluation and utilization. In: Jain, H. K., Kharkwal, M.C. (Eds), *Plant breeding: Mendelian to molecular approaches* (pp. 473–499). New Delhi, India: Narosa Publishing House.

Reddy, B. V., Ramesh, S., & Ortiz, R. (2010). Genetic and cytoplasmic-nuclear. *Plant Breeding Reviews*, *25*, 139.

Reddy, B. V., Ramesh, S., & Reddy, P. S. (2004a). Nucleus and breeder seed production in sorghum. *International Sorghum and Millets Newsletter*, *45*, 13–17.

Reddy, B. V., & Stenhouse, J. W. (1994). Sorghum improvement for semi-arid tropics region: Past current and future research thrusts in Asia. *PKV Reserach Journal*, *18*(2), 155–169.

Reddy, K.V.S., Zehr, U.B. (2004) In New directions for a diverse planet. Proceedings of the 4th International Crop Science Congress. Brisbane, Australia. pp 1-8.

Ross, W. M. (1965). Yields of *Sorghum vulgare* Pers. populations with various degrees of male sterility, *Crop Science*, 5, 480–482.

Ross, W. M., & Hackerott, H. L. (1972). Registration of seven isocytoplasmic sorghum germplasm lines, *Crop Science* 12, 720–721.

Sanjana Reddy, P., Rao, D. M., Reddy, B. V. S., & Ashok Kumar, A. (2010). Inheritance of male-fertility restoration in A1, A2, A3 and A4 (M) cytoplasmic male-sterility systems of sorghum [*Sorghum bicolor* (L.) Moench]. *Indian Journal of Genetics and Plant Breeding*, *70*(3), 240–246.

Sapkota, S., Boatwright, J. L., Kumar, N., Myers, M., Cox, A., Ackerman, A., . . . Kresovich, S. (2023). Genomic prediction of hybrid performance for agronomic traits in sorghum. *G3: Genes, Genomes, Genetics*, *13*(4), Art. jkac311.

Saxena, K. B., & Hingane, A. J. (2015). Male sterility systems in major field crops and their potential role in crop improvement. In: Bahadur, B., Venkat Rajam, M., Sahijram, L., Krishnamurthy, K.V.(Eds), *Plant biology and biotechnology. Volume I: Plant diversity, organization, function and improvement* (pp. 639–656). Springer, New Delhi.

Schaffasz, A., Windpassinger, S., Snowdon, R., & Wittkop, B. (2019). Reproductive cold stress tolerance in sorghum F1 hybrids is a heterotic trait. *Agronomy*, *9*(9), 508.

Schertz, K. F. (1963). Chromosomal, morphological, and fertility characteristics of haploids and their derivatives in *Sorghum vulgare* Pers, *Crop Science*, 3, 445–447.

Schertz, K. F. (1977). Registration of A2 Tx2753 and B Tx2753 sorghum germplasm (Reg. No. GP 30 and 31), *Crop Science*, 17, 983.

Schertz, K.F. (1994) Male sterility in sorghum: its characteristics and importance. In: Use of Molecular Markers in Sorghum and Pearl Millet Breeding for Developing Countries (J.R. Witcombe and R.R. Duncan, eds), pp. 35–37. Proceedings of the Overseas Development Administration (ODA) Plant Sciences Research Conference, 29 March–1 April 1993. Norwich: ODA.

Schertz, K. F., & Clark, L. E. (1967). Controlling dehiscence with plastic bags for hand crosses in Sorghum 1. *Crop Science*, *7*(5), 540–542.

Schertz, K. F., & Pring, D. R. (1982). Cytoplasmic sterility systems in sorghum. *Sorghum in the Eighties*, *1*, 373–383.

Schertz, K. F., & Ritchey, J. M. (1978). Cytoplasmic-Genic Male-Sterility Systems in Sorghum 1. *Crop Science*, *18*(5), 890–893.

Schertz, K. F., Sivaramakrishnan, S., Hanna, W. W., Mullet, J., Sun, Y., Murty, U. R., . . . Reddy, B. V. (1997). Alternate cytoplasms and apomixis of sorghum and pearl millet. In Proc. Intl. Conf. Genetic Improvement of Sorghum and Pearl Millet. INTSORMIL/ICRISAT, Lubbock, TX.pp 213–223.

Senthil, N., Palanisamy, S., & Sreerangasamy, S. R. (1994). Characterization of diverse cytosteriles of sorghum through fertility restoration studies. *Cereal Research Communications*, 22, 179–184.

Sharma, Y., & Sharma, S. N. (2005). Chemical hybridizing agents (CHA)—a tool for hybrid seed production—a review. *Agricultural Reviews*, *26*(2), 114–123.

Shewayrga, H., Sopade, P. A., Jordan, D. R., & Godwin, I. D. (2012). Characterisation of grain quality in diverse sorghum germplasm using a Rapid Visco-Analyzer and near infrared reflectance spectroscopy. *Journal of the Science of Food and Agriculture*, *92*(7), 1402–1410.

Sprague, G. F., & Tatum, L. A. (1942). General vs. specific combining ability in single crosses of corn, *Agronomy Journal*, 34, 923–932.

Stephens, J., & Holland, R. F. (1953). Cyto-plastic male-sterility for hybrid sorghum seed production, *Agronomy Journal*, 6, 20–2.

Stephens, J. C., & Quinby, J. R. (1952). Yield of a hand-produced hybrid sorghum, *Agronomy*, 44, 231–233.

Sukumaran, S., Xiang, W., Bean, S. R., Pedersen, J. F., Kresovich, S., Tuinstra, M. R., . . . Yu, J. (2012). Association mapping for grain quality in a diverse sorghum collection. *The Plant Genome*, *5*(3), 126–135.

Sun, Y., Niu, G., Osuna, P., Zhao, L., Ganjehunte, G., Peterson, G., . . . Gardea-Torresdey, J. L. (2014). Variability in salt tolerance of *Sorghum bicolor* L. *Agricultural Science*, *2*(1), 9.

Tang, H. V., Pedersen, J. F., Chase, C. D., & Pring, D. R. (2007). Fertility restoration of the sorghum A3 male-sterile cytoplasm through a sporophytic mechanism derived from sudangrass. *Crop Science*, *47*(3), 943–950.

Tang, H. V., & Pring, D. R. (2003). Conversion of fertility restoration of the sorghum IS1112C (A3) male-sterile cytoplasm from two genes to one gene. *Crop Science*, *43*(5), 1747–1753.

Tatum, L. A. (1971). The Southern Corn Leaf Blight Epidemic: A new race of the fungus Helminthosporium maydis threatens domestic prices and corn reserves for export. *Science*, *171*(3976), 1113–1116.

Tripathi, D. P., Rana, B. S., & Rao, N. G. P. (1985). Genetics of fertility restoration in sorghum, *Indian Journal of Genetics and Plant Breeding*, 45, 292–301.

Ullstrup, A. J. (1972). The impacts of the southern corn leaf blight epidemics of 1970–1971. *Annual Review of Phytopathology*, *10*(1), 37–50.

USDA-FAS. (2024). usda.gov. https://fas.usda.gov/data/production/commodity/0459200

Wagaw, K., Bantte, K., & Tadesse, T. (2020). Gene action, combining ability and heterotic performance of Ethiopian Sorghum (*Sorghum bicolor* (L.) Moench) lines under moisture stress areas in Ethiopia. *African Journal of Plant Science*, *14*(9), 372–394.

Wagaw, K., & Tadesse, T. (2020). Combining ability and heterosis of sorghum (*Sorghum bicolor* L. Moench) hybrids for grain and biomass yield. *American Journal of Plant Sciences*, *11*(12), 2155–2171.

Wang, L. M., Yan, H. D., Jiao, S. J., Jiang, Y., Su, D. F., & Sun, G. Q. (2020). Heterosis prediction of sweet sorghum based on combining ability and genetic distance, *Acta Agronomica Hungarica*, *53*(14), 2786–2794.

Webster, O. J., & Singh, S. P. (1964). Breeding behavior and histological structure of a nondehiscent anther character in *Sorghum vulgare* Pers, *Crop Science*, 4, 656–658.

Wei, F., Hu, J., Zhang, X., ming Cheng, L., Du, J., Si, Y. H., . . . Tian, B. M. (2012). Male sterility induced by chemical SQ-1, as an effective male specific gametocide in maize (*Zea mays*). *Maydica*, *57*(4), 244–248.

Williams-Alanís, H., Aranda, U., Cavazos, G. Á., Garcia, F. Z., Júarez, M. G., Vázquez, M. D. C. R., & Barrón, J. E. (2022). Line x tester analysis to estimate combining ability in grain sorghum (*Sorghum bicolor* L.). *Revista de la Facultad de Ciencias Agrarias UNCuyo*, *54*(2), 12–21.

Worstell, J. V., Kidd, H. J., & Schertz, K. F. (1984). Relationships Among Male-Sterility Inducing Cytoplasms of Sorghum 1. *Crop Science*, *24*(1), 186–189.

Xin, Z., Chen, J., Jiao, Y., Gladman, N., Hayes, C., Burow, G., . . . Burke, J. J. (2018). Registration of BTx623ms8, a new and easily identifiable nuclear male sterile mutant in sorghum. *Journal of Plant Registrations*, *12*(2), 278–281.

Yahaya, M. A., Shimelis, H., Laing, M., Mohammed, M. S., & Mathew, I. (2020). Methodologies for hybridization in predominantly self-pollinating crops: A review. *Journal of Crop Improvement*, *34*(2), 268–289.

Yan, W., & Hunt, L. A. (2002). Biplot analysis of diallel data. *Crop Science*, *42*(1), 21–30.

Zapata, S. D., Villanueva, R., Sekula, D., Esparza-Diaz, G., Duke, K., & Mutaleb, M. D. (2016). The Economic Impact of the Sugarcane Aphid on Sorghum Production. Southern Agricultural Economics Association, Cambridge, UK.

Zhang, Y., Chen, J., Gao, Z., Wang, H., Liang, D., Guo, Q., . . . Liu, Q. (2024). Identification of heterosis and combining ability in the hybrids of male sterile and restorer sorghum [*Sorghum bicolor* (L.) Moench] lines. *PLoS One*, *19*(1), e0296416.

Zheng, H., Dang, Y., Diao, X., & Sui, N. (2023). Molecular mechanisms of stress resistance in sorghum: Implications for crop improvement strategies. *Journal of Integrative Agriculture,* 23, 741–768.

9 Omics and Systems Biology Approaches for Developing Climate-Ready Sorghum
Opportunities and Challenges

Mumtaz Ali Saand

9.1 INTRODUCTION

Sorghum (*Sorghum bicolor* (L.) Moench), a C_4 plant, is the fifth most cultivated cereal crop globally, following maize, rice, wheat, and barley (Wang et al., 2024). Known for its rapid growth and high biomass production, sorghum serves as a versatile resource for cereal, fodder, and energy applications (Wang et al., 2024; Fu et al., 2024). Beyond its drought tolerance, sorghum exhibits tolerance to other abiotic stresses, including salinity, cold, and ion stress (Wang et al., 2022; Jia et al., 2021; Marla et al., 2023). The advent of next-generation sequencing (NGS) has revolutionized cereal crop genomics, enabling the identification of novel genes and facilitating the breeding of resilient crops under changing climatic conditions. Single-omics studies in sorghum have focused on improving yield and growth under abiotic stresses. For instance, transcriptomic analyses have identified heat shock proteins and transcription factor genes associated with combined heat and drought stress (Johnson et al., 2014).

More advanced double- and triple-omics approaches have enhanced breeding improvements in sorghum. Integrating transcriptomics and proteomics has identified genes and proteins linked to seed storage under salinity stress (Punia et al., 2020). Similarly, transcriptomics combined with metabolomics has revealed the involvement of flavonoids and hormone signal transduction pathways in response to salinity (Ma et al., 2020; Zhang et al., 2022). Other studies have demonstrated the role of transcriptomic and metabolomic integration in identifying genes and metabolites involved in dhurrin and starch biosynthesis during seed germination and seedling development (Liu et al., 2023; Khan et al., 2024). Additionally, combining transcriptomics with ionomics has highlighted changes in gene expression and ion concentrations under NPK starvation (Zhu et al., 2020).

Further integration of genomics and phenomics has uncovered complex adaptive traits associated with chilling tolerance in sorghum genotypes (Marla et al., 2023). Multi-omics approaches, spanning genomics to phenomics and integrated with systems biology, offer a holistic understanding of genetic improvements in cereal crop breeding (Saand & Yang, 2024). These strategies provide functional insights for developing resilient sorghum under diverse climatic conditions (Mukherjee et al., 2024). The integration of multi-omics technologies, ranging from genomics to phenomics, has been pivotal in elucidating gene functions and regulatory mechanisms critical for climate-resilient sorghum varieties (Liu et al., 2024). Accordingly, this chapter explores the application of omics technologies in sorghum crop research, emphasizing their role in enhancing resilience to abiotic stresses. An overview of these omics approaches is depicted in Figure 9.1.

This chapter reviews advancements in multi-omics technologies, from genomics to phenomics, for enhancing sorghum resilience. It examines integrating multi-omics with systems biology to identify key traits, genes, proteins, and metabolites under abiotic stresses like heat, drought, salinity, and

DOI: 10.1201/9781003623717-9

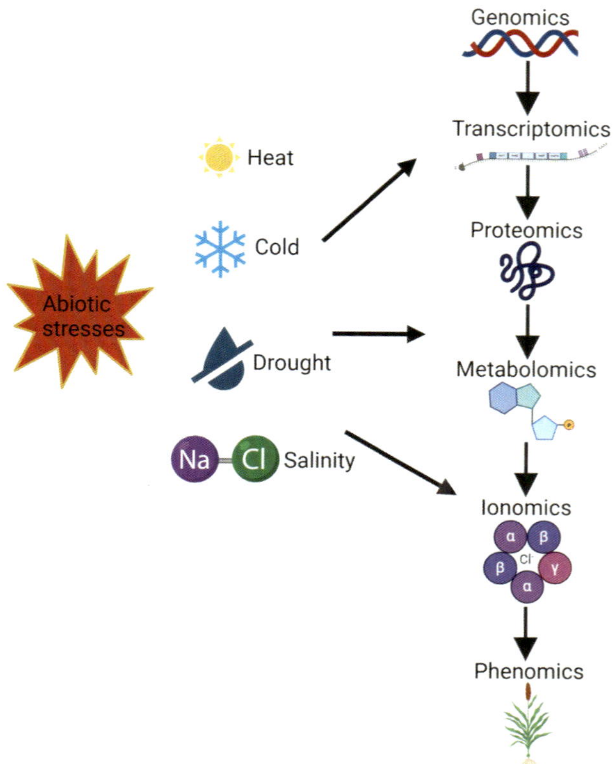

FIGURE 9.1 Overview of omics approaches for enhancing climate resilience in sorghum.

cold. Techniques for developing climate-resilient sorghum and challenges in analyzing large omics datasets are also discussed, emphasizing innovative solutions for maximizing potential of these technologies.

9.2 OMICS APPROACHES FOR DEVELOPING CLIMATE-READY SORGHUM

Developing climate-ready sorghum requires leveraging diverse omics approaches to uncover the genetic, molecular, and physiological mechanisms underlying stress resilience. Various omics approaches are discussed below-

9.2.1 Genomics

Genomics, the study of an organism's DNA focuses on the structure, function, mapping, editing, and evolution of genes. It is broadly categorized into structural genomics, which involves genetic mapping, and functional genomics, which investigates gene functions (Yang et al., 2021). As a diploid species with a small, non-duplicating genome, sorghum's genome was first sequenced using the Sanger sequencing platform for BTx623 (Paterson et al., 2009). Subsequent genome modifications identified 34,129 annotated genes and significant proportions of DNA transposons (8.7%) and retrotransposons (~59%) (McCormick et al., 2018).

The advent of NGS has revolutionized genomic research, enabling the dissection of quantitative trait loci (QTLs) for traits linked to stress tolerance and crop improvement. QTLs related to drought tolerance and other abiotic stresses have also been mapped using various markers, including simple sequence repeats (SSR) and single nucleotide polymorphisms (SNPs), aiding crop improvement (Rakkammal et al., 2022). Notably, QTL Atlas reports 150 QTL and genome-wide association studies (GWAS) conducted in sorghum since 1995, underscoring their importance in breeding programs (Mace et al., 2019).

GWAS-based studies have further identified QTLs linked to biomass, yield, and agroclimatic traits under stress conditions (Spindel et al., 2018; Morris et al., 2013). Genotyping-by-sequencing (GBS) approaches have contributed to uncovering genes related to grain pigmentation across diverse regions (Maina et al., 2018) and interactions between genetic and environmental factors influencing sorghum's adaptive responses (Olatoye et al., 2018). Comparative analyses of cultivated and wild sorghum have identified novel QTLs for grain traits, such as thousand-grain weight QTLs (Tao et al., 2018). Additionally, studies have identified SNPs associated with plant maturity, height, and seed morphology traits, including defense genes (Morris et al., 2013; Ahn et al., 2024). Marker-assisted selection (MAS) has been instrumental in leveraging these QTLs and SNPs to improve sorghum's abiotic stress tolerance, including drought and salinity (Somegowda et al., 2023).

Gene expression profiling has shed light on genomic responses of sorghum to drought and other stresses. Differential expression of genes like *SbNAC*, *SbDof*, and *SbWRKY* under drought stress and *SbARF* genes during temperature stress highlights the intricate regulatory mechanisms (Sanjari et al., 2019; Baillo et al., 2020; Chen et al., 2019). Advances in genome editing tools have further expanded functional genomic research in sorghum. Techniques like Targeting Induced Local Lesions IN Genomes (TILLING) have modified phenotypes, such as reducing lignin content to enhance digestibility (Xin et al., 2008). CRISPR-Cas9 has been tested in sorghum, alongside other tools like transcription activator-like effector nucleases (TALENs) and zinc finger nucleases (ZFNs), though their applications under stress conditions require further exploration (Jiang et al., 2013; Saand & Yang, 2024).

Pan-genomics and epigenomics also play critical roles in sorghum improvement. Pan-genomic studies have identified drought-responsive genes and SNPs linked to agronomic traits in diverse sorghum races (Ruperao et al., 2021). Epigenetic research has highlighted resilient responses to low nitrogen stress in sorghum epi-lines and identified methylation patterns and histone markers involved in photosynthesis and lignin synthesis under drought conditions (Ketumile et al., 2022; Hu et al., 2023). In summary, genomics, encompassing tools like QTL mapping, GWAS, gene expression profiling, genome editing, and epigenomics, plays a pivotal role in improving sorghum's genetic resilience to diverse abiotic stresses, offering valuable insights for crop improvement and stress adaptation strategies.

9.2.2 Transcriptomics

Earlier transcriptomics techniques, including Expressed Sequence Tags (ESTs), Serial Analysis of Gene Expression (SAGE), and cDNA-amplified fragment length polymorphism (cDNA-AFLP), were widely employed for cereal crop studies (Kaur et al., 2021). Over time, these approaches were augmented by more robust technologies such as quantitative real-time PCR (qRT-PCR) and microarrays (Kaur et al., 2021). Recently, advanced methods like digital gene expression (DGE), RNA sequencing (RNA-seq), and Next-Generation Sequencing (NGS) have revolutionized transcriptomic analyses in cereal crops (Yang et al., 2021). For instance, the single-molecule real-time long-read TAPIS (Transcriptome Analysis Pipeline for Isoform Sequencing) approach has been applied to sorghum transcriptomics, identifying numerous novel genes (Abdel-Ghany et al., 2016).

The microarray technique has identified heat shock proteins (HSPs) and transcription factor genes such as *MYB78* and *ATAF1* in sorghum under combined heat and drought stress (Johnson et al., 2014). Similarly, RNA-seq has been used to uncover differentially expressed genes (DEGs) in drought-resistant and sensitive sorghum genotypes under PEG-induced drought stress (Abdel-Ghany et al., 2020). Furthermore, RNA-seq-based transcriptomic analysis of sorghum grains revealed key pathways associated with carotenoid biosynthesis and degradation (Cruet-Burgos & Rhodes, 2023), while root, stem, and leaf profiling under saline stress identified DEGs linked to ion transporters (Ukwatta et al., 2021). Another study highlighted senescence-associated genes (SAGs) in Chinese sorghum lines (Wu et al., 2016).

Comparative transcriptomics between drought-tolerant and sensitive sorghum genotypes revealed significant differential gene expression, emphasizing the adaptive mechanisms in tolerant varieties (Fracasso et al., 2016). In seedlings, transcriptomic profiling of leaf and root tissues identified DEGs and transcription factors such as HSF, ERF, NAC, and WRKY under drought stress (Zhang et al., 2019). Leaf transcriptome studies have also identified genes enhancing sugar content under salinity stress (Sui et al., 2015) and drought-associated microRNAs (miRNAs) in multiple genotypes under water-deficit conditions (Hamza et al., 2016). Transcriptomic studies further uncovered genes involved in the flavonoid biosynthesis pathway regulating seed color morphology (Zhou et al., 2022) and miRNAs associated with cadmium accumulation in sweet sorghum (Jia et al., 2021). Additionally, ROS scavenging pathways via peroxidase (POD) and catalase (CAT) enzymes were elucidated in sorghum leaves under combined salinity and drought stress (Wang et al., 2022). Investigations into starch accumulation revealed key DEGs in seed transcriptomes of different sorghum varieties (Ke et al., 2022), while salt tolerance studies identified DEGs in both tolerant and susceptible genotypes under sodium chloride treatment (Karumanchi et al., 2023).

Emerging spatiotemporal transcriptomics has further refined gene expression analysis by providing single-cell resolution and spatial context. In sorghum, this approach has revealed DEGs involved in dhurrin biosynthesis in grains (Liu et al., 2023), secondary cell wall regulators using laser capture microdissection (LCM) techniques (Fu et al., 2024), and starch biosynthesis-related genes during seed development (Khan et al., 2024). Overall, transcriptomics has proven to be an invaluable tool for identifying critical genes and pathways that contribute to the genetic improvement of sorghum under abiotic stress conditions.

9.2.3 Proteomics

Proteomics focuses on the expression and function of proteins within cells or tissues of organisms. Numerous conventional and advanced proteomic techniques have been developed for crop proteomics analysis (Yang et al., 2021; Singh et al., 2024b). Conventional techniques include chromatography-based methods such as size exclusion, ion exchange, and western blotting. Advanced approaches, such as SDS-PAGE (sodium dodecyl sulfate-polyacrylamide gel electrophoresis) and quantitative proteomics-based techniques like isobaric tags for relative and absolute quantitation (iTRAQ), have gained prominence in cereal crop studies (Rakkammal et al., 2022; Yang et al., 2021).

Conventional proteomics methods have been applied to identify differentially expressed proteins (DEPs) in sorghum under drought stress (Jedmowski et al., 2014). A comprehensive review of sorghum grain proteomics under drought stress provided valuable insights into the utility of proteomics for improving sorghum crop performance (Labuschagne, 2018). Using the 2-DE/MS technique, Swami et al. (2011) investigated the leaf proteome of sorghum under salinity stress. Similarly, 2-DE analysis identified DEPs involved in transcriptional regulation and stress response in sorghum leaves exposed to cadmium (Cd) stress (Roy et al., 2016a). SDS-PAGE has also been employed to evaluate seed storage protein expression in sorghum under saline conditions (Punia et al., 2020).

Advanced mass spectrometry techniques like MALDI-TOF-TOF MS have enabled the identification of up- and down-regulated proteins associated with antioxidant activity and defense responses in sorghum root seedlings under drought stress (Li et al., 2020). Comparative proteomics analysis has highlighted DEPs involved in water deficit regulation and enhanced antioxidant activity in both sorghum and maize under drought stress (Ali et al., 2023). Additionally, MALDI-TOF-TOF-based sorghum leaf proteomics identified 24 DEPs under copper (Cu) stress, shedding light on mechanisms of stress tolerance (Roy et al., 2016b).

Label-free quantitative proteomics has been applied to study the leaf and root proteomes of sorghum genotypes, facilitating a deeper understanding of stress responses (Sharan et al., 2018). iTRAQ-based quantitative proteomics has revealed that tolerant sorghum lines accumulate higher levels of antioxidant proteins under aluminum (Al) stress, while sensitive lines exhibit reduced antioxidant levels, leading to increased ROS accumulation in root tips (Zhou et al., 2017). Similarly, iTRAQ analysis identified common and varietal-specific DEPs in drought-tolerant and sensitive sorghum lines under water-limiting conditions, providing insights into varietal differences in stress adaptation (Goche et al., 2020). Overall, proteomics represents a vital tool for deciphering stress-related protein dynamics and improving sorghum crop varieties to withstand drought, salinity, and metal stress conditions under abiotic stress scenarios.

9.2.4 Metabolomics

Metabolomics involves the comprehensive analysis of metabolites in cells or tissues, which can be broadly classified into primary metabolites essential for plant growth and secondary metabolites that influence the plant's quantitative and qualitative traits under environmental stresses (Yang et al., 2021; Singh et al., 2024a). Techniques such as liquid and gas chromatography-mass spectrometry (LC-MS/GC-MS), Fourier transform infrared spectroscopy (FTIR), near-infrared spectroscopy (NIR), and nuclear magnetic resonance (NMR) have been extensively employed for metabolite analysis in grain crops, including sorghum (Hamany Djande et al., 2020).

Using the GC-MS technique, drought-tolerant sorghum genotypes were found to accumulate a greater number of metabolites (17) compared to susceptible genotypes under drought stress (Ogbaga et al., 2016). Metabolic profiling using GC-MS and LC-MS approaches revealed the association of primary and secondary metabolites with biomass accumulation in sorghum (Turner et al., 2016). Furthermore, biomarkers linked to biotic stress responses were identified in sorghum through UHPLC-MS analysis (Mareya et al., 2019). Comparative metabolomics using GC-MS/MS demonstrated that sorghum lines rich in antioxidants regulated the flavonoid biosynthesis pathway (Ramalingam et al., 2021). Similarly, sweet sorghum exhibited a higher abundance of antioxidants involved in flavonoid biosynthesis compared to grain sorghum (Zhao et al., 2022). Metabolite profiling of sorghum seeds highlighted associations between various metabolites and seed pigmentation, with flavonoid biosynthesis playing a central role in color polymorphism (Zhou et al., 2020; Zhou et al., 2022). Additionally, differentially accumulated metabolites (DAMs) related to flavonoids were identified in sorghum under saline stress conditions (Ma et al., 2020).

Non-targeted metabolomics revealed variations in rhizosphere metabolites across sorghum genotypes (Miller et al., 2019). Root metabolites displayed significant alterations under nitrogen (N) stress using UPLC-MS/GC-MS (Sheflin et al., 2019), and drought stress influenced metabolic profiles in both roots and leaves analyzed by GC-MS (Pavli et al., 2013). In another study, ultrahigh-performance liquid chromatography–tandem mass spectrometry (UPLC–MS/MS) enabled the differentiation between two specific varieties of sorghum (Zhang et al., 2024). Leaf metabolome analyses revealed changes in chlorophyll content and reductions in oxidative stress under saline conditions (Punia et al., 2021). Differential metabolic adjustments involving polyamines and free amino acids were observed in sorghum leaves under salinity stress (de Oliveira et al., 2020). DAMs associated with salicylic acid-mediated salt tolerance were identified through leaf metabolic profiling (Zhang et al., 2022a). Furthermore, phenylalanine ammonia-lyase (PAL) activity was implicated in leaf metabolic responses (Simpson et al., 2021). Advanced methods have also been applied to assess sorghum metabolites during specific developmental stages. For instance, HPLC-Q-TOF-MS/MS identified lipid-driven quality control markers in two sorghum varieties (Li et al., 2024), and LC-MS/MS detected dhurrin during grain germination (Liu et al., 2023). Overall, metabolomics underscores the pivotal role of metabolites in enhancing sorghum crop resilience and breeding under various abiotic stress conditions.

9.2.5 Ionomics

Ionomics is the study of the mineral and trace element composition (ionome) in plants, focusing on their accumulation and functional roles. Ionomes play a critical role in developmental and physiological processes, influencing genetic regulation and crop responses to environmental stresses (Yang et al., 2021). Analytical techniques for ionomic studies commonly involve Inductively Coupled Plasma (ICP), which is further categorized into mass spectrometry (ICP-MS) and optical emission spectroscopy (ICP-OES) approaches. Non-destructive methods such as X-ray fluorescence (XRF) and Neutron Activation Analysis (NAA) are also widely used for plant ionome analysis (Satismruti et al., 2013).

In sorghum, ICP-OES has been applied to study cadmium (Cd) accumulation, revealing its effects on agronomic traits (Wang et al., 2021). A related ionomic study identified QTLs associated with Cd concentrations, highlighting the role of the *SbHMA3a* gene in reducing Cd levels in sorghum grains (Wahinya et al., 2022). Ionomic profiling demonstrated alterations in the root and shoot ionomes, with increased sulfur (S) and iron (Fe) concentrations, respectively, under NPK treatment in sorghum seedlings (Zhu et al., 2020). Additionally, sorghum genotypes displayed 18 elemental interactions, with phosphorus (P) showing the most significant response to nitrogen treatments (Veley et al., 2017).

Seed ionome studies have identified candidate genes linked to cadmium (Cd), nickel (Ni), manganese (Mn), and calcium (Ca) accumulation in sorghum (Shakoor, 2016). Using ICP-MS, researchers analyzed 12 elements in sorghum seeds, uncovering their nutritional significance. A positive correlation was observed between zinc (Zn) and copper (Cu) accumulation, and six promising sorghum varieties were identified as superior performers in addressing malnutrition challenges (Hacisalihoglu & Armstrong, 2022). Despite these advances, ionomic studies in sorghum remain limited, particularly in understanding micro- and macro-element interactions and their implications for plant yield, growth, and development. Nonetheless, ionomics holds substantial potential for advancing genetic improvement in sorghum under diverse abiotic stress conditions.

9.2.6 Phenomics

Phenomics involves the systematic acquisition of phenotypic data from organisms. By linking the phenome with genotypes of a certain species, it elucidates genome (G), environment (E), and phenotype (P) interactions, a concept referred to as genotype–phenotype–environment (G–P–E) relationships (Zhao et al., 2019). Advanced high-throughput phenomics approaches, leveraging sensor and imaging technologies, are now extensively applied in crop research. These techniques employ sensors for visible light, fluorescence, infrared, 3D imaging, magnetic resonance imaging (MRI), and computed tomography (CT) to analyze cereal phenomes (Yang et al., 2014; Zhao et al., 2019). LiDAR (light detection and ranging) has been employed for 3D leaf phenotyping in sorghum (Thapa et al., 2018). Techniques such as LiDAR, thermal imaging, and ultrasonic sensors have also been used to assess radiation use efficiency (RUE) in sorghum (Furbank et al., 2019). Unmanned aerial vehicles (UAVs) were utilized for canopy phenotyping in two sorghum cultivars, revealing that post-flowering canopy traits strongly correlated with stay-green breeding values compared to pre-flowering traits (Liedtke et al., 2020).

Additionally, data collected through uncrewed aircraft systems (UAS) were analyzed to identify QTLs and complex adaptive traits associated with cold tolerance via genome-to-phenome (G2P) integration (Marla et al., 2023). Semantic segmentation techniques have enabled the detection of morphological traits in sorghum, with genome-wide association studies (GWAS) linking SNPs to these traits

(Miao et al., 2020). Zhang et al. (2022b) utilized convolutional neural networks (CNNs) on sorghum biomass images to predict SNP markers and corresponding phenotypes. A genome-phenome-wide association study (GPWAS) highlighted pleiotropic effects of different mutations on phenotyping traits in sorghum (Boatwright et al., 2023). Field phenotyping has further explored transpiration responses to vapor pressure deficit, identifying two sorghum genotypes with limited transpiration traits based on stomatal conductance (Raymundo et al., 2023). The advent of next-generation phenomics, featuring advanced sensors and data-driven agronomic technologies, offers significant potential for improving cereal crops, including sorghum. Integrating phenomics with other omics layers, such as genomics, metabolomics, and ionomics, will be pivotal in advancing genetic improvement and enhancing sorghum germplasm resilience under various abiotic stresses (Roitsch et al., 2019; Yang et al., 2021).

9.3 DATABASES FOR THE ACQUISITION OF MULTI-OMICS DATA FOR SORGHUM CROP BREEDING

Saand & Yang (2024) have elucidated the roles of numerous databases in cereal crop breeding under diverse stress conditions. These databases are particularly useful for analyzing omics data in cereal crops, including sorghum (Table 9.1). For instance, Gramene and GRASSIUS are valuable resources for genome analysis, genetic variation studies, gene expression profiling, and gene mapping across cereals, including sorghum. Similarly, RGPDB and PlantNexus facilitate the analysis

TABLE 9.1
Key databases for sorghum omics analysis

	Name of database	Database focus and application	URL
1	Gramene*	Provides sorghum genome information including gene annotations, genetic variations, mapping, and pathways analysis	www.gramene.org/
2	GRASSIUS*	A comprehensive database for cereal including sorghum crops that provides information and details regarding genome-wide regulation of gene expression analysis	www.grassius.org/grasstfdb.php
3	RGPDB*	Provides information about root-associated genes and their corresponding promoters	http://sysbio.unl.edu/RGPDB/index.php
4	SorghumBase	This database is solely based on the sorghum genome and provides information related to genetic variation, gene expression, and regulatory pathways	www.sorghumbase.org/
5	SorghumFDB	Provides insights into sorghum gene family classifications, comprehensive gene annotations, miRNA-target gene relationships, orthologous pairs in model plants, gene loci conversions, and a genome browser	https://doi.org/10.1093/database/baw099
6	SorGSD	Provides information about genomic variations including SNPs and INDELs in sorghum phenotypes	https://ngdc.cncb.ac.cn/sorgsd/
7	SorbMutDB	Provides information about sorghum mutant lines	www.depts.ttu.edu/igcast/SorbMutDB.php
8	Sorghum QTL Atlas	Houses information about QTL and GWAS studies conducted in sorghum	https://aussorgm.org.au/sorghum-qtl-atlas/
10	MOROKOSHI	Provides transcriptomics data for gene expression based on RNA-Seq analysis in sorghum	https://doi.org/10.1093/pcp/pcu187.
11	PlantNexus*	Database based on gene co-expression networks analysis	https://plantnexus.ohio.edu/

TABLE 9.1 (Continued)
Key databases for sorghum omics analysis

	Name of database	Database focus and application	URL
12	Sorghum Functional genomics	Houses information about sorghum functional genomics	www.purdue.edu/sorghumgenomics/
13	GreenPhylDB*	Provides a platform for functional genomics with pan-genomic analysis	www.greenphyl.org/cgi-bin/index.cgi
14	SorghumCyc	Provides information on metabolic pathways in sorghum	https://doi.org/10.1093/nar/gkq1148

(*) in the table indicates the databases used for sorghum, as well as other cereal and non-cereal crops.

of root-associated genes, their corresponding promoters, and gene co-expression networks in cereal crops. Specifically for sorghum, several essential databases are publicly available. SorghumFDB and SorghumBase, are key platforms that support the interpretation of genetic variation, gene expression, and metabolic regulatory pathways exclusively for sorghum. GreenPhylDB provides a platform for sorghum functional genomics and pan-genomics studies, while the SorghumCyc database is tailored for metabolic pathway analysis. Additionally, SorGSD, SorbMutDB, and the Sorghum QTL Atlas are instrumental for studying SNPs, mutants, QTLs, and genome-wide association studies (GWAS) specific to sorghum (Table 9.1)

9.4 INTEGRATION OF MULTI-OMICS WITH SYSTEMS BIOLOGY FOR DEVELOPING CLIMATE-READY SORGHUM

The integration of omics approaches, spanning genomics to phenomics, with systems biology has provided transformative insights into molecular mechanisms critical for breeding climate-resilient cultivars (Saand & Yang, 2024; Saini et al., 2024a, b). The recently proposed panomics concept offers a robust platform to integrate datasets from multiple omics layers and develop predictive mathematical models. These models are essential for precision breeding, enabling the development of elite crop varieties with resistance to key biotic and abiotic stresses. By coupling multiple omics layers researchers can uncover integrated molecular mechanisms and genetic regulatory networks that underlie quantitative and qualitative traits contributing to yield improvement and stress resilience (Yang et al., 2021). For instance, combining genomics and phenomics using a genotype-to-phenotype (G2P) framework has revealed the genetic architecture of complex adaptive traits like chilling tolerance in sorghum genotypes (Marla et al., 2023). Similarly, integrative analyses have highlighted the involvement of specific molecular pathways under various stress conditions. For instance, degradome and transcriptomics integration identified miRNAs implicated in cadmium (Cd) accumulation in sweet sorghum (Jia et al., 2021), while proteomics and transcriptomics pinpointed salinity-responsive proteins and genes related to seed storage in sorghum (Punia et al., 2020).

Moreover, integrating transcriptomics with metabolomics has unveiled dynamic regulatory networks. Examples include: the identification of differentially accumulated metabolites (DAM) and differentially expressed genes (DEG) related to flavonoid biosynthesis under salinity stress (Ma et al., 2020); plant hormone signal transduction pathways under salinity stress (Zhang et al., 2022); genes involved in dhurrin biosynthesis during grain germination (Liu et al., 2023); and starch biosynthesis during seed development stages (Khan et al., 2024). Additional integrative studies have elucidated the flavonoid biosynthesis pathway regulating seed color polymorphism (Zhou et al., 2022) and the altered ionome concentration and gene expression under NPK starvation in sorghum seedlings (Zhu et al., 2020). The integration of functional genomics with phenomics, such as genome-phenome wide association studies (GPWAS), has revealed pleiotropic loci influencing multiple agronomic traits in sorghum (Boatwright et al., 2023). Furthermore, Liu et al. (2024) highlighted key genes and regulatory networks linked to sorghum's development and environmental adaptation using the G2P strategy.

Systems biology complements these integrative approaches by modeling the complexity of biological systems. This interdisciplinary framework emphasizes a holistic perspective, contrasting traditional reductionist methods, and employs computational techniques to predict system responses under varying conditions (Kumar et al., 2015). By integrating diverse datasets, systems biology captures critical information on gene expression, protein levels, and metabolic profiles, enabling a deeper understanding of regulatory networks in plant systems (Jamil et al., 2020). For instance, such integrative analyses can reveal how genetic changes influence metabolic pathways, ultimately affecting plant productivity and stress adaptation (Yang et al., 2021). As summarized in Table 9.2, studies integrating multiple omics datasets have identified key genes, proteins, and mechanisms underlying sorghum's adaptation to diverse climatic conditions and developmental processes.

TABLE 9.2
Summary of potential genes, proteins, and molecular mechanisms associated with abiotic stress tolerance identified in sorghum through integrative omics technologies

	Omics approaches	Stress or trait targeted	Key findings	Reference
1	Genomics and phenomics	Chilling	Significant SNPs and QTLs associated with chilling tolerance	Marla et al. (2023)
2	Degradome and transcriptomics	Cadmium	miRNAs and their target genes involved in Cd tolerance	Jia et al. (2021)
3	Proteomics and transcriptomics	Salinity	Salinity-responsive proteins and genes	Punia et al. (2020)
4	Transcriptomics and metabolomics	Salinity	DEGs and DAMs involved in flavonoid biosynthesis pathway	Ma et al. (2020)
5	Transcriptomics and metabolomics	Salinity	Explored hormone signal transduction pathways in sorghum genotypes under salinity stress conditions	Zhang et al. (2022a)
6	Transcriptomics and metabolomics	Seed germination	Genes and their association with the dhurrin biosynthesis pathway during sorghum grain germination	Liu et al. (2023)
7	Transcriptomics and metabolomics	Seed development	Identified regulatory networks associated with nutrition synthesis in developing seeds	Khan et al. (2024)
8	Transcriptomics and metabolomics	Flavonoid biosynthesis pathway	Identified flavonoid metabolites and the candidate genes involved in the flavonoid biosynthesis pathway	Zhou et al. (2022)
9	Transcriptomics and ionomics	NPK starvation	Observed changes in ionome concentration and dynamic gene expression under NPK nutrient starvation conditions	Zhu et al. (2020)
10	Functional genomics and phenomics	Several agronomic traits	Studied loss-of-function mutations and their functional repercussions	Boatwright et al. (2023)
11	Genomics, transcriptomics, metabolomics, and phenomics	Biomass and grain yield	Identified several key genes involved in gibberellin and cytokinin biosynthesis pathways	Breitzman, (2022)

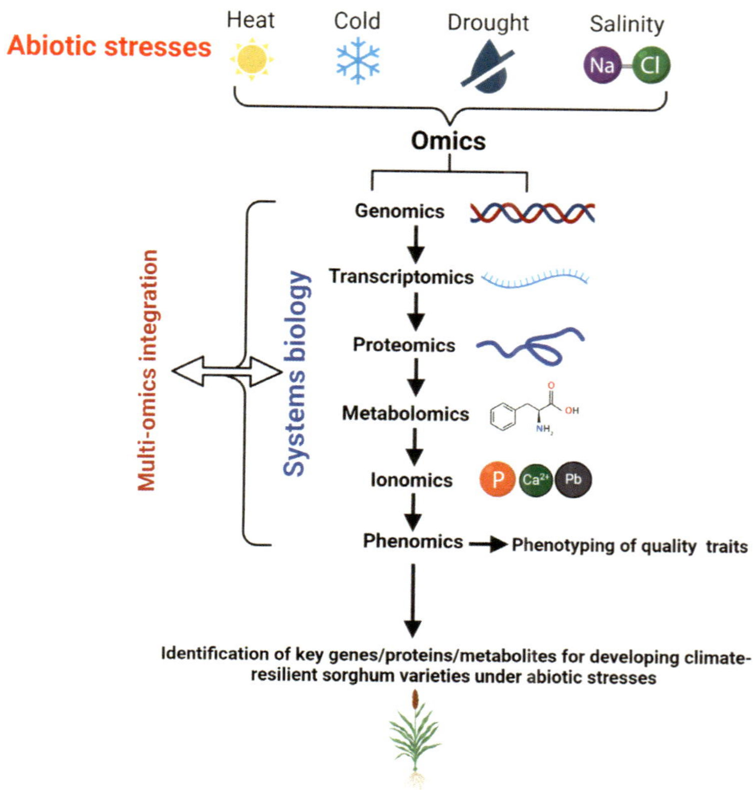

FIGURE 9.2 Proposed model for multi-omics integration and systems biology to develop resilient sorghum cultivars.

Overall, integrating diverse omics datasets along with systems biology provides a comprehensive framework for identifying key traits. This approach holds promise for developing climate-resilient sorghum varieties capable of thriving under abiotic stress conditions such as heat, cold, drought, and salinity (Figure 9.2).

9.5 CHALLENGES AND OPPORTUNITIES FOR DEVELOPING CLIMATE-READY SORGHUM

The advent of advanced functional genomics technologies has transformed sorghum breeding, enabling precision breeding to target genes linked to desirable traits for developing elite varieties. Over the last five years, high-throughput sequencing and artificial intelligence algorithms have been increasingly adopted for efficient breeding strategies in sorghum (Liu et al., 2024). Genome selection models are promising for cost-effective breeding, while novel breeding techniques, such as genome editing, have shown success in wild rice and tomato (Yu et al., 2021; Li et al., 2018). However, these tools remain underutilized in sorghum due to limited genome resources in wild species and challenges in genetic transformation. Addressing these gaps may require *de novo* domestication strategies coupled with advanced breeding technologies, synthetic biology, and targeted gene editing (Liu et al., 2024).

Mapping complex agronomic traits like yield, drought tolerance, and nutrient content remains challenging due to their polygenic nature and environmental interactions. The G–P–E model, incorporating GWAS, QTL mapping, and multi-omics approaches, offers potential solutions (Zhao et al., 2019). Transcriptomics in sorghum is hindered by the time-consuming nature and cost of advanced techniques. RNA-seq, while superior to microarrays, lacks uniform genome read coverage (Kaur et al., 2021). Emerging technologies like single-cell transcriptomics, using spatio-temporal approaches, could enhance genetic improvement under stress conditions (Wang et al., 2019). Proteomics faces limitations in precision due to the inadequacy of label-free quantification methods, which are being replaced by iTRAQ-based techniques for accuracy (Goche et al., 2020). Phosphoproteomics holds promise for understanding protein functions and regulatory mechanisms in sorghum (Saand & Yang, 2024). Metabolomics challenges include interference of secondary metabolites with primary metabolomes, identification and characterization of unknown metabolites, and high costs associated with MS and NMR techniques. Optimized metabolite extraction methods and collaborations between institutions could address these issues. Future advances, such as single-cell metabolomics, could provide a deeper understanding of sorghum metabolites (Sun et al., 2018).

Phenomics, with tools like remote sensing and UAS, has enabled high-throughput phenotyping in sorghum (Marla et al., 2023; Furbank et al., 2019). However, challenges persist in acquiring and analyzing phenotypic data. Machine learning-based phenotypic trait analysis and next-generation sensors could enhance phenomics applications (Gill et al., 2022; Roitsch et al., 2019). The integration of multi-omics approaches, from genomics to phenomics, is critical but remains a challenge due to the complexity of interpreting and mining datasets. Coupling phenomics with other omics layers could provide a holistic understanding of sorghum's adaptation to abiotic stresses and aid in tailoring resilient varieties.

REFERENCES

Abdel-Ghany, S. E., Hamilton, M., Jacobi, J. L., Ngam, P., Devitt, N., Schilkey, F., ... & Reddy, A. S. (2016). A survey of the sorghum transcriptome using single-molecule long reads. *Nature Communications*, *7*(1), 11706.

Abdel-Ghany, S. E., Ullah, F., Ben-Hur, A., & Reddy, A. S. (2020). Transcriptome analysis of drought-resistant and drought-sensitive sorghum (*Sorghum bicolor*) genotypes in response to PEG-induced drought stress. *International Journal of Molecular Sciences*, *21*(3), 772.

Ahn, E., Park, S., Hu, Z., Ellur, V., Cha, M., Lee, Y., ... & Magill, C. (2024). A genome-wide association study of seed morphology-related traits in sorghum mini-core and senegalese lines. *Crops*, *4*(2), 156–171.

Ali, A. E. E., Husselmann, L. H., Tabb, D. L., & Ludidi, N. (2023). Comparative proteomics analysis between maize and sorghum uncovers important proteins and metabolic pathways mediating drought tolerance. *Life*, *13*(1), 170.

Baillo, E. H., Hanif, M. S., Guo, Y., Zhang, Z., Xu, P., & Algam, S. A. (2020). Genome-wide Identification of WRKY transcription factor family members in sorghum (*Sorghum bicolor* (L.) moench). *PloS One*, *15*(8), e0236651.

Boatwright, J. L., Sapkota, S., & Kresovich, S. (2023). Functional genomic effects of indels using Bayesian genome-phenome wide association studies in sorghum. *Frontiers in Genetics*, *14*, 1143395.

Breitzman, M. W. (2022). *An omics approach to improving biomass and grain yield in sorghum (Sorghum bicolor (L.) Moench)* [Doctoral dissertation]. Iowa State University).

Chen, D., Wang, W., Wu, Y., Xie, H., Zhao, L., Zeng, Q., & Zhan, Y. (2019). Expression and distribution of the auxin response factors in *Sorghum bicolor* during development and temperature stress. *International Journal of Molecular Sciences*, *20*(19), 4816.

Cruet-Burgos, C., & Rhodes, D. H. (2023). Unraveling transcriptomics of sorghum grain carotenoids: A step forward for biofortification. *BMC Genomics*, *24*(1), 233.

de Oliveira, D. F., Lopes, L. D. S., & Gomes-Filho, E. (2020). Metabolic changes associated with differential salt tolerance in sorghum genotypes. *Planta*, *252*(3), 34.

Fracasso, A., Trindade, L. M., & Amaducci, S. (2016). Drought stress tolerance strategies revealed by RNA-Seq in two sorghum genotypes with contrasting WUE. *BMC Plant Biology*, *16*, 1–18.

Fu, J., McKinley, B., James, B., Chrisler, W., Markillie, L. M., Gaffrey, M. J., ... & Marshall-Colon, A. (2024). Cell-type-specific transcriptomics uncovers spatial regulatory networks in bioenergy sorghum stems. *The Plant Journal*, *118*(5), 1668–1688.

Furbank, R. T., Jimenez-Berni, J. A., George-Jaeggli, B., Potgieter, A. B., & Deery, D. M. (2019). Field crop phenomics: Enabling breeding for radiation use efficiency and biomass in cereal crops. *New Phytologist*, *223*(4), 1714–1727.

Gill, T., Gill, S. K., Saini, D. K., Chopra, Y., de Koff, J. P., & Sandhu, K. S. (2022). A comprehensive review of high throughput phenotyping and machine learning for plant stress phenotyping. *Phenomics*, *2*(3), 156–183.

Goche, T., Shargie, N. G., Cummins, I., Brown, A. P., Chivasa, S., & Ngara, R. (2020). Comparative physiological and root proteome analyses of two sorghum varieties responding to water limitation. *Scientific Reports*, *10*(1), 11835.

Hacisalihoglu, G., & Armstrong, P. R. (2022). Flax and sorghum: Multi-element contents and nutritional values within 210 varieties and potential selection for future climates to sustain food security. *Plants*, *11*(3), 451.

Hamany Djande, C. Y., Pretorius, C., Tugizimana, F., Piater, L. A., & Dubery, I. A. (2020). Metabolomics: A tool for cultivar phenotyping and investigation of grain crops. *Agronomy*, *10*(6), 831.

Hamza, N. B., Sharma, N., Tripathi, A., & Sanan-Mishra, N. (2016). MicroRNA expression profiles in response to drought stress in *Sorghum bicolor*. *Gene Expression Patterns*, *20*(2), 88–98.

Hu, Y., He, C., Shen, Y., Zeng, G., Bi, S., Huang, Q., ... & Shen, X. (2023). Dynamics of the epigenetic landscape during development and in response to drought stress in sorghum. *bioRxiv*, 2023–04.

Jamil, I. N., Remali, J., Azizan, K. A., Nor Muhammad, N. A., Arita, M., Goh, H. H., & Aizat, W. M. (2020). Systematic multi-omics integration (MOI) approach in plant systems biology. *Frontiers in Plant Science*, *11*, 944.

Jedmowski, C., Ashoub, A., Beckhaus, T., Berberich, T., Karas, M., & Brüggemann, W. (2014). Comparative analysis of *Sorghum bicolor* proteome in response to drought stress and following recovery. *International Journal of Proteomics*, *2014*(1), 395905.

Jia, W., Lin, K., Lou, T., Feng, J., Lv, S., Jiang, P., ... & Li, Y. (2021). Comparative analysis of sRNAs, degradome and transcriptomics in sweet sorghum reveals the regulatory roles of miRNAs in Cd accumulation and tolerance. *Planta*, *254*, 1–18.

Jiang, W., Zhou, H., Bi, H., Fromm, M., Yang, B., & Weeks, D. P. (2013). Demonstration of CRISPR/Cas9/sgRNA-mediated targeted gene modification in Arabidopsis, tobacco, sorghum and rice. *Nucleic Acids Research*, *41*(20), e188.

Johnson, S. M., Lim, F. L., Finkler, A., Fromm, H., Slabas, A. R., & Knight, M. R. (2014). Transcriptomic analysis of *Sorghum bicolor* responding to combined heat and drought stress. *BMC Genomics*, *15*, 1–19.

Karumanchi, A. R., Sivan, P., Kummari, D., Rajasheker, G., Kumar, S. A., Reddy, P. S., ... & Kishor, P. K. (2023). Root and leaf anatomy, ion accumulation, and transcriptome pattern under salt stress conditions in contrasting genotypes of *Sorghum bicolor*. *Plants*, *12*(13), 2400.

Kaur, B., Sandhu, K. S., Kamal, R., Kaur, K., Singh, J., Röder, M. S., & Muqaddasi, Q. H. (2021). Omics for the improvement of abiotic, biotic, and agronomic traits in major cereal crops: Applications, challenges, and prospects. *Plants*, *10*(10), 1989.

Ke, F., Zhang, K., Li, Z., Wang, J., Zhang, F., Wu, H., ... & Zhu, K. (2022). Transcriptomic analysis of starch accumulation patterns in different glutinous sorghum seeds. *Scientific Reports*, *12*(1), 11133.

Ketumile, D., Yang, X., Sanchez, R., Kundariya, H., Rajewski, J., Dweikat, I. M., & Mackenzie, S. A. (2022). Implementation of epigenetic variation in sorghum selection and implications for crop resilience breeding. *Frontiers in Plant Science*, *12*, 798243.

Khan, A., Tian, R., Bean, S. R., Yerka, M., & Jiao, Y. (2024). Transcriptome and metabolome analyses reveal regulatory networks associated with nutrition synthesis in sorghum seeds. *Communications Biology*, *7*(1), 841.

Kumar, A., Pathak, R. K., Gupta, S. M., Gaur, V. S., & Pandey, D. (2015). Systems biology for smart crops and agricultural innovation: Filling the gaps between genotype and phenotype for complex traits linked with robust agricultural productivity and sustainability. *Omics: A Journal of Integrative Biology*, *19*(10), 581–601.

Labuschagne, M. T. (2018). A review of cereal grain proteomics and its potential for sorghum improvement. *Journal of Cereal Science*, *84*, 151–158.

Li, H., Li, Y., Ke, Q., Kwak, S. S., Zhang, S., & Deng, X. (2020). Physiological and differential proteomic analyses of imitation drought stress response in *Sorghum bicolor* root at the seedling stage. *International Journal of Molecular Sciences*, *21*(23), 9174.

Li, T., Yang, X., Yu, Y., Si, X., Zhai, X., Zhang, H., ... & Xu, C. (2018). Domestication of wild tomato is accelerated by genome editing. *Nature Biotechnology*, *36*(12), 1160–1163.

Li, Y., Ma, Y., Zhu, H., Liu, Y., Pan, S., Chen, X., & Wu, T. (2024). Identifying distinct markers in two Sorghum varieties for baijiu fermentation using untargeted metabolomics and molecular network approaches. *Food Chemistry: X*, *23*, 101646.

Liedtke, J. D., Hunt, C. H., George-Jaeggli, B., Laws, K., Watson, J., Potgieter, A. B., ... & Jordan, D. R. (2020). High-throughput phenotyping of dynamic canopy traits associated with stay-green in grain sorghum. Plant Phenomics, 2020, 1–10, Article ID 4635153.

Liu, F., Baye, W., Zhao, K., Tang, S., Xie, Q., & Xie, P. (2024). Unravelling sorghum functional genomics and molecular breeding: Past achievements and future prospects. *Journal of Genetics and Genomics*.

Liu, H., Micic, N., Miller, S., Crocoll, C., & Bjarnholt, N. (2023). Species-specific dynamics of specialized metabolism in germinating sorghum grain revealed by temporal and tissue-resolved transcriptomics and metabolomics. *Plant Physiology and Biochemistry*, *196*, 807–820.

Ma, S., Lv, L., Meng, C., Zhang, C., & Li, Y. (2020). Integrative analysis of the metabolome and transcriptome of *Sorghum bicolor* reveals dynamic changes in flavonoids accumulation under saline–alkali stress. *Journal of Agricultural and Food Chemistry*, *68*(50), 14781–14789.

Mace, E., Innes, D., Hunt, C., Wang, X., Tao, Y., Baxter, J., ... & Jordan, D. (2019). The Sorghum QTL Atlas: A powerful tool for trait dissection, comparative genomics and crop improvement. *Theoretical and Applied Genetics*, *132*, 751–766.

Maina, F., Bouchet, S., Marla, S. R., Hu, Z., Wang, J., Mamadou, A., ... & Morris, G. P. (2018). Population genomics of sorghum (*Sorghum bicolor*) across diverse agroclimatic zones of Niger. *Genome*, *61*(4), 223–232.

Makita, Y., Shimada, S., Kawashima, M., Kondou-Kuriyama, T., Toyoda, T., & Matsui, M. (2015). MOROKOSHI: transcriptome database in Sorghum bicolor. *Plant and cell physiology*, *56*(1), e6-e6.

Mareya, C. R., Tugizimana, F., Piater, L. A., Madala, N. E., Steenkamp, P. A., & Dubery, I. A. (2019). Untargeted metabolomics reveal defensome-related metabolic reprogramming in *Sorghum bicolor* against infection by *Burkholderia andropogonis*. *Metabolites*, *9*(1), 8.

Marla, S., Felderhoff, T., Hayes, C., Perumal, R., Wang, X., Poland, J., & Morris, G. P. (2023). Genomics and phenomics enabled prebreeding improved early-season chilling tolerance in Sorghum. *G3: Genes, Genomes, Genetics*, *13*(8), jkad116.

McCormick, R. F., Truong, S. K., Sreedasyam, A., Jenkins, J., Shu, S., Sims, D., ... & Mullet, J. E. (2018). The *Sorghum bicolor* reference genome: Improved assembly, gene annotations, a transcriptome atlas, and signatures of genome organization. *The Plant Journal*, *93*(2), 338–354.

Miao, C., Pages, A., Xu, Z., Rodene, E., Yang, J., & Schnable, J. C. (2020). Semantic segmentation of sorghum using hyperspectral data identifies genetic associations. *Plant Phenomics,* 2020, 1–11, Article ID 4216373.

Miller, S. B., Heuberger, A. L., Broeckling, C. D., & Jahn, C. E. (2019). Non-targeted metabolomics reveals sorghum rhizosphere-associated exudates are influenced by the belowground interaction of substrate and sorghum genotype. *International Journal of Molecular Sciences*, *20*(2), 431.

Morris, G. P., Ramu, P., Deshpande, S. P., Hash, C. T., Shah, T., Upadhyaya, H. D., ... & Kresovich, S. (2013). Population genomic and genome-wide association studies of agroclimatic traits in sorghum. *Proceedings of the National Academy of Sciences*, *110*(2), 453–458.

Mukherjee, A., Maheshwari, U., Sharma, V., Sharma, A., & Kumar, S. (2024). Functional insight into multi-omics-based interventions for climatic resilience in sorghum (*Sorghum bicolor*): A nutritionally rich cereal crop. *Planta*, *259*(4), 91.

Ogbaga, C. C., Stepien, P., Dyson, B. C., Rattray, N. J., Ellis, D. I., Goodacre, R., & Johnson, G. N. (2016). Biochemical analyses of sorghum varieties reveal differential responses to drought. *PloS One*, *11*(5), e0154423.

Olatoye, M. O., Hu, Z., Maina, F., & Morris, G. P. (2018). Genomic signatures of adaptation to a precipitation gradient in Nigerian sorghum. *G3: Genes, Genomes, Genetics*, *8*(10), 3269–3281.

Paterson, A. H., Bowers, J. E., Bruggmann, R., Dubchak, I., Grimwood, J., Gundlach, H., ... & Rokhsar, D. S. (2009). The *Sorghum bicolor* genome and the diversification of grasses. *Nature*, *457*(7229), 551–556.

Pavli, O. I., Vlachos, C. E., Kalloniati, C., Flemetakis, E., & Skaracis, G. N. (2013). Metabolite profiling reveals the effect of drought on sorghum (*Sorghum bicolor* L. Moench) metabolism. *Plant Omics*, *6*(6), 371–376.

Punia, H., Tokas, J., Bhadu, S., Mohanty, A. K., Rawat, P., Malik, A., & Satpal. (2020). Proteome dynamics and transcriptome profiling in sorghum [*Sorghum bicolor* (L.) Moench] under salt stress. *3 Biotech*, *10*, 1–10.

Punia, H., Tokas, J., Malik, A., Bajguz, A., El-Sheikh, M. A., & Ahmad, P. (2021). Ascorbate–glutathione oxidant scavengers, metabolome analysis and adaptation mechanisms of ion exclusion in sorghum under salt stress. *International Journal of Molecular Sciences*, *22*(24), 13249.

Rakkammal, K., Priya, A., Pandian, S., Maharajan, T., Rathinapriya, P., Satish, L., ... & Ramesh, M. (2022). Conventional and omics approaches for understanding the abiotic stress response in cereal crops—an updated overview. *Plants*, *11*(21), 2852.

Ramalingam, A. P., Mohanavel, W., Premnath, A., Muthurajan, R., Prasad, P. V., & Perumal, R. (2021). Large-scale non-targeted metabolomics reveals antioxidant, nutraceutical and therapeutic potentials of sorghum. *Antioxidants*, *10*(10), 1511.

Raymundo, R., Wang, X., Felderhoff, T., Sexton-Bowser, S., Poland, J., Lipka, A. E., & Morris, G. P. (2023). Field phenomics reveals genetic variation for transpiration response to vapor pressure deficit in sorghum. *bioRxiv*, 2023-06.

Roitsch, T., Cabrera-Bosquet, L., Fournier, A., Ghamkhar, K., Jiménez-Berni, J., Pinto, F., & Ober, E. S. (2019). New sensors and data-driven approaches—A path to next generation phenomics. *Plant Science*, *282*, 2–10.

Roy, S. K., Cho, S. W., Kwon, S. J., Kamal, A. H. M., Kim, S. W., Oh, M. W., ... & Woo, S. H. (2016a). Morpho-physiological and proteome level responses to cadmium stress in sorghum. *PLoS One*, *11*(2), e0150431.

Roy, S. K., Kwon, S. J., Cho, S. W., Kamal, A. H. M., Kim, S. W., Sarker, K., ... & Woo, S. H. (2016b). Leaf proteome characterization in the context of physiological and morphological changes in response to copper stress in sorghum. *Biometals*, *29*, 495–513.

Ruperao, P., Thirunavukkarasu, N., Gandham, P., Selvanayagam, S., Govindaraj, M., Nebie, B., ... & Rathore, A. (2021). Sorghum pan-genome explores the functional utility for genomic-assisted breeding to accelerate the genetic gain. *Frontiers in Plant Science*, *12*, 666342.

Saand, M. A., & Yang, Y. (2024). Multi-omics assisted breeding in cereals: Applications and challenges. *Omics and System Biology Approaches for Delivering Better Cereals*, 319–349. Boca Raton, Florida: CRC Press.

Saini, D. K. (2024a). Establishing the proper links: Network-based techniques for comprehending gene functions in cereals. In: Saini, D. K., Kole, C. (Eds), *Omics and System Biology Approaches for Delivering Better Cereals* (pp. 401–430). Boca Raton, Florida: CRC Press.

Saini, D. K., Kumar, S., Saini, K., & Goyal, S. (2024b). Omics and system biology for developing climate-ready cereals: An overview. *Omics and System Biology Approaches for Delivering Better Cereals*, 1–25. Boca Raton, Florida: CRC Press.

Sanjari, S., Shirzadian-Khorramabad, R., Shobbar, Z. S., & Shahbazi, M. (2019). Systematic analysis of NAC transcription factors' gene family and identification of post-flowering drought stress responsive members in sorghum. *Plant Cell Reports*, *38*, 361–376.

Satismruti, K., Senthil, N., Vellaikumar, S., Ranjani, R. V., & Raveendran, M. (2013). Plant ionomics: A platform for identifying novel gene regulating plant mineral nutrition. *American Journal of Plant Sciences*, *4*(7), 1309–1315.

Shakoor, N., Ziegler, G., Dilkes, B. P., Brenton, Z., Boyles, R., Connolly, E. L., ... & Baxter, I. (2016). Integration of experiments across diverse environments identifies the genetic determinants of variation in *Sorghum bicolor* seed element composition. *Plant Physiology*, *170*(4), 1989–1998.

Sharan, A. A., Nikam, A. N., Jaleel, A., Tamhane, V. A., & Rao, S. P. (2018). Method for label-free quantitative proteomics for *Sorghum bicolor* L. Moench. *Tropical Plant Biology*, *11*, 78–91.

Sheflin, A. M., Chiniquy, D., Yuan, C., Goren, E., Kumar, I., Braud, M., ... & Prenni, J. E. (2019). Metabolomics of sorghum roots during nitrogen stress reveals compromised metabolic capacity for salicylic acid biosynthesis. *Plant Direct*, *3*(3), e00122.

Simpson, J. P., Olson, J., Dilkes, B., & Chapple, C. (2021). Identification of the tyrosine-and phenylalanine-derived soluble metabolomes of sorghum. *Frontiers in Plant Science*, *12*, 714164.

Singh, G., Kumar, P., Mandal, P., Aziz, D., Rafiq, S., Saini, P., ... & Saini, D. K. (2024a). Plant metabolomic profiling to better understand biotic and abiotic stress resistance in cereals. In *Omics and System Biology Approaches for Delivering Better Cereals* (pp. 268–292). Boca Raton, Florida: CRC Press.

Singh, M., Mandal, M., Kumar, V., Lahkar, C., Singh, H. P., & Saini, D. K. (2024b). Proteomics and cereals breeding: Current progress and a look ahead to the next decade. *Omics and System Biology Approaches for Delivering Better Cereals*, 232–267. Boca Raton, Florida: CRC Press.

Somegowda, V. K., Reddy, S. D., Gaddameedi, A., Kiranmayee, K. U., Naravula, J., Kishor, P. K., & Penna, S. (2023). Genomics breeding approaches for developing *Sorghum bicolor* lines with stress resilience and other agronomic traits. *Current Plant Biology*, 37, 100314.

Spindel, J. E., Dahlberg, J., Colgan, M., Hollingsworth, J., Sievert, J., Staggenborg, S. H., ... & Vogel, J. P. (2018). Association mapping by aerial drone reveals 213 genetic associations for *Sorghum bicolor* biomass traits under drought. *BMC Genomics*, *19*, 1–18.

Sui, N., Yang, Z., Liu, M., & Wang, B. (2015). Identification and transcriptomic profiling of genes involved in increasing sugar content during salt stress in sweet sorghum leaves. *BMC Genomics*, *16*, 1–18.

Sun, M., Yang, Z., & Wawrik, B. (2018). Metabolomic fingerprints of individual algal cells using the single-probe mass spectrometry technique. *Frontiers in Plant Science*, *9*, 571.

Swami, A. K., Alam, S. I., Sengupta, N., & Sarin, R. (2011). Differential proteomic analysis of salt stress response in *Sorghum bicolor* leaves. *Environmental and Experimental Botany*, *71*(2), 321–328.

Tao, Y., Mace, E., George-Jaeggli, B., Hunt, C., Cruickshank, A., Henzell, R., & Jordan, D. (2018). Novel grain weight loci revealed in a cross between cultivated and wild sorghum. *The Plant Genome*, *11*(2), 170089.

Thapa, S., Zhu, F., Walia, H., Yu, H., & Ge, Y. (2018). A novel LiDAR-based instrument for high-throughput, 3D measurement of morphological traits in maize and sorghum. *Sensors*, *18*(4), 1187.

Tian, T., You, Q., Zhang, L., Yi, X., Yan, H., Xu, W., & Su, Z. (2016). SorghumFDB: sorghum functional genomics database with multidimensional network analysis. *Database*, *2016*, baw099

Turner, M. F., Heuberger, A. L., Kirkwood, J. S., Collins, C. C., Wolfrum, E. J., Broeckling, C. D., ... & Jahn, C. E. (2016). Non-targeted metabolomics in diverse sorghum breeding lines indicates primary and secondary metabolite profiles are associated with plant biomass accumulation and photosynthesis. *Frontiers in Plant Science*, *7*, 953.

Ukwatta, J., Pabuayon, I. C. M., Park, J., Chen, J., Chai, X., Zhang, H., ... & Shi, H. (2021). Comparative physiological and transcriptomic analysis reveals salinity tolerance mechanisms in *Sorghum bicolor* (L.) Moench. *Planta*, *254*, 1–20.

Veley, K. M., Berry, J. C., Fentress, S. J., Schachtman, D. P., Baxter, I., & Bart, R. (2017). High-throughput profiling identifies resource use efficient and abiotic stress tolerant sorghum varieties. *bioRxiv*, 132787.

Wahinya, F. W., Yamazaki, K., Jing, Z., Takami, T., Kamiya, T., Kajiya-Kanegae, H., ... & Sakamoto, W. (2022). Sorghum ionomics reveals the functional *SbHMA3a* allele that limits excess cadmium accumulation in grains. *Plant and Cell Physiology*, *63*(5), 713–728.

Wang, B., Xiong, W., & Guo, Y. (2024). Dhurrin in sorghum: Biosynthesis, regulation, biological function and challenges for animal production. *Plants*, *13*(16), 2291.

Wang, S., Huang, D. Y., Zhu, Q. H., Li, B. Z., Xu, C., Zhu, H. H., & Zhang, Q. (2021). Agronomic traits and ionomics influence on Cd accumulation in various sorghum (*Sorghum bicolor* (L.) Moench) genotypes. *Ecotoxicology and Environmental Safety*, *214*, 112019.

Wang, Y., Zeng, X., Xu, Q., Mei, X., Yuan, H., Jiabu, D., ... & Nyima, T. (2019). Metabolite profiling in two contrasting Tibetan hulless barley cultivars revealed the core salt-responsive metabolome and key salt-tolerance biomarkers. *AoB Plants*, *11*(2), plz021.

Wang, Z., Wei, Y., Zhao, Y., Wang, Y., Zou, F., Huang, S., ... & Hu, H. (2022). Physiological and transcriptional evaluation of sweet sorghum seedlings in response to single and combined drought and salinity stress. *South African Journal of Botany*, *146*, 459–471.

Wu, X. Y., Hu, W. J., Luo, H., Xia, Y., Zhao, Y., Wang, L. D., ... & Jing, H. C. (2016). Transcriptome profiling of developmental leaf senescence in sorghum (*Sorghum bicolor*). *Plant Molecular Biology*, *92*, 555–580.

Xin, Z., Li Wang, M., Barkley, N. A., Burow, G., Franks, C., Pederson, G., & Burke, J. (2008). Applying genotyping (TILLING) and phenotyping analyses to elucidate gene function in a chemically induced sorghum mutant population. *BMC Plant Biology*, *8*, 1–14.

Yang, W., Guo, Z., Huang, C., Duan, L., Chen, G., Jiang, N., ... & Xiong, L. (2014). Combining high-throughput phenotyping and genome-wide association studies to reveal natural genetic variation in rice. *Nature Communications*, *5*(1), 5087.

Yang, Y., Saand, M. A., Huang, L., Abdelaal, W. B., Zhang, J., Wu, Y., ... & Wang, F. (2021). Applications of multi-omics technologies for crop improvement. *Frontiers in Plant Science*, *12*, 563953.

Youens-Clark, K., Buckler, E., Casstevens, T., Chen, C., DeClerck, G., Derwent, P., ... & Ware, D. (2010). Gramene database in 2010: updates and extensions. *Nucleic acids research*, *39*(suppl_1), D1085-D1094

Yu, H., Lin, T., Meng, X., Du, H., Zhang, J., Liu, G., ... & Li, J. (2021). A route to de novo domestication of wild allotetraploid rice. *Cell*, *184*(5), 1156–1170.

Zhang, D. F., Zeng, T. R., Liu, X. Y., Gao, C. X., Li, Y. X., Li, C. H., ... & Yu, L. I. (2019). Transcriptomic profiling of sorghum leaves and roots responsive to drought stress at the seedling stage. *Journal of Integrative Agriculture*, *18*(9), 1980–1995.

Zhang, F., Lu, F., Wang, Y., Zhang, Z., Wang, J., Zhang, K., ... & Zhu, K. (2022). Combined transcriptomic and physiological metabolomic analyses elucidate key biological pathways in the response of two sorghum genotypes to salinity stress. *Frontiers in plant science*, *13*, 880373.

Zhang, F., Lu, F., Wang, Y., Zhang, Z., Wang, J., Zhang, K., ... & Zhu, K. (2022a). Combined transcriptomic and physiological metabolomic analyses elucidate key biological pathways in the response of two sorghum genotypes to salinity stress. *Frontiers in Plant Science*, *13*, 880373.

Zhang, S., Liu, S., Yang, Y., Jiang, F., Zhao, Z., Yang, F., & Wang, L. (2024). Ultrahigh-performance liquid chromatography–tandem mass spectrometry-based untargeted metabolomics to differentiate Guizhou glutinous sorghum and Hongyingzi sorghum. *Food Analytical Methods*, 17(11), 1610–1625.

Zhang, Z., Pope, M., Shakoor, N., Pless, R., Mockler, T. C., & Stylianou, A. (2022b). Comparing deep learning approaches for understanding genotype× phenotype interactions in biomass sorghum. *Frontiers in Artificial Intelligence*, *5*, 872858.

Zhao, C., Zhang, Y., Du, J., Guo, X., Wen, W., Gu, S., ... & Fan, J. (2019). Crop phenomics: Current status and perspectives. *Frontiers in Plant Science*, *10*, 714.

Zhao, Y., Zhai, G., Li, X., Tao, H., Li, L., He, Y., ... & Zhu, Y. (2022). Metabolomics reveals nutritional diversity among six coarse cereals and antioxidant activity analysis of grain sorghum and sweet sorghum. *Antioxidants*, *11*(10), 1984.

Zhou, D., Yang, Y., Zhang, J., Jiang, F., Craft, E., Thannhauser, T. W., ... & Liu, J. (2017). Quantitative iTRAQ proteomics revealed possible roles for antioxidant proteins in sorghum aluminum tolerance. *Frontiers in Plant Science*, *7*, 2043.

Zhou, Y., Lv, J., Yu, Z., Wang, Z., Li, Y., Li, M., ... & Zhou, W. (2022). Integrated metabolomics and transcriptomic analysis of the flavonoid regulatory networks in *Sorghum bicolor* seeds. *BMC Genomics*, *23*(1), 619.

Zhou, Y., Wang, Z., Li, Y., Li, Z., Liu, H., & Zhou, W. (2020). Metabolite profiling of sorghum seeds of different colors from different sweet sorghum cultivars using a widely targeted metabolomics approach. *International Journal of Genomics*, *2020*(1), 6247429.

Zhu, Z., Li, D., Wang, P., Li, J., & Lu, X. (2020). Transcriptome and ionome analysis of nitrogen, phosphorus and potassium interactions in sorghum seedlings. *Theoretical and Experimental Plant Physiology*, *32*(4), 271–285.

10 Leveraging Physiological Responses to Abiotic Stresses for Next Generation Sorghum Breeding

Krishan Kumar Jangid, Vinay Hegde, Priyanka Negi, Aliza Pradhan, and Jagadish Rane

10.1 INTRODUCTION

Sorghum (*Sorghum bicolor* (L.) Moench), the world's fifth most important cereal crop (Kazungu et al., 2023), plays a vital role in food security, especially in semi-arid and tropical regions (Chapke & Tonapi, 2019). Its versatility in providing grain, animal feed, fuel, and other uses (Mawouma et al., 2022) underscores its adaptability to challenging environments. As a C_4 grass, sorghum can function as an annual or short-lived perennial, completing one generation per growing season. Its fibrous root system begins with seminal roots at germination, followed by nodal, crown, or adventitious roots emerging from the shoot (Kazungu et al., 2023). Nutritionally, sorghum is a powerhouse, offering essential minerals, high levels of phytochemicals with antioxidant and anti-inflammatory properties, and a lower glycemic index compared to other cereals. Its broad adaptability and nutritious profile position it as a vital crop for enhancing food security, addressing malnutrition, and improving human health (Tanwar et al., 2023). Thriving at altitudes of 800–2,000 meters with rainfall between 250 and 900 mm and temperatures ranging from 10°C to 30°C (Kazungu et al., 2023), sorghum demonstrates remarkable resilience under diverse environmental conditions.

Survival of sorghum under water scarcity is underpinned by an intricate network of physiological, molecular, and biochemical responses (Tu et al., 2023). These adaptations include water conservation mechanisms, maintenance of cellular homeostasis, and protection against environmental stress, making it particularly valuable for drought-prone regions. Specific traits like a deep root system, waxy leaves that reduce water loss, leaf angle adjustment, and the ability to maintain stomatal regulation under limited water supply enhance its drought tolerance (Abreha et al., 2022). Furthermore, the stay-green trait allows sorghum to sustain grain production under dry conditions, mitigating yield losses (Liedtke et al., 2020).

For effective breeding of abiotic stress-tolerant sorghum, understanding the plant's survival strategies under varying environmental conditions is crucial. Modern technologies such as genomics, high-throughput phenotyping, and molecular breeding offer the tools to identify genes and traits responsible for resilience (Yahaya & Shimelis, 2022). This chapter delves into the physiological responses of sorghum to abiotic stresses and explores cutting-edge strategies for mitigating these stresses to achieve better yields through improved breeding programs. Insights provided here will be instrumental in developing sorghum varieties capable of thriving amid future environmental challenges (Djanaguiraman et al., 2020). Advanced molecular and biological tools continue to expand the possibilities for breeding stress-resistant, high-yielding sorghum varieties, ensuring its role as a cornerstone crop for sustainable agriculture.

DOI: 10.1201/9781003623717-10

10.2 PHYSIOLOGICAL RESPONSES TO ABIOTIC STRESSES

Physiological responses to abiotic stresses involve a suite of mechanisms that collectively enable plants to survive and maintain productivity under challenging environmental conditions (Prasad et al., 2021) (Figure 10.1). During drought, sorghum employs stomatal regulation and osmotic adjustment to conserve water (Dourado et al., 2022). Under saline conditions, it maintains ion homeostasis and accumulates osmolytes to mitigate the adverse effects of salt stress (Mansour et al., 2021). Heat stress triggers the production of heat shock proteins which contribute to membrane stabilization and the protection of cellular integrity (Nagaraju et al., 2020), while oxidative stress is counteracted by increased antioxidant activity (Mulaudzi et al., 2022). Additionally, sorghum optimizes resource use by adjusting its root architecture and reducing leaf growth (Borrell et al., 2021). These adaptive mechanisms are complemented by the plant's ability to regulate photosynthesis and protect cellular structures, underscoring sorghum's resilience and its importance as a crop in regions prone to extreme climatic conditions (Sharma & Joshi, 2022).

10.2.1 Responses to Drought Stress

Drought, a prolonged period of water shortage, significantly impacts plant health by limiting access to atmospheric, surface, and groundwater resources essential for growth and survival (Ahluwalia et al., 2021). In sorghum, drought stress hampers key physiological processes, including cell growth, photosynthesis, and nutrient absorption, resulting in reduced productivity (Behera et al., 2022). However, sorghum exhibits remarkable drought tolerance through a variety of adaptations at the physiological, molecular, and biochemical levels. At the physiological level, drought-tolerant sorghum employs strategies such as stomatal regulation, the development of deep root systems, and osmotic adjustment to minimize water loss and optimize water uptake. Thick leaf wax serves as an additional protective barrier, reducing transpiration through the leaf surface (Sanjari et al., 2021). Stomatal conductance plays a critical role in controlling water loss and regulating leaf temperature, with plants that maintain higher stomatal conductance exhibiting cooler canopies, a key indicator of drought avoidance in sorghum (López et al., 2022).

On a biochemical level, drought tolerance mechanisms include the accumulation of protective compounds such as sucrose and proline, which aid in maintaining cellular homeostasis under water-deficit conditions. Osmoregulation, a vital metabolic process during drought stress, enables cells to accumulate solutes like proline, glycine betaine, mannitol, trehalose, and sugars such as sucrose and oligosaccharides, helping to stabilize internal conditions (Mukarram et al., 2021). Proline, in particular, has been linked to improved recovery from drought stress in sorghum leaves (de Oliveira et al., 2020). Leaf curling, a common response to decreased water potential, reflects the plant's ability to adapt osmotically under low water availability (Yahaya & Shimelis, 2022). Regulation of gene expression also plays a pivotal role in drought tolerance. Transcription factors such as DREB (Dehydration-Responsive Element-Binding) and NAC (NAM, ATAF1/2, and CUC2) are instrumental in activating drought-responsive genes (Izadi-Darbandi et al., 2023; Fan et al., 2023). Additionally, elevated levels of abscisic acid (ABA) under drought stress highlight its crucial role in enhancing drought resilience in sorghum (Abreha et al., 2022). Together, these physiological and molecular mechanisms enable sorghum to survive and thrive in water-scarce environments.

10.2.2 Responses to Heat Stress

Temperature regulation is critical for plant growth, development, and adaptation. High temperatures can severely affect a plant's structure, biochemistry, and physiology, leading to reduced growth and productivity (Ndlovu et al., 2021). Heat stress can damage proteins, disrupt synthesis, inactivate enzymes, and destroy cell membranes, impairing cell division and slowing plant growth while causing oxidative damage. Brief exposure to high temperatures during seed filling can accelerate filling, leading to reduced yield and quality (Prasad et al., 2021). Heat stress also negatively impacts plant weight, root development, and the availability of water and nutrients.

Photosystem II (PSII) is particularly sensitive to high temperatures and can partially shut down, damaging the oxygen-evolving complex and causing imbalances in electron transport (Zhu et al., 2020). Rubisco, a key enzyme in photosynthesis, becomes less effective at high temperatures due to its lower affinity for CO_2 and its tendency to bind with O_2 instead, limiting net photosynthesis. Heat stress affects various physiological processes, including canopy cooling, photosynthetic rates, stomatal conductance, chlorophyll fluorescence, and leaf curling (Behera et al., 2022). Higher temperatures also inhibit metabolic functions by denaturing proteins, increasing membrane permeability, and producing excess reactive oxygen species (Ndlovu et al., 2021). However, in C_4 plants like sorghum, the enzyme phosphoenolpyruvate carboxylase (PEP carboxylase or PEPC), which has a high affinity for CO_2, significantly reduces photorespiration, enabling sorghum to better maintain productivity and withstand elevated temperatures (Behera et al., 2022).

10.2.3 Responses to Soil Salinity

Sorghum exhibits a range of physiological responses to salt stress that can severely impact its growth, development, and yield. Understanding these responses is critical for developing strategies to enhance sorghum's resilience in saline environments (de Queiroz et al., 2023). Salt stress primarily induces osmotic imbalances and ion toxicity, particularly due to sodium accumulation, leading to developmental disruptions, physiological damage, and alterations in leaf anatomy and biochemical pathways (Liu et al., 2023). One of the most significant consequences of salt stress is a reduction in photosynthetic efficiency, which directly affects biomass production. This decline results from decreased stomatal conductance, limiting CO_2 uptake, and reduced chlorophyll content, impairing the plant's ability to perform effective photosynthesis (Rajabi Dehnavi et al., 2024).

Despite these challenges, sorghum has evolved multiple adaptive mechanisms to mitigate the effects of salinity. One key strategy involves maintaining a favorable sodium-to-potassium (Na^+/K^+) ratio, which is essential for preserving cellular functions under saline conditions. Salt-tolerant genotypes achieve this by excluding Na^+ from the cytoplasm or compartmentalizing it within vacuoles, thereby safeguarding vital cellular processes (Rajabi Dehnavi et al., 2024). In addition to ion regulation, sorghum accumulates osmoprotectants such as proline, which help maintain osmotic balance and protect cellular structures from damage (Reddy et al., 2015). Complementing this, the plant employs several ion management strategies, including salt exclusion, extrusion, dilution, and ion separation (de Queiroz et al., 2023). Salt exclusion limits Na^+ entry into roots via selective ion channels and transporters like *SOS1* preventing toxic accumulation in sensitive tissues. Salt extrusion actively transports Na^+ ions out of cells into the apoplast or rhizosphere through Na^+/H^+ antiporters, maintaining ion homeostasis. Dilution reduces ionic concentration in the cytosol through increased water uptake or tissue expansion, while ion separation involves compartmentalizing Na^+ and Cl^- into vacuoles using tonoplast transporters like *NHX* protecting enzymatic processes in the cytoplasm.

These physiological and biochemical adaptations are further supported by hormonal responses, particularly the production of abscisic acid (ABA), which modulates stress responses and growth regulation. Salt stress also triggers excessive production of reactive oxygen species (ROS), which can damage cellular components and exacerbate stress (Liu et al., 2023). To counteract oxidative damage, sorghum activates a robust antioxidant defense system comprising enzymes such as catalase (CAT), superoxide dismutase (SOD), peroxidases (POD), ascorbate peroxidase (APX), glutathione reductase (GR), and monodehydroascorbate reductase (MDAR). These enzymes play a vital role in scavenging ROS, maintaining cellular homeostasis, and enhancing the plant's overall resilience to salinity stress (Azeem et al., 2023).

10.2.4 Responses to Waterlogging Stress

Waterlogging occurs when water stagnates on the soil surface, resulting in excessive moisture in the plant's root zone. This condition disrupts key physiological processes, particularly photosynthesis, by affecting the activity of photosynthetic enzymes, reducing chlorophyll content, and leading to the accumulation of malondialdehyde (MDA), a marker of oxidative stress (Zhang et al., 2023). Studies indicate that sorghum experiences a significant decline in photosynthetic rate and efficiency under waterlogged conditions, ultimately reducing biomass production (Zhang et al., 2023). In response to waterlogging, the root system undergoes physiological alterations similar to those reported in other crops (Xuewen et al., 2014), though specific studies in sorghum are lacking. Waterlogging damages root tissues, decreases oxygen availability, and impairs nutrient uptake, further inhibiting plant growth and productivity. Prolonged exposure to waterlogged conditions can result in severe growth inhibition or even plant death (Kim et al., 2024). Symptoms such as chlorosis and excessive leaf elongation often manifest due to reduced ethylene transport.

To survive submergence, some plants have evolved adaptations like aerenchyma formation – gas-filled cavities in roots that facilitate oxygen diffusion from aerial parts to submerged roots or rhizomes (Tamaru et al., 2023). Ethylene plays a central role in the plant's response to submergence, regulating processes that mitigate damage and promote recovery (Behera et al., 2022). Additionally, a quick recovery after submergence and efficient scavenging of ROS are critical factors for post-submergence survival (Zhang et al., 2023).

10.2.5 Responses to Low Temperature Stress

Low temperatures affect plants in two major ways: chilling stress (0°C–15°C) disrupts physiological and developmental processes, while freezing stress (<0°C) causes cellular damage and dehydration (Shomo et al., 2024; Ostmeyer et al., 2020). In sorghum, cold stress impairs photosynthesis, respiration, and membrane permeability, reducing growth and yield (Rawat et al., 2021). Unlike cold-adapted C_4 species like *Miscanthus giganteus*, sorghum struggles to maintain photosynthetic

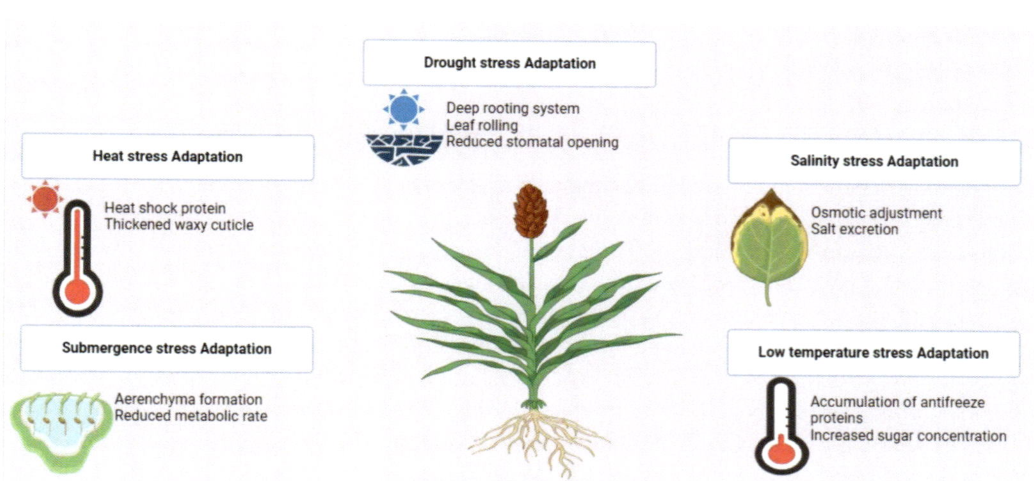

FIGURE 10.1 Physiological responses of sorghum to abiotic stresses, including drought, heat, salinity, cold, and waterlogging stress.

efficiency below 15°C, largely due to reduced activity of pyruvate phosphate dikinase (PPDK) and Rubisco, which limits carbon fixation, quantum yield (ΦPSII), and electron transport (Ortiz et al., 2017).

Cold stress also affects gas exchange and water relations. Reduced stomatal conductance, a response observed under both cold and waterlogged conditions, restricts CO_2 uptake and transpiration, impacting biomass production. Cold-tolerant plants mitigate this by adjusting stomatal closure to conserve water (Ortiz et al., 2017). At the cellular level, chilling stress compromises membrane integrity, leading to electrolyte leakage, impaired protoplasmic streaming, and metabolic slowdown. Photosystem II (PSII) is particularly vulnerable, resulting in reduced photosynthesis and nutrient uptake (Vera Hernández et al., 2023). Additionally, cold stress induces ROS accumulation, damaging membranes and enzymes (Shomo et al., 2024). However, modifications in membrane lipids and proteins help restore metabolic balance and enhance cold tolerance (Behera et al., 2022). At the molecular level, lipid homeostasis, chaperone activity, and stress signaling pathways are crucial for chilling tolerance in sorghum (Moghimi et al., 2019).

10.2.6 Responses to Multiple Abiotic Stresses

Sorghum exhibits both susceptibility and tolerance when exposed to multiple abiotic stresses, such as drought, heat, and salinity, which frequently co-occur in semi-arid regions. While studies often analyze these stresses individually, their combined impact is more detrimental, disrupting key morphological, physiological, and biochemical processes that affect growth and yield (Ndlovu et al., 2021; Zheng et al., 2024). Susceptibility to multiple stresses is evident in sorghum's impaired photosynthesis, excessive ROS accumulation, and disrupted antioxidant defense mechanisms. Under combined drought and salinity stress, despite increased superoxide dismutase (SOD) and peroxidase (POD) activity, the catalase (CAT) activity and AsA-GSH cycle are severely inhibited, failing to counteract oxidative stress, leading to membrane damage and metabolic dysfunction (Wang & Wei, 2022). Additionally, elevated temperatures intensify transpiration rates, exacerbating the effects of drought and causing severe reductions in biomass and grain yield (Ndlovu et al., 2021). Conversely, tolerance mechanisms involve stress signaling, osmotic adjustment, and metabolic reprogramming. Sorghum employs polyamine biosynthesis, unique transcription factors, and specialized signaling molecules to modulate stress responses (Johnson et al., 2014; Woldesemayat et al., 2018). It mitigates drought stress through trehalose biosynthesis, while phytohormone signaling enhances resilience to salinity and combined stress (Wang et al., 2022). Additionally, sorghum downregulates photosynthesis, activates antioxidant enzymes, and accumulates proline to sustain metabolic functions under multiple stresses (Wang et al., 2022).

10.3 GENETIC AND MOLECULAR BASIS OF STRESS PHYSIOLOGY

Understanding genetic and molecular mechanisms of stress tolerance in sorghum is key to breeding resilient crops for challenging environments. Leaf stay-green along with grain yield can serve as key indicators for drought tolerance. The stay-green trait in sorghum is controlled by multiple genes, with QTLs like *Stg1*, *Stg2*, and *Stg3* linked to this trait (Subudhi et al., 2000). *Stg2*, closely associated with chlorophyll content, is the primary driver (Subudhi et al., 2000). These QTLs have been validated in various genetic populations, showing significant variation in drought tolerance. Root length and structure, particularly the nodal root angle, also contribute to drought resistance, with narrow root angles enhancing water uptake during drought (Mace et al., 2012). Over 40% of sorghum genes are affected by drought, with heat shock proteins (HSPs) playing a vital role in stress responses. The MAPK cascade, involved in stress regulation, influences drought tolerance by modulating downstream genes. MAPK genes like *SbMAPKKK7* and *SbMAPK10* respond to drought and ABA, regulating stomatal closure (Prasad et al., 2021). Calcium-dependent protein kinases (CDPKs) are crucial in drought responses, with genes like *SbCDPK6* and *SbCDPK16* upregulated in leaves and roots, respectively. These kinases interact with ABA and MAPK pathways to regulate drought tolerance. Other genes, such as *SbSKIP* and *SbSNAC1*, have shown promise in improving drought tolerance (Li et al., 2019). Further research on genetically modifying these genes could enhance sorghum's drought tolerance (Aregawi et al., 2022).

Research on high-temperature stress-related genes in sorghum is limited, with most studies focusing on drought tolerance genes. However, some drought tolerance genes also confer heat tolerance, such as the stay-green trait QTLs (Chadalavada et al., 2021; Ndlovu et al., 2021). Genes like ARF family members and heat shock factor (HSF) genes, including *SbHSF1*, *SbHSF5*, and *SbHSF13*, are induced by both drought and heat stress (Sudhakar Reddy et al., 2016). HSPs such as *SbHSP90* and *SbHsp20* also respond to heat and drought stress, playing a role in regulating plant stress responses (Pavli et al., 2011; Nagaraju et al., 2020). Drought-tolerant sorghum varieties, like SC35, are potential sources of heat-tolerant germplasm, but their effectiveness may be compromised under heat stress, especially during flowering (Sunoj et al., 2017). Therefore, it is crucial to specifically target heat tolerance in sorghum breeding (Tack et al., 2017).

Several QTLs have been identified in sorghum seedlings, such as *qGP7-1*, *qSH1*, and *qTFW4*, which significantly influence traits like germination and biomass under salt stress (Wang et al., 2014, 2020). Importantly, QTLs like *qBrix2* and *qTB6* have shown consistent effects throughout the growth period, offering valuable targets for MAS to enhance salt tolerance (Wang et al., 2020). Salt-tolerant (M-81E) and salt-sensitive (Roma) sorghum genotypes have revealed key differences in Na^+ accumulation and transcriptomic profiles. Differentially expressed genes in M-81E are involved in pathways like ion transport and photosynthesis, contributing to its superior salt tolerance (Sui et al., 2015; Yang et al., 2018). Regulatory transcription factors such as *SbMYBHv33*, *SbWRKY50*, and *SbbHLH85* modulate salt tolerance by affecting ion transport, ABA signaling, and root traits, highlighting gene regulatory networks as pivotal mechanisms (Song et al., 2020). Beyond transcriptional regulation, post-transcriptional mechanisms like non-coding RNAs (e.g., lncRNAs and miRNAs) and epigenetic modifications (e.g., m6A methylation) further refine salt stress responses. For instance, m6A modifications dynamically adjust mRNA stability and salt-tolerance gene expression in sweet sorghum, surpassing direct transcriptional changes in sensitivity (Zheng et al., 2021).

While several studies have explored the effects of waterlogging on morphology and physiology of sorghum, genetic control of waterlogging stress in sorghum remains unexplored. In contrast, research on cold stress in sorghum has identified regulatory genes linked to cold tolerance. Moghimi et al. (2019) found three genes, (i) *Sobic.007G005400* (a DnaJ/Hsp40 protein), (ii) *Sobic.007G005500* (an m6A-binding protein), and (iii) *Sobic.007G037000* (ABHD), that influence cold tolerance, particularly by regulating lipid synthesis and signaling pathways. Additionally, QTLs like *Xtxp34*, *Xtxp88*, *Xtxp319*, and *QTL57* are associated with cold tolerance, particularly during seedling development (Parra-Londono et al., 2018). However, trade-offs exist, such as the cold-tolerant varieties from China, which have high tannin content that reduces protein digestibility but may offer benefits for ruminants and reduce bird foraging (Wu et al., 2019; Xie et al., 2019). Screening tannin-free germplasm from other cold regions, such as high-altitude African varieties, could help identify new cold tolerance-related genes and QTLs, balancing the benefits and challenges in breeding (Rutayisire et al., 2021; Ostmeyer et al., 2020).

Over the past decade, significant progress has been made in understanding the molecular responses of sorghum to various stresses (as discussed above). However, the genetic and molecular basis of combined stress physiology remains largely unexplored. Several key regulatory mechanisms contribute to sorghum's response to multiple abiotic stresses. Proline, a crucial osmoprotectant, plays a significant role in both salt and drought stress adaptation, with *SbP5CS1* and *SbP5CS2* genes actively responding to these conditions (Su et al., 2011). Similarly, aquaporins regulate osmotic balance by controlling water status under osmotic and salt stress, with sorghum displaying altered aquaporin gene expression under both salt stress and PEG-induced drought stress (Hostetler et al.,

2021). Several gene families, such as *NF- Y and trihelix* genes, exhibit responses to diverse stresses including salt, mannitol, extreme temperatures, and oxidative stress (Zheng et al., 2024). Johnson et al. (2014) highlighted the role of *MYB78, ATAF1,* heat shock proteins (HSPs), and polyamine biosynthesis in sorghum's combined stress response. Additionally, Wang et al. (2022) identified differentially expressed genes (DEGs) involved in photosynthesis, hormone signaling, trehalose metabolism, proline metabolism, and antioxidant defense under both individual and combined drought and salt stress.

An integrated multi-omics approach has been developed to identify complex trait QTLs and regulatory genes associated with multiple stresses in sorghum (Woldesemayat et al., 2018). Notably, 50% of drought-responsive genes also respond to salt, cold, heat, and oxidative stress, suggesting cross-talk between stress pathways. Some genes regulate multiple stress types simultaneously, indicating shared regulatory networks and metabolic pathways. Moreover, greenness-related QTLs show strong correlations under drought and heat stress, facilitating the maintenance of photosynthesis and overall plant performance in adverse conditions (Sukumaran et al., 2016).

10.4 TRAIT-BASED BREEDING STRATEGIES FOR ABIOTIC STRESS TOLERANCE

The integration of conventional and modern breeding approaches is crucial for developing sorghum varieties that can thrive under abiotic stress conditions (Weldemichael et al., 2024; Figure 10.2). Traditional breeding, particularly for stress tolerance, relies on methods such as pedigree selection, mass selection, and recurrent selection to develop stress-tolerant varieties (Liu et al., 2024a). These methods focus on selecting individuals with key traits that contribute to stress resilience (Wang et al., 2024), but they are labor-intensive and require multiple breeding cycles for stable improvements. To enhance the efficiency of breeding for stress tolerance, four main strategies have been proposed: (i) Breeding for high yields under optimal conditions, assuming these varieties will also perform well under suboptimal conditions; (ii) Empirical selection in field conditions targeting

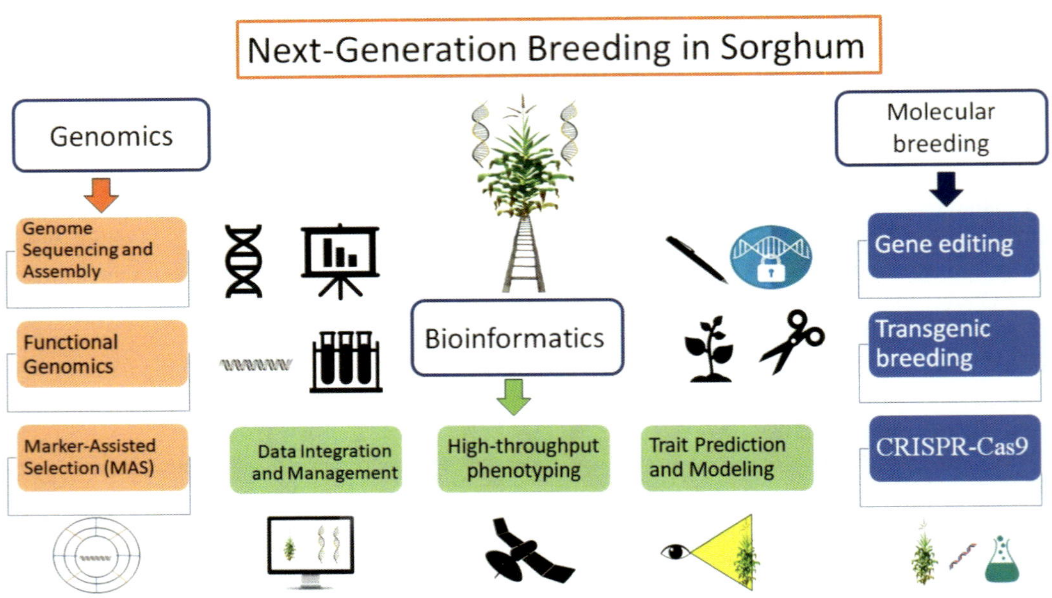

FIGURE 10.2 Next-generation trait-based breeding tools for sorghum stress tolerance, including marker-assisted breeding, targeted gene editing, transgenesis, and high-throughput phenotyping enabling the development of resilient varieties for sustainable production under abiotic stresses.

stress-prone environments to maximize yield; (iii) Integrating selected physiological and/or morphological traits that confer stress tolerance into traditional breeding programs; and (iv) Identifying key traits that provide drought tolerance at specific growth stages and incorporating them into high-yielding varieties (Reddy, 2019).

Genomic technologies, such as marker-assisted selection (MAS) and genomic selection (GS), provide greater precision in identifying and integrating stress-tolerance traits (Baloch et al., 2023). MAS is particularly effective in tracking genes linked to target traits, accelerating the breeding process and enhancing trait selection for complex stress responses (Baloch et al., 2023). It has been successful in introducing genes related to drought, heat, and salinity tolerance, especially for traits controlled by major genes (Nair & Rabbani, 2024). By validating QTLs, MAS contributes to improving sorghum and deepens our understanding of drought stress responses (Kumar et al., 2024). Furthermore, the use of donor parents with good heritability and combining ability for stress tolerance-related traits is essential for developing hybrids. For example, Knoll & Ejeta (2008) validated SSR markers (Xtxp43, Xtxp51, Xtxp211) linked to early-season cold tolerance QTLs in two sorghum populations. Xtxp43 and Xtxp51 exhibited strong effects on vigor and emergence, while Xtxp211 had smaller but significant effects. Similarly, Mwamahonje et al. (2021) employed marker-assisted backcrossing to enhance stay-green traits in Tanzanian sorghum varieties, successfully introgressing stay-green QTLs. Their work demonstrated the effectiveness of the B35xNACO Mtama 1 background for improving yield (0.127 kg/panicle), and NA316C for grain weight (0.068 kg/panicle), highlighting its potential for drought tolerance improvement.

On the other hand, GS enables breeders to predict the performance of lines based on their genetic makeup, significantly improving breeding efficiency and facilitating simultaneous selection for multiple traits (Sinha et al., 2023; Gebremedhin et al., 2024). Already successful in animal breeding, GS is now being applied increasingly in plant breeding programs. In sorghum, GS has been used alongside MAS to exploit both major and minor QTLs for key traits. Recent studies show that by combining pedigree and marker-based information, GS improves prediction accuracy and accelerates breeding cycles, especially for complex traits like grain yield, biomass, and drought adaptation. For instance, in Australia, phenotyping data from 791 sorghum hybrids and marker genotyping of 544 lines improved prediction accuracy when combined with pedigree and marker-based information, compared to pedigree-based models alone (Hunt et al., 2018). GS has been shown to increase genetic gains by accelerating breeding cycles, especially for oligogenic traits and small populations, as seen in simulations comparing GS and pedigree selection (Somegowda et al., 2023). Moreover, GS holds promise for improving drought tolerance, with multi-trait GBLUP evaluations boosting predictive ability by 16–19% for grain yield and stay-green traits under drought stress (Velazco et al., 2019a, 2020). Studies also indicate that integrating genomic data with pedigree information can optimize predictions for complex traits like grain yield in varying environmental conditions, yielding improvements of 43–72% in prediction ability (Velazco et al., 2019b). The inclusion of diverse sorghum subgroups has further enhanced prediction accuracy, underscoring the importance of a broader association panel to maximize GS potential (Velazco et al., 2019b). Overall, GS has revolutionized sorghum breeding by increasing genetic gains, reducing breeding cycles, and improving trait selection efficiency.

CRISPR/Cas9 gene editing has revolutionized plant growth and stress resilience, transitioning from initial applications in model plants like *Arabidopsis*, rice, and tobacco to widespread use in major crops, including maize, wheat, soybean, and sorghum (Tariq et al., 2023). In sorghum, studies have targeted genes such as cinnamyl alcohol dehydrogenase (*CAD*) and phytoene desaturase (*PDS*) through biolistic bombardment (Li et al., 2024). However, broader functional genomics research remains limited. In other crops, CRISPR/Cas9 has demonstrated its potential to enhance stress tolerance. For instance, knocking out *OsAnn3* in rice improved cold stress resilience (Shen et al., 2017). The integration of CRISPR/Cas9 with advanced systems like GRF4-GIF1/ternary vectors further enhances tissue regeneration and transformation efficiency, unlocking new possibilities for sorghum improvement (Li et al., 2024). Expanding these tools to abiotic stress-responsive genes in sorghum could significantly improve yield and resilience, particularly benefiting developing regions in Asia and Africa.

10.5 HIGH-THROUGHPUT PHENOTYPING FOR STRESS TOLERANCE

Phenotyping is essential in modern sorghum breeding, particularly in the context of escalating climate change and increasing abiotic stresses. By accurately measuring key physiological and agronomic traits, phenotyping helps breeders identify sorghum genotypes with enhanced tolerance to different abiotic stresses. Traits like biomass, leaf area, root architecture, and photosynthetic efficiency directly impact plant performance under stress conditions (Fiorani & Schurr, 2013). High-throughput phenotyping (HTP) platforms enable rapid, non-destructive measurement of these traits, allowing large populations to be evaluated efficiently (Gill et al., 2022; Kundu et al., 2024). In breeding for abiotic stress tolerance, phenotyping connects genotypic data with phenotypic expression under various environmental conditions (Hegde et al., 2024). This understanding helps breeders identify genetic variations that contribute to stress resilience and select traits that are quantitatively inherited and influenced by multiple genes. The integration of HTP technologies is transforming sorghum breeding by providing detailed insights into complex traits and their response to abiotic stresses (Baloch et al., 2023). These technologies generate large datasets that, when analyzed using bioinformatics tools, can identify candidate genes or metabolic pathways associated with desirable traits, thus enhancing selection efficiency (Liu et al., 2024b).

Imaging techniques play a key role in high-throughput phenotyping by enabling non-invasive capture of a wide range of plant traits (Table 10.1). In sorghum breeding, imaging technologies are

TABLE 10.1
Imaging techniques and their applications in assessing plant traits for stress tolerance and productivity

Imaging Techniques	Traits	Significance	References
Lidar	Plant Height	Important for biomass and yield estimation	Araus and Cairns (2014)
Multispectral imaging	Leaf area index	Indicator of photosynthetic capacity and drought tolerance	Furbank and Tester, (2011)
Hyperspectral imaging	Chlorophyll content	Linked to photosynthetic efficiency and overall plant health	Mishra et al. (2017)
Thermal imaging	Canopy temperature	Helps in assessing water use efficiency and drought tolerance	Cossani and Reynolds (2012)
X-ray computed tomography	Root architecture	Important for drought tolerance and nutrient uptake	Lynch (2013)
Infrared gas analyzer + imaging	Stomatal conductance	Indicator of water regulation and drought response	Jones (1999)
Time-lapse imaging	Flowering time	Key for adaptation to variable environmental conditions	Prasad and Montell, (2007)
NDVI (normalized difference vegetation index)	Stay-green trait	Important for drought tolerance and yield stability under water stress	Borrell et al. (2014)
Digital imaging and 3D reconstruction	Grain quality (size and shape)	Critical for yield traits and post-harvest processing	Pask et al. (2012)

used to monitor traits such as leaf area, plant height, biomass, and canopy temperature. Multispectral and hyperspectral imaging detect variations in chlorophyll content, water status, and photosynthetic activity, offering indirect assessments of drought tolerance (Haworth et al., 2023). These techniques are particularly useful for identifying stress-related phenotypes early in plant development, facilitating faster selection in breeding programs. Thermal imaging is commonly used to monitor canopy temperature, which is closely linked to transpiration efficiency and stomatal conductance under drought conditions (Araus & Cairns, 2014; Pradhan et al., 2022). Additionally, RGB imaging quantifies growth parameters, while fluorescence imaging assesses photosynthetic efficiency and stress-induced damage to the photosystem. By combining these imaging techniques, breeders can comprehensively evaluate sorghum cultivars' performance under various stress conditions.

Sensor-based measurements further enhance high-throughput phenotyping by providing real-time, precise data on physiological and environmental parameters. In sorghum, sensors monitor soil moisture, temperature, and light intensity, helping breeders evaluate how different genotypes perform under both controlled and field conditions. Proximal sensors, such as LiDAR, measure plant height and architecture with high accuracy, while spectrometers assess leaf reflectance properties related to water and nutrient status (Hegde et al., 2024). Integrated with automated systems, sensors enable continuous monitoring and dynamic data collection on how sorghum responds to fluctuating environmental conditions. Sensors measuring stomatal conductance and leaf water potential offer critical insights into drought tolerance mechanisms (Berger et al., 2010; Hegde et al., 2024). These tools enable breeders to select sorghum varieties with superior stress resilience more effectively.

10.6 OMICS-BASED INSIGHTS INTO PLANT RESPONSES TO ABIOTIC STRESSES AND THEIR BREEDING IMPLICATIONS

Integrating omics data into sorghum breeding enhances the understanding of plant responses to abiotic stresses. Studies on drought-responsive genes in sorghum reveal the upregulation of critical genes encoding LEA proteins, ROS detoxification enzymes, and proline biosynthesis pathways under osmotic stress (Buchanan et al., 2005). Comparative transcriptomics between drought-tolerant and sensitive genotypes have further highlighted genes like *P5CS2* (proline biosynthesis), *HKT1* (sodium transport), and genes involved in antioxidant pathways as key players in drought resistance (Fracasso et al., 2016). These findings align with broader omics analyses, which identified pathways such as starch and sucrose metabolism, phenylpropanoid biosynthesis, and phytohormone signaling as pivotal during drought response, particularly during seed germination (Zhang et al., 2024). Similarly, lipid metabolic genes are prominently upregulated in drought-tolerant genotypes, emphasizing the role of lipid metabolism in stress adaptation (Pasini et al., 2014). Insights from miRNA and proteomics studies corroborate the importance of transcription factors, antioxidant capacities, and stress signaling mechanisms, which are differentially expressed in drought-tolerant cultivars (Hamza et al., 2016; Li et al., 2020).

The mechanisms underlying salt tolerance in sorghum are equally intricate. Transcriptome analyses have identified 4,289 genes enriched in various GO terms and pathways associated with salt stress, alongside transcription factors and hormone signaling genes (Liu et al., 2024b). Comparative studies of salt-tolerant and sensitive cultivars, such as M-81E and Roma, reveal that salt tolerance is mediated by maintaining photosynthesis, sucrose metabolism, and upregulating salt-exclusion genes like *SbHKT1;5*, which enhances potassium transport (Yang et al., 2018; Tu et al., 2023). Furthermore, omics studies highlight ABA and JA signaling as crucial, with distinct expression patterns between tolerant and sensitive genotypes (Yang et al., 2017). Beyond transcriptomics, metabolomics and proteomics uncover complementary insights – e.g., osmolyte accumulation, lipidomic shifts affecting membrane integrity, and detoxification enzymes like glutathione S-transferase and peroxidase play essential roles in salinity adaptation (Ge et al., 2022; Swami et al., 2011). Notably, flavonoid metabolism emerges as a vital component, with specific flavonoids accumulating under saline-alkali conditions to mitigate oxidative damage (Punia et al., 2021). Together, these studies demonstrate how multi-omics approaches can unravel the complex interplay of pathways contributing to salt tolerance in sorghum.

Heat stress disrupts metabolism, damages chloroplasts, and impairs reproductive organ development, particularly during the reproductive stage. Heat shock proteins (HSPs) and heat stress transcription factors (HSFs) play central roles in mitigating these effects by activating stress-responsive genes (Zhang et al., 2022). Cold stress, conversely, reduces stomatal conductance, damages photosystem II, and triggers metabolic reprogramming via CBFs/DREBs and related transcription factors, such as NF-Ys, MYBs, and NACs, which regulate phenylpropanoid and amino acid biosynthesis pathways (Guo et al., 2021; Jin et al., 2021). Sorghum's sensitivity to cold during germination and early growth stages highlights the importance of oxidative stress resistance, with genes like *SbSODs* and *SbPODs* providing enhanced protection in tolerant cultivars (Shao et al., 2021). Proteomic and metabolomic studies further validate the differential regulation of stress-related proteins and metabolites under extreme temperatures, paving the way for targeted breeding strategies (Ngcala et al., 2020). Interestingly, recent research underscores the influence of diurnal rhythms on stress responses in sorghum. Approximately half of sorghum transcripts exhibit diurnal patterns, with thermo-tolerant genotypes showing pronounced rhythmic expression of key regulators, including Ca^{2+}-binding proteins (Bonnot et al., 2023).

These findings suggest that temporal regulation could be leveraged to optimize stress responses. In conclusion, significant progress has been made in elucidating the molecular responses of sorghum to diverse abiotic stresses. However, very little information is available regarding how sorghum responds to multiple concurrent stresses, and most studies have focused on individual stressors rather than their combined effects. The complexity of simultaneous abiotic stresses necessitates a deeper understanding of their interactions at physiological, biochemical, and molecular levels. Gaps remain in integrating multi-omics data across stress types and genotypes, limiting the ability to develop stress-resilient sorghum varieties effectively. Future research should prioritize combined stress studies, functional validation of candidate genes involved in multi-stress tolerance, and advancements in transformation technologies to enhance sorghum's adaptability. The establishment of unified multi-omics platforms will further facilitate large-scale data integration, enabling precise molecular breeding approaches to improve sorghum's resilience under a rapidly changing climate.

10.7 CONCLUSION

Sorghum, renowned for its inherent resilience to adverse environmental conditions, serves as a vital model for developing climate-resilient crops. Its physiological responses to abiotic stresses provide crucial insights for breeding next-generation stress-tolerant varieties/hybrids. However, the development of such varieties remains challenging due to the complexity of traits influenced by multiple environmental factors, limitations in reliable phenotyping methods, and the lengthy cycles associated with traditional breeding approaches, particularly when selecting for multiple traits simultaneously. Emerging technologies are addressing these challenges by revolutionizing the breeding landscape. CRISPR-Cas9 genome editing offers precise modifications to enhance stress tolerance without affecting other agronomic traits. High-throughput phenotyping (HTP), leveraging imaging and sensor technologies, significantly improves the efficiency and accuracy of selecting stress-tolerant traits. Moreover, the integration of multi-omics approaches unravels the molecular mechanisms underlying stress tolerance, offering targeted pathways for genetic improvement. These innovations, when combined with supportive policies, adequate funding, and collaborative efforts among research institutions and governments, hold the potential to transform sorghum breeding programs. Together, they can ensure the development of resilient varieties that meet the challenges of climate change and bolster global agricultural sustainability.

REFERENCES

Abreha, K. B., Enyew, M., Carlsson, A. S., Vetukuri, R. R., Feyissa, T., Motlhaodi, T., ... & Geleta, M. (2022). Sorghum in dryland: Morphological, physiological, and molecular responses of sorghum under drought stress. *Planta*, *255*, 1–23.

Ahluwalia, O., Singh, P. C., & Bhatia, R. (2021). A review on drought stress in plants: Implications, mitigation and the role of plant growth promoting rhizobacteria. *Resources, Environment and Sustainability*, *5*, 100032.

Araus, J. L., & Cairns, J. E. (2014). Field high-throughput phenotyping: The new crop breeding frontier. *Trends in Plant Science*, *19*(1), 52–61.

Aregawi, K., Shen, J., Pierroz, G., Sharma, M. K., Dahlberg, J., Owiti, J., & Lemaux, P. G. (2022). Morphogene-assisted transformation of *Sorghum bicolor* allows more efficient genome editing. *Plant Biotechnology Journal*, *20*(4), 748–760.

Azeem, M., Sultana, R., Mahmood, A., Qasim, M., Siddiqui, Z. S., Mumtaz, S., ... & Siddiqui, M. H. (2023). Ascorbic and salicylic acids vitalized growth, biochemical responses, antioxidant enzymes, photosynthetic efficiency, and ionic regulation to alleviate salinity stress in *Sorghum bicolor*. *Journal of Plant Growth Regulation*, *42*(8), 5266–5279.

Baloch, F. S., Altaf, M. T., Liaqat, W., Bedir, M., Nadeem, M. A., Cömertpay, G., ... & Sun, H. J. (2023). Recent advancements in the breeding of sorghum crop: Current status and future strategies for marker-assisted breeding. *Frontiers in Genetics*, *14*, 1150616.

Behera, P. P., Saharia, N., Borah, N., Devi, S. H., & Sarma, R. N. (2022). Sorghum physiology and adaptation to abiotic stresses. *International Journal of Environment and Climate Change*, *12*(10), 1005–1022.

Berger, B., Parent, B., & Tester, M. (2010). High-throughput shoot imaging to study drought responses. *Journal of Experimental Botany*, *61*(13), 3519–3528.

Bonnot, T., Somayanda, I., Jagadish, S. V. K., & Nagel, D. H. (2023). Time of day and genotype sensitivity adjust molecular responses to temperature stress in sorghum. *The Plant Journal*, *116*(4), 1081–1096.

Borrell, A. K., Mullet, J. E., George-Jaeggli, B., van Oosterom, E. J., Hammer, G. L., Klein, P. E., & Jordan, D. R. (2014). Drought adaptation of stay-green sorghum is associated with canopy development, leaf anatomy, root growth, and water uptake. *Journal of Experimental Botany*, *65*(21), 6251–6263.

Borrell, A., van Oosterom, E., George-Jaeggli, B., Rodriguez, D., Eyre, J., Jordan, D. J., ... & Hammer, G. (2021). Sorghum. In: Sadras, V.O, Calderini, D.F (Eds), *Crop Physiology Case Histories for Major Crops* (pp. 196–221). San Diego, CA, United States : Academic Press.

Buchanan, C. D., Lim, S., Salzman, R. A., Kagiampakis, I., Morishige, D. T., Weers, B. D., ... & Mullet, J. E. (2005). *Sorghum bicolor*'s transcriptome response to dehydration, high salinity and ABA. *Plant Molecular Biology*, *58*, 699–720.

Chadalavada, K., Kumari, B. R., & Kumar, T. S. (2021). Sorghum mitigates climate variability and change on crop yield and quality. *Planta*, *253*(5), 113.

Chapke, R. R., & Tonapi, V. A. (2019). Adoption and socio-economic benefits of improved post-rainy sorghum production technology. *Agricultural Research*, *8*(2), 270–278.

Cossani, C. M., & Reynolds, M. P. (2012). Physiological traits for improving heat tolerance in wheat. *Plant Physiology*, *160*(4), 1710–1718.

de Oliveira, D. F., Lopes, L. D. S., & Gomes-Filho, E. (2020). Metabolic changes associated with differential salt tolerance in sorghum genotypes. *Planta*, *252*(3), 34.

de Queiroz, G. C. M., de Medeiros, J. F., da Silva, R. R., da Silva Morais, F. M., de Sousa, L. V., de Souza, M. V. P., ... & da Silva Sá, F. V. (2023). Growth, solute accumulation, and ion distribution in sweet sorghum under salt and drought stresses in a Brazilian potiguar semiarid area. *Agriculture*, *13*(4), 803.

Djanaguiraman, M., Prasad, P. V., Ciampitti, I. A., & Talwar, H. S. (2020). Impacts of abiotic stresses on sorghum physiology. In: Tonapi, V.A., Talwar, H.S., Are, A.K., Bhat, B.V., Reddy, C.R., Dalton, T.J. (Eds), *Sorghum in the 21st Century: Food–Fodder–Feed–Fuel for a Rapidly Changing World* (pp. 157–188). Springer, Singapore.

Dourado, P. R. M., de Souza, E. R., Santos, M. A. D., Lins, C. M. T., Monteiro, D. R., Paulino, M. K. S. S., & Schaffer, B. (2022). Stomatal regulation and osmotic adjustment in sorghum in response to salinity. *Agriculture*, *12*(5), 658.

Fan, S., Chen, J., & Yang, R. (2023). Candidate genes for salt tolerance in forage sorghum under saline conditions from germination to harvest maturity. *Genes*, *14*(2), 293.

Fiorani, F., & Schurr, U. (2013). Future scenarios for plant phenotyping. *Annual Review of Plant Biology*, *64*(1), 267–291.

Fracasso, A., Trindade, L. M., & Amaducci, S. (2016). Drought stress tolerance strategies revealed by RNA-Seq in two sorghum genotypes with contrasting WUE. *BMC Plant Biology*, *16*, 1–18.

Furbank, R. T., & Tester, M. (2011). Phenomics–technologies to relieve the phenotyping bottleneck. *Trends in Plant Science*, *16*(12), 635–644.

Ge, S., Liu, D., Chu, M., Liu, X., Wei, Y., Che, X., ... & Xu, J. (2022). Dynamic and adaptive membrane lipid remodeling in leaves of sorghum under salt stress. *The Crop Journal*, *10*(6), 1557–1569.

Gebremedhin, A., Li, Y., Shunmugam, A. S., Sudheesh, S., Valipour-Kahrood, H., Hayden, M. J., ... & Kaur, S. (2024). Genomic selection for target traits in the Australian lentil breeding program. *Frontiers in Plant Science*, *14*, 1284781.

Gill, T., Gill, S. K., Saini, D. K., Chopra, Y., de Koff, J. P., & Sandhu, K. S. (2022). A comprehensive review of high throughput phenotyping and machine learning for plant stress phenotyping. *Phenomics*, *2*(3), 156–183.

Guo, Q., Li, X., Niu, L., Jameson, P. E., & Zhou, W. (2021). Transcription-associated metabolomic adjustments in maize occur during combined drought and cold stress. *Plant Physiology*, *186*(1), 677–695.

Hamza, N. B., Sharma, N., Tripathi, A., & Sanan-Mishra, N. (2016). MicroRNA expression profiles in response to drought stress in *Sorghum bicolor*. *Gene Expression Patterns*, *20*(2), 88–98.

Haworth, M., Marino, G., Atzori, G., Fabbri, A., Daccache, A., Killi, D., ... & Centritto, M. (2023). Plant physiological analysis to overcome limitations to plant phenotyping. *Plants*, *12*(23), 4015.

Hegde, V., Pradhan, A., Rathod, T., Tayade, A., & Rane, J. (2024). Application of thermal imaging for assessing desiccation stress memory in sugarcane and sorghum cultivars. *Sugar Tech*, *26*(2), 529–542.

Hostetler, A. N., Govindarajulu, R., & Hawkins, J. S. (2021). QTL mapping in an interspecific sorghum population uncovers candidate regulators of salinity tolerance. *Plant Stress*, *2*, 100024.

Hunt, C. H., van Eeuwijk, F. A., Mace, E. S., Hayes, B. J., & Jordan, D. R. (2018). Development of genomic prediction in sorghum. *Crop Science*, *58*(2), 690–700.

Izadi-Darbandi, A., Alameldin, H., Namjoo, N., & Ahmad, K. (2023). Introducing sorghum DREB2 gene in maize (*Zea mays* L.) to improve drought and salinity tolerance. *Biotechnology and Applied Biochemistry*, *70*(4), 1480–1488.

Jin, X., Long, Y., Xiong, S., Yang, Z., Chen, W., Hawar, A., ... & Sun, B. (2021). SbNAC2 enhances abiotic stress tolerance by upregulating ROS scavenging activities and inducing stress-response genes in sorghum. *Environmental and Experimental Botany*, *192*, 104664.

Johnson, S. M., Lim, F. L., Finkler, A., Fromm, H., Slabas, A. R., & Knight, M. R. (2014). Transcriptomic analysis of *Sorghum bicolor* responding to combined heat and drought stress. *BMC Genomics*, *15*, 1–19.

Jones, H. G. (1999). Use of thermography for quantitative studies of spatial and temporal variation of stomatal conductance over leaf surfaces. *Plant, Cell & Environment*, *22*(9), 1043–1055.

Kazungu, F. K., Muindi, E. M., & Mulinge, J. M. (2023). Overview of sorghum (*Sorghum bicolor.* L), its economic importance, ecological requirements and production constraints in Kenya. *International Journal of Plant & Soil Science*, *35*(1), 62–71.

Kim, Y. U., Webber, H., Adiku, S. G., Júnior, R. D. S. N., Deswarte, J. C., Asseng, S., & Ewert, F. (2024). Mechanisms and modelling approaches for excessive rainfall stress on cereals: Waterlogging, submergence, lodging, pests and diseases. *Agricultural and Forest Meteorology*, *344*, 109819.

Knoll, J., & Ejeta, G. (2008). Marker-assisted selection for early-season cold tolerance in sorghum: QTL validation across populations and environments. *Theoretical and Applied Genetics*, *116*, 541–553.

Kumar, S., Singh, A., Bist, C. M. S., & Sharma, M. (2024). Advancements in genetic techniques and functional genomics for enhancing crop traits and agricultural sustainability. *Briefings in Functional Genomics*, elae017, 23(5), 607–623.

Kundu, S., Saini, D. K., Meena, R. K., Bahuguna, R. N., & Jagadish, S. V. K. (2024). High-throughput phenotyping and AI technologies for deciphering crop resilience to heat stress. *Plant Physiology Reports*, 29, 699–715.

Li, H., Han, X., Liu, X., Zhou, M., Ren, W., Zhao, B., ... & Zhao, J. (2019). A leucine-rich repeat-receptor-like kinase gene *SbER2–1* from sorghum (*Sorghum bicolor* L.) confers drought tolerance in maize. *BMC Genomics*, *20*, 1–15.

Li, H., Li, Y., Ke, Q., Kwak, S. S., Zhang, S., & Deng, X. (2020). Physiological and differential proteomic analyses of imitation drought stress response in *Sorghum bicolor* root at the seedling stage. *International Journal of Molecular Sciences*, *21*(23), 9174.

Li, X., Liu, W., Wang, G., Sun, S. S. M., Yuan, L., & Wang, J. (2024). Improving digestibility of sorghum proteins by CRISPR/Cas9-based genome editing. *Food and Energy Security*, *13*(1), e506.

Liedtke, J. D., Hunt, C. H., George-Jaeggli, B., Laws, K., Watson, J., Potgieter, A. B., ... & Jordan, D. R. (2020). High-throughput phenotyping of dynamic canopy traits associated with stay-green in grain sorghum. Plant Phenomics, 2020, Article ID: 4635153.

Liu, F., Baye, W., Zhao, K., Tang, S., Xie, Q., & Xie, P. (2024a). Unravelling sorghum functional genomics and molecular breeding: Past achievements and future prospects. *Journal of Genetics and Genomics*.

Liu, J., Wang, X., Wu, H., Zhu, Y., Ahmad, I., Dong, G., ... & Wu, Y. (2024b). Association between reactive oxygen species, transcription factors, and candidate genes in drought-resistant sorghum. *International Journal of Molecular Sciences*, *25*(12), 6464.

Liu, J., Wu, Y., Dong, G., Zhu, G., & Zhou, G. (2023). Progress of research on the physiology and molecular regulation of sorghum growth under salt stress by gibberellin. *International Journal of Molecular Sciences*, *24*(7), 6777.

López, R., Ramírez-Valiente, J. A., & Pita, P. (2022). How plants cope with heatwaves in a drier environment. *Flora*, *295*, 152148.

Lynch, J. P. (2013). Steep, cheap and deep: An ideotype to optimize water and N acquisition by maize root systems. *Annals of Botany*, *112*(2), 347–357.

Mace, E. S., Singh, V., Van Oosterom, E. J., Hammer, G. L., Hunt, C. H., & Jordan, D. R. (2012). QTL for nodal root angle in sorghum (*Sorghum bicolor* L. Moench) co-locate with QTL for traits associated with drought adaptation. *Theoretical and Applied Genetics*, *124*, 97–109.

Mansour, M. M. F., Emam, M. M., Salama, K. H. A., & Morsy, A. A. (2021). Sorghum under saline conditions: Responses, tolerance mechanisms, and management strategies. *Planta*, *254*, 1–38.

Mawouma, S., Condurache, N. N., Turturica, M., Constantin, O. E., Croitoru, C., & Rapeanu, G. (2022). Chemical composition and antioxidant profile of sorghum (*Sorghum bicolor* (L.) moench) and pearl millet (*Pennisetum glaucum* (L.) R. Br.) grains cultivated in the far-North region of Cameroon. *Foods*, *11*(14), 2026.

Mishra, P., Asaari, M. S. M., Herrero-Langreo, A., Lohumi, S., Diezma, B., & Scheunders, P. (2017). Close range hyperspectral imaging of plants: A review. *Biosystems Engineering*, *164*, 49–67.

Moghimi, N., Desai, J. S., Bheemanahalli, R., Impa, S. M., Vennapusa, A. R., Sebela, D., ... & Jagadish, S. V. K. (2019). New candidate loci and marker genes on chromosome 7 for improved chilling tolerance in sorghum. *Journal of Experimental Botany*, *70*(12), 3357–3371.

Mukarram, M., Choudhary, S., Kurjak, D., Petek, A., & Khan, M. M. A. (2021). Drought: Sensing, signalling, effects and tolerance in higher plants. *Physiologia Plantarum*, *172*(2), 1291–1300.

Mulaudzi, T., Nkuna, M., Sias, G., Doumbia, I. Z., Njomo, N., & Iwuoha, E. (2022). Antioxidant capacity of chitosan on sorghum plants under salinity stress. *Agriculture*, *12*(10), 1544.

Mwamahonje, A., Eleblu, J. S. Y., Ofori, K., Feyissa, T., Deshpande, S., Garcia-Oliveira, A. L., ... & Tongoona, P. (2021). Introgression of QTLs for drought tolerance into farmers' preferred sorghum varieties. *Agriculture*, *11*(9), 883.

Nagaraju, M., Reddy, P. S., Kumar, S. A., Kumar, A., Rajasheker, G., Rao, D. M., & Kishor, P. K. (2020). Genome-wide identification and transcriptional profiling of small heat shock protein gene family under diverse abiotic stress conditions in *Sorghum bicolor* (L.). *International Journal of Biological Macromolecules*, *142*, 822–834.

Nair, S. K., & Rabbani, M. T. (2024). Marker-assisted selection approaches for improving quality traits. In: Chandra, K., Chand, S., Saini, R.P., Sharma, R. (Eds), *Smart Breeding* (pp. 287–328). New York: Apple Academic Press.

Ndlovu, E., Van Staden, J., & Maphosa, M. (2021). Morpho-physiological effects of moisture, heat and combined stresses on *Sorghum bicolor* [Moench (L.)] and its acclimation mechanisms. *Plant Stress*, *2*, 100018.

Ngcala, M. G., Goche, T., Brown, A. P., Chivasa, S., & Ngara, R. (2020). Heat stress triggers differential protein accumulation in the extracellular matrix of sorghum cell suspension cultures. *Proteomes*, *8*(4), 29.

Ortiz, D., Hu, J., & Salas Fernandez, M. G. (2017). Genetic architecture of photosynthesis in *Sorghum bicolor* under non-stress and cold stress conditions. *Journal of Experimental Botany*, *68*(16), 4545–4557.

Ostmeyer, T., Bheemanahalli, R., Srikanthan, D., Bean, S., Peiris, K. H., Madasamy, P., ... & Jagadish, S. V. K. (2020). Quantifying the agronomic performance of new grain sorghum hybrids for enhanced early-stage chilling tolerance. *Field Crops Research*, *258*, 107955.

Parra-Londono, S., Fiedler, K., Kavka, M., Samans, B., Wieckhorst, S., Zacharias, A., & Uptmoor, R. (2018). Genetic dissection of early-season cold tolerance in sorghum: Genome-wide association studies for seedling emergence and survival under field and controlled environment conditions. *Theoretical and Applied Genetics*, *131*, 581–595.

Pasini, L., Bergonti, M., Fracasso, A., Marocco, A., & Amaducci, S. (2014). Microarray analysis of differentially expressed mRNAs and miRNAs in young leaves of sorghum under dry-down conditions. *Journal of Plant Physiology*, *171*(7), 537–548.

Pask, A. (Ed.). (2012). *Physiological Breeding: A Field Guide to Wheat Phenotyping*. II. Cimmyt.

Pavli, O. I., Ghikas, D. V., Katsiotis, S., & Skaracis, G. N. (2011). Differential expression of heat shock protein genes in sorghum (*Sorghum bicolor* L.) genotypes under heat stress. *Australian Journal of Crop Science*, *5*(5), 511–515.

Pradhan, A., Aher, L., Hegde, V., Jangid, K. K., & Rane, J. (2022). Cooler canopy leverages sorghum adaptation to drought and heat stress. *Scientific Reports*, *12*(1), 4603.

Prasad, M., & Montell, D. J. (2007). Cellular and molecular mechanisms of border cell migration analyzed using time-lapse live-cell imaging. *Developmental Cell*, *12*(6), 997–1005.

Prasad, V. R., Govindaraj, M., Djanaguiraman, M., Djalovic, I., Shailani, A., Rawat, N., ... & Prasad, P. V. (2021). Drought and high temperature stress in sorghum: Physiological, genetic, and molecular insights and breeding approaches. *International Journal of Molecular Sciences*, *22*(18), 9826.

Punia, H., Tokas, J., Malik, A., Bajguz, A., El-Sheikh, M. A., & Ahmad, P. (2021). Ascorbate–glutathione oxidant scavengers, metabolome analysis and adaptation mechanisms of ion exclusion in sorghum under salt stress. *International Journal of Molecular Sciences*, *22*(24), 13249.

Rajabi Dehnavi, A., Zahedi, M., & Piernik, A. (2024). Understanding salinity stress responses in sorghum: Exploring genotype variability and salt tolerance mechanisms. *Frontiers in Plant Science*, *14*, 1296286.

Rawat, N., Singla-Pareek, S. L., & Pareek, A. (2021). Membrane dynamics during individual and combined abiotic stresses in plants and tools to study the same. *Physiologia Plantarum*, *171*(4), 653–676.

Reddy, P. S. (2019). Breeding for abiotic stress resistance in sorghum. In *Breeding Sorghum for Diverse End Uses* (pp. 325–340). Woodhead Publishing.

Reddy, P. S., Jogeswar, G., Rasineni, G. K., Maheswari, M., Reddy, A. R., Varshney, R. K., & Kishor, P. K. (2015). Proline over-accumulation alleviates salt stress and protects photosynthetic and antioxidant enzyme activities in transgenic sorghum [*Sorghum bicolor* (L.) Moench]. *Plant Physiology and Biochemistry*, *94*, 104–113.

Rutayisire, A., Lubadde, G., Mukayiranga, A., & Edema, R. (2021). Response of sorghum to cold stress at early developmental stage. *International Journal of Agronomy*, *2021*(1), 8875205.

Sanjari, S., Shobbar, Z. S., Ghanati, F., Afshari-Behbahanizadeh, S., Farajpour, M., Jokar, M., ... & Shahbazi, M. (2021). Molecular, chemical, and physiological analyses of sorghum leaf wax under post-flowering drought stress. *Plant Physiology and Biochemistry*, *159*, 383–391.

Shao, W., Zhang, J., Ge, S., Wei, Y., Li, J., & Wu, G. (2021). Physiological changes and related genes expression analysis of sorghum (*Sorghum bicolor*) seedlings under low temperature stress. *Journal of Agricultural Biotechnology*, *29*, 857–870.

Sharma, R., & Joshi, M. (2022). Vulnerability and resilience of sorghum to changing climatic conditions: Lessons from the past and hope for the future. In: Roy, S., Mathur, P., Chakraborty, A.P., Saha, S.P. (Eds), *Plant Stress: Challenges and Management in the New Decade* (pp. 169–181). Springer, Cham, Switzerland.

Shen, C., Que, Z., Xia, Y., Tang, N., Li, D., He, R., & Cao, M. (2017). Knock out of the annexin gene *OsAnn3* via CRISPR/Cas9-mediated genome editing decreased cold tolerance in rice. *Journal of Plant Biology*, *60*, 539–547.

Shomo, Z. D., Li, F., Smith, C. N., Edmonds, S. R., & Roston, R. L. (2024). From sensing to acclimation: The role of membrane lipid remodeling in plant responses to low temperatures. *Plant Physiology*, *196*(3), 1737–1757.

Sinha, D., Maurya, A. K., Abdi, G., Majeed, M., Agarwal, R., Mukherjee, R., ... & Chen, J. T. (2023). Integrated genomic selection for accelerating breeding programs of climate-smart cereals. *Genes*, *14*(7), 1484.

Somegowda, V. K., Reddy, S. D., Gaddameedi, A., Kiranmayee, K. U., Naravula, J., Kishor, P. K., & Penna, S. (2023). Genomics breeding approaches for developing *Sorghum bicolor* lines with stress resilience and other agronomic traits. *Current Plant Biology*, 37, 100314.

Song, Y., Li, J., Sui, Y., Han, G., Zhang, Y., Guo, S., & Sui, N. (2020). The sweet sorghum *SbWRKY50* is negatively involved in salt response by regulating ion homeostasis. *Plant Molecular Biology*, *102*, 603–614.

Su, M., Li, X. F., Ma, X. Y., Peng, X. J., Zhao, A. G., Cheng, L. Q., ... & Liu, G. S. (2011). Cloning two P5CS genes from bioenergy sorghum and their expression profiles under abiotic stresses and MeJA treatment. *Plant Science*, *181*(6), 652–659.

Subudhi, P. K., Rosenow, D. T., & Nguyen, H. T. (2000). Quantitative trait loci for the stay green trait in sorghum (*Sorghum bicolor* L. Moench): Consistency across genetic backgrounds and environments. *Theoretical and Applied Genetics*, *101*, 733–741.

Sudhakar Reddy, P., Srinivas Reddy, D., Sivasakthi, K., Bhatnagar-Mathur, P., Vadez, V., & Sharma, K. K. (2016). Evaluation of sorghum [*Sorghum bicolor* (L.)] reference genes in various tissues and under abiotic stress conditions for quantitative real-time PCR data normalization. *Frontiers in Plant Science*, *7*, 529.

Sui, N., Yang, Z., Liu, M., & Wang, B. (2015). Identification and transcriptomic profiling of genes involved in increasing sugar content during salt stress in sweet sorghum leaves. *BMC Genomics*, *16*, 1–18.

Sukumaran, S., Li, X., Li, X., Zhu, C., Bai, G., Perumal, R., ... & Yu, J. (2016). QTL mapping for grain yield, flowering time, and stay-green traits in sorghum with genotyping-by-sequencing markers. *Crop Science*, *56*(4), 1429–1442.

Sunoj, V. J., Somayanda, I. M., Chiluwal, A., Perumal, R., Prasad, P. V., & Jagadish, S. V. K. (2017). Resilience of pollen and post-flowering response in diverse sorghum genotypes exposed to heat stress under field conditions. *Crop Science*, *57*(3), 1658–1669.

Swami, A. K., Alam, S. I., Sengupta, N., & Sarin, R. (2011). Differential proteomic analysis of salt stress response in *Sorghum bicolor* leaves. *Environmental and Experimental Botany*, *71*(2), 321–328.

Tack, J., Lingenfelser, J., & Jagadish, S. V. K. (2017). Disaggregating sorghum yield reductions under warming scenarios exposes narrow genetic diversity in US breeding programs. *Proceedings of the National Academy of Sciences*, *114*(35), 9296–9301.

Tamaru, S., Goto, K., & Sakagami, J. I. (2023). Spatial O_2 Profile in *Coix lacryma-jobi* and *Sorghum bicolor* along the gas diffusion pathway under waterlogging conditions. *Plants*, *13*(1), 3.

Tanwar, R., Panghal, A., Chaudhary, G., Kumari, A., & Chhikara, N. (2023). Nutritional, phytochemical and functional potential of sorghum: A review. *Food Chemistry Advances*, 3, 100501.

Tariq, A., Mushtaq, M., Yaqoob, H., Bhat, B. A., Zargar, S. M., Raza, A., ... & Mir, R. A. (2023). Putting CRISPR-Cas system in action: A golden window for efficient and precise genome editing for crop improvement. *GM Crops & Food*, *14*(1), 1–27.

Tu, M., Du, C., Yu, B., Wang, G., Deng, Y., Wang, Y., ... & Li, Y. (2023). Current advances in the molecular regulation of abiotic stress tolerance in sorghum via transcriptomic, proteomic, and metabolomic approaches. *Frontiers in Plant Science*, *14*, 1147328.

Velazco, J. G., Jordan, D. R., Hunt, C. H., Mace, E. S., & van Eeuwijk, F. A. (2020). Genomic prediction for broad and specific adaptation in sorghum accommodating differential variances of SNP effects. *Crop Science*, *60*(5), 2328–2342.

Velazco, J. G., Jordan, D. R., Mace, E. S., Hunt, C. H., Malosetti, M., & Van Eeuwijk, F. A. (2019a). Genomic prediction of grain yield and drought-adaptation capacity in sorghum is enhanced by multi-trait analysis. *Frontiers in Plant Science*, *10*, 997.

Velazco, J. G., Malosetti, M., Hunt, C. H., Mace, E. S., Jordan, D. R., & Van Eeuwijk, F. A. (2019b). Combining pedigree and genomic information to improve prediction quality: An example in sorghum. *Theoretical and Applied Genetics*, *132*, 2055–2067.

Vera Hernández, P. F., Mendoza Onofre, L. E., & Rosas Cárdenas, F. D. F. (2023). Responses of sorghum to cold stress: A review focused on molecular breeding. *Frontiers in Plant Science*, *14*, 1124335.

Wang, H., Chen, G., Zhang, H., Liu, B., Yang, Y., Qin, L., ... & Guan, Y. (2014). Identification of QTLs for salt tolerance at germination and seedling stage of *Sorghum bicolor* L. Moench. *Euphytica*, *196*, 117–127.

Wang, H., Wang, R., Liu, B., Yang, Y., Qin, L., Chen, E., ... & Guan, Y. (2020). QTL analysis of salt tolerance in *Sorghum bicolor* during whole-plant growth stages. *Plant Breeding*, *139*(3), 455–465.

Wang, Z., & Wei, Y. (2022). Physiological and transcriptomic analysis of antioxidant mechanisms in sweet sorghum seedling leaves in response to single and combined drought and salinity stress. *Journal of Plant Interactions*, *17*(1), 1006–1016.

Wang, Z., Wei, Y., Zhao, Y., Wang, Y., Zou, F., Huang, S., ... & Hu, H. (2022). Physiological and transcriptional evaluation of sweet sorghum seedlings in response to single and combined drought and salinity stress. *South African Journal of Botany*, *146*, 459–471.

Wang, Z., Yung, W. S., Gao, Y., Huang, C., Zhao, X., Chen, Y., ... & Lam, H. M. (2024). From phenotyping to genetic mapping: Identifying water-stress adaptations in legume root traits. *BMC Plant Biology*, *24*(1), 749.

Weldemichael, M. Y., Gebremedhn, H. M., & Teklu, T. H. (2024). Advances in genome editing and future prospects for Sorghum improvement: A review. *Plant Gene*, 39, 100464.

Woldesemayat, A. A., Modise, D. M., Gemeildien, J., Ndimba, B. K., & Christoffels, A. (2018). Cross-species multiple environmental stress responses: An integrated approach to identify candidate genes for multiple stress tolerance in sorghum (*Sorghum bicolor* (L.) Moench) and related model species. *PLoS One*, *13*(3), e0192678.

Wu, Y., Guo, T., Mu, Q., Wang, J., Li, X., Wu, Y., ... & Li, X. (2019). Allelochemicals targeted to balance competing selections in African agroecosystems. *Nature Plants*, *5*(12), 1229–1236.

Xie, P., Shi, J., Tang, S., Chen, C., Khan, A., Zhang, F., ... & Xie, Q. (2019). Control of bird feeding behavior by Tannin1 through modulating the biosynthesis of polyphenols and fatty acid-derived volatiles in sorghum. *Molecular Plant*, *12*(10), 1315–1324.

Xuewen, X., Huihui, W., Xiaohua, Q., Qiang, X., & Xuehao, C. (2014). Waterlogging-induced increase in fermentation and related gene expression in the root of cucumber (*Cucumis sativus* L.). *Scientia Horticulturae*, *179*, 388–395.

Yahaya, M. A., & Shimelis, H. (2022). Drought stress in sorghum: Mitigation strategies, breeding methods and technologies—A review. *Journal of Agronomy and Crop Science*, *208*(2), 127–142.

Yang, Z., Wang, Y., Wei, X., Zhao, X., Wang, B., & Sui, N. (2017). Transcription profiles of genes related to hormonal regulations under salt stress in sweet sorghum. *Plant Molecular Biology Reporter*, *35*, 586–599.

Yang, Z., Zheng, H., Wei, X., Song, J., Wang, B., & Sui, N. (2018). Transcriptome analysis of sweet Sorghum inbred lines differing in salt tolerance provides novel insights into salt exclusion by roots. *Plant and Soil*, *430*, 423–439.

Zhang, H., Zhu, J., Gong, Z., & Zhu, J. K. (2022). Abiotic stress responses in plants. *Nature Reviews Genetics*, *23*(2), 104–119.

Zhang, R., Chang, J., Chen, H., Li, J., Huang, K., Deng, Z., ... & Feng, J. (2024). Transcriptomic insights into drought survival strategy of *Sorghum bicolor* (l.) Moench during early growth under polyethylene glycol-simulated conditions. *Agronomy*, *14*(10), 2364.

Zhang, R., Yue, Z., Chen, X., Huang, R., Zhou, Y., & Cao, X. (2023). Effects of waterlogging at different growth stages on the photosynthetic characteristics and grain yield of sorghum (*Sorghum bicolor* L.). *Scientific Reports*, *13*(1), 7212.

Zheng, H., Dang, Y., Diao, X., & Sui, N. (2024). Molecular mechanisms of stress resistance in sorghum: Implications for crop improvement strategies. *Journal of Integrative Agriculture*, *23*(3), 741–768.

Zheng, H., Sun, X., Li, J., Song, Y., Song, J., Wang, F., ... & Sui, N. (2021). Analysis of N6-methyladenosine reveals a new important mechanism regulating the salt tolerance of sweet sorghum. *Plant Science*, *304*, 110801.

Zhu, L., Cernusak, L. A., & Song, X. (2020). Dynamic responses of gas exchange and photochemistry to heat interference during drought in wheat and sorghum. *Functional Plant Biology*, *47*(7), 611–627.

11 Molecular Engineering of Sorghum Root Phenes for Increased Abiotic Stress Resilience

Kirti Bardhan, Mayank Pratap Singh Bangari, Impa Somayanda, and S. V. Krishna Jagadish

11.1 INTRODUCTION

Sorghum (*Sorghum bicolor* (L.) Moench) is a cornerstone of global food security, particularly amid climate change and the growing demand for sustainable agriculture. As the fifth most important cereal crop, sorghum serves as a staple food, animal feed, and biofuel source. Its resilience to drought and heat makes it indispensable in semi-arid regions, sustaining livelihoods of millions in Africa and Asia where other crops often fail (Takanashi, 2023; Wei, 2024; Romana et al., 2018). This adaptability, combined with its ability to thrive in resource-limited conditions like low soil fertility and water scarcity, positions sorghum as a vital crop for regions vulnerable to climate variability and food shortages (Sopandie & Wirnas, 2023; Nyoni et al., 2020). Sorghum also enhances nutritional security with its rich profile of proteins, dietary fiber, antioxidants, and phenolic compounds, offering health benefits such as protection against chronic diseases (Stefoska-Needham, 2024; Htet et al., 2022; Xiong et al., 2019). Its role in combating malnutrition and improving dietary diversity is especially significant in food-insecure areas (Istrati et al., 2019; Ratnavathi & Patil, 2014). Beyond human nutrition, sorghum supports livestock feed systems, boosts agricultural economies, and contributes to renewable energy production, enhancing its relevance in sustainable food and energy systems (Ghorai, 2022). Overall, sorghum's versatility and resilience makes it a key crop for enhancing agricultural sustainability, strengthening food security, and addressing the challenges posed by climate change (Pal, 2024; Ghorai, 2022). The economic viability of sorghum is increasingly recognized, particularly in developing countries where it supports smallholder farmers as a vital income source (Widodo et al., 2023; Mbulwe & Ajayi, 2020; Meleaku et al., 2020). Promoting sorghum as a cash crop can enhance rural livelihoods, strengthen local economies, and bolster food security.

Despite its resilience, sorghum cultivation is constrained by soil stresses such as drought, salinity, and nutrient deficiencies. Drought, the most significant abiotic stress in arid and semi-arid regions, disrupts photosynthesis, transpiration, and nutrient uptake, leading to severe yield reductions of sorghum, despite the crop's physiological adaptations including osmotic adjustment and stress-related protein production (Maharajan et al., 2021; Tari et al., 2012). Salinity further limits productivity by causing ion toxicity and osmotic stress, impairing the uptake of critical nutrients like potassium and calcium, which exacerbates growth and yield losses (da Silva, 2023). Similarly, nutrient deficiencies, particularly nitrogen, phosphorus, and potassium, stunt growth, reduce biomass, and weaken resistance to pests and diseases. Micronutrient shortages, such as iron and zinc in alkaline soils, also

DOI: 10.1201/9781003623717-11

reduce crop quality (Teixeira et al., 2020; Tari et al., 2012). Addressing these soil stresses is essential for unlocking sorghum's full potential as a resilient crop. Increasing its adaptability to drought and poor soil conditions ensures profitability and food security in regions where other crops fail (Azu et al., 2021; Hadebe et al., 2017). Understanding and improving root architecture regulation under stress conditions is critical to sustaining sorghum productivity in challenging environments.

Root phenes are specific traits of root systems that influence plant growth, development, and resource acquisition. These traits include aspects of root morphology, anatomy, and physiology, and can be classified based on their roles in resource acquisition. For example, traits like root length and branching facilitate soil exploration, while root diameter and density enhance resource exploitation (Lynch et al., 2014; York et al., 2013). Specialized anatomical features, such as aerenchyma, can improve a plant's ability to access water from drying soils (Lynch et al., 2014; Forde, 2009). The interaction between root phenes and environmental factors leads to variations in root architecture, influencing sorghum resilience to stress (Jung & McCouch, 2013; Singh et al., 2010). Identifying these traits is crucial for crop improvement, particularly under environmental stresses like drought and nutrient deficiency (Lynch et al., 2014; Lynch & Brown, 2012).

Root system architecture (RSA), defined by the depth, spread, and branching of roots, is shaped by root phenes and optimizes water and nutrient uptake (Jung & McCouch, 2013; Rich & Watt, 2013). RSA development is driven by genetic, hormonal, and environmental factors. For example, nutrient availability, especially phosphorus, can alter root architecture, prompting changes in root length and branching density to enhance nutrient acquisition (Niu et al., 2013; Williamson et al., 2001). Genetic control of root traits shows plasticity, allowing adaptation to soil conditions (Schneider et al., 2020). Understanding these traits is vital for improving crop yields in nutrient-poor soils (Schneider et al., 2020; Wang et al., 2017; Lynch & Brown, 2012). Advanced imaging and modeling techniques have enhanced the study of root phenes, enabling detailed analysis of root systems across environments. High-throughput phenotyping allows the dynamic visualization and quantification of root architecture, providing insights into how different phenes work together to optimize plant performance (Fernandez et al., 2022; Rangarajan & Lynch, 2021; Leitner et al., 2014).

Sorghum's ability to adapt its RSA is crucial for its performance under adverse conditions, such as drought and salinity, where water and nutrient acquisition are often compromised. A deeper root system, for example, enhances drought tolerance by allowing sorghum to access moisture from deeper soil layers during water scarce conditions (Gowsiga et al., 2021; Chen et al., 2020). Changes in sorghum root morphology, such as increased root length and density, improve resource exploitation (Djanaguiraman et al., 2024; Akman et al., 2021). The regulation of sorghum root growth under soil moisture stress is mediated by hormonal signals, particularly abscisic acid (ABA), which plays a key role in drought stress responses (Muthego, 2023; Dugas et al., 2011). Additionally, sorghum nodal root angle impacts root spatial distribution, influencing drought adaptation (Joshi et al., 2017; Mace et al., 2012). Quantitative trait loci (QTL) associated with nodal root angle could be targeted in breeding programs to enhance drought resilience (Rajkumar et al., 2013). The interaction between root architecture and rhizosphere microbial communities further enhances nutrient uptake and stress tolerance, with beneficial microbes improving sorghum performance under drought and salinity (Qi et al., 2022; Balasubramanian et al., 2021).

Sorghum roots also respond to drought stress through complex biochemical pathways, including the accumulation of osmolytes like proline and activation of stress-responsive genes that help maintain cellular integrity (Jin et al., 2023; Govindasamy et al., 2020). The root proteome undergoes significant changes under drought, underscoring the role of protein networks in stress tolerance (Goche et al., 2020). Sorghum root architecture also plays a vital role in salinity tolerance by minimizing salt uptake and maximizing water and nutrient absorption (Yukun et al., 2021). Sorghum cultivars with enhanced root plasticity adapt more effectively to varying salinity levels (Avci et al., 2017). Root plasticity also improves growth under nutrient-limiting conditions (Liu, 2023). Molecular breeding and genetic engineering offer promising approaches for enhancing sorghum's drought and salinity tolerance through the identification of key genes and pathways involved in root architecture and stress responses (Maina et al., 2022; Woldesemayat et al., 2017). This chapter focuses on understanding the mechanisms governing sorghum root responses to environmental stresses and their impact on aboveground growth, with the goal of developing strategies to enhance sorghum resilience and productivity in the face of climate change.

11.2 SORGHUM ROOT SYSTEM

In grain sorghum, root systems develop early, with the majority of roots concentrated in the top 40 cm of soil, although some can extend to depths of up to 80 cm. Root growth continues in the topsoil (0–20 cm) until anthesis, after which it largely ceases (Myers, 1980). However, under favorable conditions, sorghum roots can penetrate depths of 150–300 cm (Borg & Grimes, 1986), with elongation rates reaching 2–3 cm per day (Singh et al., 2010). Root depth is influenced by soil conditions, moisture availability, and management practices. For instance, under irrigated conditions, sorghum's roots quantity are maximum within the top 15 cm, while under dryland conditions, they extend deeper (Kaigama et al., 1977). Root length density increases from the topsoil to the bottom of the "A" horizon (approximately 30 cm), then decreases in deeper soil layers (Vegapareddy et al., 2010).

Sorghum's root development begins with a primary root after germination, followed by lateral roots within a week. The seminal root system, primarily vertical, lacks scutellum roots, which are found in other cereals. Nodal roots emerge at the 4–5 leaf stage and continue to develop with each leaf flush, becoming thicker and more vertically oriented with plant age. This strategy of sorghum prioritizes vertical root growth to access deeper soil moisture, especially in drought-prone areas (Singh et al., 2010). Root growth follows a sigmoidal pattern, with root density increasing until grain filling and then declining. Senescent hybrids experience greater root mass reduction compared to non-senescent sorghum hybrids (Zartman & Woyewodzic, 1979), with about two-thirds of the cumulative root length dying by heading (Blum & Arkin, 1984).

Sorghum exhibits phenotypic plasticity in response to soil and agronomic conditions. Root regrowth and death occur throughout its life cycle, influencing root mass at different growth stages. These plastic responses can enhance stress tolerance, as seen when soil compaction reduces the fine root production, minimizing carbon costs which are less efficient in nutrient and water uptake under compacted conditions (Correa et al., 2022; Correa et al., 2019). Compared to corn, sorghum's fibrous root system, with more secondary roots per unit length of primary roots, enhances water extraction efficiency (Weaver, 1926). Combined with its reduced transpiring surface, sorghum's root system provides superior drought tolerance, especially in early maturing hybrids with higher root length density and reduced leaf area (Sinclair & Muchow, 2001). Sorghum's thinner, higher-order lateral roots also improve nutrient acquisition, making it more efficient than other cereals (Rao & Ito, 1998).

11.3 ROOT RESPONSES UNDER ABIOTIC STRESS AND WHOLE PLANT TOLERANCE

Roots play a crucial role in sensing the soil condition and stresses like drought, nutrient imbalance, and salinity, besides anchorage, and root responses to these stresses and signaling modulate downstream aboveground responses which determine crop growth and productivity. Thus, understanding these root responses and signaling is crucial in the development of stress tolerant crops.

Salinity affects sorghum growth by imposing osmotic stress and ion toxicity, which can severely impair root development and overall plant health (Rajabi Dehnavi et al., 2022; Mbinda & Kimtai, 2019). In response to salinity, sorghum roots undergo a complex set of biochemical pathways and gene expression changes. One of the primary responses is the alteration of ion homeostasis, particularly in the uptake and transport of potassium (K^+) and sodium (Na^+) ions. Key transport proteins, such as ammonium transporters (AMTs) and nitrate transporters (NRTs), play critical roles in maintaining nutrient balance under saline conditions (Wang et al., 2023; Guo et al., 2023). For instance, the upregulation of genes like *SOS1* and *HKT1*, which are involved in Na^+ exclusion and K^+ retention, enhances the sorghum's ability to cope with high salt concentrations (Guo et al., 2023).

In addition to ion regulation, hormonal responses are crucial in sorghum's salinity tolerance. Hormones such as salicylic acid (SA) and gibberellic acid (GA) modulate the levels of cytokinin (CK) and auxins (IAA), influencing root growth and stress responses (Dehnavi et al., 2022; Forghani et al., 2018). For example, SA application has been shown to promote root and shoot growth under saline conditions, indicating the importance of hormonal signaling in the sorghum response to salinity (Rajabi Dehnavi et al., 2022). Morphological adaptations in the root system also play a significant role in mitigating salinity stress. Some sorghum genotypes exhibit deeper and more extensive root systems, enhancing their ability to access water and nutrients in saline environments (Mbinda & Kimtai, 2019). This root plasticity helps sorghum better withstand salinity by optimizing resource uptake and maintaining physiological functions. Moreover, the interaction between sorghum roots and soil microorganisms contributes to salinity tolerance. Beneficial root-associated bacteria can enhance nutrient uptake and improve the plant's resilience to salinity stress, playing a vital role in sorghum's adaptability in saline soils (Yukun et al., 2021).

Drought stress induces significant changes in root distribution, development, and hormonal balance in sorghum, which collectively influence the above-ground physiological responses. Root architecture and response undergo adaptations that enhance water uptake and overall plant resilience. Two key responses observed in plant roots under drought conditions are hydropatterning and xerobranching, both of which optimize the root system for efficient water acquisition in heterogeneous soil environments. Hydropatterning refers to the ability of roots to modify their growth patterns in response to localized water availability in the soil. When roots encounter moist areas, they increase branching and growth in those regions, optimizing water uptake (Orosa-Puente et al., 2018). This mechanism is particularly crucial in drought conditions, where water is unevenly distributed in the soil. Hydropatterning is regulated by various signaling pathways, including auxins. The auxin response factor *ARF7* has been identified as a key regulator in this process, facilitating root branching when roots encounter water (Orosa-Puente et al., 2018). This adaptive response allows plants to exploit available moisture more effectively, enhancing survival during periods of water deficit. Additionally, the hormone abscisic acid (ABA) plays a significant role in hydropatterning. Increased ABA levels under drought stress lead to changes in root architecture that prioritize water uptake (Rosales et al., 2019). The coordination of these hormonal signals enables plants to dynamically adjust their root systems to access water more efficiently.

Xerobranching, an adaptive modification in root architecture, enhances drought tolerance by optimizing water absorption in arid conditions. Under water stress, plants reduce overall root growth while increasing the density of lateral roots and root hairs, maximizing surface area for water uptake (Orman-Ligeza et al., 2018). Additionally, xerobranching involves thicker root tissues and increased root suberization, minimizing water loss through a reverse water potential gradient. In the Poaceae family, water deficit suppresses crown root growth, preventing excessive water extraction from dry soil. However, as moisture levels rise, new crown roots emerge, allowing root growth to resume (Sebastian et al., 2016). Sorghum's roots can detect changes in soil moisture availability through a combination of hydraulic and chemical signaling (Gowsiga et al., 2021; Chen et al., 2020). When soil moisture decreases, roots sense the changes in osmotic potential, triggering the production of stress-related hormones, particularly abscisic acid (ABA) (Muthego et al., 2023; Tiwari et al., 2017). Under drought conditions, sorghum roots undergo notable anatomical and morphological adaptations. A key adaptation is the development of a deeper root system, which enables the sorghum to access moisture from deeper soil layers that retain water for longer periods compared to surface layers (Gowsiga et al., 2021; Chen et al., 2020). This deeper rooting not only enhances water uptake but also improves nutrient absorption, which is critical for maintaining growth during water deficits. Drought-tolerant sorghum varieties typically exhibit longer and more extensive root systems than drought-sensitive varieties, enabling them to forage for water more effectively (Schittenhelm & Schroetter, 2014; Sankarapandian et al., 2013).

Furthermore, drought stress leads to an increase in the density of lateral roots and root hairs, enhancing the root surface area available for water and nutrient absorption. This morphological plasticity is an important adaptive response, allowing sorghum to optimize resource acquisition under limited water availability. Consequently, the root architecture shifts to a more fibrous structure, improving the sorghum's ability to stabilize in dry soils and enhancing its overall resilience to drought (Gowsiga et al., 2021; Chen et al., 2020).

ABA plays a crucial role in the plant's response to water deficit. Under drought conditions, ABA levels increase in the sorghum's roots, leading to stomatal closure in the leaves to reduce transpiration and conserve water (Muthego, 2023; Tiwari et al., 2017). This hormonal signaling is essential for coordinating the sorghum's physiological responses to drought, minimizing water loss while maintaining vital metabolic functions. In addition, other hormones such as auxins and cytokinins also exhibit altered levels during drought stress. Auxins are involved in root growth and development, and their balance influences root architecture. Elevated auxin levels in response to drought can promote lateral root formation, enhancing the sorghum ability to explore the soil for moisture (Ali et al., 2023). Conversely, cytokinins, which are typically associated with promoting shoot growth, may decrease under drought conditions, resulting in reduced above-ground growth to prioritize root development (Behera et al., 2022; Tiwari et al., 2017).

Changes in root structure directly influence the above-ground physiological responses of sorghum. The enhanced root system improves water uptake, which is essential for maintaining turgor pressure, a critical factor for sustaining photosynthesis and overall sorghum health during drought (Gowsiga et al., 2021; Chen et al., 2020). Drought-stressed sorghum plants often exhibit reduced leaf area and stomatal conductance, as an adaptive strategy, that help minimize water loss (Behera et al., 2022; Gowsiga et al., 2021; Chen et al., 2020). Increased ABA levels can trigger earlier leaf senescence, a mechanism that conserves resources during prolonged drought (Muthego, 2023; Tiwari et al., 2017). However, this trade-off may lead to reduced grain yield, particularly when drought coincides with critical growth stages of sorghum such as flowering and grain filling (Behera et al., 2022; Gowsiga et al., 2021; Chen et al., 2020). Moreover, beneficial microbes associated with sorghum roots can enhance drought tolerance by improving nutrient uptake and promoting root growth (Caddell et al., 2020; Carlson et al., 2020). These microbial interactions can further modulate the plant's hormonal balance and stress responses, contributing to improved drought resilience.

The availability or deficiency of nutrients plays a significant role in shaping the root architecture of sorghum plants. Nutrient levels directly influence root growth patterns, such as root length, density, and branching (Gano et al., 2021; Tari et al., 2012). For instance, under nitrogen deficiency, sorghum roots tend to elongate more to explore a larger soil volume in search of nutrients, whereas phosphorus deficiency typically results in a more extensive root system with increased lateral root development to enhance nutrient uptake (Teixeira et al., 2020; Tari et al., 2012). In contrast, potassium deficiency can reduce root biomass, compromising the sorghum's stability and nutrient absorption capacity (Teixeira et al., 2020; Tari et al., 2012). These physiological adaptations are critical for the plant's ability to adjust to varying nutrient levels, enabling sorghum to optimize growth and yield potential under suboptimal conditions.

To cope with nutrient deficiencies, sorghum employs several adaptive strategies, particularly in its root system. One such strategy is the formation of mycorrhizal associations, where sorghum roots form symbiotic relationships with arbuscular mycorrhizal fungi (AMF), enhancing nutrient uptake, especially phosphorus, and improving plant resilience (Idris et al., 2023; Gano et al., 2021). Additionally, sorghum roots can adjust their architecture in response to nutrient availability. For example, under nitrogen-limited conditions, sorghum may develop deeper roots to access moisture and nitrogen from lower soil layers (Gano et al., 2021; Tari et al., 2012). Moreover, sorghum has been shown to increase root exudation of organic acids, which help mobilize soil nutrients, particularly in nutrient-deficient soils (Gano et al., 2021; Teixeira et al., 2020). These adaptive mechanisms are essential for sustaining sorghum productivity in challenging environments, allowing it to thrive where other crops may struggle. Furthermore, the interactions between roots and soil microbes are pivotal for sorghum health and productivity, particularly under abiotic stress conditions such as drought, salinity, and nutrient imbalances. These stresses can significantly alter the composition and functionality of the root microbiome, which in turn impacts plant growth and resilience. A deeper understanding of these dynamics is crucial for developing strategies to enhance crop performance, particularly in crops like sorghum cultivated in challenging environments.

Abiotic stresses, such as drought, salinity, and nutrient imbalance, can significantly impact the microbial communities in the rhizosphere, the region of soil influenced by root secretions. Under drought conditions, plants often modify their root exudate profiles, which can alter the composition of microbial communities. For example, drought stress has been found to enrich certain beneficial bacteria that enhance sorghum stress tolerance (Qi et al., 2022; Hartman & Tringe, 2019). These bacteria can promote root growth and improve nutrient uptake, assisting the plant in coping with water scarcity. In contrast, high salinity levels typically lead to a reduction in microbial diversity, as many soil microbes are sensitive to increased salt concentrations (Zakavi et al., 2022; Gamalero et al., 2020). Nevertheless, some halotolerant microbes can thrive under saline conditions and may provide benefits to the host plant, such as improving nutrient availability and promoting osmotic adjustment (Zakavi et al., 2022; Gamalero et al., 2020). For instance, specific strains of *Bacillus* have been identified as beneficial under saline stress, enhancing maize growth and nutrient uptake (Zakavi et al., 2022).

The interactions between roots and rhizosphere microbes play a crucial role in plant growth and development. Beneficial microbes can enhance nutrient uptake, improve water retention, and promote root growth, all of which are essential for plant survival under abiotic stress. For instance, plant growth-promoting rhizobacteria (PGPR) can stimulate root elongation and branching, enabling plants to explore a larger soil volume for water and nutrients (Khan et al., 2021; Chandran et al., 2021; Pattnaik et al., 2021). This dynamic interaction not only supports plant growth but also strengthens the overall health of the soil ecosystem. In sorghum, studies have shown that drought stress leads to significant changes in root metabolite profiles and microbial communities, both of which are vital for the plant's stress response (Qi et al., 2022; Caddell et al., 2020). Under drought conditions, sorghum roots can selectively recruit beneficial microbes, such as *Actinobacteria* which enhance stress tolerance and nutrient uptake (Balasubramanian et al., 2021).

11.4 HORMONAL AND ENVIRONMENTAL REGULATION OF ROOT PHENES AND ARCHITECTURE

Plant hormones, or phytohormones, are essential regulators of root architecture, influencing various aspects of root growth and development. These hormones, which include auxins, cytokinins, ABA, strigolactones, and ethylene, interact in complex ways to modulate RSA in response to environmental cues and stress conditions. Auxins are perhaps the most widely recognized plant hormones involved in root architecture. Synthesized primarily in the shoot and transported to the root, auxins play a crucial role in regulating cell elongation and division. They promote lateral root formation by stimulating the pericycle cells, which are located just inside the root's vascular tissue (Sánchez-Bermúdez et al., 2022; Hartman & Tringe, 2019). The concentration and distribution of auxin within the root system are key factors in this process; higher auxin levels in specific regions can enhance lateral root formation, while lower concentrations can inhibit it (Neelam & Tabassum, 2023; Khan et al., 2021). Auxin influences the initiation of lateral roots by modulating gene expression and cellular responses in root tissues. Studies have demonstrated that auxin concentration gradients within the root system can dictate the spatial patterning of lateral root emergence, enabling plants to adapt their root architecture to localized soil conditions (Bellande et al., 2022; Cavallari et al., 2021; Dastidar et al., 2012). Moreover, auxin interacts with other hormones, such as ABA and cytokinins, to fine-tune root responses to environmental stresses (Meng et al., 2019; Harris, 2015; Hong et al., 2013).

The mechanism by which auxin influences lateral root development involves the activation of specific genes associated with cell division and differentiation. Auxin response factors (ARFs) are transcription factors that mediate auxin's effects on gene expression related to root development. *ARF7* and *ARF19*, in particular, play vital roles in lateral root initiation by regulating the expression of genes that promote cell division in the pericycle (Omae & Tsuda, 2022). Additionally, auxin transport proteins, such as PIN proteins, facilitate the directional flow of auxin within the root, ensuring its delivery to appropriate tissues to promote growth (Sánchez-Bermúdez et al., 2022; Hartman & Tringe, 2019). Studies have indicated that soil properties, such as compaction and texture, can affect auxin distribution within the root system, influencing lateral root formation and overall root architecture (Bellande et al., 2022; Guyomarc'h et al., 2010; McCleery et al., 2017). Furthermore, interactions between roots and soil microorganisms can alter root growth dynamics, as beneficial microbes enhance nutrient availability and influence root morphology (Ding & Bai, 2021; Galindo-Castañeda et al., 2024).

Under drought conditions, sorghum exhibits an increase in auxin levels, which is associated with enhanced root growth and development. This increase is part of a broader regulatory network that includes stress-responsive genes and metabolites. For example, drought stress has been shown to trigger the accumulation of specific metabolites, such as pipecolic acid, which correlates with changes in auxin levels and root morphology (Caddell et al., 2023; Caddell et al., 2020). Enhanced auxin signaling promotes deeper root growth, allowing sorghum to access moisture from deeper soil layers, thereby improving water use efficiency (Jin et al., 2023). The physiological mechanisms underlying auxin's role in drought response involve the modulation of reactive oxygen species (ROS) and antioxidant pathways. Drought stress induces ROS accumulation, which can be mitigated by auxin-mediated signaling pathways that enhance the scavenging ability of these reactive species (Jin et al., 2023; Govindasamy et al., 2020). This interaction not only protects sorghum root cells from oxidative damage but also promotes root elongation and branching, facilitating better water and nutrient uptake (Goche et al., 2020; Wang et al., 2010). Furthermore, the expression of auxin-related genes, such as those from the *SbGH3* and *SbLBD* families, has been shown to be upregulated under drought conditions, indicating a direct link between auxin signaling and sorghum drought stress adaptation (Chen et al., 2019; Wang et al., 2010).

In addition to drought, salinity stress also affects auxin distribution in sorghum. Salinity typically alters auxin transport and distribution within the root system, leading to changes in root morphology. Under saline conditions, auxin signaling pathways are activated to promote lateral root development, helping the plant explore a larger soil volume for water and nutrients. The interaction between auxin and other hormones, such as ABA, is critical in this context, as ABA can modulate auxin transport and distribution, further influencing root growth under salinity stress. Drought conditions have been linked to increased sorghum root thickness and density, which are adaptations that enhance the plant's ability to anchor itself in the soil while maximizing water absorption (Djanaguiraman, 2024; Gowsiga et al., 2021). Similarly, salinity stress can lead to modifications in sorghum root morphology, including changes in root hair density and length, which are essential for improving nutrient uptake efficiency (Amombo et al., 2022). These morphological root adaptations are often accompanied by changes in auxin concentration, which regulates the development of root hairs and lateral roots, thus enhancing overall sorghum RSA (Ali et al., 2023).

Cytokinins significantly influence root architecture, particularly by regulating lateral root formation and root hair development. In general, high concentrations of cytokinins inhibit lateral root development, acting antagonistically to auxin (Makaza & Kamutando, 2022; Meena et al., 2017). However, when cytokinins are present at appropriate levels, they promote root hair elongation, especially under nutrient-deficient conditions, where root hair development is crucial for maximizing nutrient and water uptake (Takatsuka et al., 2024). The balance between auxin and cytokinin levels is critical for determining whether a plant will develop lateral roots or focus on elongating the primary root (Neelam & Tabassum, 2023; Khan et al., 2021). For example, in maize, a higher auxin-to-cytokinin ratio promotes lateral root development, while a lower ratio favors the elongation of the primary root (Makaza & Kamutando, 2022; Meena et al., 2017).

Soil conditions, such as nutrient availability and moisture content, influence cytokinin biosynthesis and signaling pathways. For instance, high nitrogen levels in the soil can increase cytokinin production, leading to enhanced lateral root formation and branching in sorghum (Lamb et al., 2022; Singh et al., 2015). Conversely, nitrogen deficiency can reduce cytokinin levels, inhibiting root growth and altering root morphology (Zhang et al., 2010). Drought conditions generally result in reduced cytokinin levels, causing decreased root elongation and altered root architecture (Vedenicheva et al., 2024). However, rewatering can significantly increase cytokinin levels, promoting compensatory growth and enhanced root systems (Wang et al., 2022).

Certain beneficial bacteria, such as heterotrophic ammonia-oxidizing bacteria, can enhance cytokinin production in the rhizosphere, which improves root growth and plant resilience to stress (Wang et al., 2023; Wang, 2022). These bacteria impact nitrogen and nutrient availability, further affecting cytokinin signaling pathways. Cytokinins also play a role in regulating lateral root density by acting as negative regulators at higher concentrations (Rivas et al., 2022). This regulation is particularly important in nutrient-poor soils, where optimizing root architecture for nutrient acquisition is crucial for plant survival. Under drought conditions, roots may undergo anatomical changes to enhance water retention and nutrient uptake. Cytokinin signaling pathways mediate these changes, such as increased suberization of the endodermis and modifications in the sclerenchyma layer (Li et al., 2022; Henry et al., 2012). These structural adaptations are essential for maintaining root function under stress, allowing plants to optimize resource acquisition.

Under water deficit conditions, the accumulation of ABA plays a crucial role in modifying root architecture to enhance the plant's ability to access water (Joshi, 2018; Choudhary et al., 2011). ABA promotes deeper root growth and can stimulate lateral root formation in some species, allowing plants to explore a larger soil volume for moisture (Joshi, 2018; Choudhary et al., 2011). It also influences the expression of genes related to root development and stress responses, helping plants adapt to adverse conditions. ABA modulates auxin transport and signaling pathways, which in turn affects root growth patterns and optimizes root architecture under stress (Choudhary et al., 2011; Joshi, 2018). This crosstalk between ABA and auxin is vital for optimizing root growth and architecture under water deficit conditions.

The concentration and distribution of ABA in roots are modulated by soil moisture levels, which directly affect root morphology and function. Pérez-Pérez et al. (2020) reported that partial rootzone drying increases ABA levels in tomato roots, which is crucial for regulating water uptake and maintaining plant water status during drought conditions. Sorghum plants subjected to drought exhibit increased ABA accumulation, which is linked to enhanced root growth as a compensatory mechanism to access deeper soil moisture (Gowsiga et al., 2021; Li et al., 2020). ABA also promotes root elongation, enabling the plant to explore a larger soil volume for water and nutrients (Djanaguiraman et al., 2024; Chen et al., 2020).

In addition to drought stress, salinity also affects ABA dynamics in sorghum roots. Elevated ABA levels under saline conditions help regulate sorghum root growth and morphology by modulating stomatal closure, thereby reducing water loss through transpiration and maintaining water balance (Liu et al., 2023; Yukun et al., 2021). Arbuscular mycorrhizal fungi (AMF) enhance drought tolerance in sorghum by modulating ABA levels and improving root architecture, which facilitates better water uptake (Idris et al., 2023; Govindasamy et al., 2020). Nutrient deficiencies, particularly in nitrogen and phosphorus, can also lead to increased ABA levels, influencing sorghum root growth and nutrient uptake efficiency (Chai et al., 2024; van der Bom et al., 2024). Sorghum genotypes with higher nitrogen use efficiency show a more pronounced ABA response under nutrient stress, leading to enhanced root growth and better adaptation to low-nutrient conditions (Chai et al., 2024; Sheflin et al., 2019). Beneficial microorganisms, including bacterial endophytes, can also enhance the sorghum's ability to cope with drought and salinity by influencing ABA signaling pathways (Erlandson et al., 2021; Umapathi et al., 2021).

ABA also mediates rhizosheath formation, where soil tightly adheres to roots due to root secretions, including mucilage and exudates, under mild soil drying conditions (Zhang et al., 2021). Qin et al. (2022) reported that ABA promotes auxin biosynthesis, inhibiting primary root elongation in rice, further highlighting the critical interplay between ABA and auxin in regulating root growth under stress. Li et al. (2019) indicated that ABA receptors modulate root gravitropism and lateral root formation through the regulation of auxin transport. Under heterogeneous soil drying conditions, local root ABA accumulation is influenced by the spatial distribution of soil moisture, leading to differential growth responses in roots. This localized ABA accumulation promotes deeper root growth in search of moisture, as observed by Zegada-Lizarazu & Monti (2019) in giant reed. Manzi et al. (2015) demonstrated that sustained root ABA accumulation during prolonged water stress is vital for maintaining root hydraulic conductivity and overall plant vigor. Additionally, Rowe et al. (2016) highlighted that ABA interacts with other hormones, such as ethylene and cytokinin, to regulate root growth under osmotic stress, suggesting a complex regulatory network governing root architecture in response to environmental cues.

Strigolactones are crucial regulators of root development, particularly under nutrient deficiencies. Produced in the roots, strigolactones enhance root hair formation and lateral root development, thereby increasing the root surface area for nutrient uptake (Barnes & Tringe, 2022; Boyno, 2023). Under low phosphate conditions, strigolactones levels increase, promoting root branching and hair elongation to improve phosphorus acquisition (Barnes & Tringe, 2022; Boyno et al., 2023). Additionally, strigolactones are involved in signaling pathways that regulate plant-mycorrhizal fungi interactions, which are especially beneficial in nutrient-poor soils, helping the plant access essential nutrients through symbiosis (Boyno et al., 2023; Inbaraj, 2021). Drought stress significantly influences strigolactone production in sorghum roots. Under water-limited conditions, sorghum exhibits a marked increase in strigolactone biosynthesis, promoting deeper root growth and enhancing the plant's ability to access moisture from deeper soil layers (Hao et al., 2023; Avila et al., 2020). Drought-tolerant sorghum varieties show a more pronounced increase in strigolactone levels compared to sensitive varieties, suggesting that strigolactones may play a role in the genetic adaptation of sorghum to drought stress (Hao et al., 2023; Avila et al., 2020). Furthermore, the interaction between strigolactones and other phytohormones like ABA is crucial in modulating root architecture under drought conditions, promoting root elongation and branching.

Salinity also affects strigolactone production in sorghum. Plants exposed to saline conditions exhibit altered strigolactone profiles, promoting root adaptations, such as increased lateral root formation, which enhance the sorghum's ability to absorb water and nutrients and improve its resilience to salinity stress (Kirshibayev & Mukasheva, 2022; Guimarães et al., 2020). The presence of beneficial microorganisms in the rhizosphere of sorghum can further enhance strigolactone production and root growth under saline conditions, underscoring the importance of microbial interactions in stress tolerance (Sahib et al., 2020). Nutrient deficiencies, particularly phosphorus (P), increase strigolactone exudation from sorghum roots, promoting root growth and improving nutrient uptake efficiency (Yoneyama et al., 2013; Jamil et al., 2013). Under P-deficient conditions, strigolactones stimulate the growth of mycorrhizal fungi, which help enhance phosphorus uptake (Yoneyama et al., 2013). The genetic regulation of strigolactone biosynthesis under nutrient stress involves complex signaling pathways that integrate nutrient availability with root growth responses. For example, the expression of strigolactone biosynthesis genes is upregulated in sorghum roots under low P conditions, facilitating the plant's ability to adapt to nutrient scarcity (Maharajan et al., 2021). Overall, changes in strigolactone production and root architecture in response to drought, salinity, and nutrient limitations significantly affect sorghum root growth and distribution in the soil. Enhanced strigolactone levels promote deeper, more extensive root systems, which are critical for accessing water and nutrients in challenging environments (Hao et al., 2023; Avila et al., 2020). Increased lateral root branching, a key adaptation, helps sorghum cope with abiotic stresses (Kirshibayev & Mukasheva, 2022; Guimarães et al., 2020). These morphological changes not only improve nutrient and water uptake but also contribute to soil structure and health by promoting microbial activity in the rhizosphere (Sahib et al., 2020).

Ethylene is another hormone that plays a key role in shaping root architecture, particularly in response to environmental stresses. It promotes root hair elongation and can enhance lateral root formation under specific conditions. Ethylene signaling interacts with auxin pathways, and this crosstalk can modulate root responses to stimuli such as nutrient availability and mechanical stress (Makaza & Kamutando, 2022; Nath et al., 2016). For example, under nutrient deficiency, ethylene promotes root growth by stimulating lateral root formation, enabling the plant to explore a larger soil volume for nutrients (Makaza & Kamutando, 2022; Nath et al., 2016). During drought stress, ethylene production is often upregulated, leading to alterations in root morphology and function. This response is mediated through complex interactions with other hormones, including ABA, cytokinins, and auxins, which together coordinate the plant's adaptive responses to stress (Yan et al., 2023; Rowe et al., 2016). The interplay between these hormones can lead to a more fibrous sorghum root system, which may improve water and nutrient uptake under conditions of limited availability (Djanaguiraman et al., 2024; Li et al., 2020). Salinity also influences ethylene metabolism in sorghum roots. High salt levels trigger stress responses, including increased ethylene production, which may exacerbate root growth inhibition. Ethylene is known to promote senescence and inhibit cell elongation in wheat (Danish & Zafar-ul-Hye, 2019). The interaction between salinity and ethylene is significant because it can reduce root biomass and alter root architecture, compromising the sorghum's ability to access water and nutrients effectively (Prasad et al., 2021; Amoah & Antwi-Berko, 2020).

Nutrient limitations, particularly nitrogen deficiency, can influence ethylene production in sorghum roots. Under low nitrogen conditions, ethylene levels rise, resulting in changes in root growth patterns. Ethylene plays a critical role in regulating root hair density and length, which are essential for efficient nutrient uptake (Yan et al., 2023; Chen et al., 2020). The combined effects of nutrient limitation and drought stress pose a compounded challenge for sorghum, as both factors can lead to reduced root growth and altered architecture, ultimately affecting sorghum health and productivity (Yan et al., 2023; Behera et al., 2022). The effects of drought, salinity, and nutrient limitations on root growth and architecture are multifaceted. Drought and salinity may lead to deeper root systems as plants seek moisture, but this often comes at the expense of overall root biomass (Djanaguiraman et al., 2024; Amoah & Antwi-Berko, 2020). Conversely, nutrient limitations may lead to more branched and fibrous roots, enhancing surface area for nutrient uptake but potentially limiting root depth (Prasad et al., 2021; Chen et al., 2020).

Root architecture regulation is not solely governed by individual hormones; it results from a complex interplay of multiple hormonal signals (Figure 11.1). This hormonal crosstalk allows plants to integrate various environmental signals and adjust their root systems accordingly. The plasticity in root architecture is essential for adapting to changing environmental conditions, such as variations in soil moisture, nutrient availability, and competition with neighboring plants. By fine-tuning hormonal responses, plants can optimize root systems for efficient resource acquisition and survival (Makaza & Kamutando, 2022; Ho et al., 2017). However, the combined effects of nutrient availability, water stress, and soil conditions can lead to synergistic or antagonistic responses in root architecture. This multifaceted regulation necessitates a holistic approach to studying root plasticity, considering both intrinsic genetic factors and extrinsic environmental cues (Rellán-Álvarez et al., 2016; Sarkar et al., 2013; Rich & Watt, 2013).

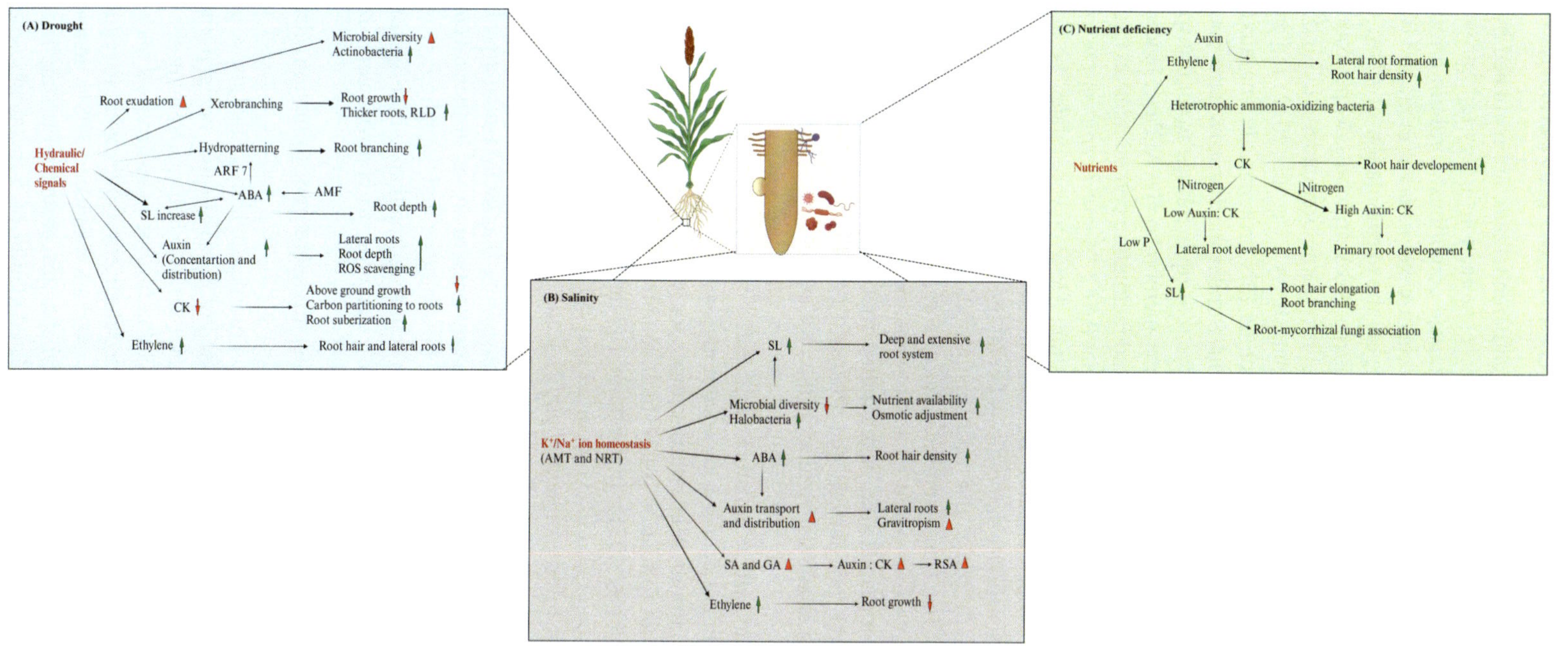

FIGURE 11.1 Schematic overview of sorghum root traits and rhizosphere under various stress conditions. (A) Drought, (B) Salinity, and (C) Nutrient deficiency. Text in red indicates the signals perceived by the roots under each stress condition, which subsequently led to hormonal crosstalk that shapes distinct root traits in sorghum. In the figure, solid red triangles represent changes, while solid green and red arrows indicate increases and decreases, respectively.

11.5 MOLECULAR REGULATION OF ROOT PHENES DEVELOPMENT

Root phenes, including length, angle, hair density, and architecture, are regulated by genetic, hormonal, and environmental factors. Understanding these interactions is key to improving sorghum's resilience to environmental stresses. Under water-limited conditions, sorghum prioritizes primary root growth over lateral root development to access deeper water reserves in the soil (Gowsiga et al., 2021). Studies have shown that moisture stress triggers the accumulation of ABA, which plays a crucial role in modulating root growth and development. ABA promotes deeper rooting by enhancing primary root elongation while inhibiting lateral root formation under PEG-induced moisture stress in sorghum (Li et al., 2022). A proteomic analysis by Li et al. (2022) indicates that ABA suppresses the activity of auxin efflux proteins, such as PIN proteins, altering auxin distribution in the root and prioritizing primary root development. Similar mechanisms have been observed in rice and Arabidopsis, where ABA accumulation in the root tip increases auxin transport (Xu et al., 2013). ABA-induced auxin transport activates plasma membrane H+ ATPase, leading to apoplast acidification in the root tip, loosening the cell wall and facilitating cell elongation (Xu et al., 2013).

Further studies in model plants have revealed that *ARF2* (*Auxin Response Factor 2*) acts as a repressor of lateral root formation under ABA signaling. *ARF2* downregulates the homeobox gene *HB33*, which is essential for root branching. Research in Arabidopsis has shown that overexpressing *HB33* leads to ABA hypersensitivity and inhibits lateral root emergence (Wang et al., 2011). These findings could shed light on the molecular mechanisms underlying primary root development in sorghum under drought stress. In a transcriptomic atlas of sorghum, McCormick et al. (2018) reported enrichment of *YODA*, a kinase involved in auxin signaling that modulates root development. In Arabidopsis, *YODA* plays a crucial role in regulating root growth, with *yda1* mutants exhibiting short roots, increased lateral root density, and abnormal cellular arrangements (Smekalova et al., 2014). *YODA* may regulate auxin distribution in the root and play a similar role in regulating root architecture in sorghum.

In agricultural conditions where moisture is primarily stored in the subsoil and production is heavily dependent on rainfall, a deeper root system may not necessarily be advantageous. In such situations, lateral root proliferation becomes critical for efficient moisture and nutrient uptake. Conditions where lateral root formation is key, include nutrient deficiencies (Adu et al., 2022) and salt stress (Özyazıcı & Açıkbaş, 2021). In sorghum, lignin biosynthesis plays a significant role in lateral root formation. Saluja et al. (2021) reported that caffeic acid O-methyltransferase (COMT) activity in sorghum brown midrib mutant (*bmr12*) leads to reduced syringyl lignin deposition, altering root anatomy and reducing lateral root density. The reduced lateral root density is attributed to impaired lateral root primordia formation, demonstrating that COMT-mediated lignin biosynthesis is essential for root system architecture. Transcriptomic analyses further reveal that the loss of COMT activity increases gibberellic acid (GA) biosynthesis and signaling, which negatively regulates lateral root density. This highlights a molecular mechanism where lignin biosynthesis intersects with hormonal regulation, specifically GA signaling, to control lateral root formation (Saluja et al., 2021).

Under low P conditions, sorghum adapts by enhancing lateral root formation, characterized by longer and thinner lateral roots, a strategy that increases the root surface area for more efficient P uptake from the soil (Galdman et al., 2022). This morphological adaptation is governed by a combination of transcriptional and epigenetic remodeling processes that occur in the root apices. Sorghum employs a dual strategy: activating beneficial genes while maintaining the repression of non-essential pathways. Galdman et al. (2022) found that histone modifications play a pivotal role in this process, with *H3K4me3*, associated with active transcription, increasing in regulatory regions and correlating with the activation of phosphorus starvation-induced (PSI) genes. Conversely, *H3K27me3*, linked to gene repression, shows minimal changes, ensuring the stability of gene silencing in non-essential genes. Additionally, the hypomethylation of 5-methylcytosine, particularly around key genes and promoters, further contributes to epigenetic remodeling. This remodeling landscape becomes enriched with DNA-binding motifs for transcription factors such as AP2/ERF, NAC, WRKY, and LOB, which orchestrate key pathways, including auxin signaling. In barley, the *LOB* gene is expressed at the boundaries between stem cells and developing leaf primordia at the shoot apical meristem, as well as at the base of lateral roots and the junction between the primary root and lateral root primordia (Janiak et al., 2019). Similarly, in rice, the overexpression of WRKY genes like *OsWRKY31* and in cotton, *GmWRKY13*, influences lateral root development by modulating auxin response or transport (Wang et al., 2018; Yu et al., 2012). These gene upregulations are localized predominantly to the lateral root region, suggesting that primary transcriptomic remodeling of the RSA is concentrated within the lateral root apices. This epigenetic shift likely supports the changes in RSA, enabling sorghum to maximize root growth in low phosphorus environments.

In addition to P uptake, sorghum also benefits from the expression of the bHLH transcription factor *SbbHLH85*, which plays a crucial role in regulating root hair development and enhancing salt stress tolerance. Overexpression of *SbbHLH85* leads to longer and more abundant root hairs by modulating auxin and ABA signaling pathways. It regulates key genes, such as *SbPYL4* (ABA signaling) and *SbPIN3* (auxin transport), influencing root system architecture under saline conditions (Song et al., 2022). The upper ground root system of sorghum is equally important, particularly the nodal roots, which not only exhibit photosynthetic activity but are also associated with nitrogen-fixing microbes that enhance nutrient acquisition. This multifunctional adaptation is crucial for improving sorghum's sustainability and yield (Wojciechowski & Kant, 2021; Mace et al., 2012). Nodal root bud development is essential for nodal root formation, and at the molecular level, genes like *WOX4*, *PFA5*, and *NGAL2* regulate meristem maintenance, auxin signaling, and cell proliferation, establishing the foundation for nodal root bud formation. As development progresses, mid to late-stage genes, including *WOX11*, *WOX5*, and LOB-domain proteins like *LBD29*, promote the establishment of root meristem niches and the formation of lateral roots. Furthermore, genes such as *SOMBRERO* play a key role in regulating root cap formation and programmed cell death, ensuring a functional and adaptive root system (Lamb et al., 2022).

11.6 DESIGNING ROOTS USING MOLECULAR TOOLS AND TECHNIQUES

Advancements in molecular biology have enabled the precise manipulation of the root traits. Molecular tools such as genomics, transcriptomics, and genome editing technologies allow for the identification and modification of genes associated with stress responses, paving the way for the development of crops with improved tolerance to adverse environmental conditions (Joshi et al., 2022).

11.6.1 Genomic Approaches to Improving Root Traits in Sorghum

Genomic approaches, such as genome-wide association studies (GWAS) and quantitative trait locus (QTL) mapping, are powerful tools for identifying genetic regions associated with key root traits, which are essential for enhancing crop adaptation to environmental stresses. These techniques help uncover stress-responsive genes and their regulatory networks, providing valuable targets for improving root architecture and function. In sorghum, a multi-locus GWAS conducted on 129 Ethiopian accessions identified 49 stable quantitative trait nucleotides (QTNs) across ten chromosomes, linked to important root traits such as root angle, root length, and root-to-shoot ratio, key attributes for drought adaptation. This study also pinpointed candidate genes, including *ALUMINUM-ACTIVATED MALATE TRANSPORTER* (*ALMT*), which influences root elongation, and an F-BOX domain protein that affects root angle (Kebede et al., 2024). These findings suggest that these QTNs and candidate genes can serve as critical targets for breeding drought-tolerant sorghum varieties with optimized root systems. Further, GWAS in diverse sorghum germplasm identified significant candidate genes associated with root traits. For instance, *Sobic.005G104100*, a peroxidase 13-related gene, was linked to root length and was shown to play a vital role in promoting root hair growth and elongation under stress conditions. Another gene, *Sobic.005G141801*, encodes a cadmium-induced protein and is associated with root thickness. Additionally, QTNs related to root traits were located near QTLs for root angle, root number, and root dry weight, emphasizing their role in improving root architecture for enhanced water and nutrient uptake (Ramalingam et al., 2023).

Beyond sorghum, studies in other cereal crops provide further insights into root trait genetic regulation. In rice, a GWAS on 529 accessions identified 21 RSA traits, including root length, volume, and weight, under both drought and normal conditions. Key genes such as *DEEPER ROOTING 1* (*DRO1*), *NARROW LEAF1* (*NAL1*), and *JASMONATE ZIM-domain 1* (*OsJAZ1*) were found to be linked to root growth traits under stress, suggesting their potential application in rice root improvement under drought conditions. Similarly, in maize, a GWAS revealed loci associated with root volume and transpiration efficiency under water deficit (Zaidi et al., 2016), while wheat studies identified root depth QTLs under osmotic stress (Ayalew et al., 2018). In Arabidopsis, key genes like *CYP79B2* and *HKT1* were implicated in lateral root adaptation to salinity stress (Julkowska et al., 2017; Kobayashi et al., 2016). These findings underscore the utility of GWAS in identifying genetic targets that can be leveraged for breeding stress-resilient crops with optimized root systems. In the case of sorghum, such genomic insights are invaluable for accelerating the development of varieties with enhanced drought tolerance and resource-efficient root architectures.

11.6.2 Genome Editing to Modify Root Traits in Sorghum

CRISPR/Cas9 technology has revolutionized plant biotechnology, enabling precise genome editing that can modify root traits to enhance crop performance under varying environmental conditions. This system uses the Cas9 endonuclease, directed by a guide RNA (gRNA), to introduce double-strand breaks at specific genomic loci. The subsequent repair processes can result in targeted mutations, facilitating the alteration of genes that regulate root development and function. In sorghum, genome editing has been effectively applied using the microprojectile transformation system. Immature embryos from the sorghum genotype Tx430 were transformed with plasmids containing the *SbPIN1*, *SbPIN2*, and *SbPIN4* genes, driven by their native promoters (Borrell et al., 2022). Notably, overexpression of *SbPIN2* led to significant improvements in root traits, including enhanced root biomass, an increased number of nodal roots, and changes in root length, all of which directly promoted root growth. These findings demonstrate that modifying auxin transport genes, such as *SbPIN2*, can effectively improve root architecture in sorghum.

Building on this, further research could explore how other genes involved in root growth, such as those regulating cytokinin metabolism and gravitropism, influence root development in sorghum. For example, studies in other crops like rice and barley have shown that genes like *HvPIN2* (Aldiss et al., 2024) and *OsERA1* (Ogata et al., 2020), which regulate auxin transport and ABA sensitivity respectively, play crucial roles in root architecture. While such studies provide valuable insights, further exploration of these pathways in sorghum could yield important strategies for improving root traits, especially in response to abiotic stresses like drought and nutrient deficiency. In barley, for instance, CRISPR/Cas9-mediated knockout of *HvPIN2* resulted in wider, shallower roots with a loss of gravitropic response, which emphasizes the role of auxin regulation in root architecture (Aldiss et al., 2024). Similarly, the knockout of *HvCKX1* in barley led to longer roots and increased root surface area, highlighting the potential for modifying cytokinin pathways to enhance root growth (Gasparis et al., 2019). Such findings further underscore the importance of understanding and manipulating root trait regulatory networks, a concept that could be applied to sorghum for improving root development under stress conditions. Although CRISPR/Cas9 research in sorghum is still evolving, the modification of key genes involved in root development offers a promising avenue for improving root architecture in sorghum. By leveraging the insights gained from other species and focusing on specific genes like *SbPIN2*, future studies could lead to more resilient and productive sorghum cultivars better suited for challenging environments.

11.6.3 Designing the Root Architecture Using Synthetic Biology

Root architecture is a complex trait controlled by gene regulatory networks. Manipulating these networks through genetic engineering can enhance sorghum's adaptability to soil conditions and environmental stresses, such as drought, by optimizing root systems for improved water and nutrient uptake. Gene expression modification in sorghum can be achieved using specific promoters that serve as regulatory switches to control when and where genes are expressed. For example, tissue-specific promoters, which drive expression patterns of genes relevant to root development, can be identified from available databases to generate reporter genes for sorghum. However, to achieve precise regulation of gene expression at spatial and magnitude levels, advanced techniques like synthetic transcriptional regulators are necessary. Brophy et al. (2022) designed synthetic transcriptional activators and repressors that could be utilized in sorghum. These regulators, based on bacterial DNA domains, VP15 activation domains, and SV40 nuclear localization signals, allow for the precise modulation of gene expression. Their orthogonal properties, minimal cross-reactivity with other promoters, allow for the construction of genetic circuits capable of executing Boolean logic functions such as AND, OR, and NOT gates, enabling complex regulation of gene expression (Brophy et al., 2022). Such circuits can be used to fine-tune root architecture, providing the flexibility needed to optimize sorghum root systems for drought resistance and other environmental stresses.

In sorghum, genetic modifications that alter root architecture sometimes result in pleiotropic effects on other root traits, similar to those observed in other species. For instance, a gain-of-function mutation in the *INDOLE-3-ACETIC ACID INDUCIBLE 14* (*IAA14*) gene, which regulates root branching, could result in reductions in root branching, gravitropism, root hair development, and primary root growth (Fukaki et al., 2002). However, expressing this mutant gene using tissue-specific promoters in sorghum could limit its effect to lateral root branching, enabling precise modification of root traits without disrupting other aspects of root development. By using synthetic genetic circuits such as BUFFER gates, which regulate the expression of the *IAA14* mutant gene under the control of tissue-specific promoters, sorghum plants could be engineered to exhibit specific root traits that optimize water and nutrient uptake in water-limited environments.

A novel system for monitoring gene activity in lateral root development has also been developed using recombinase-based synthetic circuits in *Arabidopsis thaliana* (Guiziou et al., 2023), and this system could be adapted for use in sorghum. By using integrase enzymes, researchers can track the activity of specific genes expressed in lateral root founder cells, allowing for a detailed understanding of root development. This technique, which enables tracing of gene activity using fluorescent markers, could provide new insights into the regulation of root traits and help identify key genes involved in root system architecture. To further refine root architecture engineering, researchers could improve this system by dividing the coding sequence of the integrase enzyme into two fragments, as was done in *Arabidopsis thaliana* (Guiziou et al., 2023). These fragments could be expressed separately and reconstituted into a functional enzyme only when both fragments meet in the same cell. This system provides temporal control over gene expression, allowing for the activation of specific root-related genes at different stages of development or in response to environmental signals. For sorghum, this approach could be used to optimize lateral root development, potentially enhancing drought tolerance by improving root depth and water uptake capacity.

While significant progress has been made in the field of synthetic biology, challenges remain in adapting these tools for monocots and particularly for sorghum. The use of tissue-specific promoters and the development of efficient transformation systems for sorghum will be crucial for successfully applying synthetic biology to modify root architecture. Transient transformation techniques offer promising solutions but require further optimization. Nonetheless, the ongoing advancements in synthetic biology and genetic engineering hold great potential for creating sorghum varieties with improved root systems that are better equipped to withstand environmental stresses and support sustainable agricultural practices in the face of climate change.

11.7 CONCLUSION

The root system architecture and phenes play a pivotal role in optimizing resource acquisition and overall abiotic stress tolerance in sorghum. However, the regulation of individual root phenes and the overall architecture is a complex phenomenon, influenced by a combination of soil environmental cues, hormonal balance, and genetic factors. Understanding the molecular mechanisms that govern root phenotypic plasticity is key to unlocking sorghum's potential for better adaptation to these stresses. The molecular engineering of sorghum root phenes presents a transformative opportunity to mitigate the impacts of abiotic stresses such as drought, salinity, and poor soil conditions.

A more refined understanding of the genetic and hormonal-environmental regulatory mechanisms that control root development can significantly enhance sorghum's tolerance to these stresses. Tools like GWAS, QTL mapping, and CRISPR/Cas9 genome editing provide powerful means to precisely manipulate critical root traits, allowing for the targeted improvement of root architecture. Additionally, the exploitation of beneficial root-microbe interactions opens new avenues for enhancing stress resilience, as these interactions can influence nutrient uptake, drought tolerance, and overall plant health. Given these advancements, there is an urgent need for an integrative approach that combines molecular engineering with agronomic practices. By leveraging these innovative tools and approaches, we can develop climate-resilient sorghum varieties that are better equipped to withstand the challenges posed by a changing climate. Such efforts will contribute significantly to the creation of sustainable food systems and enhance global food security in the face of increasing environmental challenges.

REFERENCES

Adu, M. O., Asare, P. A., Yawson, D. O., Amoah, K. K., Atiah, K., Duah, M. K., & Graham, A. (2022). Root system traits contribute to variability and plasticity in response to phosphorus fertilization in 2 field-grown sorghum [*Sorghum bicolor* (L.) Moench] Cultivars. *Plant Phenomics*, *2022*, 0002.

Akman, H., Zhang, C., & Ejeta, G. (2021). Physio-morphological, biochemical, and anatomical traits of drought-tolerant and susceptible sorghum cultivars under pre-and post-anthesis drought. *Physiologia Plantarum*, *172*(2), 912–921.

Aldiss, Z., Robinson, H., Lam, Y., Dixon, R., Crisp, P. A., Godwin, I. D., ... & Massel, K. (2024). The PIN2 ortholog in barley modifies root gravitropism and architecture without impacting the shoot. *bioRxiv*, 2024–03.

Ali, A. E. E., Husselmann, L. H., Tabb, D. L., & Ludidi, N. (2023). Comparative proteomics analysis between maize and sorghum uncovers important proteins and metabolic pathways mediating drought tolerance. *Life*, *13*(1), 170.

Amoah, J. N., & Antwi-Berko, D. (2020). Comparative physiological, biochemical and transcript response to drought in sorghum genotypes. *Biotechnology Journal International*, *24*(3), 1–14.

Amombo, E., Ashilenje, D. S., Lazaar, K., Hirich, A., Kouisni, L., Devkota, K. P., ... & Nilahyane, A. (2022). Will roots play a decisive role in forage sorghum production under salt stress?. *Environmental Sciences Proceedings*, *16*(1), 61.

Avci, S., İleri, O., & Demirkaya, M. (2017). Determination of genotypic variation among sorghum cultivars for seed vigor, salt and drought stresses. *Journal of Agricultural Sciences*, *23*, 335–343.

Avila, R. G., Magalhães, P. C., da Silva, E. M., Gomes Júnior, C. C., de Paula Lana, U. G., de Alvarenga, A. A., & de Souza, T. C. (2020). Silicon supplementation improves tolerance to water deficiency in sorghum plants by increasing root system growth and improving photosynthesis. *Silicon*, *12*, 2545–2554.

Ayalew, H., Liu, H., Börner, A., Kobiljski, B., Liu, C., & Yan, G. (2018). Genome-wide association mapping of major root length QTLs under PEG induced water stress in wheat. *Frontiers in Plant Science*, *9*, 1759.

Azu, E., Elegba, W., Asare, A. T., Blege, P. K., Amoatey, H. M., & Danquah, E. Y. (2021). Responses of small-holder farmers on sorghum production preferences and constraints in the Upper East Region of Ghana. *Journal of Agricultural Extension and Rural Development*, *13*(3), 202–216.

Balasubramanian, V. K., Dampanaboina, L., Cobos, C. J., Yuan, N., Xin, Z., & Mendu, V. (2021). Induced secretion system mutation alters rhizosphere bacterial composition in *Sorghum bicolor* (L.) Moench. *Planta*, *253*, 1–18.

Barnes, E. M., & Tringe, S. G. (2022). Exploring the roles of microbes in facilitating plant adaptation to climate change. *Biochemical Journal*, *479*(3), 327–335.

Behera, P. P., Saharia, N., Borah, N., Devi, S. H., & Sarma, R. N. (2022). Sorghum physiology and adaptation to abiotic stresses. *International Journal of Environment and Climate Change*, *12*(10), 1005–1022.

Bellande, K., Laplaze, L., & Guyomarc'h, S. (2022). Lateral root formation. *eLS*, *3*, 1–10.

Blum, A., & Arkin, G. F. (1984). Sorghum root growth and water-use as affected by water supply and growth duration. *Field Crops Research*, *9*, 131–142.

Borg, H., & Grimes, D. W. (1986). Depth development of roots with time: An empirical description. *Transactions of the ASAE*, *29*(1), 194–197.

Borrell, A. K., Wong, A. C., George-Jaeggli, B., van Oosterom, E. J., Mace, E. S., Godwin, I. D., ... & Jordan, D. R. (2022). Genetic modification of PIN genes induces causal mechanisms of stay-green drought adaptation phenotype. *Journal of Experimental Botany*, *73*(19), 6711–6726.

Boyno, G., Rezaee Danesh, Y., Demir, S., Teniz, N., Mulet, J. M., & Porcel, R. (2023). The Complex interplay between arbuscular mycorrhizal fungi and strigolactone: Mechanisms, sinergies, applications and future directions. *International Journal of Molecular Sciences*, *24*(23), 16774.

Brophy, J. A., Magallon, K. J., Duan, L., Zhong, V., Ramachandran, P., Kniazev, K., & Dinneny, J. R. (2022). Synthetic genetic circuits as a means of reprogramming plant roots. *Science*, *377*(6607), 747–751.

Caddell, D. F., Louie, K., Bowen, B., Sievert, J. A., Hollingsworth, J., Dahlberg, J., ... & Coleman-Derr, D. (2020). Drought shifts sorghum root metabolite and microbiome profiles and enriches the stress response factor pipecolic acid. *BioRxiv*, 2020–11.

Caddell, D. F., Pettinga, D., Louie, K., Bowen, B. P., Sievert, J. A., Hollingsworth, J., ... & Coleman-Derr, D. (2023). Drought shifts sorghum root metabolite and microbiome profiles and enriches for pipecolic acid. *Phytobiomes Journal*, *7*(4), 449–463.

Carlson, R., Tugizimana, F., Steenkamp, P. A., Dubery, I. A., Hassen, A. I., & Labuschagne, N. (2020). Rhizobacteria-induced systemic tolerance against drought stress in *Sorghum bicolor* (L.) Moench. *Microbiological Research*, *232*, 126388.

Cavallari, N., Artner, C., & Benkova, E. (2021). Auxin-regulated lateral root organogenesis. *Cold Spring Harbor Perspectives in Biology*, *13*(7), a039941.

Chai, Y. N., Qi, Y., Goren, E., Chiniquy, D., Sheflin, A. M., Tringe, S. G., ... & Schachtman, D. P. (2024). Root-associated bacterial communities and root metabolite composition are linked to nitrogen use efficiency in sorghum. *Msystems*, *9*(1), e01190–e01223.

Chandran, H., Meena, M., & Swapnil, P. (2021). Plant growth-promoting rhizobacteria as a green alternative for sustainable agriculture. *Sustainability*, *13*(19), 10986.

Chen, D., Wang, W., Wu, Y., Xie, H., Zhao, L., Zeng, Q., & Zhan, Y. (2019). Expression and distribution of the auxin response factors in *Sorghum bicolor* during development and temperature stress. *International Journal of Molecular Sciences*, *20*(19), 4816.

Chen, X., Wu, Q., Gao, Y., Zhang, J., Wang, Y., Zhang, R., ... & Huang, R. (2020). The role of deep roots in sorghum yield production under drought conditions. *Agronomy*, *10*(4), 611.

Choudhary, D. K., Sharma, K. P., & Gaur, R. K. (2011). Biotechnological perspectives of microbes in agro-ecosystems. *Biotechnology Letters*, *33*, 1905–1910.

Correa, J., Postma, J. A., & Wojciechowski, T. (2022). Phenotypic response to soil compaction varies among genotypes and correlates with plant size in sorghum. *Plant and Soil*, 472, 59–76.

Correa, J., Postma, J. A., Watt, M., & Wojciechowski, T. (2019). Soil compaction and the architectural plasticity of root systems. *Journal of Experimental Botany*, *70*(21), 6019–6034.

da Silva, J. O. N., Pessoa, L. G. M., da Silva, E. M., da Silva, L. R., dos Santos Freire, M. B. G., de Souza, E. S., ... & do Nascimento Alencar, E. L. (2023). Effects of silicon alone and combined with organic matter and *Trichoderma harzianum* on sorghum yield, ions accumulation and soil properties under saline irrigation. *Agriculture*, *13*(11), 1–23.

Danish, S., & Zafar-ul-Hye, M. (2019). Co-application of ACC-deaminase producing PGPR and timber-waste biochar improves pigments formation, growth and yield of wheat under drought stress. *Scientific Reports*, *9*(1), 5999.

Dastidar, M. G., Jouannet, V., & Maizel, A. (2012). Root branching: Mechanisms, robustness, and plasticity. *Wiley Interdisciplinary Reviews: Developmental Biology*, *1*(3), 329–343.

Ding, Z., & Bai, Y. (2021). The current and future studies on plant root development and root microbiota. *Scientia Sinica Vitae*, *51*, 1447–1456.

Djanaguiraman, M., Gowsiga, S., Govindaraj, M., Habyarimana, E., Senthil, A., Thavaprakaash, N., ... & Chellammal, C. (2024). Impact of root architecture and transpiration rate on drought tolerance in stay-green sorghum. *Crop Science*, *64*(5), 2612–2629.

Dugas, D. V., Monaco, M. K., Olson, A., Klein, R. R., Kumari, S., Ware, D., & Klein, P. E. (2011). Functional annotation of the transcriptome of *Sorghum bicolor* in response to osmotic stress and abscisic acid. *BMC Genomics*, *12*, 1–21.

Erlandson, S., Bheemanahalli, R., Narayana, N. K., Johnson, M., Graham, C., Jagadish, S. V. K., & Subramanian, S. (2021). Evidence for microbiome-dependent chilling tolerance in sorghum. *bioRxiv*, 2021–03.

Fernandez, R., Crabos, A., Maillard, M., Nacry, P., & Pradal, C. (2022). High-throughput and automatic structural and developmental root phenotyping on Arabidopsis seedlings. *Plant Methods*, *18*(1), 127.

Forde, B. G. (2009). Is it good noise? The role of developmental instability in the shaping of a root system. *Journal of Experimental Botany*, *60*(14), 3989–4002.

Forghani, A. H., Almodares, A., & Ehsanpour, A. A. (2018). Potential objectives for gibberellic acid and paclobutrazol under salt stress in sweet sorghum (*Sorghum bicolor* [L.] Moench cv. Sofra). *Applied Biological Chemistry*, *61*, 113–124.

Fukaki, H., Tameda, S., Masuda, H., & Tasaka, M. (2002). Lateral root formation is blocked by a gain-of-function mutation in the SOLITARY-ROOT/IAA14 gene of Arabidopsis. *The Plant Journal*, *29*(2), 153–168.

Galindo-Castañeda, T., Hartmann, M., & Lynch, J. P. (2024). Location: Root architecture structures rhizosphere microbial associations. *Journal of Experimental Botany*, *75*(2), 594–604.

Gamalero, E., Bona, E., Todeschini, V., & Lingua, G. (2020). Saline and arid soils: Impact on bacteria, plants, and their interaction. *Biology*, *9*(6), 116.

Gano, B., Dembele, J. S. B., Tovignan, T. K., Sine, B., Vadez, V., Diouf, D., & Audebert, A. (2021). Adaptation responses to early drought stress of West Africa sorghum varieties. *Agronomy*, *11*(3), 443.

Gasparis, S., Przyborowski, M., Kała, M., & Nadolska-Orczyk, A. (2019). Knockout of the *HvCKX1* or *HvCKX3* gene in barley (*Hordeum vulgare* L.) by RNA-guided Cas9 nuclease affects the regulation of cytokinin metabolism and root morphology. *Cells*, *8*(8), 782.

Ghorai, A. K. (2022). Zero till water use efficient rabi sorghum *(Sorghum bicolor* L. Moench), an effective alternative to water guzzling summer rice (*Oryza sativa*), under changing climatic scenario. *Asian Journal of Agricultural Extension, Economics & Sociology*, *40*(11), 497–502.

Gladman, N., Hufnagel, B., Regulski, M., Liu, Z., Wang, X., Chougule, K., ... & Ware, D. (2022). Sorghum root epigenetic landscape during limiting phosphorus conditions. *Plant Direct*, *6*(5), e393.

Goche, T., Shargie, N. G., Cummins, I., Brown, A. P., Chivasa, S., & Ngara, R. (2020). Comparative physiological and root proteome analyses of two sorghum varieties responding to water limitation. *Scientific Reports*, *10*(1), 11835.

Govindasamy, V., George, P., Kumar, M., Aher, L., Raina, S. K., Rane, J., ... & Minhas, P. S. (2020). Multi-trait PGP rhizobacterial endophytes alleviate drought stress in a senescent genotype of sorghum [*Sorghum bicolor* (L.) Moench]. *3 Biotech*, *10*, 1–14.

Gowsiga, S., Djanaguiraman, M., Thavaprakaash, N., Jeyakumar, P., & Govindaraj, M. (2021). Sorghum drought tolerance is associated with deeper root system and decreased transpiration rate. *Preprints*, 2021110144.

Guimarães, M. J. M., Simões, W. L., Barros, J. R. A., Willadino, L. G., & Erickson, R. (2020). Salinity decreases transpiration of sorghum plants. *Experimental Results*, *1*, e14.

Guiziou, S., Maranas, C. J., Chu, J. C., & Nemhauser, J. L. (2023). An integrase toolbox to record gene-expression during plant development. *Nature Communications*, *14*(1), 1844.

Guo, H., Nie, C. Y., Li, Z., Kang, J., Wang, X. L., & Cui, Y. N. (2023). Physiological and transcriptional analyses provide insight into maintaining ion homeostasis of sweet sorghum under salt stress. *International Journal of Molecular Sciences*, *24*(13), 11045.

Guyomarc'h S Lucas M Laplaze L . 2010. Lateral/secondary roots. In: Encyclopedia of Life Sciences (eLS). Chichester: John Wiley & Sons. doi: 10.1002/9780470015902.a0002060.pub2.

Hadebe, S. T., Modi, A. T., & Mabhaudhi, T. (2017). Drought tolerance and water use of cereal crops: A focus on sorghum as a food security crop in sub-Saharan Africa. *Journal of Agronomy and Crop Science*, *203*(3), 177–191.

Hao, J., Yang, Y., Futrell, S., Kelly, E. A., Lorts, C. M., Nebie, B., ... & Schachtman, D. P. (2023). CRISPR/Cas9-mediated mutagenesis of carotenoid cleavage dioxygenase (CCD) genes in sorghum alters strigolactone biosynthesis and plant biotic Interactions. *Phytobiomes Journal*, *7*(3), 339–351.

Harris, J. M. (2015). Abscisic acid: Hidden architect of root system structure. *Plants*, *4*(3), 548–572.

Hartman, K., & Tringe, S. G. (2019). Interactions between plants and soil shaping the root microbiome under abiotic stress. *Biochemical Journal*, *476*(19), 2705–2724.

Henry, A., Cal, A. J., Batoto, T. C., Torres, R. O., & Serraj, R. (2012). Root attributes affecting water uptake of rice (*Oryza sativa*) under drought. *Journal of Experimental Botany*, *63*(13), 4751–4763.

Ho, Y. N., Mathew, D. C., & Huang, C. C. (2017). Plant-microbe ecology: Interactions of plants and symbiotic microbial communities. *Plant Ecology-Traditional Approaches to Recent Trends*, *93*, 119.

Hong, J. H., Seah, S. W., & Xu, J. (2013). The root of ABA action in environmental stress response. *Plant Cell Reports*, *32*, 971–983.

Htet, M. N. S., Feng, B., Wang, H., Tian, L., & Yadav, V. (2022). Comparative assessment of nutritional and functional properties of different sorghum genotypes for ensuring nutritional security in dryland agro-ecosystem. *Frontiers in Nutrition*, *9*, 1048789.

Idris, I., Fefirenta, A. D., Sari, V. K., & Sudiana, I. M. (2023). Arbuscular mycorrhizal fungi induced different proline accumulations in two sorghum accessions in a response to drought stress. *Agriculture (Pol'nohospodárstvo)*, *68*(3), 127–142.

Inbaraj, M. P. (2021). Plant-microbe interactions in alleviating abiotic stress—A mini review. *Frontiers in Agronomy*, *3*, 667903.

Istrati, D. I., Constantin, O. E., Vizireanu, C., Dinică, R., & Furdui, B. (2019). Sorghum as source of functional compounds and their importance in human nutrition. *The Annals of the University Dunarea de Jos of Galati. Fascicle VI-Food Technology*, *43*(2), 189–205.

Jamil, M., Van Mourik, T. A., Charnikhova, T., & Bouwmeester, H. J. (2013). Effect of diammonium phosphate application on strigolactone production and *Striga hermonthica* infection in three sorghum cultivars. *Weed Research*, *53*(2), 121–130.

Janiak, A., Kwasniewski, M., Sowa, M., Kuczyńska, A., Mikołajczak, K., Ogrodowicz, P., & Szarejko, I. (2019). Insights into barley root transcriptome under mild drought stress with an emphasis on gene expression regulatory mechanisms. *International Journal of Molecular Sciences*, *20*(24), 6139.

Jin, X., Zheng, Y., Wang, J., Chen, W., Yang, Z., Chen, Y., ... & Sun, B. (2023). *SbNAC9* improves drought tolerance by enhancing scavenging ability of reactive oxygen species and activating stress-responsive genes of sorghum. *International Journal of Molecular Sciences*, *24*(3), 2401.

Joshi, D. C., Singh, V., Hunt, C., Mace, E., van Oosterom, E., Sulman, R., ... & Hammer, G. (2017). Development of a phenotyping platform for high throughput screening of nodal root angle in sorghum. *Plant Methods*, *13*, 1–12.

Joshi, R. (2018). Tripartite plant microbe and nutrients interaction in the below ground soil ecosystems. *Journal of Medicinal Plants*, *6*(4), 36–40.

Joshi, S., Chinnusamy, V., & Joshi, R. (2022). Root system architecture and omics approaches for belowground abiotic stress tolerance in plants. *Agriculture*, *12*(10), 1677.

Julkowska, M. M., Koevoets, I. T., Mol, S., Hoefsloot, H., Feron, R., Tester, M. A., ... & Testerink, C. (2017). Genetic components of root architecture remodeling in response to salt stress. *The Plant Cell*, *29*(12), 3198–3213.

Jung, J. K., & McCouch, S. (2013). Getting to the roots of it: Genetic and hormonal control of root architecture. *Frontiers in Plant Science*, *4*, 186.

Kaigama, B. K., Teare, I. D., Stone, L. R., & Powers, W. L. (1977). Root and top growth of irrigated and nonirrigated grain sorghum 1. *Crop Science*, *17*(4), 555–559.

Kebede, A., Barka, G. D., Kebede, M., Tadesse, T., Girma, G., & Menamo, T. M. (2024). Multi-locus genome-wide association analysis for root and shoot traits at seedling stage in Ethiopian sorghum (*Sorghum bicolor* (L.) Moench) accessions. *Genetic Resources and Crop Evolution*, 72, 1289–1311.

Khan, N., Ali, S., Shahid, M. A., Mustafa, A., Sayyed, R. Z., & Curá, J. A. (2021). Insights into the interactions among roots, rhizosphere, and rhizobacteria for improving plant growth and tolerance to abiotic stresses: A review. *Cells*, *10*(6), 1551.

Kirshibayev, Y., & Mukasheva, D. (2022). Effect of NaCl toxicity on development and anatomical structure of varieties of Sorghum saccharatum (L.) in Kazakhstan, *Asian Journal of Plant Sciences*, 21, (1), 119–129.

Kobayashi, Y., Sadhukhan, A., Tazib, T., Nakano, Y., Kusunoki, K., Kamara, M., ... & Koyama, H. (2016). Joint genetic and network analyses identify loci associated with root growth under NaCl stress in Arabidopsis thaliana. *Plant, Cell & Environment*, *39*(4), 918–934.

Lamb, A., Weers, B., McKinley, B., Rooney, W., Morgan, C., Marshall-Colon, A., & Mullet, J. (2022). Bioenergy sorghum's deep roots: A key to sustainable biomass production on annual cropland. *GCB Bioenergy*, *14*(2), 132–156.

Leitner, D., Felderer, B., Vontobel, P., & Schnepf, A. (2014). Recovering root system traits using image analysis exemplified by two-dimensional neutron radiography images of lupine. *Plant Physiology*, *164*(1), 24–35.

Li, H., Li, Y., Ke, Q., Kwak, S. S., Zhang, S., & Deng, X. (2020). Physiological and differential proteomic analyses of imitation drought stress response in *Sorghum bicolor* root at the seedling stage. *International Journal of Molecular Sciences*, *21*(23), 9174.

Li, Y., Tan, B., Wang, D., Mu, Y., Li, G., Zhang, Z., ... & Zhu, L. (2022). Proteomic analysis revealed different molecular mechanisms of response to PEG stress in drought-sensitive and drought-resistant sorghums. *International Journal of Molecular Sciences*, *23*(21), 13297.

Li, Y., Wang, Y., Tan, S., Li, Z., Yuan, Z., Glanc, M., ... & Zhang, J. (2020). Root growth adaptation is mediated by PYLs ABA receptor-PP2A protein phosphatase complex. *Advanced Science*, *7*(3), 1901455.

Li, Y., Zhu, J., Wu, L., Shao, Y., Wu, Y., & Mao, C. (2019). Functional divergence of PIN1 paralogous genes in rice. *Plant and Cell Physiology*, *60*(12), 2720–2732.

Liu, Z., Qin, T., Atienza, M., Zhao, Y., Nguyen, H., Sheng, H., ... & Kochian, L. V. (2023). Constitutive basis of root system architecture: Uncovering a promising trait for breeding nutrient-and drought-resilient crops. *Abiotech*, *4*(4), 315–331.

Lynch, J. P., & Brown, K. M. (2012). New roots for agriculture: Exploiting the root phenome. *Philosophical Transactions of the Royal Society B: Biological Sciences*, *367*(1595), 1598–1604.

Lynch, J. P., Chimungu, J. G., & Brown, K. M. (2014). Root anatomical phenes associated with water acquisition from drying soil: Targets for crop improvement. *Journal of Experimental Botany*, *65*(21), 6155–6166.

Mace, E. S., Singh, V., Van Oosterom, E. J., Hammer, G. L., Hunt, C. H., & Jordan, D. R. (2012). QTL for nodal root angle in sorghum (*Sorghum bicolor* L. Moench) co-locate with QTL for traits associated with drought adaptation. *Theoretical and Applied Genetics*, *124*, 97–109.

Maharajan, T., Krishna, T. A., Kiriyanthan, R. M., Ignacimuthu, S., & Ceasar, S. A. (2021). Improving abiotic stress tolerance in sorghum: Focus on the nutrient transporters and marker-assisted breeding. *Planta*, *254*, 1–16.

Maina, F., Harou, A., Hamidou, F., & Morris, G. P. (2022). Genome-wide association studies identify putative pleiotropic locus mediating drought tolerance in sorghum. *Plant Direct*, *6*(6), e413.

Makaza, W., & Kamutando, C. N. (2022). Harnessing beneficial plant-microbe interactions for enhanced plant adaptation to abiotic stresses. *Plant Defense Mechanisms*, *143*.

Manzi, M., Lado, J., Rodrigo, M. J., Zacarías, L., Arbona, V., & Gómez-Cadenas, A. (2015). Root ABA accumulation in long-term water-stressed plants is sustained by hormone transport from aerial organs. *Plant and Cell Physiology*, *56*(12), 2457–2466.

Mbinda, W., & Kimtai, M. (2019). Evaluation of morphological and biochemical characteristics of sorghum [*Sorghum bicolor* [L.] Moench] varieties in response salinity stress. *Annual Research & Review in Biology*, *33*(1), 1–9.

Mbulwe, L., & Ajayi, O. C. (2020). Case study–sorghum improvement in Zambia: Promotion of sorghum open pollinated varieties (SOPVs). *European Journal of Agriculture and Food Sciences*, *2*(5), 1–12.

McCleery, W. T., Mohd-Radzman, N. A., & Grieneisen, V. A. (2017). Root branching plasticity: Collective decision-making results from local and global signalling. *Current Opinion in Cell Biology*, *44*, 51–58.

McCormick, R. F., Truong, S. K., Sreedasyam, A., Jenkins, J., Shu, S., Sims, D., ... & Mullet, J. E. (2018). The Sorghum bicolor reference genome: improved assembly, gene annotations, a transcriptome atlas, and signatures of genome organization. *The Plant Journal*, *93*(2), 338–354.

Meena, K. K., Sorty, A. M., Bitla, U. M., Choudhary, K., Gupta, P., Pareek, A., ... & Minhas, P. S. (2017). Abiotic stress responses and microbe-mediated mitigation in plants: The omics strategies. *Frontiers in Plant Science*, *8*, 172.

Meleaku, T., Goshu, D., & Tegegne, B. (2020). Determinants sorghum market among smallholder farmers in Kafta Humera district Tigeray Ethiopia. *South Asian Journal of Social Studies and Economics*, 8, 1–13.

Meng, F., Xiang, D., Zhu, J., Li, Y., & Mao, C. (2019). Molecular mechanisms of root development in rice. *Rice*, *12*(1), 1.

Muthego, D., Moloi, S. J., Brown, A. P., Goche, T., Chivasa, S., & Ngara, R. (2023). Exogenous abscisic acid treatment regulates protein secretion in sorghum cell suspension cultures. *Plant Signaling & Behavior*, *18*(1), 2291618.

Myers, R. J. K. (1980). The root system of a grain sorghum crop. *Field Crops Research*, *3*, 53–64.

Nath, M., Bhatt, D., Prasad, R., Gill, S. S., Anjum, N. A., & Tuteja, N. (2016). Reactive oxygen species generation-scavenging and signaling during plant-arbuscular mycorrhizal and *Piriformospora indica* interaction under stress condition. *Frontiers in Plant Science*, *7*, 1574.

Neelam, A., & Tabassum, S. (2023). Optical sensing technologies to elucidate the interplay between plant and microbes. *Micromachines*, *14*(1), 195.

Niu, Y. F., Chai, R. S., Jin, G. L., Wang, H., Tang, C. X., & Zhang, Y. S. (2013). Responses of root architecture development to low phosphorus availability: A review. *Annals of Botany*, *112*(2), 391–408.

Nyoni, N., Dube, M., Bhebhe, S., Sibanda, B., Maphosa, M., & Bombom, A. (2020). Understanding biodiversity in sorghums to support the development of high value bio-based products in sub-Saharan Africa. *Journal of Cereals and Oilseeds*, *11*(2), 37–43.

Ogata, T., Ishizaki, T., Fujita, M., & Fujita, Y. (2020). CRISPR/Cas9-targeted mutagenesis of *OsERA1* confers enhanced responses to abscisic acid and drought stress and increased primary root growth under nonstressed conditions in rice. *PloS One*, *15*(12), e0243376.

Omae, N., & Tsuda, K. (2022). Plant-microbiota interactions in abiotic stress environments. *Molecular Plant-Microbe Interactions*, *35*(7), 511–526.

Orman-Ligeza, B., Morris, E. C., Parizot, B., Lavigne, T., Babé, A., Ligeza, A., ... & Draye, X. (2018). The xerobranching response represses lateral root formation when roots are not in contact with water. *Current Biology*, *28*(19), 3165–3173

Orosa-Puente, B., Leftley, N., Von Wangenheim, D., Banda, J., Srivastava, A. K., Hill, K., ... & Bennett, M. J. (2018). Root branching toward water involves posttranslational modification of transcription factor *ARF7*. *Science*, *362*(6421), 1407–1410.

Özyazıcı, M. A., & Açıkbaş, S. (2021). Effects of different salt concentrations on germination and seedling growth of some sweet sorghum [*Sorghum bicolor* var. saccharatum (L.) Mohlenbr.] cultivars. *Türkiye Tarımsal Araştırmalar Dergisi*, *8*(2), 133–143.

Pal, B., Pramanik, S., & Kishore, A. (2024). Scented sorghum (*Sorghum bicolor* L. Moench): A novel avenue to boost the millet's popularity. *AgroEnvironmental Sustainability*, *2*(1), 51–61.

Pattnaik, S., Mohapatra, B., & Gupta, A. (2021). Plant growth-promoting microbe mediated uptake of essential nutrients (Fe, P, K) for crop stress management: Microbe–soil–plant continuum. *Frontiers in Agronomy*, *3*, 689972.

Pérez-Pérez, J. G., Puertolas, J., Albacete, A., & Dodd, I. C. (2020). Alternation of wet and dry sides during partial rootzone drying irrigation enhances leaf ethylene evolution. *Environmental and Experimental Botany*, *176*, 104095.

Prasad, V. R., Govindaraj, M., Djanaguiraman, M., Djalovic, I., Shailani, A., Rawat, N., ... & Prasad, P. V. (2021). Drought and high temperature stress in sorghum: Physiological, genetic, and molecular insights and breeding approaches. *International Journal of Molecular Sciences*, *22*(18), 9826.

Qi, M., Berry, J. C., Veley, K. W., O'Connor, L., Finkel, O. M., Salas-González, I., ... & Bart, R. S. (2022). Identification of beneficial and detrimental bacteria impacting sorghum responses to drought using multi-scale and multi-system microbiome comparisons. *The ISME Journal*, *16*(8), 1957–1969.

Qin, T., Kazim, A., Wang, Y., Richard, D., Yao, P., Bi, Z., ... & Bai, J. (2022). Root-related genes in crops and their application under drought stress resistance—A review. *International Journal of Molecular Sciences*, *23*(19), 11477.

Rajabi Dehnavi, A., Zahedi, M., Ludwiczak, A., & Piernik, A. (2022). Foliar application of salicylic acid improves salt tolerance of sorghum (*Sorghum bicolor* (L.) Moench). *Plants*, *11*(3), 368.

Rajkumar, Fakrudin, B., Kavil, S. P., Girma, Y., Arun, S. S., Dadakhalandar, D., ... & Kamatar, M. Y. (2013). Molecular mapping of genomic regions harbouring QTLs for root and yield traits in sorghum (*Sorghum bicolor* L. Moench). *Physiology and Molecular Biology of Plants*, *19*, 409–419.

Ramalingam, A. P., Mohanavel, W., Kambale, R., Rajagopalan, V. R., Marla, S. R., Prasad, P. V., ... & Perumal, R. (2023). Pilot-scale genome-wide association mapping in diverse sorghum germplasms identified novel genetic loci linked to major agronomic, root and stomatal traits. *Scientific Reports*, *13*(1), 21917.

Rangarajan, H., & Lynch, J. P. (2021). A comparative analysis of quantitative metrics of root architecture. *Plant Phenomics,* 2021, 6953197.

Rao, T. P., & Ito, O. (1998). Differences in root system morphology and root respiration in relation to nitrogen uptake among six crop species. *Japan Agricultural Research Quarterly*, *32*, 97–104.

Ratnavathi, C. V., & Patil, J. V. (2014). Sorghum utilization as food, *Nutrition & Food Sciences*, 4(1), 1–8.

Rellán-Álvarez, R., Lobet, G., & Dinneny, J. R. (2016). Environmental control of root system biology. *Annual Review of Plant Biology*, *67*(1), 619–642.

Rich, S. M., & Watt, M. (2013). Soil conditions and cereal root system architecture: Review and considerations for linking Darwin and Weaver. *Journal of Experimental Botany*, *64*(5), 1193–1208.

Rivas, M. Á., Friero, I., Alarcón, M. V., & Salguero, J. (2022). Auxin-cytokinin balance shapes maize root architecture by controlling primary root elongation and lateral root development. *Frontiers in Plant Science*, *13*, 836592.

Romana, K. K., Chander, G., Deshpande, S., & Gupta, R. (2018). Genomic-assisted enhancement in stress tolerance for productivity improvement in sorghum. *Biotechnologies of Crop Improvement, Volume 3: Genomic Approaches*, 3, 265–288.

Rosales, M. A., Maurel, C., & Nacry, P. (2019). Abscisic acid coordinates dose-dependent developmental and hydraulic responses of roots to water deficit. *Plant Physiology*, *180*(4), 2198–2211.

Rowe, J. H., Topping, J. F., Liu, J., & Lindsey, K. (2016). Abscisic acid regulates root growth under osmotic stress conditions via an interacting hormonal network with cytokinin, ethylene and auxin. *New Phytologist*, *211*(1), 225–239.

Sahib, M. R., Pervaiz, Z. H., Williams, M. A., Saleem, M., & DeBolt, S. (2020). Rhizobacterial species richness improves sorghum growth and soil nutrient synergism in a nutrient-poor greenhouse soil. *Scientific Reports*, *10*(1), 15454.

Saluja, M., Zhu, F., Yu, H., Walia, H., & Sattler, S. E. (2021). Loss of *COMT* activity reduces lateral root formation and alters the response to water limitation in sorghum brown midrib (bmr) 12 mutant. *New Phytologist*, *229*(5), 2780–2794.

Sánchez-Bermúdez, M., Del Pozo, J. C., & Pernas, M. (2022). Effects of combined abiotic stresses related to climate change on root growth in crops. *Frontiers in Plant Science*, *13*, 918537.

Sankarapandian, R., Audilakshmi, S., Sharma, V., Ganesamurthy, K., Talwar, H. S., & Patil, J. V. (2013). Effect of morpho-physiological traits on grain yield of sorghum grown under stress at different growth stages, and stability analysis. *The Journal of Agricultural Science*, *151*(5), 630–647.

Sarkar, A. K., Mayandi, K., Gautam, V., Barik, S., & Sarkar Das, S. (2013). Improving the plant root system architecture to combat abiotic stresses incurred as a result of global climate changes. In: Tuteja, N., Gill, S.S (Eds), *Climate Change and Plant Abiotic Stress Tolerance*, (pp.305–324). Wiley-VCH Verlag GmbH & Co. KGaA.

Schittenhelm, S., & Schroetter, S. (2014). Comparison of drought tolerance of maize, sweet sorghum and sorghum-sudangrass hybrids. *Journal of Agronomy and Crop Science*, *200*(1), 46–53.

Schneider, H. M., Klein, S. P., Hanlon, M. T., Kaeppler, S., Brown, K. M., & Lynch, J. P. (2020). Genetic control of root anatomical plasticity in maize. *The Plant Genome*, *13*(1), e20003.

Sebastian, J., Yee, M. C., Goudinho Viana, W., Rellán-Álvarez, R., Feldman, M., Priest, H. D., ... & Dinneny, J. R. (2016). Grasses suppress shoot-borne roots to conserve water during drought. *Proceedings of the National Academy of Sciences*, *113*(31), 8861–8866.

Sheflin, A. M., Chiniquy, D., Yuan, C., Goren, E., Kumar, I., Braud, M., ... & Prenni, J. E. (2019). Metabolomics of sorghum roots during nitrogen stress reveals compromised metabolic capacity for salicylic acid biosynthesis. *Plant Direct*, *3*(3), e00122.

Sinclair, T. R., & Muchow, R. C. (2001). System analysis of plant traits to increase grain yield on limited water supplies. *Agronomy Journal*, *93*(2), 263–270.

Singh, P., Mohanta, T. K., & Sinha, A. K. (2015). Unraveling the intricate nexus of molecular mechanisms governing rice root development: *OsMPK3/6* and auxin-cytokinin interplay. *PLoS One*, *10*(4), e0123620.

Singh, V., van Oosterom, E. J., Jordan, D. R., Messina, C. D., Cooper, M., & Hammer, G. L. (2010). Morphological and architectural development of root systems in sorghum and maize. *Plant and Soil*, *333*, 287–299.

Smékalová, V., Luptovčiak, I., Komis, G., Šamajová, O., Ovečka, M., Doskočilová, A., ... & Šamaj, J. (2014). Involvement of YODA and mitogen activated protein kinase 6 in Arabidopsis post-embryogenic root development through auxin up-regulation and cell division plane orientation. *New Phytologist*, *203*(4), 1175–1193.

Song, Y., Li, S., Sui, Y., Zheng, H., Han, G., Sun, X., ... & Sui, N. (2022). SbbHLH85, a bHLH member, modulates resilience to salt stress by regulating root hair growth in sorghum. *Theoretical and Applied Genetics*, 1–16.

Sopandie, D., & Wirnas, D. (2023). Phosphorus deficiency tolerance in sorghum. *Jurnal Agronomi Indonesia (Indonesian Journal of Agronomy)*, *51*(1), 121–133.

Stefoska-Needham, A. (2024). Sorghum and health: An overview of potential protective health effects. *Journal of Food Science,* 89, 30–41.

Takanashi, H. (2023). Genetic control of morphological traits useful for improving sorghum. *Breeding Science*, *73*(1), 57–69.

Takatsuka, H., Amari, T., & Umeda, M. (2024). Cytokinin signaling is involved in root hair elongation in response to phosphate starvation. *Plant Signaling & Behavior*, *19*(1), 2305030.

Tari, I., Laskay, G., Takács, Z., & Poór, P. (2012). Response of sorghum to abiotic stresses: A review. *Journal of Agronomy and Crop Science*, *199*(4), 264–274.

Teixeira, G. C. M., de Mello Prado, R., Oliveira, K. S., D'Amico-Damião, V., & da Silveira Sousa Junior, G. (2020). Silicon increases leaf chlorophyll content and iron nutritional efficiency and reduces iron deficiency in sorghum plants. *Journal of Soil Science and Plant Nutrition*, *20*, 1311–1320.

Tiwari, S., Lata, C., Singh Chauhan, P., Prasad, V., & Prasad, M. (2017). A functional genomic perspective on drought signalling and its crosstalk with phytohormone-mediated signalling pathways in plants. *Current Genomics*, *18*(6), 469–482.

Umapathi, M., Chandrasekhar, C. N., Senthil, A., Kalaiselvi, T., Santhi, R., & Ravikesavan, R. (2021). Effect of bacterial endophytes inoculation on morphological and physiological traits of sorghum (*Sorghum bicolor* (L.) Moench) under drought. *Journal of Pharmacognosy and Phytochemistry*, *10*(2), 1021–1028.

van der Bom, F., Williams, A., Raymond, N., Sulman, R., McLean, G., & Bell, M. (2024). Spatio-temporal distribution of water and phosphorus determine growth of sorghum genotypes with contrasting nodal root angle. *Plant and Soil*, *499*(1), 37–54.

Vedenicheva, N. P., Shcherbatiuk, M. M., & Kosakivska, I. V. (2024). Endogenous cytokinins in plants of Secale cereale (Poaceae) under the effects of soil drought. *Ukrainian Botanical Journal*, *81*(3), 242–250.

Vegapareddy, M., Richter, G. M., & Goulding, K. W. T. (2010). Using digital image analysis to quantify the architectural parameters of roots grown in thin rhizotrons. *Plant Biosystems*, *144*(2), 499–506.

Wang, H., Ye, L., Zhou, L., Yu, J., Pang, B., Zuo, D., ... & Wang, H. (2023). Co-expression network analysis of the transcriptome identified hub genes and pathways responding to saline–alkaline stress in *Sorghum bicolor* L. *International Journal of Molecular Sciences*, *24*(23), 16831.

Wang, L., Hua, D., He, J., Duan, Y., Chen, Z., Hong, X., & Gong, Z. (2011). Auxin Response Factor2 (*ARF2*) and its regulated homeodomain gene *HB33* mediate abscisic acid response in Arabidopsis. *PLoS Genetics*, *7*(7), e1002172.

Wang, P., Xu, X., Tang, Z., Zhang, W., Huang, X. Y., & Zhao, F. J. (2018). OsWRKY28 regulates phosphate and arsenate accumulation, root system architecture and fertility in rice. *Frontiers in Plant Science*, *9*, 1330.

Wang, S., Bai, Y., Shen, C., Wu, Y., Zhang, S., Jiang, D., ... & Qi, Y. (2010). Auxin-related gene families in abiotic stress response in *Sorghum bicolor*. *Functional & Integrative Genomics*, *10*, 533–546.

Wang, X. L., Ma, K., Qi, L., Liu, Y. H., Shi, J., Li, X. L., ... & Song, P. (2022). Effect of ammonia-oxidizing bacterial strain that survives drought stress on corn compensatory growth upon post-drought rewatering. *Frontiers in Plant Science*, *13*, 947476.

Wang, X. L., Si, Z. Q., Yu, H., Qi, L., Liu, W., Shi, J., & Song, P. (2023). Unveiling the dual role of heterotrophic ammonia-oxidizing bacteria: Enhancing plant regrowth through modulating cytokinin delivery. *Frontiers in Microbiology*, *14*, 1268442.

Wang, X., Chen, Y., Thomas, C. L., Ding, G., Xu, P., Shi, D., ... & Shi, L. (2017). Genetic variants associated with the root system architecture of oilseed rape (*Brassica napus* L.) under contrasting phosphate supply. *DNA Research*, *24*(4), 407–417.

Weaver, J. E. (1926). *Root development of field crops*. McGraw-Hill.

Wei, C., Gao, L., Xiao, R., Wang, Y., Chen, B., Zou, W., ... & Tao, Y. (2024). Complete telomere-to-telomere assemblies of two sorghum genomes to guide biological discovery. *iMeta*, *3*(2), e193.

Widodo, S., Triastono, J., Sahara, D., Pustika, A. B., Kristamtini, Purwaningsih, H., ... & Muslimin. (2023). Economic value, farmers perception, and strategic development of sorghum in Central Java and Yogyakarta, Indonesia. *Agriculture*, *13*(3), 516.

Williamson, L. C., Ribrioux, S. P., Fitter, A. H., & Leyser, H. O. (2001). Phosphate availability regulates root system architecture in Arabidopsis. *Plant Physiology*, *126*(2), 875–882.

Wojciechowski, T., & Kant, J. (2021). How sorghum root traits can contribute to cereal yield increase. In: Goyal, A.K (Ed.), *Cereal Grains-Volume 2*. IntechOpen.

Woldesemayat, A. A., Van Heusden, P., Ndimba, B. K., & Christoffels, A. (2017). An integrated and comparative approach towards identification, characterization and functional annotation of candidate genes for drought tolerance in sorghum (*Sorghum bicolor* (L.) Moench). *BMC Genetics*, *18*, 1–24.

Xiong, Y., Zhang, P., Warner, R. D., & Fang, Z. (2019). Sorghum grain: From genotype, nutrition, and phenolic profile to its health benefits and food applications. *Comprehensive Reviews in Food Science and Food Safety*, *18*(6), 2025–2046.

Xu, W., Jia, L., Shi, W., Liang, J., Zhou, F., Li, Q., & Zhang, J. (2013). Abscisic acid accumulation modulates auxin transport in the root tip to enhance proton secretion for maintaining root growth under moderate water stress. *New Phytologist*, *197*(1), 139–150.

Yan, J., Wang, D., Kang, S., He, Z., Li, X., Tang, W., ... & Pang, C. (2023). Ethylene regulates combined drought and low nitrogen tolerance in wheat: Proteomic analysis. *Agronomy*, *13*(7), 1950.

Yoneyama, K., Xie, X., Kisugi, T., Nomura, T., & Yoneyama, K. (2013). Nitrogen and phosphorus fertilization negatively affects strigolactone production and exudation in sorghum. *Planta*, *238*, 885–894.

York, L. M., Nord, E. A., & Lynch, J. P. (2013). Integration of root phenes for soil resource acquisition. *Frontiers in Plant Science*, *4*, 355.

Yu, F., Huaxia, Y., Lu, W., Wu, C., Cao, X., & Guo, X. (2012). *GhWRKY15*, a member of the WRKY transcription factor family identified from cotton (*Gossypium hirsutum* L.), is involved in disease resistance and plant development. *BMC Plant Biology*, *12*, 1–18.

Yukun, G., Jianghui, C., Genzeng, R., Shilin, W., Puyuan, Y., Congpei, Y., ... & Jinhua, C. (2021). Changes in the root-associated bacteria of sorghum are driven by the combined effects of salt and sorghum development. *Environmental Microbiome*, *16*, 1–15.

Zaidi, P. H., Seetharam, K., Krishna, G., Krishnamurthy, L., Gajanan, S., Babu, R., ... & Vivek, B. S. (2016). Genomic regions associated with root traits under drought stress in tropical maize (*Zea mays* L.). *PloS One*, *11*(10), e0164340.

Zakavi, M., Askari, H., & Shahrooei, M. (2022). Maize growth response to different Bacillus strains isolated from a salt-marshland area under salinity stress. *BMC Plant Biology*, *22*(1), 367.

Zartman, R. E., & Woyewodzic, R. T. (1979). Root distribution patterns of two hybrid grain sorghums under field conditions 1. *Agronomy Journal*, *71*(2), 325–328.

Zegada-Lizarazu, W., & Monti, A. (2019). Deep root growth, ABA adjustments and root water uptake response to soil water deficit in giant reed. *Annals of Botany*, *124*(4), 605–615.

Zhang, H., Chen, T., Wang, Z., Yang, J., & Zhang, J. (2010). Involvement of cytokinins in the grain filling of rice under alternate wetting and drying irrigation. *Journal of Experimental Botany*, *61*(13), 3719–3733.

Zhang, Y., Xu, F., Ding, Y., Du, H., Zhang, Q., Dang, X., ... & Xu, W. (2021). Abscisic acid mediates barley rhizosheath formation under mild soil drying by promoting root hair growth and auxin response. *Plant, Cell & Environment*, *44*(6), 1935–1945.

12 Redesigning the Sorghum Root Microbiome for Improved Water and Nutrient Use Efficiency

Ali Yetgin, Fatma Azgın, and Sadık Dinçer

12.1 INTRODUCTION

Sorghum (*Sorghum bicolor* (L.) Moench) is a crucial cereal crop cultivated extensively in arid and semi-arid regions due to its remarkable tolerance to adverse climatic conditions. The primary reason for this resilience lies in its physiological and morphological adaptations, including a deep and extensive root system, waxy leaf cuticle, and the capacity to decrease metabolic activity during water-deficit stress (Abreha et al., 2022). These characteristics empower sorghum to maintain growth and yield in conditions that are unfavorable for other cereals like maize and wheat. Hence, sorghum plays a vital role as a food and fodder source for millions of individuals and cattle, respectively, in drought-prone regions such as sub-Saharan Africa (Khalifa & Eltahir, 2023). Its significance extends beyond human consumption to biofuel production and industrial uses, enhancing its economic value. The crop's ability to thrive in poor soils with minimal inputs positions it as a crucial element of sustainable agriculture and food security strategies in water-limited environments.

Despite its benefits, sorghum cultivation encounters numerous obstacles that impede its productivity and broader acceptance. One significant obstacle is its vulnerability to various biotic stresses, including pests and diseases like sorghum midge, stem borers, and anthracnose, resulting in significant yield reductions (Madhusudhana, 2020). Additionally, sorghum frequently faces nutrient deficiencies in the marginal soils where it is commonly grown, leading to suboptimal growth and reduced yields. Another crucial challenge is the crop's relatively limited genetic diversity, constraining the potential for breeding programs to create improved varieties with enhanced stress resistance and higher nutritional quality. Moreover, socio-economic factors, such as restricted access to high-quality seeds, insufficient extension services, and inadequate market infrastructure, hinder the adoption of optimal agronomic practices and technologies among smallholder farmers. Overcoming these challenges necessitates an integrated approach that combines advancements in breeding, agronomy, and biotechnological interventions, including optimizing the sorghum root microbiome to enhance nutrient and water use efficiency (Yahaya & Shimelis, 2022).

The root microbiome, comprising a diverse community of bacteria, fungi, archaea, and other microorganisms, plays a vital role in plant health and development. These microorganisms engage in complex interactions with plant roots, influencing key physiological processes. Beneficial microbes, such as mycorrhizal fungi and nitrogen-fixing bacteria, enhance nutrient uptake by increasing the availability of essential elements like nitrogen, phosphorus, and micronutrients (Yuan et al., 2024). Additionally, root-associated microbes bolster plant defenses against pathogens through competitive exclusion, the production of antimicrobial compounds, and the activation of systemic resistance (Ku et al., 2024). The root microbiome also aids in stress tolerance by producing phytohormones and osmolytes that help plants cope with abiotic stresses, including drought and salinity (Trivedi,

DOI: 10.1201/9781003623717-12

2021). Collectively, these functions improve plant growth, resilience, and productivity, underscoring the importance of a robust root microbiome for sustainable agriculture.

Optimization of the root microbiome emerges as a promising approach to enhance WUE and NUE in agricultural crops, such as sorghum. Through targeted stimulation of advantageous microbial populations, there exists the potential to ameliorate root functionality and overall plant performance amidst resource constraints. Notably, specific rhizobacteria can synthesize exopolysaccharides that foster soil aggregation and water retention, consequently enhancing WUE (Dar et al., 2021). Similarly, microbes like arbuscular mycorrhizal fungi extend the root surface area via hyphal networks, facilitating increased water and nutrient absorption (Miransari, 2011). Moreover, phosphate-solubilizing bacteria have the ability to convert insoluble phosphorus into plant-accessible forms, thereby enhancing PUE (Rezakhani et al., 2020). The use of tailored microbial inoculants or biofertilizers customized to sorghum's needs can improve fertilizer and water efficiency, reducing both costs and environmental impact (Diaz Franco et al., 2015). Understanding root microbiome dynamics under varying environmental conditions enables strategies to maintain microbial stability and functionality, ensuring consistent crop yields. Harnessing and optimizing the root microbiome thus represents a crucial step in advancing sorghum cultivation, supporting food security, and promoting sustainable agricultural practices (Chai et al., 2024).

In recent times, the field of microbiome research has witnessed significant advancements, reshaping our comprehension of plant-microbe interactions and their implications in agriculture. High-throughput sequencing methodologies, such as metagenomics and metatranscriptomics, have yielded comprehensive insights into the constitution and functional capacities of root-associated microbial communities (Latorre-Pérez et al., 2020; Kang et al., 2016). Advances in cutting-edge techniques have illuminated the intricate networks of microbial consortia essential for plant health and productivity. Progress in bioinformatics and systems biology now enables the integration of multi-omics data, providing a comprehensive understanding of the microbiome's role in nutrient cycling, pathogen control, and stress resilience (Lucaciu et al., 2019). Synthetic biology and genetic engineering studies have demonstrated the potential to modify microbial genomes to enhance beneficial traits (Ziegler & Takors, 2020). The development of microbiome-based products, such as biofertilizers and biopesticides, highlights the practical applications of this research.

Restructuring the sorghum root microbiome requires a strategic approach to enhance WUE and NUE (Xu et al., 2018; Chai et al., 2024). A key strategy involves identifying and incorporating beneficial microbial strains with traits such as nitrogen fixation, phosphate solubilization, and drought tolerance. These microbes can be applied as inoculants, either individually or in consortia, to establish a favorable root microbiome. Advanced genetic engineering techniques, including CRISPR-Cas systems, enable precise modifications of microbial genomes to enhance their functional capabilities (Ding et al., 2020). Additionally, agronomic practices like crop rotation, organic amendments, and reduced chemical inputs can promote beneficial microbial proliferation while suppressing harmful ones (Das et al., 2022). An innovative approach employs synthetic biology to engineer microbial communities with optimized interactions and functionalities (Liu et al., 2019). This chapter comprehensively examines these strategies, providing a solid framework for leveraging the root microbiome to advance sorghum cultivation. By integrating microbiome science with agricultural practices, this chapter aims to chart a course toward more resilient and productive sorghum cropping systems.

12.2 SORGHUM ROOT MICROBIOME COMPOSITION

The rhizosphere, defined as the soil region influenced by plant roots, harbors a diverse assortment of microbial communities, encompassing bacteria, fungi, archaea, and protists. The composition and abundance of microbial communities is highly variable between different host plants, although it is also influenced by abiotic factors such as the physico-chemical structure of the soil and the climatic characteristics of the environment (Shi et al., 2024). Within the sorghum root microbiome, bacterial populations are abundant and fulfill crucial functions in nutrient cycling, disease suppression and growth stimulation (Lopes et al., 2021). Noteworthy bacterial taxa consist of nitrogen-fixing genera like *Rhizobium* and *Azospirillum*, phosphate-solubilizing bacteria such as *Pseudomonas* and *Bacillus*, and plant growth-promoting rhizobacteria (PGPR) that synthesize phytohormones. Fungal communities, especially mycorrhizal fungi, establish symbiotic partnerships with sorghum roots, enhancing water and nutrient absorption, notably phosphorus. Other fungal categories can serve as decomposers or agents for biological control (Afridi et al., 2022). Furthermore, archaea, despite being less studied, participate in nutrient conversions and methane metabolism (Afridi et al., 2022). Protists and viruses also reside in the rhizosphere, contributing to microbial predation and horizontal gene transfer (Soucy et al., 2015; Guo et al., 2022). This microbial diversity and their interactions with plant roots are imperative for preserving soil quality and augmenting sorghum growth and resilience (Liu et al., 2023).

Progress in molecular methodologies has substantially enhanced the capacity to scrutinize and comprehend the constitution of root microbiomes. Metagenomics, entailing the direct sequencing of DNA retrieved from environmental specimens, enables the recognition and portrayal of microbial communities devoid of the necessity for cultivation. This approach yields insights into the taxonomic diversity and potential functional competencies of the microbiome (Maropola et al., 2015). Advanced sequencing technologies like Illumina, Oxford Nanopore and PacBio yield extensive datasets, streamlining comprehensive analyses of microbial communities (Sevim et al., 2019). Metatranscriptomics, which assesses the RNA profiles of microbial communities, unveils the active metabolic pathways and gene expression patterns, providing a dynamic perspective on microbial functions. Alternative techniques, such as amplicon sequencing (e.g., 16S rRNA for bacteria, ITS for fungi), furnish targeted insights into specific microbial assemblages (De Filippis et al., 2017). State-of-the-art bioinformatics tools and repositories, such as QIIME, SILVA, and Greengenes, facilitate the processing, annotation, and elucidation of sequencing data (Fung et al., 2021). Additionally, methodologies like fluorescence in situ hybridization (FISH) and stable isotope probing (SIP) furnish spatial and functional details concerning microbial populations in the rhizosphere (Cao et al., 2023). Collectively, these approaches increase our understanding of the intricate and dynamic attributes of the sorghum root microbiome.

Nitrogen-fixing bacteria assume a pivotal role in boosting nitrogen accessibility for sorghum, a crop frequently grown in nitrogen-deficient soils. These bacteria, encompassing symbiotic species like *Rhizobium* and free-living or associative bacteria such as *Azospirillum* and *Azotobacter*, possess the enzyme nitrogenase, which expedites the conversion of atmospheric nitrogen (N_2) into ammonia (NH_3), a form assimilable by plants. In sorghum, associative nitrogen-fixers colonize the rhizosphere or root surface, forging advantageous associations that enhance plant nitrogen levels without forming root nodules (Chai et al., 2024). This biological nitrogen fixation diminishes the dependence on synthetic nitrogen fertilizers, which are expensive and ecologically harmful. Augmented nitrogen availability fosters sorghum growth, escalates biomass output, and enhances grain yield, rendering nitrogen-fixing bacteria indispensable for sustainable sorghum agriculture, particularly in resource-constrained environments (Ostmeyer et al., 2022).

Phosphate-solubilizing microorganisms (PSMs) play a crucial role in enhancing phosphorus (P) availability in soil, particularly in regions where a significant portion of phosphorus exists in insoluble forms, inaccessible to plants. These microorganisms, such as *Pseudomonas*, *Bacillus*, *Rhizobium*, *Aspergillus*, and *Penicillium*, secrete organic acids, enzymes, and siderophores that facilitate the solubilization of bound phosphates, converting them into plant-absorbable forms (Mattos et al., 2020). In the context of sorghum farming, PSMs boost the uptake of phosphorus by roots, thereby promoting improved root growth, increased nutrient assimilation, and enhanced overall plant vigor and productivity (de Oliveira et al., 2021). Their effectiveness is particularly valuable in regions with low phosphorus content in the soil, which is a common characteristic of many sorghum-growing areas. By promoting the solubilization and uptake of phosphorus, these microorganisms play a key role in decreasing reliance on chemical phosphorus fertilizers, thereby advocating for sustainable agricultural practices.

Mycorrhizal fungi, notably arbuscular mycorrhizal fungi (AMF), establish symbiotic relationships with sorghum roots and various other plant species, significantly augmenting the uptake of nutrients and water (Wang et al, 2024). These fungi extend their hyphae beyond the root system into the surrounding soil, effectively expanding the root surface area and facilitating the absorption of nutrients and water from a larger soil volume. AMF enhances phosphorus uptake by accessing phosphorus reserves beyond the typical root depletion zone (Chandra et al., 2022). Furthermore, they improve the absorption of other vital nutrients like nitrogen, potassium, and micronutrients. The mutualistic association between sorghum and AMF also enhances resistance to drought conditions by improving water uptake and bolstering the plant's ability to withstand water stress through enhanced osmotic regulation and root hydraulic conductivity (Birhanu & Negussie, 2024). In addition to nutrient and water uptake benefits, AMF play a role in maintaining soil structure stability by producing glomalin, a glycoprotein that aids in soil aggregation (Zhang et al., 2023). This symbiotic bond is essential for the vitality and productivity of sorghum, particularly in environments characterized by nutrient scarcity and susceptibility to drought.

The constitution of the sorghum root microbiome is significantly impacted by soil characteristics and environmental circumstances (Sun et al., 2021). Various soil attributes, such as pH, consistency, amount of organic material, and nutrient accessibility, play a pivotal role in molding microbial populations. For example, acidic soils tend to support specific fungal groups, while neutral to alkaline soils may facilitate a more diverse bacterial community (Rousk et al., 2009). Soil consistency, which influences aeration and water retention, also affects the suitability of microbial habitats and the availability of resources. Environmental factors like temperature, moisture, and precipitation patterns additionally affect microbial dynamics. In dry regions where sorghum is commonly grown, the availability of moisture is a crucial factor determining microbial composition and activity (Enciso et al., 2019). Microorganisms adapted to drought conditions, like certain mycorrhizal fungi and bacteria tolerant to drought, are more abundant in these areas. Furthermore, seasonal changes and agricultural methods, such as watering and fertilization, can modify the structure of microbial communities by altering the microenvironment of the soil and nutrient flows. Understanding these influences from soil and the environment is vital for the management of the sorghum root microbiome to enhance plant health and productivity (Chai et al., 2024).

In order to maximize sorghum's overall performance in a variety of environmental circumstances, microbial activities play crucial roles, as shown in Figure 12.1. The method for maximizing WUE takes into account the activities of microbes that produce phytohormones, exopolysaccharides (EPS), deaminase activity for 1-aminocyclopropane-1-carboxylate (ACC), and osmoprotectants (Gamayunova et al., 2023). Together, these systems increase water retention, control stress ethylene levels, and preserve cellular integrity during osmotic stress, all of which help plants become more resilient to drought. Biochemical nitrogen fixation, phosphate solubilization, hydrolytic enzyme activity, and siderophore secretion are examples of microbial processes that are included in an efficient nutrient acquisition and utilization approach. The aforementioned actions promote greater accessibility and assimilation of vital nutrients, specifically nitrogen, phosphorus, and iron, hence augmenting plant development and yield (Hassan, 2020). These microbial consortia work with the sorghum plant to develop a mutually beneficial relationship that not only supports a sustainable method of crop production but also enhances the plant's resistance to abiotic stresses, maximizing its potential for agricultural use.

The genetic makeup of sorghum itself is a key determinant of the composition of the root microbiome. Different varieties of sorghum display unique patterns of root exudation, which can selectively attract and sustain specific microbial communities (Wang et al., 2021). Root exudates, like sugars, amino acids, organic acids, and secondary metabolites, act as nutrients and signaling molecules for microbes in the rhizosphere. Variations in these exudates among diverse sorghum

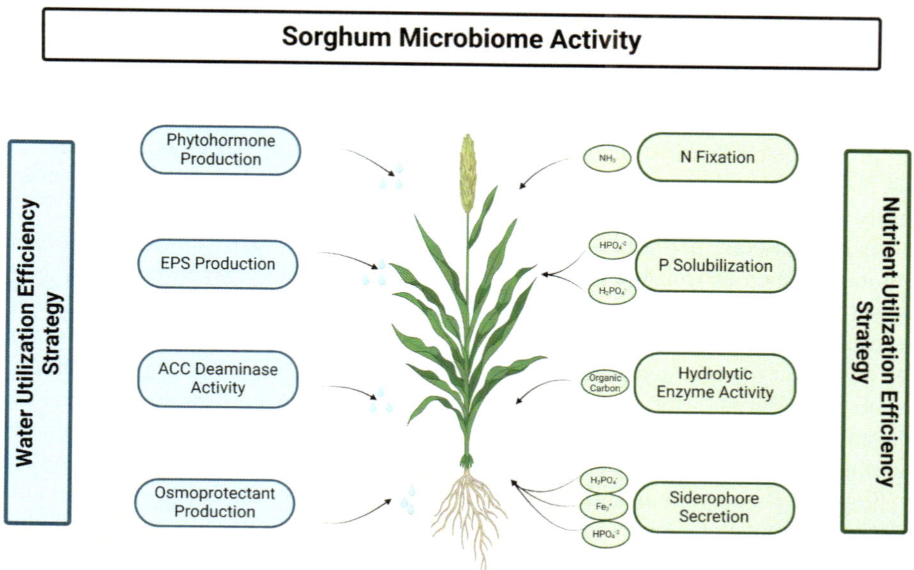

FIGURE 12.1 Strategies for optimizing the sorghum root microbiome to enhance water and nutrient use efficiency.

genotypes can lead to the formation of distinct microbial communities (Mechan-Llontop et al., 2023). Moreover, genetic disparities in root structure and growth patterns impact the spatial arrangement and density of microbial populations. For instance, genotypes with deeper or more expansive root systems may host different microbial communities compared to those with shallower roots (Abera et al., 2022). Additionally, genetic characteristics linked to disease resistance or tolerance to stress can indirectly influence the microbiome by changing the plant's interaction with pathogens and beneficial microbes. Breeding initiatives that consider interactions with the root microbiome can select sorghum genotypes that encourage beneficial microbial connections, thus improving nutrient uptake and resilience to stress (Lopes et al., 2021). This interplay between genotype and microbiome presents a promising approach to optimize sorghum performance through targeted microbiome management.

12.3 ROOT MICROBIOME AND WATER USE EFFICIENCY (WUE)

Microbial interactions play a significant role in shaping root morphology and function, ultimately enhancing WUE in sorghum. The presence of beneficial microbes, such as AMF and specific rhizobacteria, can bring about alterations in root architecture to promote increased water uptake. AMF, for instance, extend the root system by forming extensive hyphal networks that can penetrate deeper soil layers, accessing water beyond the reach of roots alone (Busso & Busso, 2022). This symbiotic association contributes to enlarging the root surface area, thereby enhancing water absorption especially in drought conditions. Moreover, rhizobacteria like *Azospirillum* spp. have been found to stimulate root hair development and lateral root formation, further enhancing the plant root's capacity to explore the soil for water. The production of phytohormones such as auxins and cytokinins by microbes leads to enhanced root branching and elongation, stimulating overall root growth. These morphological adaptations enable sorghum plants to achieve better hydration by tapping into water sources from a larger soil volume, consequently improving WUE and enhancing the plant's ability to withstand water scarcity (Goche et al., 2020). Furthermore, the presence of microbes in association with sorghum roots contributes significantly to the enhancement of WUE by producing stress hormones and osmolytes. In situations of water stress, specific rhizobacteria are known to produce stress-related phytohormones like abscisic acid (ABA) and ethylene. ABA, in particular, plays a vital role in regulating stomatal closure, reducing water loss through transpiration while ensuring sufficient CO_2 intake for photosynthesis.

Ethylene is a gaseous hormone synthesized by higher plants. This hormone is essential for various processes such as fruit ripening and flowering, root initiation when produced at less than the optimum amount of 1.0 $\mu l\ L^{-1}$, but exposure to stresses such as drought causes an increase in ethylene levels and undesirable effects for the plant such as leaf abscission, loss of chlorophyll pigments, inhibition of root, and nodulation (Carlson et al. 2020). The strategy of inducing plant resistance to stress through the degradation of 1-aminocyclopropane-1-carboxylic acid (ACC), a precursor molecule in ethylene production, is important for sorghum grown in water-limited areas. ACC deaminase activity, the enzyme responsible for ACC degradation, has been frequently reported in many *Bacillus*, *Pseudomonas* and *Rhizobium* species (Yuan et al. 2023). This modulation of ethylene enhances root elongation and increases the rhizosphere area, thereby facilitating the plant's uptake of water and nutrients. Undoubtedly, one of the tools to increase the WUE of sorghum could be the induction of microbiota related to ACC deaminase enzyme activity.

Additionally, microbes have the capacity to trigger the plant's synthesis of osmoprotectants such as proline, glycine betaine, and trehalose (Jawahar et al., 2019). These osmolytes play a crucial role in preserving cell turgor and enzyme activity under conditions of dehydration stress, thereby safeguarding cellular integrity and metabolic functions. Moreover, the microbial synthesis of exopolysaccharides (EPS) serves to improve soil structure and water retention in the vicinity of the root zone, establishing a microenvironment that shields the plant against abrupt changes in soil moisture levels (Panda et al., 2020). Through these physiological and biochemical adaptations, beneficial microbes empower sorghum plants to better endure drought conditions, ultimately resulting in enhanced WUE.

Certain bacteria and microbial consortia have been found in numerous studies to considerably increase sorghum's WUE (Govindasamy et al., 2017). For instance, studies have demonstrated that sorghum's resistance to drought can be increased by inoculating it with the AMF *Rhizophagus irregularis* (Idris et al., 2023). This fungus helps the plant absorb more water from the soil and improves the hydraulic conductivity of roots, enabling the plant to continue its physiological processes even in times of water scarcity. An additional investigation revealed that the introduction of *Azospirillum brasilense*, a rhizobacterium that promotes plant growth, into sorghum resulted in enhanced root length and surface area, which in turn led to better water uptake and utilization efficiency. It has also been discovered that bacterial consortiums, such as those consisting of *Bacillus* and *Pseudomonas* species, work in concert to improve WUE by generating exopolysaccharides that promote soil moisture retention and by triggering systemic resistance mechanisms in the host plant (Radhakrishnan et al., 2017; Zboralski & Filion, 2023). In addition to increasing WUE, these microbial treatments support general plant health and drought-adapted production.

Studies that draw comparisons with other crops can offer important information about the possible advantages and functioning of microbial improvements in WUE. For example, rhizobacteria and AMF have been shown to have comparable beneficial effects on WUE in different cereal crops (Brar et al., 2024). It has been demonstrated that introducing AMF species, like *Glomus mosseae*, into maize increases its tolerance to drought by improving root water uptake and lowering transpiration rates (Thangaraj et al., 2022). *Bacillus megaterium* has been shown to enhance root development and soil moisture retention in wheat, resulting in increased drought resistance. These investigations highlight how advantageous plant-microbe interactions occur in many crop species. Comparative studies also emphasize how crucial it is to choose suitable microbial strains that are well-suited to particular crops and environmental circumstances. These kinds of insights are essential for creating microbial inoculants that are specifically designed for sorghum, utilizing the success of other crops to maximize sorghum's water use efficiency (Santana et al., 2020). Researchers and farmers can improve the resilience and production of sorghum in water-limited environments by comprehending and putting these principles into practice.

12.4 ROOT MICROBIOME AND NUTRIENT USE EFFICIENCY (NUE)

Microbial enhancement of nutrient uptake plays a pivotal role in improving NUE in sorghum (Chai et al., 2024). Certain soil microorganisms, including bacteria and fungi, specialize in converting inaccessible nutrients into plant-available forms, thus facilitating nutrient absorption. For example, nitrogen-fixing bacteria such as *Rhizobium* and *Azospirillum* use the enzyme nitrogenase to transform atmospheric nitrogen into ammonia (Heiniger & Harwood, 2015). This biologically fixed nitrogen supports sorghum growth and reduces dependence on synthetic nitrogen fertilizers (Kamtchoum et al., 2019). Similarly, phosphate-solubilizing bacteria (PSB) like *Pseudomonas* and *Bacillus* release organic acids and phosphatases that solubilize insoluble phosphate compounds, making phosphorus available for plant uptake. Mycorrhizal fungi, especially AMF, enhance nutrient acquisition by extending their hyphal networks into the soil, increasing the effective root surface area and improving the uptake of immobile nutrients such as phosphorus, zinc, and copper (Busso & Busso, 2022). By facilitating the assimilation of these critical nutrients, beneficial microbes not only boost plant growth and yield but also improve fertilizer efficiency, thereby mitigating environmental impacts.

The symbiotic associations between sorghum and soil microorganisms are pivotal for nutrient cycling and the enhancement of NUE. Mycorrhizal partnerships serve as a prime example of such symbiosis, where AMF colonizes sorghum roots and, in exchange for carbohydrates derived from photosynthesis, enhance nutrient and water absorption capacities (Wipf et al., 2019). This mutually beneficial interaction significantly boosts phosphorus uptake, a nutrient often scarce in numerous soils. Moreover, these fungi can tap into nutrient reservoirs beyond the root depletion zones, thereby ensuring sustained nutrient availability. Likewise, endophytic bacteria residing within root tissues establish symbiotic connections that bolster nutrient uptake. These bacteria can synthesize plant hormones like auxins and cytokinins, which promote root growth and enhance nutrient absorption efficiency (Renugadevi et al., 2021).

Furthermore, microbial communities in the rhizosphere contribute to nutrient cycling by decomposing organic matter, converting nutrients into plant-accessible forms. Decomposer bacteria and fungi decompose intricate organic compounds, releasing nitrogen, phosphorus, and other vital nutrients back into the soil in forms accessible to plants (Thepbandit & Athinuwat, 2024). These processes not only recycle nutrients but also enhance soil quality and structure over time. By preserving a dynamic and interconnected microbial community, nutrient cycling processes are optimized, ensuring that sorghum plants receive a consistent supply of essential nutrients (Lopes et al., 2021). This comprehensive improvement of nutrient dynamics through microbial involvement is essential for attaining high NUE, supporting sustainable agricultural practices, and reducing reliance on chemical fertilizers.

Numerous investigations have identified particular microorganisms that enhance the NUE in sorghum, thus significantly contributing to improved growth and yield. *Azospirillum brasilense* is a notable nitrogen-fixing bacterium renowned for its ability to enhance NUE in sorghum (Skonieski et al., 2019). The inoculation with *A. brasilense* has been proven to elevate nitrogen absorption, increase biomass production, and improve grain yield by transforming atmospheric nitrogen into plant-usable forms. Additionally, *Bacillus megaterium*, a bacterium that solubilizes phosphate, has exhibited substantial effectiveness in enhancing phosphorus availability. Field experiments involving

Bacillus megaterium inoculation have led to increased phosphorus uptake, enhanced root and shoot growth, and overall improved plant health. AMF such as *Glomus intraradices* also play a pivotal role in enhancing NUE. These fungi establish symbiotic associations with sorghum roots, extending their hyphal networks into the soil to access phosphorus and other nutrients outside the root zone. Research has demonstrated that sorghum plants inoculated with AMF display improved nutrient uptake, enhanced resistance to drought, and higher yields compared to non-inoculated counterparts. These instances underscore the potential of harnessing beneficial microbes to optimize nutrient uptake and utilization in sorghum cultivation (Schlemper et al., 2018).

The utilization of biofertilizers and microbial inoculants has exhibited substantial potential in enhancing NUE in sorghum, providing a sustainable alternative to traditional chemical fertilizers. Biofertilizers containing nitrogen-fixing bacteria, phosphate-solubilizing microbes, and mycorrhizal fungi offer a targeted strategy for enhancing nutrient availability and uptake. For example, commercially available biofertilizers containing *A. brasilense* have been effectively applied to sorghum fields, resulting in increased nitrogen levels in plant tissues and enhanced yield performance (de Asevedo Soares et al., 2021). Similarly, biofertilizers containing *B. megaterium* have proven effective in solubilizing soil phosphorus, thereby improving phosphorus availability and uptake by sorghum plants (Vijayalakshmi et al., 2020).

Microbial inoculants, whether single strains or consortia of beneficial microbes, further enhance these advantages. In field trials, inoculants consisting of a combination of nitrogen-fixing and phosphate-solubilizing bacteria have demonstrated synergistic impacts, significantly enhancing both nitrogen and phosphorus uptake. The application of AMF inoculants has also been observed to enhance sorghum's nutrient absorption, particularly under conditions of low inputs and stress (Kamali & Mehraban et al., 2020). Furthermore, these biofertilizers and inoculants contribute to soil health by enhancing soil structure, increasing microbial diversity, and improving nutrient cycling processes. The collective effect of biofertilizers and microbial inoculants is a more robust and effective nutrient management system, leading to sustainable productivity enhancements in sorghum cultivation. By diminishing dependence on synthetic fertilizers, these biological inputs offer an eco-friendly solution for achieving high NUE and promoting sustainable agricultural practices.

12.5 STRATEGIES FOR DESIGNING THE SORGHUM ROOT MICROBIOME

Selecting advantageous microbial strains to enhance water and nutrient utilization efficiency in sorghum encompasses various crucial factors. Initially, it is imperative to validate the effectiveness of the microbes in improving nutrient absorption and stress resilience. This involves demonstrating their capacity to fix nitrogen, solubilize phosphorus, or generate growth-promoting hormones. Subsequently, ensuring compatibility with the sorghum genotype is vital. The chosen microbes must effectively colonize sorghum roots and establish a stable symbiotic association. Environmental adaptability is also essential, with the microbes needing to thrive in the specific soil types and climatic conditions where sorghum is grown (Yukun et al., 2021). Moreover, resilience to biotic and abiotic stresses is crucial for the persistence and functionality of the microbes under challenging circumstances like drought or pathogen pressure. Lastly, emphasizing the safety and non-pathogenic attributes of the microbes is paramount to prevent any negative impacts on the plant, soil ecosystem, or human health (Tokas et al., 2023). Considering scalability and cost-effectiveness in production and application are significant aspects for the practical implementation of agricultural practices. Adhering to these criteria enables researchers to pinpoint and utilize microbial strains that notably enhance sorghum's water and nutrient use efficiency.

Figure 12.2 illustrates a systematic approach to optimizing the sorghum microbiome for improved growth, productivity, and stress tolerance. The strategies for microbial inoculation are categorized into three methods: seed treatment, root treatment, and phyllosphere treatment. Each of these techniques has a specific function in microbiome engineering and targets different plant compartments (Trabelsi & Mhamdi, 2013; O'Callaghan, 2016). Before being sown, seeds are treated by covering them with advantageous microorganisms. The objectives of this approach are to increase early seedling vigor, encourage quick germination and emergence, and provide a solid basis for endophytic microbial colonization. Early microbial establishment promotes resilience against soil-borne diseases, aids in the support of the plant's early growth phases and creates the conditions for a symbiotic connection that can last the entire plant's life cycle. The main goal of root treatment is to use irrigation systems to introduce microbial inoculants into the root zone. This method improves the colonization of rhizospheric microbes, which is important for root health, water absorption, and nutrient uptake. This treatment has the potential to improve plant health and productivity by stimulating root growth, increasing NUE and WUE, and providing biocontrol against root-associated diseases by fostering beneficial microbial communities around the root. Using foliar sprays, phytosphere treatment involves introducing advantageous microorganisms to the plant's aerial sections. The goal of this technique is to colonize the phyllosphere, which will improve plant immunity to foliar diseases, encourage development, and alter the plant's overall physiology (Shinde et al., 2017). Beneficial microorganisms can interact with the metabolic processes of plants by colonizing their above-ground surfaces. This can enhance the plant's ability to assimilate nutrients and defend against environmental stresses (see Figure 12.2).

The integration of these strategies enables the comprehensive engineering of the sorghum microbiome, facilitating the establishment of a balanced microbial consortium that can adapt to varying environmental conditions and optimize plant performance. This holistic approach not only improves the plant's ability to utilize available resources more efficiently but also enhances its resilience to abiotic and biotic stresses, contributing to sustainable agricultural practices and increased crop yields. Genetic manipulation provides potent mechanisms to amplify the beneficial

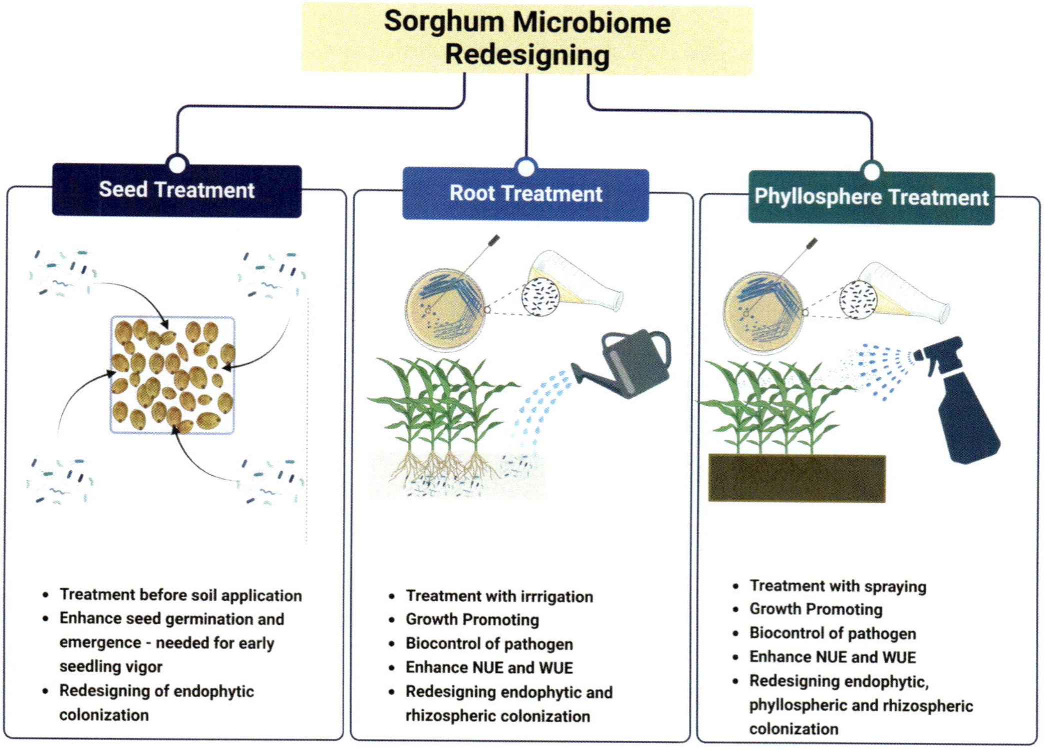

FIGURE 12.2 Sorghum microbiome engineering strategies for enhanced plant performance and resilience.

characteristics of microbes, thereby optimizing their efficacy in enhancing water and nutrient utilization efficiency in sorghum (Li et al., 2024). One strategy involves incorporating specific genes that encode enzymes or proteins facilitating nutrient uptake. For instance, transferring nitrogen fixation genes (*nif* genes) to non-nitrogen-fixing bacteria can yield engineered strains capable of converting atmospheric nitrogen into usable forms for sorghum (Guo et al., 2023). Similarly, introducing genes related to phosphate solubilization or potassium mobilization can enhance the nutrient solubilizing capabilities of microbial strains (Li et al., 2023). The utilization of CRISPR-Cas9 technology permits precise editing of microbial genomes to amplify advantageous traits while eliminating undesirable features (Croteau et al., 2018).

Another tactic entails engineering microbes to produce elevated levels of plant growth-promoting substances like auxins, cytokinins, or gibberellins, which can stimulate root growth and enhance nutrient absorption. Additionally, microbes can be genetically modified to generate stress-responsive compounds such as osmoprotectants or antifungal agents, thereby bolstering plant resilience to various stresses. Synthetic biology methodologies enable the development of synthetic microbial communities with complementary functions, establishing a resilient system that can effectively support sorghum growth in diverse conditions (Ibrahim et al., 2021). These communities can be customized to encompass multiple engineered strains, each optimized for specific roles such as nitrogen fixation, phosphate solubilization, and stress alleviation. Through leveraging genetic engineering techniques, it becomes feasible to create highly efficient microbial inoculants that significantly enhance water and nutrient utilization efficiency in sorghum, consequently leading to increased yields and more sustainable agricultural practices.

Table 12.1 provides a comprehensive overview of various microbial inoculants, application methods, observed effects, and their implications for enhancing water and nutrient utilization

TABLE 12.1
List of microbial inoculants used as a strategy for regulating the sorghum root microbiome to enhance water and nutrient utilization efficiency

Microbial inoculant	Observed effects	Implications for water and nutrient use efficiency	References
Azospirillum brasilense	Enhanced root growth, increased nitrogen uptake	Improved nitrogen utilization, better root architecture for water absorption	Fernandes Souza et al. (2023)
Bacillus subtilis	Reduced disease incidence, improved plant health	Enhanced nutrient uptake and water use efficiency due to healthier plants	Rodríguez-Hernández et al. (2020)
Pseudomonas fluorescens	Increased phosphate availability, disease resistance	Better phosphate utilization, improved overall plant vigor and water uptake	Shaik et al. (2019)
Glomus intraradices	Improved phosphorus and micronutrient uptake	Enhanced water and nutrient use efficiency through mycorrhizal associations	Nakmee et al. (2016)
Rhizobium leguminosarum	Increased nitrogen fixation, improved plant growth	Enhanced nitrogen utilization, improved growth leading to better water use	Pelletier et al. (2020)
Azotobacter chroococcum	Increased nitrogen availability, stimulated plant growth	Improved nitrogen uptake, enhanced root development for efficient water use	Jat et al. (2013)
Enterobacter cloacae	Increased phosphorus availability, improved plant growth	Better phosphorus uptake, enhanced root growth for improved water and nutrient efficiency	Singh et al. (2017)
Streptomyces spp.	Suppressed soil pathogens, enhanced nutrient availability	Improved plant health and growth, leading to better water and nutrient utilization	Gopalakrishnan et al. (2021)

efficiency in sorghum. Seed, soil, and foliar applications of microbial inoculants can be utilized to restructure the sorghum microbiome. The microbial inoculants can be formulated in solid or liquid forms, depending on the carrier material. A key consideration is maintaining the stability of viable cell numbers throughout the product's shelf life, while ensuring the organisms remain active. Encapsulation technologies have been developed to address these challenges, allowing microorganisms to survive longer in the soil and extending the shelf life of the formulations (Fadiji et al., 2024).

12.6 CONCLUSION

Improving the WUE and NUE of sorghum through root microbiome optimization is a promising strategy to enhance both productivity and sustainability. Key to this approach is understanding the diversity of microbial communities in sorghum roots, the roles of critical microbial groups such as phosphate-solubilizing microbes, mycorrhizal fungi, and nitrogen-fixing bacteria, and the factors influencing microbiome composition, including soil type and sorghum genotype. Advanced techniques like synthetic biology and CRISPR are crucial for developing beneficial bacteria and engineering artificial microbial consortia tailored for sorghum. The use of microbial inoculants and biofertilizers, alongside sustainable practices like crop rotation and organic amendments, represents effective microbiome management strategies that significantly boost microbial interventions. However, long-term field studies and a deeper understanding of plant-microbiome interactions are essential to address challenges such as the stability and persistence of introduced microbes and their interactions with native soil microbiota. The potential for scaling microbial technologies and integrating them with sustainable agriculture offers promising opportunities for widespread adoption.

CONFLICT OF INTEREST STATEMENT

The authors have declared no conflict of interest.

REFERENCES

Abera, S., Shimels, M., Tessema, T., Raaijmakers, J. M., & Dini-Andreote, F. (2022). Back to the roots: Defining the core microbiome of *Sorghum bicolor* in agricultural field soils from the centre of origin. *FEMS Microbiology Ecology*, *98*(12), fiac136.

Abreha, K. B., Enyew, M., Carlsson, A. S., Vetukuri, R. R., Feyissa, T., Motlhaodi, T., ... & Geleta, M. (2022). Sorghum in dryland: Morphological, physiological, and molecular responses of sorghum under drought stress. *Planta*, *255*, 1–23.

Afridi, M. S., Fakhar, A., Kumar, A., Ali, S., Medeiros, F. H., Muneer, M. A., ... & Saleem, M. (2022). Harnessing microbial multitrophic interactions for rhizosphere microbiome engineering. *Microbiological Research*, *265*, 127199.

Birhanu, M. W., & Negussie, Z. T. (2024). Arbuscular Mycorrhizal (AM) Fungi Symbiosis in Sustainable production of sorghum (*Sorghum bicolor* L. Moench) under drought stress: An emerging biofertilizer in dryland areas. *Advances in Agriculture*, *2024*(1), 8814543.

Brar, B., Bala, K., Saharan, B. S., Sadh, P. K., & Duhan, J. S. (2024). Bio-boosting agriculture: Harnessing the potential of fungi-bacteria-plant synergies for crop improvement. *Discover Plants*, *1*(1), 21.

Busso, M., & Busso, M. (2022). Arbuscular mycorrhizal fungi and common mycorrhizal networks benefit plants through morphological, physiological and productive traits and soil quality. *Lilloa*, *59*(2), 301–317.

Cao, Z., Zuo, W., Wang, L., Chen, J., Qu, Z., Jin, F., & Dai, L. (2023). Spatial profiling of microbial communities by sequential FISH with error-robust encoding. *Nature Communications*, *14*(1), 1477.

Carlson, R., Tugizimana, F., Steenkamp, P. A., Dubery, I. A., Hassen, A. I., & Labuschagne, N. (2020). Rhizobacteria-induced systemic tolerance against drought stress in *Sorghum bicolor* (L.) Moench. *Microbiological Research*, *232*, 126388.

Chai, Y. N., Qi, Y., Goren, E., Chiniquy, D., Sheflin, A. M., Tringe, S. G., ... & Schachtman, D. P. (2024). Root-associated bacterial communities and root metabolite composition are linked to nitrogen use efficiency in sorghum. *Msystems*, *9*(1), e01190–e01223.

Chandra, P., Singh, A., Prajapat, K., Rai, A. K., & Yadav, R. K. (2022). Native arbuscular mycorrhizal fungi improve growth, biomass yield, and phosphorus nutrition of sorghum in saline and sodic soils of the semi–arid region. *Environmental and Experimental Botany*, *201*, 104982.

Croteau, F. R., Rousseau, G. M., & Moineau, S. (2018). The CRISPR-Cas system: Beyond genome editing. *Medecine Sciences: M/S*, *34*(10), 813–819.

Dar, A., Zahir, Z. A., Iqbal, M., Mehmood, A., Javed, A., Hussain, A., ... & Ahmad, M. (2021). Efficacy of rhizobacterial exopolysaccharides in improving plant growth, physiology, and soil properties. *Environmental Monitoring and Assessment*, *193*, 1–15.

Das, P. P., Singh, K. R., Nagpure, G., Mansoori, A., Singh, R. P., Ghazi, I. A., ... & Singh, J. (2022). Plant-soil-microbes: A tripartite interaction for nutrient acquisition and better plant growth for sustainable agricultural practices. *Environmental Research*, *214*, 113821.

de Asevedo Soares, D., Andreotti, M., Vicentini, M. E., Freitas, L. A., Modesto, V. C., Nakao, A. H., ... & Teixeira Filho, M. C. M. (2021). Grain sorghum grown as second crop and inoculated with *Azospirillum brasilense* associated with nitrogen fertilization. *Revista de Agricultura Neotropical*, *8*(3), e5117.

De Filippis, F., Laiola, M., Blaiotta, G., & Ercolini, D. (2017). Different amplicon targets for sequencing-based studies of fungal diversity. *Applied and Environmental Microbiology*, *83*(17), e00905–e00917.

de Oliveira, I. F., Simeone, M. L. F., de Guimarães, C. C., Garcia, N. S., Schaffert, R. E., & de Sousa, S. M. (2021). Sorgoleone concentration influences mycorrhizal colonization in sorghum. *Mycorrhiza*, *31*, 259–264.

Diaz Franco, A., Galvez Lopez, D., & Ortiz Chairez, F. E. (2015). Bio-inoculation and reduced chemical fertilization associated with plant growth and productivity of sorghum. *Revista Internacional de Contaminación Ambiental*, *31*(3), 245–252.

Ding, W., Zhang, Y., & Shi, S. (2020). Development and application of CRISPR/Cas in microbial biotechnology. *Frontiers in Bioengineering and Biotechnology*, *8*, 711.

Enciso, J., Chavez, J. C., Ganjegunte, G., & Zapata, S. D. (2019). Energy sorghum production under arid and semi-arid environments of Texas. *Water*, *11*(7), 1344.

Fadiji, A. E., Xiong, C., Egidi, E., & Singh, B. K. (2024). Formulation challenges associated with microbial biofertilizers in sustainable agriculture and paths forward. *Journal of Sustainable Agriculture and Environment*, *3*(3), e70006.

Fernandes Souza, J., Rigueira, J. P. S., Brant Albuquerque, C. J., Júnior, V. R. R., Soares dos Santos, A., da Cunha Siqueira Carvalho, C., ... & Pinto Monção, F. (2023). Yield and nutritional value of silage of different sorghum hybrids inoculated with *Azospirillum brasilense*. *Journal of Applied Animal Research*, *51*(1), 424–433.

Fung, C., Rusling, M., Lampeter, T., Love, C., Karim, A., Bongiorno, C., & Yuan, L. L. (2021). Automation of QIIME2 metagenomic analysis platform. *Current Protocols*, *1*(9), e254.

Gamayunova, V., Honenko, L., Baklanova, T., & Pilipenko, T. (2023). Current trends in sorghum use, grain yield and water consumption depending on the hybrid composition. *Ecological Engineering & Environmental Technology*, *24*, 211–220.

Goche, T., Shargie, N. G., Cummins, I., Brown, A. P., Chivasa, S., & Ngara, R. (2020). Comparative physiological and root proteome analyses of two sorghum varieties responding to water limitation. *Scientific Reports*, *10*(1), 11835.

Gopalakrishnan, S., Srinivas, V., Naresh, N., Pratyusha, S., Ankati, S., Madhuprakash, J., ... & Sharma, R. (2021). Deciphering the antagonistic effect of Streptomyces spp. and host-plant resistance induction against charcoal rot of sorghum. *Planta*, *253*, 1–12.

Govindasamy, V., Raina, S. K., George, P., Kumar, M., Rane, J., Minhas, P. S., & Vittal, K. P. R. (2017). Functional and phylogenetic diversity of cultivable rhizobacterial endophytes of sorghum [*Sorghum bicolor* (L.) Moench]. *Antonie Van Leeuwenhoek*, *110*, 925–943.

Guo, K., Yang, J., Yu, N., Luo, L., & Wang, E. (2023). Biological nitrogen fixation in cereal crops: Progress, strategies, and perspectives. *Plant Communications*, *4*(2), 100499 .

Guo, S., Tao, C., Jousset, A., Xiong, W., Wang, Z., Shen, Z., ... & Geisen, S. (2022). Trophic interactions between predatory protists and pathogen-suppressive bacteria impact plant health. *The ISME Journal*, *16*(8), 1932–1943.

Hassan, O. H. A. (2020). Phytochemical screening and antibacterial activities of *Sorghum bicolor* leaves derived from in vitro culture. *GSC Biological and Pharmaceutical Sciences*, *10*(1), 065–072.

Heiniger, E. K., & Harwood, C. S. (2015). Posttranslational modification of a vanadium nitrogenase. *MicrobiologyOpen*, *4*(4), 597–603.

Ibrahim, M., Raajaraam, L., & Raman, K. (2021). Modelling microbial communities: Harnessing consortia for biotechnological applications. *Computational and Structural Biotechnology Journal*, *19*, 3892–3907.

Idris, I., Fefirenta, A. D., Sari, V. K., & Sudiana, I. M. (2023). Arbuscular mycorrhizal fungi induced different proline accumulations in two sorghum accessions in a response to drought stress. *Agriculture (Pol'nohospodárstvo)*, *68*(3), 127–142.

Jat, M. K., Purohit, H. S., Singh, B., Garhwal, R. S., & Choudhary, M. (2013). Effect of integrated nutrient management on yield and nutrient uptake in sorghum (*Sorghum bicolor*). *Indian Journal of Agronomy*, *58*(4), 543–547.

Jawahar, G., Rajasheker, G., Maheshwari, P., Punita, D. L., Jalaja, N., Kumari, P. H., ... & Kishor, P. B. K. (2019). Osmolyte diversity, distribution, and their biosynthetic pathways. In: Khan, M.I.R., Reddy, P.S., Ferrante, A., Khan, N.A. (Eds), *Plant Signaling Molecules* (pp. 449–458). Woodhead Publishing.

Kamali, S., & Mehraban, A. (2020). Effects of Nitroxin and arbuscular mycorrhizal fungi on the agro-physiological traits and grain yield of sorghum (*Sorghum bicolor* L.) under drought stress conditions. *Plos One*, *15*(12), e0243824.

Kamtchoum, S. M., Nchoutnji, I., Temegne, C. N., Tonfack, L. B., Fofe, L., Seutchueng, T. G. T., ... & Youmbi, E. (2019). Comparative effect of biological fixation of nitrogen and chemical fertilizer on yield optimization of two sorghum varieties in the Western Highlands. *Asian Journal of Agricultural and Horticultural Research*, *4*(3), 1–10.

Kang, D. D., Rubin, E. M., & Wang, Z. (2016). Reconstructing single genomes from complex microbial communities. *IT-Information Technology*, *58*(3), 133–139.

Khalifa, M., & Eltahir, E. A. (2023). Assessment of global sorghum production, tolerance, and climate risk. *Frontiers in Sustainable Food Systems*, *7*, 1184373.

Ku, Y. S., Liao, Y. J., Chiou, S. P., Lam, H. M., & Chan, C. (2024). From trade-off to synergy: Microbial insights into enhancing plant growth and immunity. *Plant Biotechnology Journal, 22*(9), 2461–2471.

Latorre-Pérez, A., Pascual, J., Porcar, M., & Vilanova, C. (2020). A lab in the field: Applications of real-time, in situ metagenomic sequencing. *Biology Methods and Protocols*, *5*(1), bpaa016.

Li, F., Sun, A., Jiao, X., Yu, D. T., Ren, P., Wu, B. X., ... & Hu, H. W. (2024). Nitrogenous fertilizer plays a more important role than cultivars in shaping sorghum-associated microbiomes. *Science of The Total Environment*, 943, 173831.

Li, H. P., Han, Q. Q., Liu, Q. M., Gan, Y. N., Rensing, C., Rivera, W. L., ... & Zhang, J. L. (2023). Roles of phosphate-solubilizing bacteria in mediating soil legacy phosphorus availability. *Microbiological Research*, *272*, 127375.

Liu, Q., Wang, J., Wu, A., Sun, A., Dong, E., Wang, Y., ... & Jiao, X. (2023). Plant-associated microorganisms during the reproductive period best predict sorghum yield and quality. *Field Crops Research*, *304*, 109167.

Liu, Y. X., Qin, Y., & Bai, Y. (2019). Reductionist synthetic community approaches in root microbiome research. *Current Opinion in Microbiology*, *49*, 97–102.

Lopes, L. D., Chai, Y. N., Marsh, E. L., Rajewski, J. F., Dweikat, I., & Schachtman, D. P. (2021). Sweet sorghum genotypes tolerant and sensitive to nitrogen stress select distinct root endosphere and rhizosphere bacterial communities. *Microorganisms*, *9*(6), 1329.

Lucaciu, R., Pelikan, C., Gerner, S. M., Zioutis, C., Köstlbacher, S., Marx, H., ... & Rattei, T. (2019). A bioinformatics guide to plant microbiome analysis. *Frontiers in Plant Science*, *10*, 1313.

Madhusudhana, R. (2020). Breeding for resistance to biotic stresses. In: Tonapi, V.A., Talwar, H.S., Are, A.K., Bhat, B.V., Reddy, C.R., Dalton, T.J. (Eds), *Sorghum in the 21st Century: Food–Fodder–Feed–Fuel for a Rapidly Changing World* (pp. 369–392). Springer, Singapore.

Maropola, M. K. A., Ramond, J. B., & Trindade, M. (2015). Impact of metagenomic DNA extraction procedures on the identifiable endophytic bacterial diversity in *Sorghum bicolor* (L. Moench). *Journal of Microbiological Methods*, *112*, 104–117.

Mattos, B. B., Marriel, I. E., De Sousa, S. M., Lana, U. G. D. E. P., Schaffert, R. E., & Gomes, E. A. (2020). Sorghum genotypes response to inoculation with phosphate solubilizing bacteria. *Brazilian Journal Maize Sorghum*, *19*, 14.

Mechan-Llontop, M. E., Mullet, J., & Shade, A. (2023). Phyllosphere exudates select for distinct microbiome members in sorghum epicuticular wax and aerial root mucilage. *Phytobiomes Journal*, *7*(2), 184–197.

Miransari, M. (2011). Arbuscular mycorrhizal fungi and nitrogen uptake. *Archives of Microbiology*, *193*(2), 77–81.

Nakmee, P. S., Techapinyawat, S., & Ngamprasit, S. (2016). Comparative potentials of native arbuscular mycorrhizal fungi to improve nutrient uptake and biomass of *Sorghum bicolor* Linn. *Agriculture and Natural Resources*, *50*(3), 173–178.

O'Callaghan, M. (2016). Microbial inoculation of seed for improved crop performance: Issues and opportunities. *Applied Microbiology and Biotechnology*, *100*(13), 5729–5746.

Ostmeyer, T. J., Bahuguna, R. N., Kirkham, M. B., Bean, S., & Jagadish, S. V. K. (2022). Enhancing sorghum yield through efficient use of nitrogen–challenges and opportunities. *Frontiers in Plant Science*, *13*, 845443.

Panda, R., Das, M., & Nayak, S. K. (2020). Estimation and optimization of exopolysaccharide production from rice rhizospheric soil and its interaction with soil carbon pools. *Rhizosphere*, *14*, 100206.

Pelletier, D. A., Li, Z., Lu, T. Y. S., Zhang, L., Hu, Z., Morris, G. P., ... & Pan, C. (2020). Genome sequences of 42 bacteria isolated from *Sorghum bicolor* roots. *Microbiology Resource Announcements*, *9*(37), 10–1128.

Radhakrishnan, R., Hashem, A., & Abd_Allah, E. F. (2017). Bacillus: A biological tool for crop improvement through bio-molecular changes in adverse environments. *Frontiers in Physiology*, *8*, 293128.

Renugadevi, R., Ayyappadas, M. P., Priya, V. S., Shobana, M. F., & Vivekanandhan, K. (2021). Applications of bacterial endophytes and their advanced identification methodologies. *Journal of Applied Biology and Biotechnology*, *9*(6), 51–55.

Rezakhani, L., Motesharezadeh, B., Tehrani, M. M., Etesami, H., & Mirseyed Hosseini, H. (2020). Effect of silicon and phosphate-solubilizing bacteria on improved phosphorus (P) uptake is not specific to insoluble P-fertilized sorghum (*Sorghum bicolor* L.) plants. *Journal of Plant Growth Regulation*, *39*, 239–253.

Rodríguez-Hernández, M. G., Gallegos-Robles, M. Á., Rodríguez-Sifuentes, L., Fortis-Hernández, M., Luna-Ortega, J. G., & González-Salas, U. (2020). Native *Bacillus spp.* strains as sustainable alternative in the yield of corn forage. *Terra Latinoamericana*, *38*(2), 313–321.

Rousk, J., Brookes, P. C., & Bååth, E. (2009). Contrasting soil pH effects on fungal and bacterial growth suggest functional redundancy in carbon mineralization. *Applied and Environmental Microbiology*, *75*(6), 1589–1596.

Santana, S. R. A., Voltolini, T. V., Antunes, G. D. R., da Silva, V. M., Simões, W. L., Morgante, C. V., ... & Fernandes-Júnior, P. I. (2020). Inoculation of plant growth-promoting bacteria attenuates the negative effects of drought on sorghum. *Archives of Microbiology*, *202*, 1015–1024.

Schlemper, T. R., Dimitrov, M. R., Gutierrez, F. A. S., van Veen, J. A., Silveira, A. P., & Kuramae, E. E. (2018). Effect of *Burkholderia tropica* and *Herbaspirillum frisingense* strains on sorghum growth is plant genotype dependent. *PeerJ*, *6*, e5346.

Sevim, V., Lee, J., Egan, R., Clum, A., Hundley, H., Lee, J., ... & Woyke, T. (2019). Shotgun metagenome data of a defined mock community using Oxford Nanopore, PacBio and Illumina technologies. *Scientific Data*, *6*(1), 285.

Shaik, A. B., Ahil, S. B., Govardhan, R., & Pal, P. P. (2019). Management of sorghum grain mold through induced resistance mechanism using foliar spray of Pseudomonas fluorescens strains and non-conventional chemicals: A field study. *Archives of Phytopathology and Plant Protection*, *52*(11–12), 969–988.

Shi, X., Guo, P., Chen, Y., Liu, C., Liu, C., Yu, H., ... & Zou, H. (2024). Integrated analysis of soil metagenome and soil metabolome reveals the differential responses of sorghum and peanut rhizosphere microbes to salt stress. *Journal of Soil Science and Plant Nutrition*, 24(2), 2959–2971.

Shinde, N. K., Seema, M. N., Ghorade, R. B., & Mohod, N. B. (2017). Effect of integrated nutrient management on yield and quality of parching sorghum varieties. *International Journal of Current Microbiology and Applied Sciences*, *6*(12), 4268–4272.

Singh, R. P., Nalwaya, S., & Jha, P. N. (2017). The draft genome sequence of the plant growth promoting rhizospheric bacterium *Enterobacter cloacae* SBP-8. *Genomics Data*, *12*, 81–83.

Skonieski, F. R., Viégas, J., Martin, T. N., Mingotti, C. C. A., Naetzold, S., Tonin, T. J., ... & Meinerz, G. R. (2019). Effect of nitrogen topdressing fertilization and inoculation of seeds with *Azospirillum brasilense* on corn yield and agronomic characteristics. *Agronomy*, *9*(12), 812.

Soucy, S. M., Huang, J., & Gogarten, J. P. (2015). Horizontal gene transfer: Building the web of life. *Nature Reviews Genetics*, *16*(8), 472–482.

Sun, A., Jiao, X. Y., Chen, Q., Wu, A. L., Zheng, Y., Lin, Y. X., ... & Hu, H. W. (2021). Microbial communities in crop phyllosphere and root endosphere are more resistant than soil microbiota to fertilization. *Soil Biology and Biochemistry*, *153*, 108113.

Thangaraj, K., Li, J., Mei, H., Hu, S., Han, R., Zhao, Z., ... & Kamatchi Reddiar, D. (2022). Mycorrhizal colonization enhanced *Sorghum bicolor* tolerance under soil water deficit conditions by coordination of proline and reduced glutathione (GSH). *Journal of Agricultural and Food Chemistry*, *70*(14), 4243–4255.

Thepbandit, W., & Athinuwat, D. (2024). Rhizosphere microorganisms supply availability of soil nutrients and induce plant defense. *Microorganisms*, *12*(3), 558.

Tokas, D., Singh, S., Yadav, R., & Singh, A. N. (2023). Plant-microbe interactions: Role in sustainable agriculture and food security in a changing climate. In: Swapnil, P., Meena, M., Harish, Marwal, A., Vijayalakshmi, S., Zehra, A. (Eds), *Plant-Microbe Interaction-Recent Advances in Molecular and Biochemical Approaches* (pp. 363–391). Academic Press.

Trabelsi, D., & Mhamdi, R. (2013). Microbial inoculants and their impact on soil microbial communities: A review. *BioMed Research International*, *2013*(1), 863240.

Trivedi, N. (2021). Improved plant resistance by phytomicrobiome community towards biotic and abiotic stresses. In A. Verma, J. K. Saini, A. E. Hesham, & H. B. Singh (Eds.), *Phytomicrobiome Interactions and Sustainable Agriculture* (pp. 207–216). Wiley.

Vijayalakshmi, V., Pradeep, S., Manjunatha, H., Krishna, V., & Jyothi, V. (2020). The impact of nitrogen fixers and phosphate solubilizing microbes on sorghum (*Sorghum bicolor*) yield. *Current Biotechnology*, *9*(3), 198–208.

Wang, P., Chai, Y. N., Roston, R., Dayan, F. E., & Schachtman, D. P. (2021). The sorghum bicolor root exudate sorgoleone shapes bacterial communities and delays network formation. *MSystems*, *6*(2), 10–1128.

Wang, Q., Liu, M., Wang, Z., Li, J., Liu, K., & Huang, D. (2024). The role of arbuscular mycorrhizal symbiosis in plant abiotic stress. *Frontiers in Microbiology*, *14*, 1323881.

Wipf, D., Krajinski, F., van Tuinen, D., Recorbet, G., & Courty, P. E. (2019). Trading on the arbuscular mycorrhiza market: From arbuscules to common mycorrhizal networks. *New Phytologist*, *223*(3), 1127–1142.

Xu, L., Naylor, D., Dong, Z., Simmons, T., Pierroz, G., Hixson, K. K., ... & Coleman-Derr, D. (2018). Drought delays development of the sorghum root microbiome and enriches for monoderm bacteria. *Proceedings of the National Academy of Sciences*, *115*(18), E4284–E4293.

Yahaya, M. A., & Shimelis, H. (2022). Drought stress in sorghum: Mitigation strategies, breeding methods and technologies—A review. *Journal of Agronomy and Crop Science*, *208*(2), 127–142.

Yuan, A., Kumar, S. D., Wang, H., Wang, S., Impa, S., Wang, H., ... & Shao, R. (2024). Dynamic interplay among soil nutrients, rhizosphere metabolites, and microbes shape drought and heat stress responses in summer maize. *Soil Biology and Biochemistry*, *191*, 109357.

Yuan, Y., Shi, Y., Liu, Z., Fan, Y., Liu, M., Ningjing, M., & Li, Y. (2023). Promotional properties of ACC deaminase-producing bacterial strain DY1–3 and its enhancement of maize resistance to salt and drought stresses. *Microorganisms*, *11*(11), 2654.

Yukun, G., Jianghui, C., Genzeng, R., Shilin, W., Puyuan, Y., Congpei, Y., ... & Jinhua, C. (2021). Changes in the root-associated bacteria of sorghum are driven by the combined effects of salt and sorghum development. *Environmental Microbiome*, *16*, 1–15.

Zboralski, A., & Filion, M. (2023). Pseudomonas spp. can help plants face climate change. *Frontiers in Microbiology*, *14*, 1198131.

Zhang, Y., Dong, L., & Shangguan, Z. (2023). Appropriate N addition improves soil aggregate stability through AMF and glomalin-related soil proteins in a semiarid agroecosystem. *Land Degradation & Development*, *34*(3), 710–722.

Ziegler, M. Takors, R. (2020). Reduced and Minimal Cell Factories in Bioprocesses: Towards a Streamlined Chassis. In *Minimal Cells: Design, Construction, Biotechnological Applications*; Lara, A.R., Gosset, G., Eds.; Springer Nature: Basel, Switzerland, pp. 1–44..

13 Omics Approaches to Modulate Sorghum Seed Compositional Dynamics for Improved Health Alternatives

Mohammad Jafar Tanin, Praveenkumar Alagappan, Jatinder Paul Kaur, Arshvir Kaur Boparai, Shivani Upadhyay, and Himanshu Sharma

13.1 INTRODUCTION

Sorghum, a member of the Poaceae family and widely cultivated as a cereal crop, was domesticated approximately 4,000 years ago and is native to Africa (Ananda et al., 2020). As the fifth most produced cereal crop globally, following barley, wheat, rice, and maize, it has proven to be an elite resilient C_4 plant, capable of tolerating drought conditions and commonly grown in arid and semi-arid regions (Kazungu et al., 2023). In addition, sorghum can tolerate low temperatures ranging from 12 to 15°C and thrive in areas with high precipitation and waterlogging conditions (Muller et al., 2020), making it a crucial staple food for millions of people in developing countries. Its resilience also positions it as a potential solution to changing climate and food security challenges.

Sorghum offers numerous health benefits due to its rich nutrient content, serving as a significant source of polysaccharides, lipids, proteins, and essential minerals such as iron and zinc, which are vital for meeting daily human nutritional needs (Tanwar et al., 2023). Moreover, sorghum contains bioactive compounds like phenolic compounds, flavonoids, stilbenes, phytosterols, and condensed tannins, which contribute to its potent antioxidant capacity (Liangli et al., 2012). These bioactive compounds are linked to various health benefits, including reducing inflammation, inhibiting cancer, and promoting cardiovascular health (Oluwole et al., 2022). Sorghum's polysaccharides provide energy (Jaworski et al., 2015), while its proteins and lipids support human growth and development (de Morais Cardoso et al., 2017). Furthermore, being gluten-free, sorghum is an excellent choice for individuals with celiac disease or gluten intolerance (Moreno et al., 2014). With higher levels of protein, iron, and other vitamins and minerals than common grains such as maize, wheat, and rice, sorghum is a highly nutrient-dense grain (Tanwar et al., 2023). Thus, optimizing these components is essential for enhancing sorghum's role in nutrition and health.

Despite its robust nature and favorable nutritional profile, sorghum's content of certain bioactive compounds is considered lower or less diverse compared to other major cereals, such as wheat, rice, and maize (Ananda et al., 2020). Therefore, improving the nutritional composition of sorghum grain is critical to maximizing its potential as a functional food ingredient and enhancing its value for human consumption. This is particularly vital given the crop's significant role in ensuring nutritional security and addressing its importance under climate change scenarios. To achieve this, it is essential to have a reliable method for developing elite biofortified, climate-resilient cultivars with high

DOI: 10.1201/9781003623717-13

yields. Plant breeding is widely recognized as a key strategy for agricultural advancement, aiming to ensure a sufficient supply of nutritious food for the growing global population (Qaim, 2020; Hasan et al., 2021). To tackle the challenges posed by climate change and to meet the nutritional demands of the expanding population, innovative technologies are necessary to enhance plant breeding efficiency (Rial-Lovera et al., 2017). Recent advancements in omics technologies, such as genomics, transcriptomics, proteomics, metabolomics, epigenomics, and microbiomics, have significantly improved the understanding of sorghum's genetic, molecular, and biochemical foundations (Abera et al., 2022; Draye et al., 2001). Multi-omics studies have effectively analyzed omics data, illustrating the molecular regulators and metabolic mechanisms governing the nutritional features and yield of sorghum (Mahmood et al., 2022). Furthermore, progress in DNA sequencing technologies has facilitated high-throughput genotyping and streamlined data analysis for omics studies, enhancing the efficiency of the genotyping process. As a result, these technologies enable the identification of key genes, regulatory networks, and metabolic pathways responsible for sorghum's nutritional composition, offering opportunities to manipulate these factors for biofortification purposes.

Several studies have highlighted various breeding and genetic methods for biofortification and nutritional improvement in sorghum (Elkonin et al., 2018; Are et al., 2019), but none have provided comprehensive insights into the importance and application of omics approaches for biofortification. This chapter aims to fill this gap by offering a detailed discussion on sorghum biofortification. It explores the chemical composition and nutritional value of sorghum seeds, examining the roles of polysaccharides, proteins, lipids, minerals, vitamins, and bioactive compounds. The chapter also highlights various omics approaches for sorghum biofortification, including genomic strategies like QTL mapping, GWAS, and GS studies; transcriptomics techniques such as RNAseq, scRNAseq, and candidate gene validation using qRT-PCR; and proteomics, metabolomics, epigenomics, and microbiomics focusing on seed endophytes. Finally, the integration of these multi-omics approaches for a holistic understanding and enhancement of sorghum seed composition is discussed, addressing the challenges and future perspectives in the field. By harnessing the power of omics technologies, plant breeders can significantly improve the nutritional quality and health benefits of sorghum, ultimately making it a more valuable crop for tackling global nutritional challenges. This chapter provides a comprehensive overview of the current state of knowledge and recent advancements in this area, emphasizing the potential of omics approaches to modulate sorghum seed compositional dynamics for better health alternatives.

13.2 CHEMICAL COMPOSITION AND NUTRITIONAL VALUE OF SORGHUM GRAINS

As shown in Figure 13.1, sorghum grain contains key nutritional components, including essential polysaccharides, proteins, vitamins, minerals, and bioactive compounds, which contribute significantly to its nutritional value and health benefits.

13.2.1 Polysaccharides

Sorghum grains include polysaccharides such as starch, fructose, raffinose, sucrose, maltose, and glucose. The polysaccharides are mainly divided into starch and non-starch. Starch, the primary component, is predominantly in the endosperm, constituting 55–70% of the grain's weight (Khoddami et al., 2023). Similar to other starches, sorghum starch comprises amylose and amylopectin, although waxy sorghum may contain less amylose (Xiong et al., 2019). Recent research indicates that sorghum starches typically exhibit higher gelatinization temperatures compared to starches from wheat, maize, and barley, due to fewer short branch chains in sorghum amylopectin (Ali & Hasnain, 2013). Lower gelatinization temperature, higher water uptake, and viscosity are

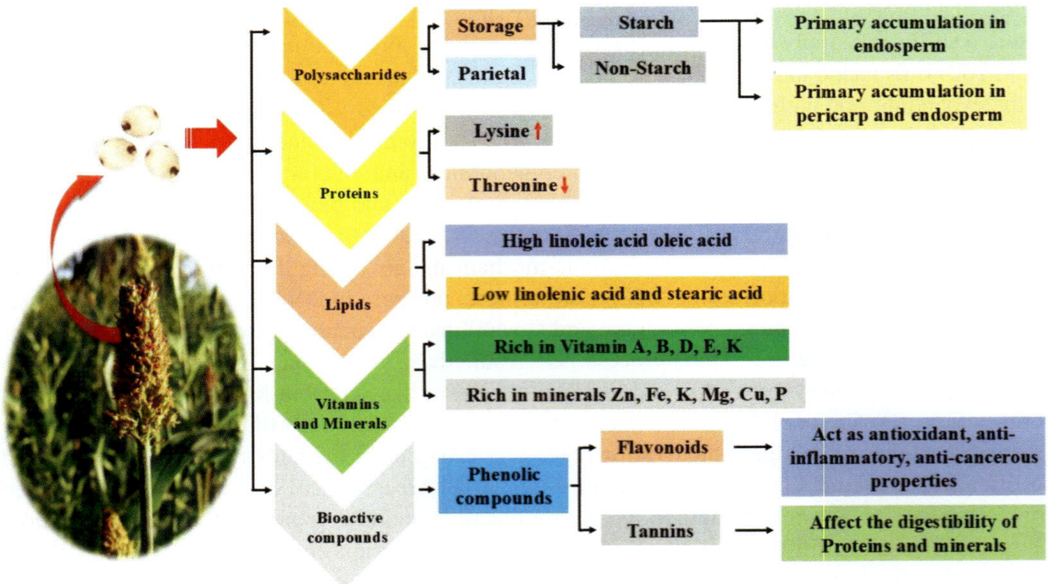

FIGURE 13.1 Illustrating the key nutritional components of sorghum grain, highlighting its rich content of essential nutrients such as polysaccharides, proteins, vitamins, minerals, and bioactive compounds which contribute to its nutritional value and health benefits.

advantageous for baking sorghum bread, whereas higher gelatinization temperature and low viscosity favor stiff porridge baking (Saeed Omer et al., 2023).

Sorghum grains contain free sugar at a concentration of 1–2% (Khoddami et al., 2023). The non-starch polysaccharides available in the sorghum grains, specifically in the pericarp as well as endosperm cell walls, potentially make the endosperm carbohydrates more water-insoluble (Miafo et al., 2019). Unlike wheat endosperm, which contains water-soluble arabinoxylans, sorghum endosperm cell walls are rich in water-insoluble glucuronoarabinoxylans, which can negatively impact bread-making and brewing processes (Dahir et al., 2015; Zhurlova, 2017). Additionally, sorghum starch has a lower *in vitro* digestibility at 33–48%, than that of maize starch at 53–58% (Ezeogu et al., 2005). Floury sorghum grains, characterized by an amorphous structure, are more susceptible to enzymatic breakdown during milling, resulting in higher starch digestibility compared to corneous grains (Khoddami et al., 2023). Non-waxy sorghum is characterized by much lower starch digestibility than that of waxy sorghum (Elhassan et al., 2015).

13.2.2 Proteins

Sorghum grains exhibit varying levels of protein, ranging from 6% to 25%, depending on the variety and growing conditions (Bean et al., 2019). The protein in sorghum is primarily found in the endosperm, and its content is influenced by factors such as variety and environmental conditions. Sorghum's protein is rich in essential amino acids, but it is relatively low in threonine compared to other grains such as wheat and rice (Taylor & Taylor, 2024). As a result, sorghum may not be the best sole protein source and is often recommended to be complemented with other protein-rich foods. The primary protein types in sorghum include albumin, globulins, glutelins, and kafirins (De Mesa-Stonestreet et al., 2010). Notably, sorghum lacks gliadin and glutenin, the two main components of gluten, making it a naturally gluten-free grain (Tanwar et al., 2023). This characteristic makes sorghum an appropriate food resource for individuals with celiac disease or those following a gluten-free diet.

Kafirins, which account for 50–80% of the total protein content in sorghum, are hydrophobic and protease-resistant (De Mesa-Stonestreet et al., 2010). Glutelins make up the second largest protein fraction (Tanwar et al., 2023). Despite being a valuable protein source, sorghum's amino acid profile is not entirely balanced. Its lysine content is relatively low, which is why it is often recommended to consume sorghum in combination with legumes or other lysine-rich foods to ensure a balanced diet (Henley et al., 2010). The kafirins, which are the predominant proteins in sorghum, are particularly low in lysine. However, both β- and γ-kafirins are rich in cysteine, another essential amino acid, while δ-kafirin is relatively high in methionine (Xiao et al., 2017). The protein in both the outer and inner layers of the endosperm primarily consists of α-kafirin, with the inner layer containing higher concentrations of β- and γ-kafirins compared to the outer layer (Khoddami et al., 2023).

Studies on the protein components in sorghum grain have shown that γ- and β-kafirins surround the more digestible α-kafirin within a disulfide bond polymer network, which helps protect it from protease exposure (Shull et al., 1992). *In vivo* and *in vitro* studies have reported that the digestibility of sorghum protein ranges between 30% and 72%, which is lower than that of wheat (81%) and maize (73%) (Axtell et al., 1981). However, the digestibility of whole-grain sorghum can be improved by 35% after decortication and extrusion, reaching 46% (MacLean et al., 1981). Despite this improvement, sorghum proteins are unable to form the gas-entrapping viscoelastic dough essential for making flatbread and baked goods (Taylor & Dewar, 2001). Additionally, the digestibility of sorghum protein may be further reduced during processing, such as cooking (Arbab & El Tinay, 1997).

13.2.3 Lipids

Sorghum grain contains approximately 3% lipids, a level lower than that of millet and maize but higher than that of rice and wheat (Dykes & Rooney, 2006). The lipids in sorghum are primarily concentrated in the scutellum and have a characteristic fatty acid profile, which includes 49% linoleic acid, 31% oleic acid, 14% palmitic acid, 2.7% linolenic acid, and 2.1% stearic acid (Dykes & Rooney, 2006). This lipid content can influence the flavor and shelf life of sorghum-based foods (Khoddami et al., 2023).

The fatty acid profile of sorghum is similar to that of maize, with linoleic acid and oleic acid making up 80% of the total fatty acids, although sorghum contains a lower overall lipid content. Sorghum also tends to have relatively high levels of lipid waxes, including fatty alcohols, wax esters, and fatty aldehydes, compared to maize and other grains. For example, the wax content in sorghum is approximately 2.0 mg/g of grain, which is about 50 times higher than that of maize (Weller & Hwang, 2005). Among these waxes, polycosanols are notable for their potential health benefits, such as providing protection against cardiovascular disease (Awika, 2017). Sorghum contains about 75 mg of policosanols per 100 g of grain (Aruna & Visarada, 2019).

13.2.4 Minerals and vitamins

Sorghum grain is a significant source of β-carotene (vitamin A precursor), tocopherols, and both fat-soluble (A, D, K) and water-soluble (B) vitamins (viz., thiamine, riboflavin, pyridoxine) (Martino et al., 2012). It is rich in minerals such as magnesium, iron, zinc, copper, calcium, phosphorus, and potassium, primarily concentrated in the pericarp, aleurone layer, and germ (Anglani, 1998). However, decortication removes these essential nutrients. Sorghum's bioavailability of minerals is hindered by polyphenols and phytates, with zinc and iron availability ranging from 9.7% to 17.1% and 6.6% to 15.7%, respectively (Kruger et al., 2014).

13.2.5 Bioactive Compounds

13.2.5.1 Phenolic Compounds

Phenolic compounds, a major group of phytochemicals in plants, are characterized by an aromatic ring with one or more hydroxyl groups (Chirinos et al., 2009). They are classified as simple phenols or polyphenols, which include coumarins, lignins, lignans, tannins, phenolic acids, and flavonoids (Soto-Vaca et al., 2012). These compounds are synthesized in response to ecological and physiological stresses such as pathogen attacks, insect damage, UV radiation, and wounding (Pratyusha, 2022). As antioxidants, phenolic compounds are believed to help prevent heart disease (Dai & Mumper, 2010), reduce inflammation (Zhang et al., 2019), lower cancer and diabetes risks, and inhibit mutagenesis in human cells (Sawadogo et al., 2012). The health benefits of consuming plant-based foods like fruits, vegetables, and legumes are largely attributed to their phenolic content. Phytochemicals, especially phenolics, have also gained recognition for their antioxidant properties and potential as food colorants, especially with increasing restrictions on synthetic antioxidants (Zhang et al., 2021).

Sorghum grain is particularly rich in phenolic compounds, such as flavonoids, phenolic acids, and tannins, which exhibit strong antioxidant activity (Awika, 2011) and are more effective than those from other plants (Rhodes & Price, 1997). Sorghum also contains anthocyanins and flavonoids with anti-inflammatory properties (Shim et al., 2013) and offers a range of health benefits, including anti-cardiovascular, anti-diabetic, anti-cancer, anti-lipidemic, immunomodulatory, anti-anemic, neuroprotective, anti-diarrheal, anti-microbial, and anthelminthic effects (Massey et al., 2014; Ademiluyi et al., 2014; Mukkun et al., 2024). Whole-grain sorghum has been shown to lower plasma glucose in diabetics compared to whole wheat and decorticated sorghum (Lakshmi & Vimala, 1996). Furthermore, non-tannin sorghum lipid extract reduces cholesterol absorption in hamsters, likely due to policosanols and sterols in the lipid fraction (Carr et al., 2005).

13.2.5.2 Tannins

Sorghum tannins are known for forming poorly digestible protein complexes, which can reduce its feed value, but they also offer health benefits. In addition to flavan-3-ols, 3-deoxyflavans contribute to the polymeric structure of sorghum tannins, with glycosylated 3-deoxyflavan polymers and terminal flavanones identified in sorghum (Krueger et al., 2003). Although 3-deoxyflavonoids are common as monomers in sorghum, they play a minor role in polymeric tannin synthesis, indicating that they are less involved in flavonoid polymerization. Along with the classical 4→8 B-type flavan linkages, A-type interflavan linkages are also present, highlighting the structural diversity of sorghum tannins. Condensed tannins, the most studied flavonoids in sorghum, bind strongly to proteins, a characteristic that not only reduces sorghum's digestibility by over 70% when cooked but is also exploited in leather production. Additionally, tannins can bind micronutrients like iron and zinc, further compromising the nutritional value of sorghum. A study on guinea pigs showed that those fed with a diet containing 58% low-tannin sorghum had significantly lower cholesterol levels compared to those fed wheat, rolled oats, or pearl millet (Yang et al., 2023). Tannins also inhibit digestive enzymes, particularly in type III high-tannin sorghums (Awika & Rooney, 2004). Despite these anti-nutritional effects, tannins serve a protective role against bird pests, reducing predation on maturing sorghum grain (Mofokeng & Shargie, 2016). In regions with severe bird pest issues, such as parts of Eastern and Southern Africa, tannin-rich sorghum varieties are more commonly grown, as birds tend to prefer non-tannin varieties.

13.2.5.3 Flavonoids

Flavonoids in sorghum are a unique group of phenolic compounds that contribute to the diverse coloration of sorghum grains (Xiong et al., 2019). As the most abundant and diverse group of plant phenolics, flavonoids are characterized by a flavan structure, with classification based on substitutions around the heterocyclic (C) ring. While flavonoids are minor components in most cereals such as rice, wheat, maize, and oats, except in some pigmented varieties (Rusu et al., 2023), sorghum stands out for its high levels and wide range of flavonoids, many of which are rare in other cereals and food plants. This makes sorghum flavonoids a valuable area for research in health and nutrition.

Sorghum flavonoids are classified into monomeric and polymeric types. Monomeric flavonoids in sorghum include 3-deoxyanthocyanins, flavones, flavanones, and flavan-4-ols. Notably, sorghum's monomeric flavonoids are predominantly 3-deoxyflavonoids, which lack a substitution at position 3 of the heterocyclic (C) ring. This structural characteristic influences the chemical and biochemical properties that are significant for food processing and health (Yang et al., 2009). Polymeric flavonoids in sorghum are mainly tannins, which are condensed polymers of flavan-3-ols with an average polymerization degree of around 15, connected by 4→8 interflavan linkages, forming organized linear structures (Hammouda et al., 2013).

13.2.5.4 Phytosterols and Polycosanols

Sorghum grains are rich in lipids, particularly polycosanols and phytosterols, which offer various health benefits. Polycosanols, accounting for 33.4% to 44% of the total lipid content, have been shown to reduce cholesterol levels and promote cardiovascular health. Unpolished sorghum grains contain 74.5 mg/100 g of polycosanols, whereas polished grains contain only 9.8 mg/100 g due to the removal of the outer layers during the polishing process (Tanwar et al., 2023). In addition to polycosanols, sorghum is also a significant source of phytosterols such as sitosterol, campesterol, and stigmasterol. These plant-based compounds help lower cholesterol levels and, in some cases, sorghum's phytosterol content exceeds that of fruits, vegetables, and other cereals, with the levels varying across different varieties and influenced by cultivation practices (Chung et al., 2013).

However, despite these beneficial compounds, sorghum also contains anti-nutritional factors, including phytates, lectins, and protease inhibitors (e.g., trypsin, chymotrypsin, and amylase), which can negatively impact protein and carbohydrate digestibility, as well as mineral bioavailability.

13.3 GENOMIC APPROACHES FOR SORGHUM BIOFORTIFICATION

Since 2005, sorghum breeding has been transformed by advancements in sequencing technologies, trait mapping, genome editing, genome selection, and high-throughput phenotyping. These innovations have enabled large-scale identification of polygenic and oligo-genic effects, enhancing the use of both natural and artificial variation. Biofortification efforts for future climates are also advancing through omics integration, revealing gene-protein-trait relationships. While QTL analysis remains crucial, challenges like marker availability and recombination rates limit its resolution (Tanin et al., 2024). Advances in NGS have made genotyping-by-sequencing (GBS) a cost-effective tool for species with large genomes, facilitating SNP discovery and genotyping without prior genomic knowledge. GBS and GWAS have successfully utilized for trait mapping in sorghum, linking genomic regions to traits and identifying causal SNPs for agronomic and phytochemical traits (Elshire et al., 2011; Kim et al., 2016). Zhang et al. (2013) mapped flanking markers of 604 QTL from 35 studies onto the sorghum genome assembly (v1.4), although they did not adjust QTL confidence intervals based on factors influencing the reliability of the original studies. A search on PubMed Central database using terms like crop name, QTL, and GWAS reveals 2.7 times more hits for maize and 3.6 times more for rice compared to sorghum.

A major concern in sorghum is the limited bioavailability of micronutrients due to the presence of polyphenols and tannins. Phuke et al. (2017) identified QTLs for grain Fe and Zn content in a recombinant inbred line (RIL) population from the cross 296B × PVK 801, which showed significant variation in these micronutrients. A significant G×E interaction was observed, with Fe being more influenced by the environment than Zn. The genetic map for this RIL population covered

1,355.52 cM with 2,088 markers, identifying 11 QTLs for Fe and Zn (individual environments) and 3 QTLs across environments (Kotla et al., 2019). Key studies have also identified important QTLs for grain composition. Mutations regulating amylose production (Karper, 1933), sucrose content (Boyer and Liu, 1983), and lysine content (Singh & Axtell, 1973) have been linked to specific QTLs, with two major QTLs for high protein digestibility found on chromosome 1 (Winn et al., 2009). Furthermore, QTLs for grain Fe and Zn concentrations offer new opportunities for early generation selection (Kotla et al., 2016).

Linkage and association studies have mapped genes underlying sorghum grain composition, including the waxy mutation located on chromosome 10 (McIntyre et al., 2008; mace & Jordan, 2011). The first association mapping study for sorghum grain quality identified significant MTAs related to starch synthesis and grain storage proteins (de Alencar Figueiredo et al., 2010). GWAS has further identified QTLs for mineral characteristics, protein, fat, and starch content, with some genes, such as alpha-amylase-3, strongly associated with protein and fat variation (Rhodes et al., 2014; 2017). Additionally, a robust QTL linked to crude fat was found on chromosome SBI-10, explaining up to 28.1% of the phenotypic variation in one environment (Boyles, 2017). Other studies have identified SNPs associated with carotenoid content, as well as genes regulating the accumulation of Zn, Mn, and other elements (Cruet-Burgos & Rhodes, 2023). The joint linkage-association mapping approach has been applied in sorghum, but meta-analyses across studies are needed for more comprehensive identification of QTLs influencing complex traits (Gudi et al., 2022; Halladakeri et al., 2023; Sethi et al., 2023). Overall, improving the nutritional quality of sorghum relies on leveraging QTLs/genes identified through GWAS and mutation studies, which can be introgressed into elite cultivars via genomics-assisted breeding (Varshney et al., 2021).

13.4 TRANSCRIPTOMIC ADVANCES FOR SORGHUM BIOFORTIFICATION

Advances in transcriptomics, including RNA sequencing (RNAseq), single-cell RNA sequencing (scRNAseq), and quantitative real-time PCR (qRT-PCR), have greatly facilitated this effort by enabling detailed studies of gene expression and regulatory mechanisms. RNAseq, a high-throughput technique, provides insights into gene expression patterns and regulatory networks, identifying genes involved in nutrient metabolism and transport. It has revealed pathways linked to iron (Kabir & Bennetzen, 2024), zinc (Abou-Elwafa et al., 2019), starch and phenolic compounds (Wang et al., 2023), carbohydrates and flavonoids (Zhou et al., 2020), proteins and fats (Rhodes et al., 2017), and vitamins (Cruet-Burgos et al., 2023). For instance, genes associated with carotenoid biosynthesis, degradation, and precursors were found to be differentially expressed across developmental stages in sorghum grains, demonstrating the feasibility of breeding for higher carotenoid content (Cruet-Burgos et al., 2020). RNAseq has also been used to analyze genes related to cell wall biosynthesis and remodeling, providing insights crucial for improving grain quality (Costes et al., 2023). Differentially expressed genes identified via RNAseq serve as markers for breeding programs and help reconstruct metabolic pathways, advancing our understanding of nutrient accumulation.

scRNAseq enables gene expression analysis at the single-cell level, offering a high-resolution view of cellular heterogeneity and tissue-specific regulation. This method reveals cell-type-specific expression patterns and uncovers regulatory networks controlling nutrient biosynthesis, aiding in targeted biofortification strategies (Serrano-Ron et al., 2021). scRNAseq has also provided temporal insights into gene expression during seed development, essential for optimizing nutrient accumulation (Li et al., 2024). qRT-PCR validates candidate genes identified through RNAseq or scRNAseq, quantifying their expression across tissues and developmental stages (Chaudhari and Fakrudin 2017). It has been used to investigate genes such as *SbDMAS2*, *SbNAS2*, and *SbNAS3*, which play significant roles in enhancing micronutrient content in sorghum grains (Guleria et al., 2023). By overexpressing or silencing these genes in transgenic sorghum lines, researchers confirm their function and establish causative links to biofortification traits (Khan et al., 2024). Integrating these transcriptomic approaches provides a robust framework for understanding the genetic basis of biofortification and accelerating the development of nutritionally enriched sorghum varieties.

13.5 PROTEOMIC APPROACHES FOR SORGHUM BIOFORTIFICATION

Studies reveal significant genetic variation in sorghum for mineral concentration (Ng'uni et al., 2012; Susmita & Selvi, 2014; Ng'Uni et al., 2016; Djanaguiraman et al., 2024). However, uneven nutrient distribution among plant parts poses a major challenge to achieving biofortification goals (Tang et al., 2008). In gramineous grains like wheat and sorghum, mineral accumulation and distribution are heavily influenced by the source-sink relationship, which remains inadequately understood (Kirkby et al., 2023). In biofortified wheat, iron concentrates in the aleurone layer and embryo's scutellum, while zinc is distributed across the embryonic axis and scutellum epithelium (Wan et al., 2022). Sorghum exhibits similar patterns, though with notable cell-to-cell variability in iron and zinc content (Gaddameedi et al., 2022). Understanding the mechanisms of seed development, nutrient composition, and accumulation is critical for effective biofortification (Yang et al., 2007).

Transporter proteins embedded in cell membranes play a vital role in nutrient uptake, distribution, and storage, making their identification and characterization essential for improving biofortified crops (Singh et al., 2020). Advances in proteomics, including gel-based and gel-free approaches, have enabled the profiling of such proteins, facilitating the understanding of biochemical pathways and linking gene products with their functional responses (Sharma et al., 2023). For instance, proteomics has identified zinc-regulated transporters and iron-regulated transporter-like proteins (ZIP) in various crops, including sorghum, contributing to enhanced nutrient uptake and remobilization (Li et al., 2019).

Proteomics also aids in elucidating nutrient assimilation pathways and transport mechanisms, which can help increase mineral concentrations in grains (Sohail et al., 2022). African biofortified sorghum demonstrates the potential of transgenic technology by enhancing grain digestibility, improving iron and zinc bioavailability, and boosting levels of essential amino acids and vitamins (Zhao, 2007). However, genetic modifications can result in unintended effects, such as toxins, antinutrients, or allergens, raising safety concerns about their impact on human health (Gong et al., 2010). Systematic proteomic analyses are therefore essential for identifying these trade-offs and ensuring the biological safety of genetically modified crops (Gong & Wang, 2013).

Proteomics bridges the gap between genotype and phenotype by identifying biomarkers associated with nutrient traits and tracking protein expression changes influencing complex traits like zinc and iron content in grains (Barkla, 2016). Such biomarkers enable the development of molecular tools to enhance nutrient uptake efficiency, tolerance to nutrient stress, and overall crop productivity (Cremer et al., 2014). While genomic sequencing offers valuable insights, proteomics provides a direct link to protein function and plant physiology, making it indispensable for comprehensive molecular characterization and biofortification strategies (Sharan et al., 2018).

13.6 METABOLOMIC ADVANCES TO SUPPORT SORGHUM BIOFORTIFICATION

Metabolomics involves the quantitative analysis of metabolites, small organic molecules vital for biochemical processes in cells, tissues, and organisms (Daviss, 2005). Metabolites are classified as primary (e.g., carbohydrates, proteins, lipids) for growth and development or secondary (e.g., phenolics, alkaloids) for defense and interactions (Tissier et al., 2014). The study of all metabolites, the metabolome, offers insights beyond proteomics by measuring metabolic expression and identifying key biochemical events influencing gene function (Jamaloddin et al., 2021). Techniques like gas chromatography-mass spectroscopy (GC-MS), capillary electrophoresis mass spectroscopy (CE-MS), liquid chromatography-mass spectroscopy (LC-MS), high-resolution mass spectroscopy

(HRMS), fourier transform ion cyclotron resonance mass spectroscopy (FI-ICR-MS) mass spectroscopy (MS), nuclear magnetic resonance spectroscopy (NMR), direct infusion mass spectroscopy (DIMS), and direct flow injection (DFI) mass spectrometry are pivotal in plant metabolomics, enhancing our understanding of biological processes, phenotypes, and plant nutritional value (Tahir et al., 2019).

Recent metabolomic analyses further underscore sorghum's potential as a functional food. For instance, Ramalingam et al. (2021) conducted a large-scale metabolomic study on diverse sorghum accessions, revealing bioactive compounds such as flavonoids, phytosterols, condensed tannins, and polyphenols with anti-cancer, anti-inflammatory, and antioxidant properties. Dark red pericarp sorghums exhibited higher concentrations of these beneficial metabolites, suggesting their potential as biomarkers for genotype differentiation. Such findings pave the way for biofortified varieties with enhanced nutraceutical properties through metabolic engineering and molecular breeding. Similarly, Zhou et al. (2020) identified 651 metabolites across white, red, and purple seeds in sweet sorghum, highlighting the potential for biofortification using metabolic profiling.

Advanced approaches like metabolomic genome-wide association studies (mGWAS) and metabolomic quantitative trait loci (mQTL) mapping have proven effective in dissecting metabolic networks in other crops, aiding in stress resistance and enhancing food quality (Peng et al., 2017). Extending these methods to sorghum could identify genetic markers linked to beneficial metabolites, facilitating breeding programs aimed at improving stress tolerance, nutritional quality, and flavor. Despite limited comprehensive metabolomic analyses in sorghum, existing studies lay the groundwork for identifying key metabolites and their biosynthetic pathways. Analyzing the metabolome at different seed development stages will enable the discovery of candidate genes associated with nutrient accumulation. Such insights will enhance our understanding of sorghum's metabolic networks, supporting its role in sustainable agriculture and food security, particularly in developing regions.

13.7 EPIGENOMIC APPROACHES FOR SORGHUM BIOFORTIFICATION

Epigenomics, the study of heritable changes in gene expression without altering DNA sequence, offers innovative strategies for enhancing sorghum's nutritional quality (Selma & Orzáez, 2021). Mechanisms such as DNA methylation, histone modifications, and non-coding RNAs play key roles in regulating gene expression, influencing nutrient biosynthesis and stress adaptation.

DNA methylation, which involves adding methyl groups to cytosines, regulates gene activity by silencing or activating genes. This modification is essential for gene expression, transposon activity, and genomic imprinting, particularly in regulatory regions like promoters and terminators (Zhang et al., 2018). For biofortification, CRISPR/dCas9-mediated DNA methyltransferases or demethylases can target genes such as *NAS* to enhance iron and zinc accumulation in grains. Similar strategies in rice and wheat have successfully increased micronutrient content, demonstrating their potential for improving sorghum (Vafadarshamasbi et al., 2022). Additionally, studies like Gladman et al. (2022) on phosphorus deficiency adaptation emphasize the role of epigenomics in improving nutrient use efficiency in sorghum.

Histones, which structure chromatin, undergo modifications like methylation, acetylation, and phosphorylation, influencing gene expression (Shilatifard, 2006). Enzymes such as histone acetyltransferases (HATs) and deacetylases (HDACs) modulate chromatin accessibility, affecting nutrient biosynthesis pathways (Etier et al., 2022). Targeted acetylation of histones can, for example, activate the *PSY* gene to enhance provitamin A biosynthesis. Small molecule inhibitors or activators targeting these enzymes could help breed nutritionally enhanced sorghum varieties (Anwar et al., 2021). Furthermore, microRNAs (miRNAs) and long non-coding RNAs (lncRNAs) regulate gene expression by affecting mRNA stability and translation (Bunch, 2018). Engineering miRNAs to suppress negative regulators of nutrient biosynthesis or using lncRNAs to enhance the expression of beneficial genes can optimize nutrient profiles in sorghum.

Epigenetic techniques complement traditional breeding for biofortification. Integrating metabolomic and epigenomic data helps identify candidate genes and molecular markers, accelerating the development of nutrient-rich sorghum varieties (Chandana et al., 2022). While genetic modifications, such as transgenic sorghum enriched in provitamin A carotenoids, have shown success (Cruet-Burgos et al., 2023), regulatory challenges have hindered their widespread adoption. Epigenomic interventions can lead to off-target effects, requiring advanced precision tools and rigorous field trials for stability across environments. Additionally, regulatory and ethical concerns demand the establishment of robust frameworks. Despite these challenges, combining epigenomics with conventional breeding holds the potential to revolutionize sorghum biofortification, enhancing global nutrition and food security. Further research into sorghum's epigenetic response to environmental stresses, such as drought, is essential for developing resilient, nutrient-rich varieties (https://jgi.doe.gov/studying-drought-tolerance-sorghum/). Overall, epigenomics offers transformative potential for sorghum improvement through DNA methylation, histone modifications, and non-coding RNA regulation. Leveraging these tools alongside precision breeding technologies can address malnutrition and enhance food security in sorghum-dependent regions.

13.8 MICROBIOMIC ADVANCES FOR SORGHUM BIOFORTIFICATION

The plant microbiome, comprising endophytic and rhizospheric microbial communities, plays a key role in shaping a plant's nutritional status by influencing its health, nutrient dynamics, stress tolerance, and overall fitness (Trivedi et al., 2020). Microbes offer a sustainable method for enhancing mineral, vitamin, and secondary metabolite content in crops through natural processes. Microbe-mediated biofortification, though underappreciated, has shown significant potential. Introducing rhizospheric or endophytic microbes can boost micronutrient levels in plant tissues like seeds, roots, and fruits, with endophytes emerging as important contributors to biofortification (Ku et al., 2019). These microorganisms colonize various plant organs, promoting growth through mechanisms like nutrient acquisition and developmental modulation (de Fretes et al., 2021). Some endophytes, for instance, solubilize phosphates, positioning them as potential biofertilizers for crop improvement (de Fretes et al., 2021).

Several plant growth-promoting microbes have been identified as beneficial for sorghum growth. For example, *Streptomyces* spp. have been shown to promote sorghum growth and biomass production (Alekhya & Gopalakrishnan, 2016). Additionally, the seed-endophytic bacterium *Bacillus pumilus*, isolated from sorghum seeds, has been found to protect seedlings from the fungal pathogen *Rhizoctonia solani*. This bacterium inhibits fungal growth, produces beneficial lipopeptides, and enhances the levels and gene expression of protective enzymes in seedlings, underscoring the importance of seed endophytes in enhancing plant defense against pathogens (Kumar et al., 2023). Moreover, co-inoculation with *Bacillus* strains exhibiting plant growth-promoting abilities has been shown to enhance iron content in sorghum, providing a bio-based, chemical-free approach to iron biofortification (Manasa et al., 2021). This study highlighted the synergistic effects of *Bacillus* spp. isolated from various sources, demonstrating their potential to improve sorghum growth and increase iron uptake. Though not yet tested in field conditions, the results suggest that these microbial inoculants could reduce malnutrition by improving the nutritional quality of sorghum grains (Manasa et al., 2021).

Research has also isolated and characterized a diverse range of endophytic bacteria, including *Pantoea*, *Enterobacter*, and *Pseudomonas*, from surface-sterilized sorghum stems and roots. These bacteria show promise for future research into their potential effects on sorghum growth and biofortification (Mareque et al., 2015). Furthermore, the use of *Bacillus* bioformulations under various environmental conditions could address iron deficiency in both plants and humans, providing an alternative method for biofortification through microbial inoculants. Endophytic fungi also play an important role in plant stress tolerance. For instance, the endophytic fungus *Pseudomonas brassicacearum* (*SVB6R1*) enhances the expression of ACC deaminase, which contributes to improved salt tolerance in sorghum (Gamalero et al., 2020). This highlights the potential of microbial communities, particularly endophytes, in promoting plant health and improving the nutritional profile of crops. Further research into these microbial interactions is essential for optimizing their use in sustainable agriculture and addressing nutrient deficiencies globally.

Microbe-host interactions play a fundamental role in shaping evolutionary biology and understanding them requires both classical and cutting-edge "omics" approaches (Rodriguez et al., 2019). This multi-omics approach will provide a holistic view of plant-microbe interactions, transcending the limitations of studying individual components (Kimotho & Maina, 2024). Combining host and microbial omics data improves precision in plant microbiome research, enhancing the understanding of microbiota's influence on plant fitness and nutrient dynamics. However, the impact of climate change on this ecological niche and its effects on crop nutrient uptake and utilization remain underexplored (Tripathi et al., 2022).

13.9 MULTI-OMICS APPROACHES TO CATALYZE SORGHUM BIOFORTIFICATION

Multi-omics approaches have been extensively utilized in sorghum research to enhance nutritional quality through biofortification, offering valuable insights into the genetic and molecular pathways of nutritional traits (Figure 13.2). The integration of genomics, transcriptomics, proteomics, and metabolomics has identified key genes, secondary metabolites, and protein complexes associated with flavonoids, driving the nutritional enhancement of biofortified sorghum (Ren et al., 2022). This holistic approach has supported the development of biofortified sorghum cultivars with higher

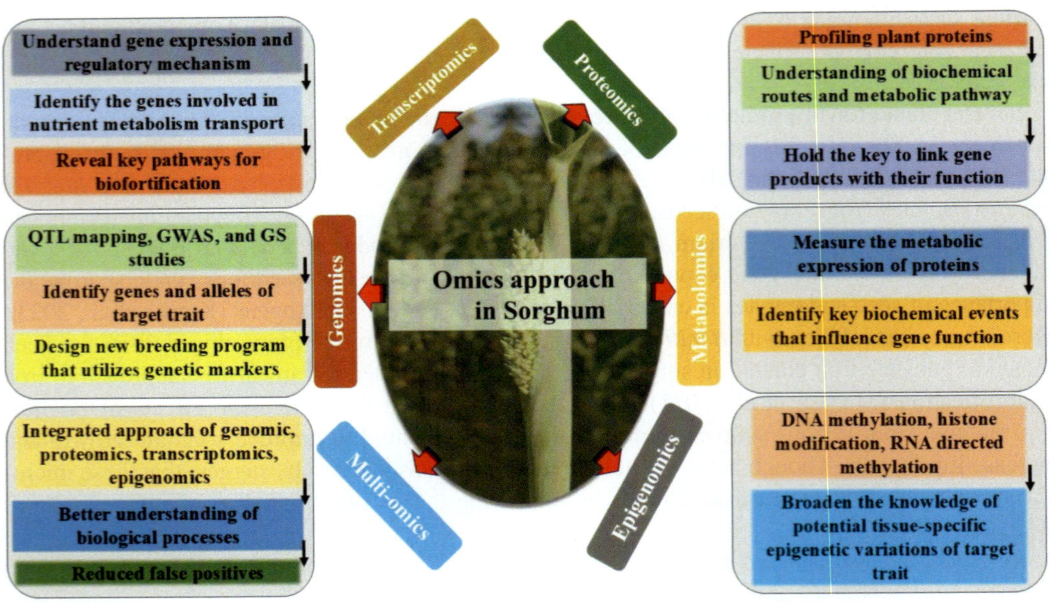

FIGURE 13.2 Comprehensive omics approaches for sorghum nutritional improvement. This figure illustrates the integration of genomics, transcriptomics, proteomics, metabolomics, and epigenomics to enhance the nutritional profile of sorghum.

nutritional value, accelerating the creation of climate-resilient varieties to address global concerns about food security and climate change (Mukherjee et al., 2024).

Building on these efforts, Cruet-Burgos and Rhodes (2023) characterized carotenoids in 446 sorghum accessions, uncovering high-carotenoid lines that were previously unknown. Genome-wide association studies identified zeaxanthin epoxidase as a key gene influencing zeaxanthin, lutein, and β-carotene. High-carotenoid lines showed limited genetic diversity, mostly from one country, while genomic predictions in 2,495 accessions revealed novel genetic diversity. Both oligogenic and polygenic variations were confirmed, supporting marker-assisted and genomic selection for breeding. Similarly, Kudapa et al. (2023) applied GWAS to identify QTLs for micronutrient concentrations, integrating proteomics and metabolomics to explore nutrient accumulation pathways. These integrated multi-omics approaches offer a comprehensive understanding of sorghum's genome, advancing efforts to address climate change and sustainable agriculture. However, challenges in data integration, protocol standardization, result interpretation, high technology costs, and environmental variability remain obstacles to improving nutritional quality.

13.10 CHALLENGES AND FUTURE PERSPECTIVES

Sorghum's complex genome, with abundant repetitive sequences and polyploidy in some species, complicates the identification and manipulation of genes related to seed composition. Although significant progress has been made in sorghum genomics, including a well-annotated reference genome and available transcriptomic resources, it still lags behind crops like rice and maize in terms of functional genomics tools and high-throughput resources for gene characterization. Environmental factors such as soil type, climate, and agricultural practices also significantly influence seed composition, requiring extensive field trials for better understanding. Integrating data from multi-omics approaches is challenging due to differing data formats and the need for advanced computational tools and bioinformatics pipelines. Traits like protein, starch, and lipid content are controlled by multiple genes and are quantitatively inherited, making the identification of stable and major QTLs a complex task. The use of genetic modification techniques, such as CRISPR/Cas9, faces regulatory and public acceptance challenges. However, declining sequencing costs and advancements in genome editing and bioinformatics tools promise to improve our understanding of trait genetics in sorghum. The availability of diverse germplasm collections and development of high-density genetic maps, and QTL mapping can support breeding programs and seed composition improvements. Further, incorporating precision agriculture technologies like remote sensing and drones can allow for real-time monitoring and better environmental management. Multi-omics integration, metabolic engineering, and synthetic biology may enhance the nutritional value of sorghum. Further, raising consumer awareness and expanding market opportunities are essential for promoting the adoption of biofortified sorghum. Although challenges remain, advances in genomic technologies, systems biology, and precision agriculture offer promising prospects for the future of sorghum biofortification.

ACKNOWLEDGEMENT

M. J. T. extends sincere gratitude to the Institute of International Education for generously awarding the Scholar Rescue Fund fellowship.

REFERENCES

Abera, S., Shimels, M., Tessema, T., Raaijmakers, J. M., & Dini-Andreote, F. (2022). Back to the roots: Defining the core microbiome of *Sorghum bicolor* in agricultural field soils from the centre of origin. *FEMS Microbiology Ecology*, *98*(12), fiac136.

Abou-Elwafa, S. F., Amin, A. E. E. A. Z., & Shehzad, T. (2019). Genetic mapping and transcriptional profiling of phytoremediation and heavy metals responsive genes in sorghum. *Ecotoxicology and Environmental Safety*, *173*, 366–372.

Ademiluyi, A. O., Oboh, G., Agbebi, O. J., Boligon, A. A., & Athayde, M. L. (2014). Sorghum [*Sorghum bicolor* (L.) Moench] leaf sheath dye protects against cisplatin-induced hepatotoxicity and oxidative stress in rats. *Journal of Medicinal Food*, *17*(12), 1332–1338.

Alekhya, G., & Gopalakrishnan, S. (2016). Plant growth-promotion by Streptomyces spp. in sorghum (*Sorghum bicolor* L.). *African Journal of Biotechnology*, *15*(33), 1781–1788.

Ali, T. M., & Hasnain, A. (2013). Effect of emulsifiers on complexation and retrogradation characteristics of native and chemically modified White sorghum (*Sorghum bicolor*) starch. *Thermochimica Acta*, *552*, 46–53.

Ananda, G. K., Myrans, H., Norton, S. L., Gleadow, R., Furtado, A., & Henry, R. J. (2020). Wild sorghum as a promising resource for crop improvement, *Frontiers in Plant Science*, *11*, 1108.

Anglani, C. (1998). Sorghum for human food–A review. *Plant Foods for Human Nutrition*, *52*, 85–95.

Anwar, S., Brenya, E., Alagoz, Y., & Cazzonelli, C. I. (2021). Epigenetic control of carotenogenesis during plant development. *Critical Reviews in Plant Sciences*, *40*(1), 23–48.

Arbab, M. E., & El Tinay, A. H. (1997). Effect of cooking and treatment with sodium bisulphite or ascorbic acid on the in vitro protein digestibility of two sorghum cultivars. *Food Chemistry*, *59*(3), 339–343.

Are, A. K., Srivastava, R. K., Mahalingam, G., Gorthy, S., Gaddameedi, A., Kunapareddy, A., ... & Jaganathan, J. (2019). Application of plant breeding and genomics for improved sorghum and pearl millet grain nutritional quality. In: Taylor, J.R.N., Duodu, K.G. (Eds), *Sorghum and Millets* (pp. 51–68). Cambridge, MA, United States: AACC International Press.

Aruna, C., & Visarada, K. B. R. S. (2019). Other industrial uses of sorghum. In: Aruna. C, Visarada, K.B.R.S, Bhat, B.V., Tonapi, V.A. (Eds), *Breeding Sorghum for Diverse End Uses* (pp. 271–292). Cambridge, MA, United States: Woodhead Publishing.

Awika, J. M. (2011). Sorghum flavonoids: Unusual compounds with promising implications for health. In: Awika, J.M., Piironen, V., Bean, S. (Eds), *Advances in Cereal Science: Implications to Food Processing and Health Promotion*, ACS Symposium Series, Vol. 1089, 171–200. American Chemical Society, Washington, DC Distributed in print by Oxford University Press, Inc.

Awika, J. M. (2017). Sorghum: Its unique nutritional and health-promoting attributes. In: Taylor, J.R.N., Awika, J.M. (Eds), *Gluten-Free Ancient Grains* (pp. 21–54). Cambridge, MA, United States: Woodhead Publishing.

Awika, J. M., & Rooney, L. W. (2004). Sorghum phytochemicals and their potential impact on human health. *Phytochemistry*, *65*(9), 1199–1221.

Axtell, J. D., Kirleis, A. W., Hassen, M. M., D'Croz Mason, N. A., Mertz, E. T., & Munck, L. (1981). Digestibility of sorghum proteins. *Proceedings of the National Academy of Sciences*, *78*(3), 1333–1335.

Barkla, B. J. (2016). Identification of abiotic stress protein biomarkers by proteomic screening of crop cultivar diversity. *Proteomes*, *4*(3), 26.

Bean, S. R., Wilson, J. D., Moreau, R. A., Galant, A., Awika, J. M., Kaufman, R. C., ... & Ioerger, B. P. (2019). Structure and composition of the sorghum grain. *Sorghum: A State of the Art and Future Perspetives*, *58*, 173–214.

Boyer, C. D., & Liu, K. C. (1983). Starch and water-soluble polysaccharides from sugary endosperm of sorghum. *Phytochemistry*, *22*(11), 2513-2515.

Boyles, R. E., Pfeiffer, B. K., Cooper, E. A., Rauh, B. L., Zielinski, K. J., Myers, M. T., ... & Kresovich, S. (2017). Genetic dissection of sorghum grain quality traits using diverse and segregating populations. *Theoretical and Applied Genetics*, *130*, 697–716.

Bunch, H. (2018). Gene regulation of mammalian long non-coding RNA. *Molecular Genetics and Genomics*, *293*(1), 1–15.

Carr, T. P., Weller, C. L., Schlegel, V. L., Cuppett, S. L., Guderian Jr, D. M., & Johnson, K. R. (2005). Grain sorghum lipid extract reduces cholesterol absorption and plasma non-HDL cholesterol concentration in hamsters. *The Journal of Nutrition*, *135*(9), 2236–2240.

Chandana, B. S., Mahto, R. K., Singh, R. K., Ford, R., Vaghefi, N., Gupta, S. K., ... & Kumar, R. (2022). Epigenomics as potential tools for enhancing magnitude of breeding approaches for developing climate resilient chickpea. *Frontiers in Genetics*, *13*, 900253.

Chaudhari, G. N., & Fakrudin, B. (2017). Candidate gene prediction and expression profiling of near isogenic lines (NILs) carrying stay-green QTLs in rabi sorghum. *Journal of Plant Biochemistry and Biotechnology*, *26*, 64–72.

Chirinos, R., Betalleluz-Pallardel, I., Huamán, A., Arbizu, C., Pedreschi, R., & Campos, D. (2009). HPLC-DAD characterisation of phenolic compounds from Andean oca (*Oxalis tuberosa* Mol.) tubers and their contribution to the antioxidant capacity. *Food Chemistry*, *113*(4), 1243–1251.

Chung, I. M., Yong, S. J., Lee, J., & Kim, S. H. (2013). Effect of genotype and cultivation location on β-sitosterol and α-, β-, γ-, and δ-tocopherols in sorghum. *Food Research International*, *51*(2), 971–976.

Costes, C., Mameri, H., Terrier, N., Navarro Sanz, S., Calatayud, C., Soriano, A., ... & Francin-Allami, M. (2023). *Analysis of Polysaccharide Composition during Sorghum Grain Development*. CIRAD.

Cremer, J. E., Bean, S. R., Tilley, M. M., Ioerger, B. P., Ohm, J. B., Kaufman, R. C., ... & Godwin, I. D. (2014). Grain sorghum proteomics: Integrated approach toward characterization of endosperm storage proteins in kafirin allelic variants. *Journal of Agricultural and Food Chemistry*, *62*(40), 9819–9831.

Cruet-Burgos, C., & Rhodes, D. H. (2023). Unraveling transcriptomics of sorghum grain carotenoids: A step forward for biofortification. *BMC Genomics*, *24*(1), 233.

Cruet-Burgos, C., Cox, S., Ioerger, B. P., Perumal, R., Hu, Z., Herald, T. J., ... & Rhodes, D. H. (2020). Advancing provitamin A biofortification in sorghum: Genome-wide association studies of grain carotenoids in global germplasm. *The Plant Genome*, *13*(1), e20013.

Dahir, M., Zhu, K. X., Guo, X. N., Aboshora, W., & Peng, W. (2015). Possibility to utilize sorghum flour in a modern bread making industry. *Journal Academia and Industrial Research*, *4*(4), 128–135.

Dai, J., & Mumper, R. J. (2010). Plant phenolics: Extraction, analysis and their antioxidant and anticancer properties. *Molecules*, *15*(10), 7313–7352.

Daviss, B. (2005). Growing pains for metabolomics: The newest'omic science is producing results--and more data than researchers know what to do with. *The Scientist*, *19*(8), 25–29.

de Alencar Figueiredo, L. F., Sine, B., Chantereau, J., Mestres, C., Fliedel, G., Rami, J. F., ... & Courtois, B. (2010). Variability of grain quality in sorghum: Association with polymorphism in *Sh2*, *Bt2*, *SssI*, *Ae1*, *Wx* and *O2*. *Theoretical and Applied Genetics*, *121*, 1171–1185.

de Fretes, C. E., Widianto, D., Purwestri, Y. A., & Nuringtyas, T. R. (2021). Plant growth-promoting activity of endophytic bacteria from sweet sorghum (*Sorghum bicolor* (L.) Moench). *Indonesian Journal of Biotechnology*, *26*(4), 190–196.

De Mesa-Stonestreet, N. J., Alavi, S., & Bean, S. R. (2010). Sorghum proteins: The concentration, isolation, modification, and food applications of kafirins. *Journal of Food Science*, *75*(5), R90–R104.

de Morais Cardoso, L., Pinheiro, S. S., Martino, H. S. D., & Pinheiro-Sant'Ana, H. M. (2017). Sorghum (*Sorghum bicolor* L.): Nutrients, bioactive compounds, and potential impact on human health. *Critical Reviews in Food Science and Nutrition*, *57*(2), 372–390.

Djanaguiraman, M., Vimala, K., Sofi, P. A., Perumal, R., & Prasad, P. V. V. (2024). Genetic variation for grain iron and zinc concentration in the US sorghum [*Sorghum bicolor* (L.) Moench] association panel. *Crop Science*, 64(5), 2652–2665.

Draye, X., Lin, Y. R., Qian, X. Y., Bowers, J. E., Burow, G. B., Morrell, P. L., ... & Paterson, A. H. (2001). Toward integration of comparative genetic, physical, diversity, and cytomolecular maps for grasses and grains, using the sorghum genome as a foundation. *Plant Physiology*, *125*(3), 1325–1341.

Dykes, L., & Rooney, L. W. (2006). Sorghum and millet phenols and antioxidants. *Journal of Cereal Science*, *44*(3), 236–251.

Elhassan, M. S., Emmambux, M. N., Hays, D. B., Peterson, G. C., & Taylor, J. R. (2015). Novel biofortified sorghum lines with combined waxy (high amylopectin) starch and high protein digestibility traits: Effects on endosperm and flour properties. *Journal of Cereal Science*, *65*, 132–139.

Elkonin, L., Italyanskaya, J., & Panin, V. (2018). Genetic modification of sorghum for improved nutritional value: State of the problem and current approaches. *Journal of Investigative Genomics*, *5*, 39–48.

Elshire, R. J., Glaubitz, J. C., Sun, Q., Poland, J. A., Kawamoto, K., Buckler, E. S., & Mitchell, S. E. (2011). A robust, simple genotyping-by-sequencing (GBS) approach for high diversity species. *PloS One*, *6*(5), e19379.

Etier, A., Dumetz, F., Chéreau, S., & Ponts, N. (2022). Post-translational modifications of histones are versatile regulators of fungal development and secondary metabolism. *Toxins*, *14*(5), 317.

Ezeogu, L. I., Duodu, K. G., & Taylor, J. R. N. (2005). Effects of endosperm texture and cooking conditions on the in vitro starch digestibility of sorghum and maize flours. *Journal of Cereal Science*, *42*(1), 33–44.

Gaddameedi, A., Sheraz, S., Kumar, A., Li, K., Pellny, T., Gupta, R., ... & Shewry, P. R. (2022). The location of iron and zinc in grain of conventional and biofortified lines of sorghum. *Journal of Cereal Science*, *107*, 103531.

Gamalero, E., Favale, N., Bona, E., Novello, G., Cesaro, P., Massa, N., ... & Lingua, G. (2020). Screening of bacterial endophytes able to promote plant growth and increase salinity tolerance. *Applied Sciences*, *10*(17), 5767.

Gladman, N., Hufnagel, B., Regulski, M., Liu, Z., Wang, X., Chougule, K., ... & Ware, D. (2022). Sorghum root epigenetic landscape during limiting phosphorus conditions. *Plant Direct*, *6*(5), e393.

Gong, C. Y., & Wang, T. (2013). Proteomic evaluation of genetically modified crops: Current status and challenges. *Frontiers in Plant Science*, *4*, 41.

Gong, X., Wang, P., Yang, F., Chang, S., Liu, B., He, H., ... & Wang, C. (2010). Protein–protein docking with binding site patch prediction and network-based terms enhanced combinatorial scoring. *Proteins: Structure, Function, and Bioinformatics*, *78*(15), 3150–3155.

Gudi, S., Saini, D. K., Singh, G., Halladakeri, P., Kumar, P., Shamshad, M., ... & Sharma, A. (2022). Unravelling consensus genomic regions associated with quality traits in wheat using meta-analysis of quantitative trait loci. *Planta*, *255*(6), 115.

Guleria, G., Rana, M., Priyadarshini, P., Kaldate, R., Kumar, N., Rana, R., ... & Ahmed, S. (2023). Biofortification of Sorghum (*Sorghum bicolor*). In: Deshmukh, R., Nadaf, A., Ansari, W.A., Singh, K., Sonah, H. (Eds), *Biofortification in Cereals: Progress and Prospects* (pp. 259–284). Springer Nature Singapore.

Halladakeri, P., Gudi, S., Akhtar, S., Singh, G., Saini, D. K., Hilli, H. J., ... & Mir, R. R. (2023). Meta-analysis of the quantitative trait loci associated with agronomic traits, fertility restoration, disease resistance, and seed quality traits in pigeonpea (Cajanus cajan L.). *The Plant Genome*, *16*(3), e20342.

Hammouda, H., Chérif, J. K., Trabelsi-Ayadi, M., Baron, A., & Guyot, S. (2013). Detailed polyphenol and tannin composition and its variability in Tunisian dates (Phoenix dactylifera L.) at different maturity stages. *Journal of Agricultural And Food Chemistry*, *61*(13), 3252–3263.

Hasan, N., Choudhary, S., Naaz, N., Sharma, N., & Laskar, R. A. (2021). Recent advancements in molecular marker-assisted selection and applications in plant breeding programmes. *Journal of Genetic Engineering and Biotechnology*, *19*(1), 128.

Henley, E. C., Taylor, J. R. N., & Obukosia, S. D. (2010). The importance of dietary protein in human health: Combating protein deficiency in sub-Saharan Africa through transgenic biofortified sorghum. *Advances in Food and Nutrition Research*, *60*, 21–52.

Jamaloddin, M., Maliha, A., Gokulan, C. G., Gaur, N., & Patel, H. K. (2021). Metabolomics-assisted breeding for crop improvement: An emerging approach. *Omics Technologies for Sustainable Agriculture and Global Food Security*, *1*, 241–279.

Jaworski, N. W., Lærke, H. N., Bach Knudsen, K. E., & Stein, H. H. (2015). Carbohydrate composition and in vitro digestibility of dry matter and nonstarch polysaccharides in corn, sorghum, and wheat and coproducts from these grains. *Journal of Animal Science*, *93*(3), 1103–1113.

Kabir, A. H., & Bennetzen, J. L. (2024). Molecular insights into the mutualism that induces iron deficiency tolerance in sorghum inoculated with Trichoderma harzianum. *Microbiological Research*, *281*, 127630.

Karper, R.E. (1933). Inheritance of waxy endosperm in sorghum. Journal of Heredity 24,257–262.

Kazungu, F. K., Muindi, E. M., & Mulinge, J. M. (2023). Overview of sorghum (*Sorghum bicolor*. L), its economic importance, ecological requirements and production constraints in Kenya. *International Journal of Plant & Soil Science*, *35*(1), 62–71.

Khan, A., Tian, R., Bean, S. R., Yerka, M., & Jiao, Y. (2024). Transcriptome and metabolome analyses reveal regulatory networks associated with nutrition synthesis in sorghum seeds. *Communications Biology*, *7*(1), 841.

Khoddami, A., Messina, V., Vadabalija Venkata, K., Farahnaky, A., Blanchard, C. L., & Roberts, T. H. (2023). Sorghum in foods: Functionality and potential in innovative products. *Critical Reviews in Food Science and Nutrition*, *63*(9), 1170–1186.

Kim, C., Guo, H., Kong, W., Chandnani, R., Shuang, L. S., & Paterson, A. H. (2016). Application of genotyping by sequencing technology to a variety of crop breeding programs. *Plant Science*, *242*, 14–22.

Kimotho, R. N., & Maina, S. (2024). Unraveling plant–microbe interactions: Can integrated omics approaches offer concrete answers?. *Journal of Experimental Botany*, *75*(5), 1289–1313.

Kirkby, E. A., Nikolic, M., White, P. J., & Xu, G. (2023). Mineral nutrition, yield, and source–sink relationships. In: Rengel, Z., Cakmak, I., White, P.J. (Eds), *Marschner's Mineral Nutrition of Plants* (pp. 131–200). San Diego, CA, United States: Academic Press.

Kotla, A., Phuke, R., Hariprasanna, K., Mehtre, S. P., Rathore, A., Gorthy, S., ... & Kumar, A. A. (2019). Identification of QTLs and candidate genes for high grain Fe and Zn concentration in sorghum [*Sorghum bicolor* (L.) Moench]. *Journal of Cereal Science*, *90*, 102850.

Kotla, A., Phuke, R., Hariprasanna, K., Mehtre, S. P., Rathore, A., Radhika, K., … & Kumar, A. (2016). Identification of QTLs and underlying candidate genes controlling grain Fe and Zn concentration in sorghum [*Sorghum bicolor* (L). Moench]. In *Plant and Animal Genome Conference XXIV, January 08-13, 2016, San Diego, CA*.

Krueger, C. G., Vestling, M. M., & Reed, J. D. (2003). Matrix-assisted laser desorption/ionization time-of-flight mass spectrometry of heteropolyflavan-3-ols and glucosylated heteropolyflavans in sorghum [*Sorghum bicolor* (L.) Moench]. *Journal of Agricultural and Food Chemistry*, *51*(3), 538–543.

Kruger, J., Oelofse, A., & Taylor, J. R. (2014). Effects of aqueous soaking on the phytate and mineral contents and phytate: mineral ratios of wholegrain normal sorghum and maize and low phytate sorghum. *International Journal of Food Sciences and Nutrition*, *65*(5), 539-546.

Ku, Y. S., Rehman, H. M., & Lam, H. M. (2019). Possible roles of rhizospheric and endophytic microbes to provide a safe and affordable means of crop biofortification. *Agronomy*, *9*(11), 764.

Kudapa, H., Barmukh, R., Vemuri, H., Gorthy, S., Pinnamaneni, R., Vetriventhan, M., ... & Govindaraj, M. (2023). Genetic and genomic interventions in crop biofortification: Examples in millets. *Frontiers in Plant Science*, *14*, 1123655.

Kumar, K., Pal, G., Verma, A., Kumar, D., Shukla, P., & Verma, S. K. (2023). Seed vectored bacterial endophyte Bacillus pumilus protect sorghum (*Sorghum bicolor* L.) seedlings from a fungal pathogen Rhizoctonia solani. *Biological Control*, *183*, 105249.

Lakshmi, K. B., & Vimala, V. (1996). Hypoglycemic effect of selected sorghum recipes. *Nutrition Research*, *16*(10), 1651–1658.

Li, J., Wang, Y., Zheng, L., Li, Y., Zhou, X., Li, J., ... & Zhang, W. (2019). The intracellular transporter *AtNRAMP6* is involved in Fe homeostasis in Arabidopsis. *Frontiers in Plant Science*, *10*, 1124.

Li, Y., Ma, H., Wu, Y., Ma, Y., Yang, J., Li, Y., ... & Min, L. (2024). Single-Cell Transcriptome Atlas and Regulatory Dynamics in Developing Cotton Anthers. *Advanced Science*, *11*(3), 2304017.

Liangli, L. Y., Tsao, R., & Shahidi, F. (Eds.). (2012). *Cereals and Pulses: Nutraceutical Properties and Health Benefits*. John Wiley.

Mace, E. S., & Jordan, D. R. (2011). Integrating sorghum whole genome sequence information with a compendium of sorghum QTL studies reveals uneven distribution of QTL and of gene-rich regions with significant implications for crop improvement. *Theoretical and Applied Genetics*, *123*(1), 169–191.

Maclean Jr, W. C., de RomaÑa, G. L., Placko, R. P., & Graham, G. G. (1981). Protein quality and digestibility of sorghum in preschool children: balance studies and plasma free amino acids. *The Journal of nutrition*, *111*(11), 1928–1936.

Mahmood, U., Li, X., Fan, Y., Chang, W., Niu, Y., Li, J., ... & Lu, K. (2022). Multi-omics revolution to promote plant breeding efficiency. *Frontiers in Plant Science*, *13*, 1062952.

Manasa, M., Ravinder, P., Gopalakrishnan, S., Srinivas, V., Sayyed, R. Z., El Enshasy, H. A., ... & Hameeda, B. (2021). Co-inoculation of *Bacillus spp.* for growth promotion and iron fortification in sorghum. *Sustainability*, *13*(21), 12091.

Mareque, C., Taulé, C., Beracochea, M., & Battistoni, F. (2015). Isolation, characterization and plant growth promotion effects of putative bacterial endophytes associated with sweet sorghum (*Sorghum bicolor* (L) Moench). *Annals of Microbiology*, *65*, 1057–1067.

Martino, H. S. D., Tomaz, P. A., Moraes, E. A., Conceição, L. D., Oliveira, D. D. S., Queiroz, V. A. V., ... & Ribeiro, S. M. R. (2012). Chemical characterization and size distribution of sorghum genotypes for human consumption. *Rev Inst Adolfo Lutz*, *71*(2), 337-344

Massey, A. R., Reddivari, L., & Vanamala, J. (2014). The dermal layer of sweet sorghum (*Sorghum bicolor*) stalk, a byproduct of biofuel production and source of unique 3-deoxyanthocyanidins, has more antiproliferative and proapoptotic activity than the pith in p53 variants of HCT116 and colon cancer stem cells. *Journal of Agricultural and Food Chemistry*, *62*(14), 3150–3159.

McIntyre, C. L., Drenth, J., Gonzalez, N., Henzell, R. G., & Jordan, D. R. (2008). Molecular characterization of the waxy locus in sorghum. *Genome*, *51*(7), 524–533.

Miafo, A. P. T., Koubala, B. B., Kansci, G., & Muralikrishna, G. (2019). Free sugars and non-starch polysaccharides–phenolic acid complexes from bran, spent grain and sorghum seeds. *Journal of Cereal Science*, *87*, 124–131.

Mofokeng, M. A., & Shargie, N. G. (2016). Bird damage and control strategies in grain sorghum produciton. *International Journal of Agricultural and Environmental Research*, 2(4), 264–269.

Moreno, M. D. L., Comino, I., & Sousa, C. (2014). Alternative grains as potential raw material for gluten-free food development in the diet of celiac and gluten-sensitive patients. *Austin Journal of Nutrition and Food Sciences*, *2*(3), 1016.

Mukherjee, A., Maheshwari, U., Sharma, V., Sharma, A., & Kumar, S. (2024). Functional insight into multi-omics-based interventions for climatic resilience in sorghum (*Sorghum bicolor*): A nutritionally rich cereal crop. *Planta*, *259*(4), 91.

Mukkun, L., Arung, E. T., Ismayati, M., Kleden, Y. L., Pakan, P. D., Nguru, O., ... & Naibaho, N. (2024). Antibacterial, antioxidant and anti-inflammatory activities of red, brown, and black sorghum (*Sorghum bicolor*) cultivated in a dry land area. *Biodiversitas: Journal of Biological Diversity*, *25*(8), 2560–2569.

Müller, M., Dembélé, S., Zougmoré, R. B., Gaiser, T., & Partey, S. T. (2020). Performance of three sorghum cultivars under excessive rainfall and waterlogged conditions in the Sudano-Sahelian Zone of West Africa: A case study at the climate-smart village of Cinzana in Mali. *Water*, *12*(10), 2655.

Ng'Uni, D., Shargie, N. G., Andersson, S. C., Van Biljon, A., & Labuschagne, M. T. (2016). Genetic variation and trait associations of yield, protein and grain micronutrients for identification of promising sorghum varieties. *Cereal Research Communications*, *44*(4), 681–693.

Ng'uni, D., Geleta, M., Hofvander, P., Fatih, M., & Bryngelsson, T. (2012). Comparative genetic diversity and nutritional quality variation among some important Southern African sorghum accessions [*'Sorghum bicolor'*(L.) Moench]. *Australian Journal of Crop Science*, *6*(1), 56–64.

Oluwole, O., Fernando, W. B., Lumanlan, J., Ademuyiwa, O., & Jayasena, V. (2022). Role of phenolic acid, tannins, stilbenes, lignans and flavonoids in human health–A review. *International Journal of Food Science & Technology*, *57*(10), 6326–6335.

Peng, M., Shahzad, R., Gul, A., Subthain, H., Shen, S., Lei, L., ... & Luo, J. (2017). Differentially evolved glucosyltransferases determine natural variation of rice flavone accumulation and UV-tolerance. *Nature Communications*, *8*(1), 1975.

Phuke, R. M., Anuradha, K., Radhika, K., Jabeen, F., Anuradha, G., Ramesh, T., ... & Kumar, A. A. (2017). Genetic variability, genotype× environment interaction, correlation, and GGE biplot analysis for grain iron and zinc concentration and other agronomic traits in RIL population of sorghum (*Sorghum bicolor* L. Moench). *Frontiers in Plant Science*, *8*, 712.

Pratyusha, S. (2022). Phenolic compounds in the plant development and defense: An overview. In: Hasanuzzaman, M., Nahar, K. *(Eds), Plant Stress Physiology-Perspectives in Agriculture*, 125–140.

Qaim, M. (2020). Role of new plant breeding technologies for food security and sustainable agricultural development. *Applied Economic Perspectives and Policy*, *42*(2), 129–150. IntechOpen, London, United Kingdom.

Ramalingam, A. P., Mohanavel, W., Premnath, A., Muthurajan, R., Prasad, P. V., & Perumal, R. (2021). Large-scale non-targeted metabolomics reveals antioxidant, nutraceutical and therapeutic potentials of sorghum. *Antioxidants*, *10*(10), 1511.

Ren, G., Yang, P., Cui, J., Gao, Y., Yin, C., Bai, Y., ... & Chang, J. (2022). Multiomics analyses of two sorghum cultivars reveal the molecular mechanism of salt tolerance. *Frontiers in Plant Science*, *13*, 886805.

Rhodes, D. H., Hoffmann Jr, L., Rooney, W. L., Ramu, P., Morris, G. P., & Kresovich, S. (2014). Genome-wide association study of grain polyphenol concentrations in global sorghum [*Sorghum bicolor* (L.) Moench] germplasm. *Journal of Agricultural and Food Chemistry*, *62*(45), 10916–10927.

Rhodes, D. H., Hoffmann, L., Rooney, W. L., Herald, T. J., Bean, S., Boyles, R., ... & Kresovich, S. (2017). Genetic architecture of kernel composition in global sorghum germplasm. *BMC Genomics*, *18*, 1–8.

Rhodes, M. J. C., & Price, K. R. (1997). Identification and analysis of plant phenolic antioxidants. *European Journal of Cancer Prevention*, *6*(6), 518–521.

Rial-Lovera, K., Davies, W. P., & Cannon, N. D. (2017). Implications of climate change predictions for UK cropping and prospects for possible mitigation: A review of challenges and potential responses. *Journal of the Science of Food and Agriculture*, *97*(1), 17–32.

Rodriguez, P. A., Rothballer, M., Chowdhury, S. P., Nussbaumer, T., Gutjahr, C., & Falter-Braun, P. (2019). Systems biology of plant-microbiome interactions. *Molecular Plant*, *12*(6), 804–821.

Rusu, A. V., Socol, C. T., Bangar, S. P., Coşier, V., & Trif, M. (2023). Colored cereals: Genetics and chemistry of pigments. In: Bangar, S.P., Kumar, M. (Eds), *Functionality and Application of Colored Cereals* (pp. 111–134). Cambridge, MA, United States: Academic Press.

Saeed Omer, S. H., Hong, J., Zheng, X., & Khashaba, R. (2023). Sorghum flour and sorghum flour enriched bread: Characterizations, challenges, and potential improvements. *Foods*, *12*(23), 4221.

Sawadogo, W. R., Maciuk, A., Banzouzi, J. T., Champy, P., Figadere, B., Guissou, I. P., & Nacoulma, O. G. (2012). Mutagenic effect, antioxidant and anticancer activities of six medicinal plants from Burkina Faso. *Natural Product Research*, *26*(6), 575–579.

Selma, S., & Orzáez, D. (2021). Perspectives for epigenetic editing in crops. *Transgenic Research*, *30*(4), 381–400.

Serrano-Ron, L., Cabrera, J., Perez-Garcia, P., & Moreno-Risueno, M. A. (2021). Unraveling root development through single-cell omics and reconstruction of gene regulatory networks. *Frontiers in Plant Science*, *12*, 661361.

Sethi, M., Saini, D. K., Devi, V., Kaur, C., Singh, M. P., Singh, J., ... & Chaudhary, D. P. (2023). Unravelling the genetic framework associated with grain quality and yield-related traits in maize (Zea mays L.). *Frontiers in Genetics*, *14*, 1248697.

Sharan, A. A., Nikam, A. N., Jaleel, A., Tamhane, V. A., & Rao, S. P. (2018). Method for label-free quantitative proteomics for *Sorghum bicolor* L. Moench. *Tropical Plant Biology*, *11*, 78–91.

Sharma, A., Sharma, D., & Verma, S. K. (2023). A systematic in silico report on iron and zinc proteome of *Zea mays*. *Frontiers in Plant Science*, *14*, 1166720.

Shilatifard, A. (2006). Chromatin modifications by methylation and ubiquitination: Implications in the regulation of gene expression. *Annual Review of Ciochemistry*, *75*(1), 243–269.

Shim, T. J., Kim, T. M., Jang, K. C., Ko, J. Y., & Kim, D. J. (2013). Toxicological evaluation and anti-inflammatory activity of a golden gelatinous sorghum bran extract. *Bioscience, Biotechnology, and Biochemistry*, *77*(4), 697–705.

Shull, J. M., Watterson, J. J., & Kirleis, A. W. (1992). Purification and immunocytochemical localization of kafirins in *Sorghum bicolor* (L. Moench) endosperm. *Protoplasma*, *171*, 64–74.

Singh, P. K., Pratap, S. G., & Tandon, P. K. (2020). The mechanisms of trace element uptake and transport up to grains of crop plants. In: Mishra, K., Tandon, P.K., Srivastava, S. (Eds), *Sustainable Solutions for Elemental Deficiency and Excess in Crop Plants*, 119–133. Springer, Singapore.

Singh, R., & Axtell, J. D. (1973). High lysine mutant gene (hl that improves protein quality and biological value of grain sorghum 1. *Crop Science*, *13*(5), 535–539.

Sohail, M. I., ur Rehman, M. Z., Aziz, T., Akmal, F., Azhar, M., Nadeem, F., ... & Khalid, M. A. (2022). Iron bio-fortification and heavy metal/(loid)s contamination in cereals: Successes, issues, and challenges. *Crop and Pasture Science*, *73*(8), 877–895.

Soto-Vaca, A., Gutierrez, A., Losso, J. N., Xu, Z., & Finley, J. W. (2012). Evolution of phenolic compounds from color and flavor problems to health benefits. *Journal of Agricultural and Food Chemistry*, *60*(27), 6658–6677.

Susmita, C., & Selvi, B. (2014). Genetic variability for grain iron, zinc, other nutrients and yield related traits in sorghum (*Sorghum bicolor* (L.) Moench). *International Journal of Agricultural Science Research*, *4*(3), 91–100.

Tahir, A. T., Fatmi, Q., Nosheen, A., Imtiaz, M., & Khan, S. (2019). Metabolomic approaches in plant research. In: Hakeem, K., Shaik, N., Banaganapalli, B., Elango, R. (Eds), *Essentials of Bioinformatics, Volume III: In Silico Life Sciences: Agriculture* (pp. 109–140). Springer, Cham, Switzerland.

Tang, J., Zou, C., He, Z., Shi, R., Ortiz-Monasterio, I., Qu, Y., & Zhang, Y. (2008). Mineral element distributions in milling fractions of Chinese wheats. *Journal of Cereal Science*, *48*(3), 821–828.

Tanin, M. J., Saini, D. K., Kumar, P., Gudi, S., Sharma, H., Kaur, J. P., ... & Sharma, A. (2024). Iron biofortification in wheat: Past, present, and future. *Current Plant Biology*, 38, 100328.

Tanwar, R., Panghal, A., Chaudhary, G., Kumari, A., & Chhikara, N. (2023). Nutritional, phytochemical and functional potential of sorghum: A review. *Food Chemistry Advances*, 3, 100501.

Taylor, J. R., & Dewar, J. (2001). Developments in sorghum food technologies. *Advances in Food and Nutrition Research, 43,* 217–264. https://doi.org/10.1016/s1043-4526(01)43006-3

Taylor, J. R., & Taylor, J. (2024). Proteins from sorghum and millets. In: Nadathur, S., Wanasundara, J.P.D., Scanlin, L. (Eds), *Sustainable Protein Sources* (pp. 93–119). Cambridge, MA, United States: Academic Press.

Tissier, A., Ziegler, J., & Vogt, T. (2014). Specialized plant metabolites: Diversity and biosynthesis. In: Kraussand, G.J., Nies, D.H. (Eds), *Ecological Biochemistry: Environmental and Interspecies Interactions*, 14–37. Boschstr, Weinheim, Germany: Wiley-VCH Verlag GmbH & Co.KGaA.

Tripathi, S., Bahuguna, R. N., Shrivastava, N., Singh, S., Chatterjee, A., Varma, A., & Jagadish, S. V. K. (2022). Microbial biofortification: A sustainable route to grow nutrient-rich crops under changing climate. *Field Crops Research, 287*, 108662.

Trivedi, P., Leach, J. E., Tringe, S. G., Sa, T., & Singh, B. K. (2020). Plant–microbiome interactions: From community assembly to plant health. *Nature Reviews Microbiology, 18*(11), 607–621.

Vafadarshamasbi, U., Mace, E., Jordan, D., & Crisp, P.A. (2022). Decoding the sorghum methylome: Understanding epigenetic contributions to agronomic traits. *Biochemical Society Transactions, 50*(1), 583–596.

Varshney, R. K., Bohra, A., Yu, J., Graner, A., Zhang, Q., & Sorrells, M. E. (2021). Designing future crops: Genomics-assisted breeding comes of age. *Trends in Plant Science, 26*(6), 631–649.

Wan, Y., Stewart, T., Amrahli, M., Evans, J., Sharp, P., Govindan, V., ... & Shewry, P. R. (2022). Localisation of iron and zinc in grain of biofortified wheat. *Journal of Cereal Science, 105*, 103470.

Wang, L., Ni, D. R., Yang, F., Lin, L., Yang, Y. B., Sun, C. D., ... & Kong, X. L. (2023). Transcriptome profiling of *Sorghum bicolor* reveals cultivar-specific molecular signatures associated with starch and phenolic compounds biosyntheses and accumulation during sorghum grain development. *Czech Journal Genetics and Plant Breeding, 59(4),* 235–252.

Weller, C. L., & Hwang, K. T. (2005). Extraction of lipids from grain sorghum DDG. *Transactions of the ASAE, 48*(5), 1883–1888.

Winn, J. A., Mason, R. E., Robbins, A. L., Rooney, W. L., & Hays, D. B. (2009). QTL mapping of a high protein digestibility trait in *Sorghum bicolor. International Journal of Plant Genomics, 2009*(1), 471853.

Xiao, J., Chen, Y., & Huang, Q. (2017). Physicochemical properties of kafirin protein and its applications as building blocks of functional delivery systems. *Food & Function, 8*(4), 1402–1413.

Xiong, Y., Zhang, P., Warner, R. D., & Fang, Z. (2019). Sorghum grain: From genotype, nutrition, and phenolic profile to its health benefits and food applications. *Comprehensive Reviews in Food Science and Food Safety, 18*(6), 2025–2046.

Yang, D. L., Jing, R. L., Chang, X. P., & Li, W. (2007). Identification of quantitative trait loci and environmental interactions for accumulation and remobilization of water-soluble carbohydrates in wheat (*Triticum aestivum* L.) stems. *Genetics, 176*(1), 571–584.

Yang, L., Browning, J. D., & Awika, J. M. (2009). Sorghum 3-deoxyanthocyanins possess strong phase II enzyme inducer activity and cancer cell growth inhibition properties. *Journal of Agricultural and Food Chemistry, 57*(5), 1797–1804.

Yang, Q., Van Haute, M., Korth, N., Sattler, S., Rose, D., Juritsch, A., ... & Benson, A. K. (2023). The waxy mutation in sorghum and other cereal grains reshapes the gut microbiome by reducing levels of multiple beneficial species. *Gut Microbes, 15*(1), 2178799.

Zhang, H., Lang, Z., & Zhu, J. K. (2018). Dynamics and function of DNA methylation in plants. *Nature Reviews Molecular Cell Biology, 19*(8), 489–506.

Zhang, W., Chen, L., Xiong, Y., Panayi, A. C., Abududilibaier, A., Hu, Y., ... & Liu, G. (2021). Antioxidant therapy and antioxidant-related bionanomaterials in diabetic wound healing. *Frontiers in Bioengineering and Biotech,. 9*, 707479.

Zhang, Y., Li, M., Gao, H., Wang, B., Tongcheng, X., Gao, B., & Yu, L. (2019). Triacylglycerol, fatty acid, and phytochemical profiles in a new red sorghum variety (Ji Liang No. 1) and its antioxidant and anti-inflammatory properties. *Food Science & Nutrition, 7*(3), 949–958.

Zhao, Z. Y. (2007, December). The Africa biofortified sorghum project–applying biotechnology to develop nutritionally improved sorghum for Africa. In *Biotechnology and Sustainable Agriculture 2006 and*

Beyond: Proceedings of the 11th IAPTC&B Congress, August 31-18, 2006 Beijing, China (pp. 273–277). Springer, Netherlands.

Zhou, Y., Wang, Z., Li, Y., Li, Z., Liu, H., & Zhou, W. (2020). Metabolite profiling of sorghum seeds of different colors from different sweet sorghum cultivars using a widely targeted metabolomics approach. *International Journal of Genomics*, *2020*(1), 6247429.

Zhurlova, O. D. (2017). The current trends and future perspectives of arabinoxylans prebiotics research: A review. *Grain Products and Mixed Fodder's*, *17*(4), 4–11.

14 Leveraging High-Throughput Phenotyping and Artificial Intelligence for Accelerating Sorghum Improvement

Aliva Bakshi, Swaraj Pramanik, Chaoxin Wang, Md Abdullah Al Bari, Josef Aranjo, Mayank Pratap Singh Bangari, Ritesh Kumar, Impa Somayanda, Dinesh K. Saini, Terry Felderhoff, Scott R. Bean, Chad Hayes, Yves Emendack, S. V. Krishna Jagadish, and Doina Caragea

14.1 INTRODUCTION

Sorghum (*Sorghum bicolor* (L.) Moench) is one of the top five cereal crops worldwide, with an annual global production of 60.1 million tons (Ciampitti & Prasad, 2020; USDA, 2024). Ten countries contribute 78.6% of this output, including the United States, Nigeria, Brazil, India, among others (USDA, 2024). The U.S. leads with 14% of global production (8.07 million tons), followed by Nigeria at 11% (6.4 million tons), and Brazil and India each at 8% (USDA, 2024). The leading U.S. states in sorghum production, such as Kansas, Texas, Oklahoma, South Dakota, Nebraska, and Colorado, form the "Sorghum Belt", where the crop is mostly rainfed. In the U.S., sorghum is primarily used for livestock feed and energy production (Ciampitti & Prasad, 2020), whereas in Asia and Africa, it is mainly consumed by humans (Rakshit et al., 2014; Khoddami et al., 2023).

Yield improvements in sorghum for semi-arid regions could enhance food security and stabilize the global food system (Demarco et al., 2023; Mayor et al., 2023). Sorghum yield improvements in the U.S. have resulted from advancements in genotype, environment, and agronomic practices (Assefa & Staggenborg, 2010; Ciampitti & Prasad, 2020; Demarco et al., 2023). The introduction of hybrid sorghum between 1950 and 1980 boosted yields by 39% (Miller and Kebede, 1984). Due to its drought tolerance, sorghum strengthens the U.S. dryland production system (Staggenborg et al., 2008; Demarco et al., 2023). However, its full genetic potential remains untapped. There is a pressing need to enhance the productivity, nutritional quality, and stress resilience of sorghum to meet global food demands and cope with climate change. Sorghum improvement has traditionally relied on breeding methods like mass selection, pedigree breeding, and hybridization, resulting in varieties with better yield, stress tolerance, disease resistance, and grain quality attributes (Rakshit & Bellundagi, 2019). However, these methods are slow and labor-intensive, especially for complex traits like grain yield and stress tolerance involving multiple genes with enormous environmental influence. Molecular breeding, combined with physiological insights, has accelerated genetic gains in crops like maize (Lee and Tollenaar, 2007; Cooper et al., 2014). Similarly, understanding the physiological determinants of sorghum yield stability can enhance breeding decisions and accelerate genetic gains.

 DOI: 10.1201/9781003623717-14

Given the advances in plant genotyping over the last decade, plant phenotyping is currently the limiting factor in determining the genetic regulation of essential phenotypic features (Araus et al., 2022; Gill et al., 2022; Baloch et al., 2023). High-throughput plant phenotyping (HTPP) enables rapid plant measurements, allowing for indirect selection of primary traits like grain yield through vegetation indices (Gill et al., 2022; Kundu et al., 2024). In sorghum, various indices correlate with overall plant health and yield (Galli et al., 2020; Santiago et al., 2024). The dynamic nature of the phenome, shaped by both predictable and unpredictable factors, highlights the need for high temporal resolution in data collection (Cobb et al., 2013). While this generates vast amounts of data requiring sophisticated storage and analysis, it enhances the ability to capture critical traits like yield and stress responses. Integrating HTPP into sorghum breeding is vital for developing improved cultivars that can withstand climate change and contribute to food security. Developments in AI and machine learning technologies, combined with advances in remote sensing and imaging technologies, have resulted in tools that have expanded our ability to rapidly and accurately assess plant traits across large (spatio-temporal) populations. Such tools have enabled researchers to capture a wide range of genetic diversity (Singh et al., 2016; Ubbens & Stavness, 2017; Taghavi Namin et al., 2018; Jiang and Li, 2020; Li et al., 2020; van Dijk et al., 2021; Araus et al., 2022; Khan et al., 2022), a task which has been hardly possible in the past, given the amount of time and effort involved in manual analysis (Singh et al., 2016).

The full potential of AI-based HTPP can be achieved using data acquired in the field through remote sensing based on Unmanned Aerial Vehicles (UAVs) equipped with optical RGB cameras, multispectral and hyperspectral cameras, thermal cameras or LiDAR (Light Detection and Ranging) (Alsadik et al., 2023), or through robotic-assisted imaging platforms (Fahlgren et al., 2015). Many plant traits can be assessed using remote sensing, including seedling and plant counts, inflorescence and fruit counts, leaf area, plant health (diseases, pests, nutrient deficiencies), height, yield, biomass, etc. (Araus et al., 2022). Different data acquisition technologies produce more appropriate data for predicting some plant traits than others. For instance, multispectral and/or hyperspectral data can lead to more accurate early predictions of plant diseases, pest infestations, or nutrient deficiencies. In contrast, plant height can be more accurately predicted based on 3D reconstruction from LiDAR point cloud data (Waliman & Zakhor, 2020). Traits such as seedling/plant counts or inflorescence/fruit counts can be predicted from RGB images. It is also worth noting that different modalities of data used together through multimodal models can lead to more accurate predictions of some plant traits (Maimaitijiang et al., 2020).

While AI-based tools for assisting with HTPP in the wild using field data are highly desirable and greatly needed to enhance modern plant breeding (Ninomiya, 2022), advances in plant breeding can also be made based on close-up, proximal imaging in more controlled settings, e.g., greenhouse or laboratory settings (Saini et al., 2024). From a machine learning and computer vision perspective, RGB images collected in controlled environments can lead to more accurate models that can be subsequently transferred or adapted to more complex field environments (Zhao et al., 2024). In what follows, we review recent works that leverage HTPP using AI-based tools for accelerating sorghum improvement. Specifically, we focus on studies that use deep learning models for sorghum trait phenotyping based on both proximal and remote sensing RGB images, as RGB images can be used to assess a variety of traits by performing image-based object detection or instance segmentation at a low cost, as compared with other data acquisition technologies (Alsadik et al., 2023; Kior et al., 2024).

Deep learning, a key technology behind many recent advances in artificial intelligence (AI), is a sub-area of machine learning that involves using deep neural networks to uncover complex patterns and relationships in data (LeCun et al., 2015). Deep learning models have been used extensively to assess sorghum panicle detection/segmentation and counting based on UAV images (Guo et al.,

2018; Ghosal et al., 2019; Malambo et al., 2019; Oh et al., 2019; Lin & Guo, 2020; Gonzalo-Martı́n et al., 2021; Li et al., 2022; Santiago et al., 2024). In addition, deep learning has been used for various phenotyping tasks including sorghum grain counting (Santiago et al., 2024; James et al., 2024), sorghum seedling/plant counting (Ribera et al., 2017), leaf counting (Miao et al., 2021), estimating flowering time (Cai et al., 2021), predicting yield (Zannou & Houndji, 2019), stomata counting (Bheemanahalli et al., 2021), etc.

Most HTPP tasks are addressed with convolutional neural networks (CNNs), which are deep learning models designed for processing structured grid-like data, such as images (Lecun et al., 1989). CNNs are highly effective in tasks like image classification and object detection/segmentation due to their ability to automatically learn spatial hierarchies of features. A CNN consists of convolutional layers followed by non-linear activations, pooling layers and fully connected layers. The fully connected layers are generally used to combine the features learned by convolutional layers into a final classification or a regression output. For a classification task, the network assigns a categorical label to its input, while for a regression task, the network predicts a continuous numerical value based on input data. CNNs are also commonly used as a backbone for object detection and instance segmentation models. In object detection, the model detects objects of interest (e.g., panicles), localizes them using rectangular bounding boxes and classifies the objects into one of several predefined categories (Zaidi et al., 2022). As opposed to object detection, where objects of interest are marked using bounding boxes, in instance segmentation, the model further identifies precise segmentation masks for each object instance by performing category classification at pixel level (Hafiz & Bhat, 2020).

Given the prevalence of object detection and instance segmentation approaches in the literature on AI-based HTPP for advancing sorghum improvement, these two approaches are reviewed in detail, along with the generic workflow for training and evaluating models for HTPP in sorghum. Sample tools available for data labeling in object detection and instance segmentation tasks are discussed, as well as evaluation metrics commonly used in such tasks. The application of these approaches to assess several sorghum selection traits, including panicle count, grain count (as a proxy for yield estimation), and stomata density estimation, is also explored. Key conclusions are presented, highlighting the limitations of current approaches and proposing directions for future research to enhance the utility of these methodologies.

14.2 OBJECT DETECTION AND INSTANCE SEGMENTATION MODELS

This section reviews the generic architecture of object detection and segmentation models, differentiating between one-stage and two-stage detectors. Popular object detection and image segmentation approaches are briefly described, with a focus on Faster R-CNN, Mask R-CNN, and several YOLO (You Only Look Once) variants commonly used in HTPP, particularly for trait phenotyping in sorghum.

14.2.1 Generic Model Architecture

The architecture of deep neural networks for object detection typically consists of two primary components: a *backbone* and a *detection head*. The backbone, usually a pre-trained CNN, generates feature representations or maps from the input images. The detection head is responsible for identifying objects by predicting bounding boxes around objects (defined by their coordinates) and classifying the objects into various categories of interest (Bochkovskiy et al., 2020). Object detection approaches can be categorized as one-stage detectors (e.g., YOLO family of detectors) and two-stage detectors (e.g., Faster R-CNN).

One-stage detectors such as the YOLO models perform object detection and classification in a single step, through the means of a dense region-free prediction head. In contrast, two-stage detectors, such as Faster R-CNN, perform the object detection and classification tasks in two separate

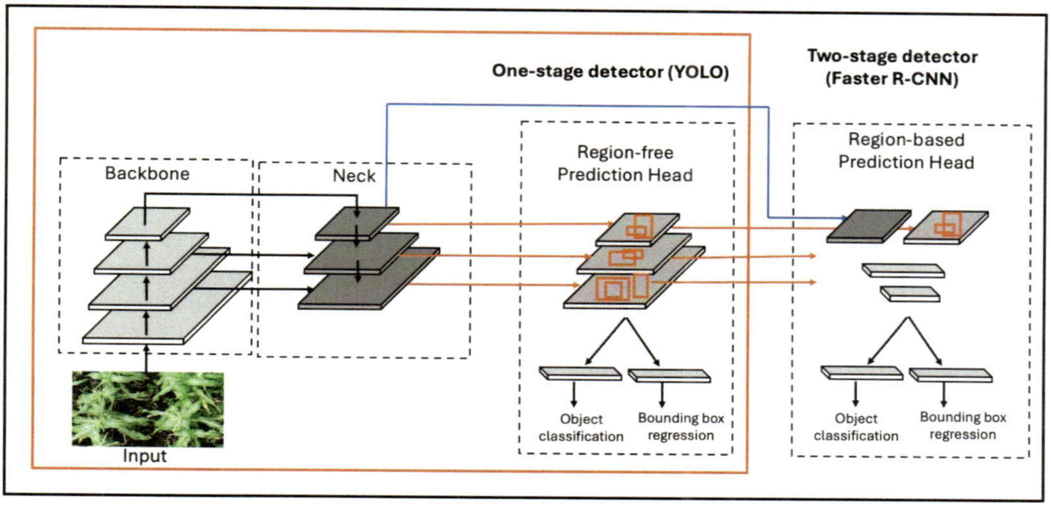

FIGURE 14.1 Generic architecture for one-stage and two-stage object detection approaches.

stages: in the first stage, regions of interest (RoIs) that might contain objects are identified using a dense prediction head similar to the head used by YOLO. In the second stage, the bounding boxes of the RoI proposals are refined and classified into different categories using a sparse regression head.

More recent object detection models often include an additional component called a *neck*, which is inserted between the backbone and the head. The neck combines features from different layers of the backbone network, enhancing the model's ability to detect objects of varying scales (Bochkovskiy et al., 2020). A widely-used neck structure is the Feature Pyramid Network (FPN) (Lin et al., 2017). FPN uses a top-down path with lateral connections to extract and fuse multi-scale feature maps, with the goal of improving detection performance across different object sizes. The basic architecture of the one-stage and two-stage detectors is shown in Figure 14.1.

14.2.2 Faster R-CNN Model

The Faster R-CNN model (Ren et al., 2016) is a popular two-stage object detection model that uses a pre-trained CNN backbone along with an FPN network to generate feature maps at various scales. In the first stage, the extracted feature maps are fed into a region proposal network (RPN), which acts as the dense prediction head. The RPN identifies Regions of Interest (RoIs; areas likely to contain objects) and their locations as bounding boxes. RoIs are labeled as object or background based on their overlap with the ground truth boxes, and non-maximum suppression (NMS) is used to remove overlapping regions likely representing the same object. In the second stage, the selected RoIs and feature maps are passed to a sparse prediction head, which classifies the RoIs into specific categories (e.g., panicle) and adjusts their bounding box coordinates through a regression task. The model is trained end-to-end using a multi-task loss that combines cross-entropy for classification with L2 loss for bounding box regression (Girshick, 2015).

14.2.3 Mask R-CNN Model

Mask R-CNN (He et al., 2017), a widely used instance segmentation model, builds on the Faster R-CNN framework (Ren et al., 2016). Like Faster R-CNN, Mask R-CNN can detect objects of interest (e.g., panicles) in an image and localize them using bounding boxes. Additionally, Mask R-CNN produces accurate segmentation masks for each object instance. It extends Faster R-CNN by adding layers dedicated to predicting instance masks for each RoI, while simultaneously performing object classification and bounding box regression. Mask R-CNN also uses an FPN (Lin et al., 2017) with a ResNet backbone to detect objects at various scales. Unlike Faster R-CNN, it replaces the RoI Pool layer with a RoI Align layer, ensuring precise pixel-level alignment between the input and output, allowing for the generation of accurate instance masks.

14.2.4 YOLO Models

As previously explained, two-stage detectors perform object detection by first using an RPN to identify RoIs that may contain objects, followed by a second stage that classifies these RoIs into specific categories. In contrast, one-stage detectors directly process the input image to simultaneously classify objects and determine their bounding box coordinates. Originally, one-stage detectors such as YOLO (Redmon, 2016) were designed to improve inference time and enable real-time applications where speed is critical. While the first iterations of the YOLO models traded performance for speed, later versions, from YOLOv5 (Jocher, 2020) to the most recent YOLOv11 model (released as of September 30, 2024) (Jocher and Qiu, 2024), have continued to improve performance and expand the capabilities of the models. Currently, YOLO models achieve state-of-the-art performance in terms of both accuracy and speed on benchmark datasets (Jocher et al., 2023). Furthermore, models such as YOLOv8 (Jocher et al., 2023) and YOLOv11 (Jocher and Qiu, 2024) have expanded their capabilities from object detection and segmentation to other tasks such as pose/keypoint detection, oriented detection and classification (Jocher et al., 2023) opening new exciting doors for a variety of real-time computer science applications.

While the YOLO family of models has evolved very fast, research on AI-based HTPP approaches for sorghum improvement has also followed the trend, with models such as YOLOv5 (Jocher, 2020), YOLOv7 (Wang et al., 2023), YOLOv8 (Jocher et al., 2023), YOLOv9 (Wang et al., 2025) already being documented in the sorghum literature as having high performance in terms of HTPP sorghum phenotyping (Santiago et al., 2024). These models are discussed in the following sections.

14.2.4.1 YOLOv5

The backbone of YOLOv5 (Jocher, 2020) is based on a combination of CSP-Darknet53 (Redmon, 2018) and CSPNet (Wang et al., 2020). Darknet53, inspired by the ResNet architecture, was specifically designed for object detection. The neck in YOLOv5 includes two components: SPPF and CSP-PAN. SPPF is a faster variant of Spatial Pyramid Pooling (SPP) (He et al., 2015), which aids in detecting small objects and objects at various scales. The PANet (Path Aggregation Network; Liu et al., 2018), combined with CSP (Cross Stage Partial connections), further enhances computational speed. YOLOv5 also uses several data augmentation techniques, including mosaic augmentation (Bochkovskiy et al., 2020), which stitches four images together to train the model to detect objects in different areas of the image, not just the center. Additionally, YOLOv5 uses automatically generated anchors with varying scales and aspect ratios to predict bounding boxes directly from the input image. The model has a dense prediction head with three detection layers for different scales, predicting bounding boxes with three aspect ratios at each scale. Similar to Faster R-CNN-FPN, the NMS technique filters out bounding boxes that represent the same object.

14.2.4.2 YOLOv7

The YOLOv7 (Wang et al., 2023) architecture is built on the "bag-of-freebies" concept introduced by Bochkovskiy et al. (2020), which prioritizes inference speed while allowing slower training if it enhances model accuracy. YOLOv7 incorporates several innovations in both network architecture and training strategies. A key innovation is the extended efficient layer aggregation network (E-ELAN) (Wang et al., 2022), used in both the backbone and neck/head networks. E-ELAN prevents performance degradation when scaling the network and helps the model learn more diverse features. The network's neck is based on a PAFPN structure (combining PAN and FPN) and utilizes E-ELAN for feature extraction and fusion. YOLOv7 also introduces an "auxiliary head" in the middle of the network to support the "lead head", which may be too deep in the network. Similar to YOLOv5, YOLOv7 employs an automated anchor selection approach and data augmentation techniques, along with non-maximum suppression (NMS) to filter bounding boxes. The model is trained from scratch on the COCO dataset using a multi-task learning loss, combining classification loss (binary cross-entropy), objectness loss (binary cross-entropy), and location loss.

14.2.4.3 YOLOv8

YOLOv8 (Jocher et al., 2023) is highly popular and effective model which was designed as a direct successor of YOLOv5 (Jocher, 2020) to further balance speed and accuracy. It was the first model to expand its capabilities beyond object detection and instance segmentation. The backbone in YOLOv8 is a custom CSPDarknet53 network, which combines Darknet and CSPNet architectures. By incorporating cross-stage partial (CSP) connections, the backbone enhances information flow between layers (Jocher et al., 2023), which helps reduce speed while maintaining accuracy. YOLOv8's neck consists of a novel module (called C2f) which combines high-level semantic features with low-level spatial information through the use of PANet (Liu et al., 2018). As another innovation, YOLOv8 introduces dynamic anchor assignment, which enables the adaptation of the anchor box dimensions during training. Together, these innovations help improve detection accuracy, especially for small objects (such as sorghum seeds). As for the head, YOLOv8 is similar to its predecessors in terms of how it leverages different modules to predict bounding boxes, objectness scores, and specific categories.

14.2.4.4 YOLOv9

YOLOv9 (Wang et al., 2025) is a successor of YOLOv7. Two main innovations were introduced in YOLOv9, the Programmable Gradient Information (PGI) and the Generalized Efficient Layer Aggregation Network (GELAN), which together help improve feature extraction, gradient flow, and network efficiency. The PGI component was designed to prevent data loss and ensure accurate gradient updates through an auxiliary reversible branch that enables gradient information propagation at different semantic levels. On the other hand, the GELAN architecture optimizes lightweight models with gradient path planning by simultaneously considering things like number of parameters, computational complexity, accuracy, and inference speed.

14.3 Generic Workflow for Training and Evaluating Models

Figure 14.2 shows the generic workflow that is followed when training and evaluating object detection and instance segmentation approaches.

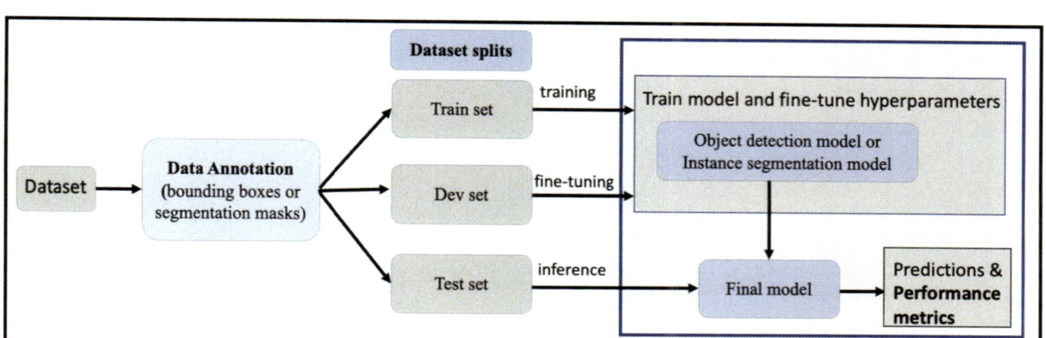

FIGURE 14.2 Generic workflow for object detection/segmentation tasks.

14.3.1 Data Annotation

To be able to train a model, one needs an image dataset that is representative of the objects targeted (e.g., seeds or panicles). Furthermore, target objects in the image dataset need to be annotated and used as ground truth for training and evaluating detection/segmentation (and counting) models. For object detection, this means drawing rectangular boxes (called bounding boxes) around each of the target objects in an image. For instance, in segmentation, the annotations are in the form of precise polygonal masks. There are many tools available for image data annotation, including the popular LabelMe (Russell et al., 2008) and LabelBox (LabelBox, 2024) toolboxes. A sample image of sorghum seeds annotated with bounding boxes using LabelMe is shown in Figure 14.3.

The annotations performed with various tools can be exported as JSON files. These JSON files have the location information for each object in an image provided as (x-max, y-max) and (x-min, y-min) coordinates, where these coordinates represent the end points of the main diagonal of the bounding box. The JSON files also contain the label of the object (e.g., seed) and the shape of the object (e.g., rectangle or polygon). As different models require the annotations in different formats, tools that convert one format to another need to be used. For instance, Roboflow (https://roboflow.com/) can convert the JSON files from LabelMe to the proper format required by different object detection models (e.g., COCO format for Faster R-CNN, or YOLO format for YOLO models).

14.3.2 Dataset Splits

Annotated images are split into three disjoint subsets, specifically a training (train) subset, a development (dev) subset, and a test subset. The training subset is used to train the model, while the development subset is used as a proxy for the test subset to fine-tune the hyperparameters of the models (that need to be fixed prior to training). Once a model is trained, the performance is evaluated based on the test subset.

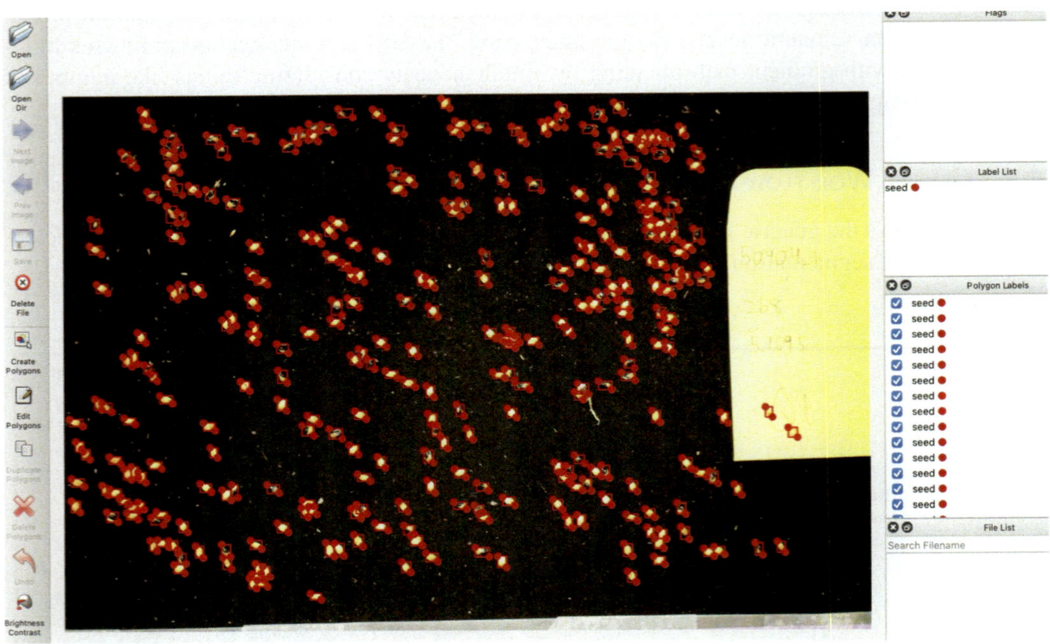

FIGURE 14.3 Sample image of sorghum seeds annotated with bounding boxes using LabelMe.

14.3.3 Performance Metrics

Several metrics are commonly used to evaluate models trained for object detection and instance segmentation. First, the Intersection-over-Union (IoU) metric is used to measure how well the predicted bounding boxes (or alternatively, segmentation masks) overlap with the ground truth bounding boxes (or ground truth segmentation masks). Using a threshold on the IoU (e.g., 50%), one can identify the number of true positives (TP), true negatives (TN), false positives (FP), and false negatives (FN) for each category of interest in the dataset. This information can be further used to calculate the mean average precision over all objects in an image and all images in the test set, denoted by $mAP@n$ for an IoU threshold of n, which identifies a prediction to be correct if the IoU between the detected object and the ground truth object annotation is greater or equal than n. Both IoU and $mAP@n$ metrics range between 0 and 1, with high values representing more accurate models.

When the detection/segmentation results are ultimately used for counting different objects, error metrics such as the Root Mean Squared Error (RMSE), Mean Absolute Error (MAE), or mean absolute percentage error (MAPE) are used for evaluation. In the case of error metrics, values close to zero correspond to accurate models. Finally, when evaluating models in terms of their ability to detect the right number of objects in an image, the R^2 (coefficient of determination) can be used to identify the fit of the predicted counts to the ground truth counts.

14.4 SORGHUM SELECTION TRAITS AND PHENOTYPING APPROACHES

We focus the discussion in this section on three important selection traits for sorghum breeding, including two key yield-related traits, panicle counts and grain density, which are crucial for improving plant productivity, and one stress-related trait, stomata density (and the related stomatal complex area), as stomata have an essential role in regulating gas exchange and abiotic stress (e.g., drought) response in plants.

14.4.1 Sorghum Panicle Detection and Density Estimation

Panicle count, or the number of grain-bearing panicles of a sorghum plant, is a vital determinant of total grain yield (Maman et al., 2004). Each panicle holds seeds, and a higher number of panicles can potentially lead to higher yields for plants, assuming the panicles have similar architecture. Also, growing conditions and abiotic stresses can influence the panicle count and its seed density (Ndlovu et al., 2021). Given its high impact on yield, the panicle count represents an important selection criterion for breeding programs (Kumar et al., 2013). Traditionally, the panicle count has been obtained by manually counting the panicles on each plant in the field. However, the manual counting approach is expensive, laborious, time-demanding, subjective, and prone to errors. Thus, it cannot be a feasible option to assess panicle counts on a large number of sorghum genotypes across environmental conditions and abiotic stresses, particularly for large-scale trials. Therefore, tools for effective, non-destructive HTPP of panicle count are essential, as evidenced by several studies focusing on deep learning approaches for panicle detection, segmentation, and counting.

Several studies have focused on developing HTPP tools for counting sorghum panicles from field images. For instance, Olsen et al. (2018) proposed automated sorghum panicle detection and counting from aerial RGB images. This study has primarily developed tools for panicle annotation, detection, and counting. The annotation tool uses superpixels for efficient labeling. The detection model is based on 3D color histograms followed by logistic regression, and the counting model uses superpixel-level detection scores with regression. This system was evaluated on aerial imagery from 18 sorghum varieties and had high accuracy with area under the curve (AUC) = 0.98 for detection. It also achieved an MAE value of 1.88 for the counting task when the variety-specific information was considered and an MAE value of 2.66 when it was not. These results demonstrated the system's ability to contribute towards high-throughput phenotyping while notably advancing sorghum breeding programs.

Malambo et al. (2019) used an early deep-learning segmentation model called SegNet (Badrinarayanan et al., 2017) (which uses VGG-19 as the backbone network) for segmenting UAV images into panicles, foliage, and ground. Subsequently, the model was applied to field orthomosaic to generate semantic segmentation at the field level. Evaluation based on 60 randomly sampled plots showed that the SegNet model achieved an accuracy of 94% for panicle detection, while evaluation at plot level showed that the model's predictions had a strong 0.88 Spearman correlation with the manually annotated panicle counts.

Oh et al. (2019) proposed a two-stage network based on the counting CNN (CCNN) architecture (Onoro-Rubio & López-Sastre, 2016) to predict sorghum panicle density maps, which are subsequently integrated into panicle counts. In the first stage, the model performs panicle segmentation, while in the second stage, it uses the panicle segmentation output to infer density maps. Experimental results using UAV images show that the proposed model outperforms baselines for crowd counting, such as CCNN (Onoro-Rubio & López-Sastre, 2016), MCNN (Zhang et al., 2016), and CSRNet (Li et al., 2018), and achieves an MAE value of 1.06. James et al. (2023) introduced a complete pipeline for panicle detection from field plots at a commercial scale, deploying deep learning protocol using UAV images. However, James et al. (2023) used extensive, laborious hand-held image acquisition protocol to capture and develop training datasets. The pipeline contains several components, from data collection and annotations to model training, performance evaluation, model adaptation, and density map estimation. Models trained include two detection models, Faster R-CNN and YOLOv5, and one regression model, CSRnet (Li et al., 2018). The best results in terms of count MAE and RMSE were obtained with the CSRnet model.

The high-resolution head density maps produced at the final step of the pipeline can help breeders estimate crop yield. Santiago et al. (2024) explored deep learning approaches for panicle segmentation based on images captured using a smartphone. Specifically, they trained a Mask R-CNN model using the Detectron2 framework and a YOLOv8 model for panicle segmentation. Experimental results showed that the two models achieved an average precision *AP* of 0.75 and 0.89, respectively. Overall, the studies mentioned above have highlighted the potential of deep learning to transform crop management, improve yields, and enhance food security by optimizing critical breeding decisions and relevant agricultural processes.

Our group leverages the advancement of deep learning in computer vision applications in the sorghum space and is working on developing sorghum yield forecasting using deep learning in a computer vision framework. The aim is to forecast sorghum yield four to six weeks before physical harvesting by extracting yield attributes from the imagery. As part of this research, we optimized the protocol for detecting sorghum panicles in the field plots. The field trial was conducted in Ashland Bottoms, Kansas, with 36 distinct sorghum genotypes arranged in a randomized complete block design with three replicates, using 18-foot, 4-row plots in 2023. We captured images on September 27, 2023, on a clear sunny day, with a windspeed less than 10 mph. We used a DJI M300 (DJI, Shenzhen, China) Uncrewed Aerial Vehicle (UAV) equipped with a ZenmuseP1 sensor together termed as Uncrewed Aerial System (UAS), capturing imagery at a ground sample distance of 0.08–0.1 cm/pixel. This setup allowed for the development of a robust dataset. Close-range imagery was captured from six meters above ground level from three angles: nadir (ND) at 90°, a 45° oblique angle facing north (FN), and a 45° oblique angle facing south (FS) (Figure 14.4).

To develop our training dataset, we used UAS imagery, in contrast to handheld devices described by James et al. (2023). The large image dimensions of approximately 8,100 x 4,600 pixels were cropped to 1,280 x 1,280 pixels, resized to 640 x 640 pixels, and annotated using custom tools derived from the base Segment Anything Model (SAM) (Kirillov et al., 2023). The images were further corrected in LabelMe (Russell et al., 2008). We annotated 200 images each from the ND, FN, and FS angles, which were augmented using Roboflow. The augmentations included flipping,

FIGURE 14.4 Image capture and representation of field panicles: A. DJI M300 Drone during image acquisition, B. Nadir perspective of field panicles, C. Facing North view of panicles, D. Facing South view of panicles.

rotation, grayscale, hue, and saturation modifications, yielding a total of 600 images across ND, FN, and FS groups. We then used a training, validation, and test split of 90:5:5.

In terms of model architectures, we applied several open-source algorithms, including Faster R-CNN and three YOLO versions, YOLOv7, YOLOv8, and YOLOv9. Each of the YOLO models was trained for 100 iterations, while Faster R-CNN was trained for 300 iterations. The YOLO models showed similar performance across versions and view angles, with AP values ranging from 0.64 to 0.86, and AP@50 values between 0.95 and 0.98. YOLOv7 consistently achieved the highest AP and AP@50 across all angles. Our findings highlight the strong potential of all YOLO models, especially in detecting sorghum panicles from the nadir view, which consistently yielded the highest precision. The detections from YOLOv7, YOLOv8, and YOLOv9 are exhibited in Figure 14.5 across view angles ND, FN, and FS.

In comparison, the Faster R-CNN model achieved an AP of 0.59 and an AP@50 of 0.89. While this performance is lower than the performance of the YOLO models, it still demonstrates the utility of the model in panicle detection tasks. These results collectively offer promising avenues for optimizing panicle detection in sorghum fields and contribute to the ongoing development of methods to digitize sorghum yield forecasting four to six weeks before harvesting.

14.4.2 SORGHUM GRAIN DETECTION AND DENSITY ESTIMATION

Grain count, which refers to the total number of seeds per panicle, is another significant trait that directly influences the overall yield (Van Oosterom & Hammer, 2008), together with other grain characteristics, such as grain size and weight. Higher grain counts per panicle are associated with higher productivity, making the grain count one of the prime targets in breeding programs. Traditional

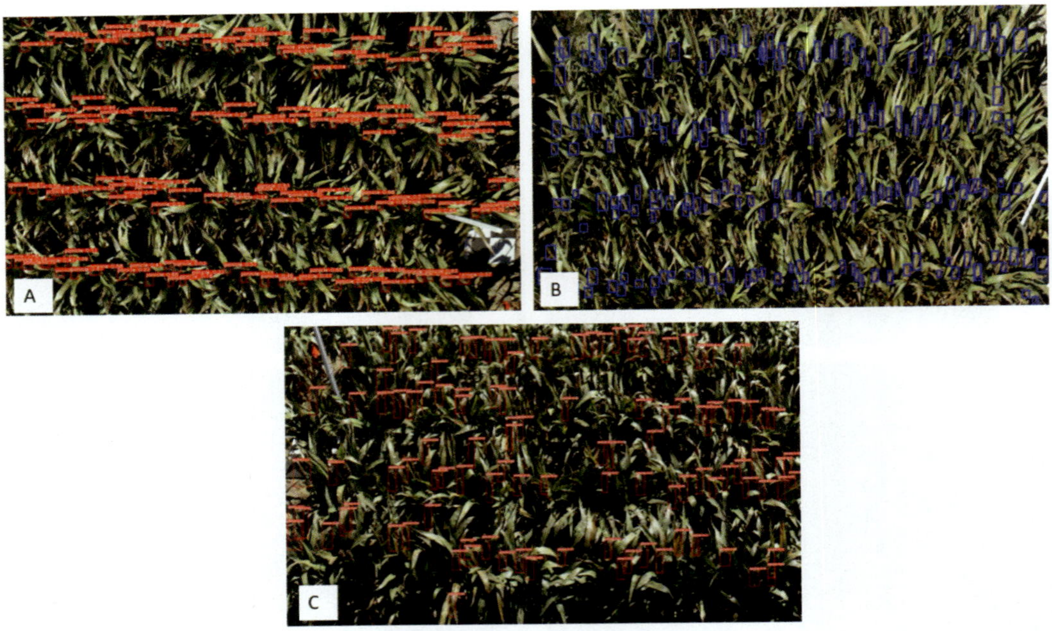

FIGURE 14.5 Detected panicles with A. YOLOv7 for Nadir view, B. YOLOv8 for Facing North, and C. YOLOv9 for Facing South.

grain count phenotyping involves manually sampling panicles at plant level from genotypes grown under different conditions, separating the seeds from the panicle either manually or with a thresher, using a cleaning system to separate the grains from the waste, and finally using a grain counting machine to count the grains (James et al., 2024). The manual sampling process is labor intensive and error-prone (Santiago et al., 2024), while the sensing methods used by grain counting machines can over-count or under-count the grains (e.g., grains with smaller volumes may not be detected by capacitance methods) (Liu et al., 2021). Furthermore, non-destructive methods that can be used for HTPP in the field are highly desirable and greatly needed.

Some recent studies have integrated AI and deep learning with traditional field phenotyping to reduce labor costs and improve the efficiency of sorghum grain count phenotyping. For instance, Santiago et al. (2024) utilized smartphone imagery from the field to first detect and then segment sorghum panicles (using the Mask R-CNN model from Detectron2 and YOLOv8 models) and subsequently estimate the grain count per panicle using three variants of the density-based MCNN (Zhang et al., 2016), initially designed for crowd counting. The panicle images used in the experiments for grain count estimation were manually cropped to ensure perfect segmentation. The panicle images were split into three fragments to reduce the number of counts in an image. The best result in terms of MAPE was 17%, and was obtained using a custom CNN network with three columns. While this is an encouraging result, the approach has the disadvantage of working on fragmented panicle images. Furthermore, the total number of images and genotypes included in the study was rather small and did not ensure diversity. Finally, the use of a smartphone to capture images, while cheaper and more readily available than UAVs, does not scale up well to field level, and it is still labor-intensive and prone to selection bias.

In another significant contribution towards phenotyping grain count, James et al. (2024) introduced a framework called GrainPointNet, which combines RGB images and point cloud models constructed via proximal structure-from-motion-based photogrammetry workflow for grain counting. YOLOv5 was used to detect grains in an image, and PointNet was used to extract features from the point cloud models. The features extracted are used together with the image-based grain counts to infer the panicle grain count. The GrainPointNet model achieved MAPE values of 6.5% and 6.8% for high- and low-resolution point clouds, respectively, showcasing its potential for both lab and field conditions. One main disadvantage of this approach is the fact that a large number of images need to be taken for each panicle to be able to perform point cloud reconstruction of the panicle. In addition, images were taken in a controlled laboratory setting, and thus they are not similar to those that would be taken in the field, despite the fact that the authors showed that their approach works well with lower scale images (which would resemble the scale of the field images).

To address prior work limitations related to dataset size and genotype diversity, our group has assembled a large benchmark dataset consisting of two types of sorghum grain images from approximately 350 diverse sorghum genotypes from the Sorghum Association Panel (SAP) (Casa et al., 2008). Panicles in the dataset contain as few as 30 grains and as many as 3,775 grains, with an average of approximately 1,400 grains per panicle. Using the benchmark dataset, we have also developed counting-by-detection approaches for predicting grain count based on the assembled dataset. A total of four panicles were manually collected from each genotype and subsequently subjected to imaging. The imaging process was performed in a lab environment and encompassed both the entire panicles containing grains (front and back), as well as threshed grains that were partitioned into four quadrants on a piece of black cloth. Machine counts were also obtained for each panicle. The four quadrant images were used as an alternative source of ground truth for grain counts (and to validate the machine counts), while the panicle images were used to train and evaluate models for grain count estimation directly from panicle images.

We trained Faster R-CNN and YOLOv7, YOLOv8, YOLOv9 models for both types of images using a subset of carefully annotated datasets containing images from 70 highly versatile genotypes with different grain color, panicle shape, and grain density. The models were evaluated on a test dataset with a diverse set of images from 25 genotypes, most of which do not appear in the train set (which allows us to evaluate the robustness of the models to new genotypes).

The predictions of these object detection models are compared with the machine and/or manually annotated ground truth of our test data to assess performance, which is measured in terms of *AP* values, *MAPE* and R^2. Our best results for grain detection and counting in quadrant images were obtained using the YOLOv8 model, which achieved an *AP@50* value of 98.3, an *MAPE* value of 1.089% and an R^2 value of 0.997. The results of this model also suggested that the machine generally under-counts the grains on a panicle, for dense panicles with a large number of seed, possibly due to some grains with smaller volume (Liu et al., 2021). The best results for grain detection and counting in panicle images were obtained using the YOLOv7 model, which achieved an *AP@50* value of 84.62, and an *MAPE* value of 9.26%. As for R^2, the best model achieved an R^2 value of 0.81 on test genotypes that were not included in the training set. Figure 14.6 shows examples of seed detection on a panicle image (left) and on a quadrant image (right).

Overall, our results show that modern object detection models in the YOLO family can achieve good performance for counting sorghum grains per panicle, even when the panicle contains a large number of seeds. However, we should note that counting grains on a panicle by detection results in lower counts, as not all grains are visible in an image. Approaches based on density maps and regression are needed to estimate the actual counts better. While our results are obtained with images taken in a lab setting, they can serve as a rich source of data for transfer to field settings.

14.4.3 Stomatal Density and Stomatal Complex Area

Stomatal density (SD) and stomatal complex area (SCA) are key traits that regulate gas exchange and plant responses to abiotic stresses. While these stomatal characteristics have been studied in crops exposed to different environments using manual, low-throughput methods (Gitz & Baker, 2009), the genetic basis controlling these traits and their environmental responses in sorghum

FIGURE 14.6 Example of seed detection on panicle and quadrant images. The panicle image shows an annotated seed count of 442 and a predicted seed count of 420. The quadrant image shows an annotated seed count of 1,852 and a predicted seed count of 1,801.

insufficiently understood. Additionally, the diversity in stomatal traits is largely unexplored and underutilized in breeding programs due to the labor-intensive phenotyping process, which requires significant resource investment. For instance, manual phenotyping involves collecting stomatal imprints, imaging specimens, and manually counting stomata, with counting being the most time-consuming aspect (Fetter et al., 2019; Sakoda et al., 2019). Sorghum is commonly grown in arid and semi-arid regions, making its productivity heavily reliant on rainfall patterns, posing challenges in regions like the U.S.A., sub-Saharan Africa, and India (Leff et al., 2004). Despite their adaptation to arid environments, sorghum hybrids remain vulnerable to harsh conditions during various growth stages (Tack et al., 2017). As a C_4 crop adapted to hot and arid climates (Osborne & Freckleton, 2009), sorghum presents a valuable opportunity to investigate natural variation in SD and SCA under field conditions. Identifying the loci controlling stomatal traits is crucial for developing sorghum varieties with optimized stomatal regulation.

In recent years, several computer vision-based tools have been developed to address the need for automated high-throughput detection of stomata in microscopic images. These tools include the Cascade object detection algorithm (Higaki et al., 2014; Laga et al., 2014; Duarte et al., 2017; Jayakody et al., 2017), deep convolutional neural networks like AlexNet (Fetter et al., 2019), and YOLO models (Casado & Heras, 2018). Additionally, several methods and tools for quantifying stomatal variations from images have been developed (Dittberner et al., 2018; Sakoda et al., 2019). Most of these techniques have employed object detection rather than the more accurate semantic object segmentation approach.

To overcome this limitation, our recent study (Bheemanahalli et al., 2021) used a Mask R-CNN model to automatically detect and segment stomata in images, enabling the counting and measurement of stomata to estimate SD and SCA. Both traditional and deep learning methods were employed to assess genetic diversity in stomatal traits across 311 grain sorghum accessions in two field environments in Kansas, U.S.A. The analysis of nearly 12,000 images from abaxial (Ab) and adaxial (Ad) leaf surfaces revealed significant variation in stomatal traits. Examples of stomata images, including segmentation results, are presented in Figure 14.7.

Our study demonstrated significant agreement between manual and deep learning methods in predicting SD and SCA, as well as agreement when mapping genomic regions associated with

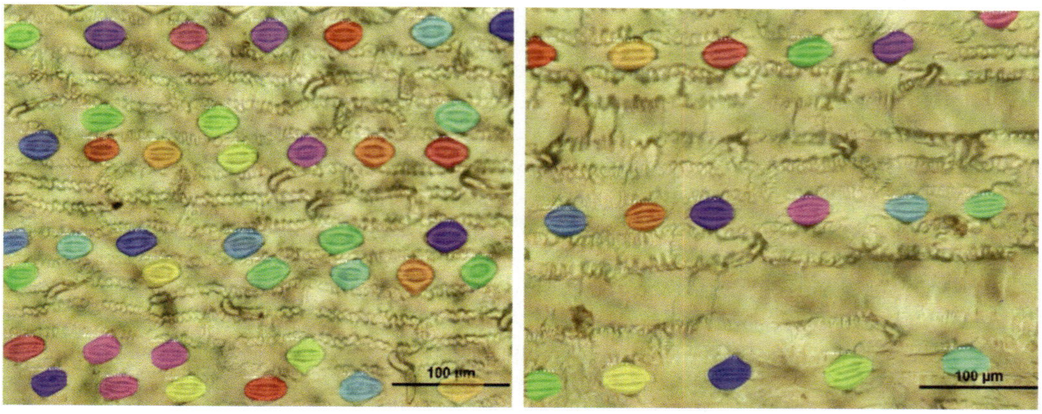

FIGURE 14.7 Samples of stomata images, where stomata are annotated using our Mask R-CNN model.

FIGURE 14.8 Web-Server for stomata density and stomatal complex area detection. Users can check the model predictions on sample images or can upload their own images to obtain estimates for SD and SCA traits (Bheemanahalli et al., 2021).

stomatal number and area. Specifically, the R^2 computed by comparing the predicted counts to manual counts (ground truth) was 0.96 for two different growing environments, while the R^2 for stomatal complex area was around 0.90. A Web-based tool was developed with our trained stomata segmentation model in the backend. The interface of the tool is shown in Figure 14.8. From a breeding perspective, leveraging a genome-wide association analysis of the genetic variants led to the identification of genomic regions associated with SD and SCA traits, and promising candidate genes controlling those traits in the detected genomic regions (Bheemanahalli et al., 2021).

14.5 LIMITATIONS OF CURRENT RESEARCH AND FUTURE DIRECTIONS

Many of the existing HTPP phenotyping tools, e.g., for grain density (Walker et al., 2013) and stomata density estimation (Bheemanahalli et al., 2021), are designed to work with images taken in a lab setting, under controlled conditions. As discussed above, manual sampling of data to be imaged in the lab (e.g., panicles or sorghum leaves) is time-consuming, expensive and prone to bias and error.

Technologies and tools that can be used for HTPP based on images taken in the field hold great promise, as shown by recent studies on panicle count (Olsen et al., 2018; Malambo et al., 2019; James et al., 2023; Santiago et al., 2024). However, there are unique challenges that need to be addressed when working with field images. As UAVs become a primary data collection tool (Kim et al., 2019), future work must focus on adapting deep learning models to the complexities of dynamic field conditions. Unlike controlled environments, field images present more differences caused by variable lighting conditions, weather conditions, camera angles, etc. Furthermore, field images may contain numerous occlusions of the objects of interest, which can distort data collection. Robust deep learning models capable of handling field data without compromising accuracy need to be developed.

Another drawback of the current research in the area of HTPP which limits model training is related to the availability of training data. As with any other AI applications, accurate model training begins with data annotation. However, this step presents significant challenges, especially in agricultural domains, due to its laborious and expensive nature. Tools such as AnyLabeling (Nguyen, 2024), integrated with sophisticated models like the SAM (Kirillov et al., 2023), have simplified the annotation process by automating pixel-level labeling. SAM's ability to generalize across various objects and domains enhances the accuracy of annotations, yet limitations persist. In highly specific datasets, such as those involving small, overlapping sorghum seeds, SAM may struggle to deliver consistent results. Additionally, the pixel-level precision it offers can become computationally expensive when applied to large datasets, making it difficult to scale for extensive agricultural use. The model's performance is also reliant on the quality of the prompts used during annotation, necessitating continuous human oversight. However, looking ahead, there is potential to overcome the annotation challenges by developing models that are specialized for agricultural imagery. Integrating more crop-specific models into tools like AnyLabeling could enhance such tools, help produce labeled data at a low cost, and ultimately make the models more effective in the field. Further advancements could include models that require minimal human input (e.g., prompts) for generating annotations, and models that are capable of generalizing to various crop types and environmental conditions. Collaborating with agricultural scientists to create large, domain-specific datasets such as (Zhang et al., 2024; Cai et al., 2022) could further enhance existing annotation tools and thus help reduce the reliance on labor-intensive manual annotation.

In particular, prompt-based phenotyping, where models are guided by specific inputs to recognize traits of interest, is gaining traction as an effective method for object identification (Du et al., 2022). However, its success largely depends on the quality of the prompts used. In agricultural datasets, variability in crop appearance caused by different growth stages, environmental conditions, or camera angles can make crafting effective prompts a challenge. Future research should focus on optimizing these prompts for agricultural applications. Additionally, developing methods to automatically generate or refine prompts could reduce the need for expert input and broaden the applicability of prompt-based models (Wang et al., 2024; Minderer et al., 2022; Radford et al., 2021) in the agricultural domain.

Once data is annotated, data augmentation becomes essential to make machine learning models more robust (Zoph et al., 2020). This technique, which involves manipulating existing data through random cropping, rotations, scaling, and color adjustments, is especially important in agricultural applications where annotated datasets are often limited. In the case of phenotyping, however, random augmentations can sometimes distort the shape and size of objects of interest, negatively affecting model accuracy. Future research could address this by using context-aware augmentation techniques (Kim et al., 2024; Gordienko et al., 2022; Hu et al., 2023). For instance, generative adversarial networks (GANs) (Goodfellow et al., 2020; Bowles et al., 2018) or diffusion models (Nichol et al., 2021; Rombach et al., 2022; Croitoru et al., 2023) could create synthetic datasets that are both realistic and preserve the structural integrity of the object of interest. Furthermore, exploring 3D data augmentation could provide different perspectives of objects (e.g., panicles), thus improving the robustness of existing models in real-world agricultural environments.

Video data analysis has grown important in crop phenotyping (Yang & Han, 2020) and yield estimation (Chaivivatrakul et al., 2010) as it enables continuous observation of crop growth and tracking the developmental stages, identifying stress responses, and evaluating plant health in real time. Videos capture variations in plant morphology and behavior under different conditions leading to more accuracy in terms of crop phenotype. Moreover, video-data analysis allows faster processing of large volumes of data. Thus, another direction to be addressed in the future is HTPP phenotyping based on video analysis.

Finally, explainability helps understanding of the behavior and predictions of the models (Hassija et al., 2024; Nazir et al., 2023; Kinger & Kulkarni, 2021). Explainability can help with acceptance of the model outputs by sorghum breeders by providing insights into which features influence the outcomes (Coulibaly et al., 2022). It can also facilitate iterative improvement of models by diagnosing model error. Overall, explainability enhances transparency and reliability, and could increase the adoption of data-driven approaches in crop management, eventually leading to improved agricultural productivity and sustainability.

14.6 CONCLUSIONS

AI-based approaches are transforming sorghum improvement, offering powerful tools for high-throughput phenotyping. By integrating advanced imaging technologies and powerful deep learning models, AI has accelerated the development of sorghum varieties that can meet the challenges of climate change, disease pressure, and food security. Ongoing advancements in AI and deep learning techniques will continue to drive innovation in sorghum breeding, helping to optimize crop performance and resilience.

ACKNOWLEDGEMENTS

Mention of trade names or commercial products in this publication is solely for the purpose of providing specific information and does not imply recommendation or endorsement by the U.S. Department of Agriculture. The USDA is an equal opportunity provider and employer. This work was supported in part by the U.S. Department of Agriculture, Agricultural Research Service.

REFERENCES

Alsadik, B., Javan, F. D., & Nex, F. C. (2023). UAV remote sensing for smart agriculture. *GIM International*, *2023*(1), 14–17.

Araus, J. L., Kefauver, S. C., Vergara-Díaz, O., Gracia-Romero, A., Rezzouk, F. Z., Segarra, J., ... & Bort, J. (2022). Crop phenotyping in a context of global change: What to measure and how to do it. *Journal of Integrative Plant Biology*, *64*(2), 592–618.

Assefa, Y., & Staggenborg, S. A. (2010). Grain sorghum yield with hybrid advancement and changes in agronomic practices from 1957 through 2008. *Agronomy Journal*, *102*(2), 703–706.

Badrinarayanan, V., Kendall, A., & Cipolla, R. (2017). Segnet: A deep convolutional encoder-decoder architecture for image segmentation. *IEEE Transactions on Pattern Analysis and Machine Intelligence*, *39*(12), 2481–2495.

Baloch, F. S., Altaf, M. T., Liaqat, W., Bedir, M., Nadeem, M. A., Cömertpay, G., ... & Sun, H. J. (2023). Recent advancements in the breeding of sorghum crop: Current status and future strategies for marker-assisted breeding. *Frontiers in Genetics*, *14*, 1150616.

Bheemanahalli, R., Wang, C., Bashir, E., Chiluwal, A., Pokharel, M., Perumal, R., ... & Jagadish, S. V. K. (2021). Classical phenotyping and deep learning concur on genetic control of stomatal density and area in sorghum. *Plant Physiology*, *186*(3), 1562–1579.

Bochkovskiy, A., Wang, C. Y., & Liao, H. Y. M. (2020). Yolov4: Optimal speed and accuracy of object detection. *arXiv preprint arXiv:2004.10934*.

Bowles, C., Chen, L., Guerrero, R., Bentley, P., Gunn, R., Hammers, A., ... & Rueckert, D. (2018). Gan augmentation: Augmenting training data using generative adversarial networks. *arXiv preprint arXiv:1810.10863*.

Cai, E., Baireddy, S., Yang, C., Delp, E. J., & Crawford, M. (2021, July). Panicle counting in UAV images for estimating flowering time in sorghum. In *2021* IEEE International Geoscience and Remote Sensing Symposium IGARSS (pp. 6280–6283). IEEE.

Cai, E., Luo, Z., Baireddy, S., Guo, J., Yang, C., & Delp, E. J. (2022). High-resolution uav image generation for sorghum panicle detection. In Proceedings of the IEEE/CVF Conference on Computer Vision and Pattern Recognition (pp. 1676–1685). IEEE.

Casa, A. M., Pressoir, G., Brown, P. J., Mitchell, S. E., Rooney, W. L., Tuinstra, M. R., ... & Kresovich, S. (2008). Community resources and strategies for association mapping in sorghum. *Crop Science*, *48*(1), 30–40.

Casado, Á., & Heras, J. (2018). Guiding the creation of deep learning-based object detectors. *arXiv preprint arXiv:1809.03322*.

Chaivivatrakul, S., Moonrinta, J., & Dailey, M. N. (2010, May). Towards automated crop yield estimation-detection and 3D reconstruction of pineapples in Video sequences. In Proceedings of the Fifth International Conference on Computer Vision Theory and Applications (pp. 180–183). Angers: INSTICC Press.

Ciampitti, I. A., & Prasad, P. V. (2020). *Sorghum: State of the Art and Future Perspectives*. Monograph 58. Madison, WI: American Society of Agronomy.

Cobb, J. N., DeClerck, G., Greenberg, A., Clark, R., & McCouch, S. (2013). Next-generation phenotyping: Requirements and strategies for enhancing our understanding of genotype–phenotype relationships and its relevance to crop improvement. *Theoretical and Applied Genetics*, *126*, 867–887.

Cooper, M., Gho, C., Leafgren, R., Tang, T., & Messina, C. (2014). Breeding drought-tolerant maize hybrids for the US corn-belt: Discovery to product. *Journal of Experimental Botany*, *65*(21), 6191–6204.

Coulibaly, S., Kamsu-Foguem, B., Kamissoko, D., & Traore, D. (2022). Explainable deep convolutional neural networks for insect pest recognition. *Journal of Cleaner Production*, *371*, 133638.

Croitoru, F. A., Hondru, V., Ionescu, R. T., & Shah, M. (2023). Diffusion models in vision: A survey. *IEEE Transactions on Pattern Analysis and Machine Intelligence*, *45*(9), 10850–10869.

Demarco, P. A., Mayor, L., Rotundo, J. L., Prasad, P. V., Morris, G. P., Fernandez, J. A., ... & Ciampitti, I. A. (2023). Retrospective study in US commercial sorghum breeding: II. Physiological changes associated to yield gain. *Crop Science*, *63*(2), 867–878.

Dittberner, H., Korte, A., Mettler-Altmann, T., Weber, A. P., Monroe, G., & de Meaux, J. (2018). Natural variation in stomata size contributes to the local adaptation of water-use efficiency in Arabidopsis thaliana. *Molecular Ecology*, *27*(20), 4052–4065.

Du, Y., Wei, F., Zhang, Z., Shi, M., Gao, Y., & Li, G. (2022). Learning to prompt for open-vocabulary object detection with vision-language model. In *Proceedings of the IEEE/CVF Conference on Computer Vision and Pattern Recognition* (pp. 14084–14093). IEEE.

Duarte, K. T., De Carvalho, M. A., & Martins, P. S. (2017, February). Segmenting high-quality digital images of stomata using the wavelet spot detection and the watershed transform. In *International Conference on Computer Vision Theory and Applications* (Vol. 5, pp. 540–547). SciTePress.

Fahlgren, N., Gehan, M. A., & Baxter, I. (2015). Lights, camera, action: High-throughput plant phenotyping is ready for a close-up. *Current Opinion in Plant Biology*, *24*, 93–99.

Fetter, K. C., Eberhardt, S., Barclay, R. S., Wing, S., & Keller, S. R. (2019). StomataCounter: A neural network for automatic stomata identification and counting. *New Phytologist*, *223*(3), 1671–1681.

Galli, G., Horne, D. W., Collins, S. D., Jung, J., Chang, A., Fritsche-Neto, R., & Rooney, W. L. (2020). Optimization of UAS-based high-throughput phenotyping to estimate plant health and grain yield in sorghum. *The Plant Phenome Journal*, *3*(1), e20010.

Ghosal, S., Zheng, B., Chapman, S. C., Potgieter, A. B., Jordan, D. R., Wang, X., ... & Guo, W. (2019). A weakly supervised deep learning framework for sorghum head detection and counting. *Plant Phenomics, 2019,* Article ID: 1525874.

Gill, T., Gill, S. K., Saini, D. K., Chopra, Y., de Koff, J. P., & Sandhu, K. S. (2022). A comprehensive review of high throughput phenotyping and machine learning for plant stress phenotyping. *Phenomics*, *2*(3), 156–183.

Girshick, R. (2015). Fast r-cnn. *arXiv preprint arXiv:1504.08083.*

Gitz, D. C., & Baker, J. T. (2009). Methods for creating stomatal impressions directly onto archivable slides. *Agronomy Journal, 101*(1), 232–236.

Gonzalo-Martín, C., García-Pedrero, A., & Lillo-Saavedra, M. (2021). Improving deep learning sorghum head detection through test time augmentation. *Computers and Electronics in Agriculture*, *186*, 106179.

Goodfellow, I., Pouget-Abadie, J., Mirza, M., Xu, B., Warde-Farley, D., Ozair, S., ... & Bengio, Y. (2020). Generative adversarial networks. *Communications of the ACM*, *63*(11), 139–144.

Gordienko, Y., Rokovyi, O., Alienin, O., & Stirenko, S. (2022, June). Context-aware data augmentation for efficient object detection by UAV surveillance. In 2022 10th International Symposium on Digital Forensics and Security (ISDFS) (pp. 1–6). IEEE.

Guo, W., Zheng, B., Potgieter, A. B., Diot, J., Watanabe, K., Noshita, K., ... & Chapman, S. C. (2018). Aerial imagery analysis–quantifying appearance and number of sorghum heads for applications in breeding and agronomy. *Frontiers in Plant Science*, *9*, 1544.

Hafiz, A. M., & Bhat, G. M. (2020). A survey on instance segmentation: State of the art. *International Journal of Multimedia Information Retrieval*, *9*(3), 171–189.

Hassija, V., Chamola, V., Mahapatra, A., Singal, A., Goel, D., Huang, K., ... & Hussain, A. (2024). Interpreting black-box models: A review on explainable artificial intelligence. *Cognitive Computation*, *16*(1), 45–74.

He, K., Gkioxari, G., Dollár, P., & Girshick, R. (2017). Mask r-cnn. In *Proceedings of the IEEE International Conference on Computer Vision* (pp. 2961–2969). IEEE.

He, K., Zhang, X., Ren, S., & Sun, J. (2015). Spatial pyramid pooling in deep convolutional networks for visual recognition. *IEEE Transactions on Pattern Analysis and Machine Intelligence*, *37*(9), 1904–1916.

Higaki, T., Kutsuna, N., & Hasezawa, S. (2014). CARTA-based semi-automatic detection of stomatal regions on an Arabidopsis cotyledon surface. *Plant Morphology*, *26*(1), 9–12.

Hu, X., Duan, Z., Huang, X., Xu, Z., Ming, D., & Ma, J. (2023, October). Context-aware data augmentation for lidar 3d object detection. In 2023 IEEE International Conference on Image Processing (ICIP) (pp. 11–15). IEEE.

James, C., Gu, Y., Potgieter, A., David, E., Madec, S., Guo, W., ... & Chapman, S. (2023). From prototype to inference: A pipeline to apply deep learning in sorghum panicle detection. *Plant Phenomics*, *5*, 0017.

James, C., Smith, D., He, W., Chandra, S. S., & Chapman, S. C. (2024). GrainPointNet: A deep-learning framework for non-invasive sorghum panicle grain count phenotyping. *Computers and Electronics in Agriculture*, *217*, 108485.

Jayakody, H., Liu, S., Whitty, M., & Petrie, P. (2017). Microscope image based fully automated stomata detection and pore measurement method for grapevines. *Plant Methods*, *13*, 1–12.

Jiang, Y., & Li, C. (2020). Convolutional neural networks for image-based high-throughput plant phenotyping: A review. *Plant Phenomics*.

Jocher, G., Stoken, A., Borovec, J., Changyu, L., Hogan, A., Diaconu, L., ... & Rai, P. (2020). Ultralytics/yolov5: v3. 1-bug fixes and performance improvements. *Zenodo*. https://zenodo.org/records/4154370#.Y4LNkHZBxPY

Jocher, G., Chaurasia, A., & Qiu, J. (2023). Ultralytics yolov8. https://github.com/ultralytics/ultralytics. (accessed on September 30, 2024).

Jocher, G., & Qiu, J. (2024). Ultralytics yolo11. https://github.com/ultralytics/ultralytics. (accessed on September 30, 2024).

Jocher, G., Qiu, J., & Chaurasia, A. (2023, January). Ultralytics YOLO. https://github.com/ultralytics/ultralytics. (accessed on September 30, 2024).

Khan, M. H. U., Wang, S., Wang, J., Ahmar, S., Saeed, S., Khan, S. U., ... & Feng, X. (2022). Applications of artificial intelligence in climate-resilient smart-crop breeding. *International Journal of Molecular Sciences*, *23*(19), 11156.

Khoddami, A., Messina, V., Vadabalija Venkata, K., Farahnaky, A., Blanchard, C. L., & Roberts, T. H. (2023). Sorghum in foods: Functionality and potential in innovative products. *Critical Reviews in Food Science and Nutrition*, *63*(9), 1170–1186.

Kim, H., Kim, D., Ahn, P., Suh, S., Cho, H., & Kim, J. (2024). ContextMix: A context-aware data augmentation method for industrial visual inspection systems. *Engineering Applications of Artificial Intelligence*, *131*, 107842.

Kim, J., Kim, S., Ju, C., & Son, H. I. (2019). Unmanned aerial vehicles in agriculture: A review of perspective of platform, control, and applications. *IEEE Access*, *7*, 105100–105115.

Kinger, S., & Kulkarni, V. (2021, August). Explainable ai for deep learning based disease detection. In Proceedings of the 2021 Thirteenth International Conference on Contemporary Computing (pp. 209–216). Association for Computing Machinery.

Kior, A., Yudina, L., Zolin, Y., Sukhov, V., & Sukhova, E. (2024). RGB imaging as a tool for remote sensing of characteristics of terrestrial plants: A review. *Plants*, *13*(9), 1262.

Kirillov, A., Mintun, E., Ravi, N., Mao, H., Rolland, C., Gustafson, L., ... & Girshick, R. (2023). Segment anything. In *Proceedings of the IEEE/CVF International Conference on Computer Vision* (pp. 4015–4026). IEEE.

Kundu, S., Saini, D. K., Meena, R. K., Bahuguna, R. N., & Jagadish, S. V. K. (2024). High-throughput phenotyping and AI technologies for deciphering crop resilience to heat stress. *Plant Physiology Reports*, *29*, 1–17.

Saini, D. K., Rana, A., Halder, J., Billah, M. M., Gill, H. S., Zhang, J., ... & Sehgal, S. K. (2024). Rapid estimation of DON content in wheat flour using close-range hyperspectral imaging and machine learning. *The Plant Phenome Journal*, *7*(1), e70001.

Kumar, A. A., Sharma, H. C., Sharma, R., Blummel, M., Reddy, P. S., & Reddy, B. V. (2013). Phenotyping in sorghum [*Sorghum bicolor* (L.) Moench]. In Panguluri, S. and Kumar, A. (eds.), *Phenotyping for Plant Breeding: Applications of Phenotyping Methods for Crop Improvement* (pp. 73–109). New York, NY: Springer.

LabelBox. (2024). LabelBox. https://labelbox.com. (accessed on September 30, 2024).

Laga, H., Shahinnia, F., & Fleury, D. (2014, December). Image-based plant stornata phenotyping. In 2014 13th International Conference on Control Automation Robotics & Vision (ICARCV) (pp. 217–222). IEEE.

LeCun, Y., Bengio, Y., & Hinton, G. (2015). Deep learning. *Nature*, *521*(7553), 436–444.

LeCun, Y., Boser, B., Denker, J. S., Henderson, D., Howard, R. E., Hubbard, W., & Jackel, L. D. (1989). Backpropagation applied to handwritten zip code recognition. *Neural Computation*, *1*(4), 541–551.

Lee, E. A., & Tollenaar, M. (2007). Physiological basis of successful breeding strategies for maize grain yield. *Crop Science*, *47*, S-202.

Leff, B., Ramankutty, N., & Foley, J. A. (2004). Geographic distribution of major crops across the world. *Global Biogeochemical Cycles*, *18*(1), 1–27.

Li, H., Wang, P., & Huang, C. (2022). Comparison of deep learning methods for detecting and counting sorghum heads in UAV imagery. *Remote Sensing*, *14*(13), 3143.

Li, Y., Zhang, X., & Chen, D. (2018). Csrnet: Dilated convolutional neural networks for understanding the highly congested scenes. In *Proceedings of the IEEE Conference on Computer Vision and Pattern Recognition* (pp. 1091–1100). IEEE.

Li, Z., Guo, R., Li, M., Chen, Y., & Li, G. (2020). A review of computer vision technologies for plant phenotyping. *Computers and Electronics in Agriculture*, *176*, 105672.

Lin, T. Y., Dollár, P., Girshick, R., He, K., Hariharan, B., & Belongie, S. (2017). Feature pyramid networks for object detection. In *Proceedings of the IEEE Conference on Computer Vision and Pattern Recognition* (pp. 2117–2125). IEEE.

Lin, Z., & Guo, W. (2020). Sorghum panicle detection and counting using unmanned aerial system images and deep learning. *Frontiers in Plant Science*, *11*, 534853.

Liu, S., Qi, L., Qin, H., Shi, J., & Jia, J. (2018). Path aggregation network for instance segmentation. In *Proceedings of the IEEE Conference on Computer Vision and Pattern Recognition* (pp. 8759–8768). IEEE.

Liu, Z., Xia, J., Hu, M., Du, J., Luo, C., & Zheng, K. (2021). Design and analysis of a performance monitoring system for a seed metering device based on pulse width recognition. *PloS One*, *16*(12), e0261593.

Maimaitijiang, M., Sagan, V., Sidike, P., Hartling, S., Esposito, F., & Fritschi, F. B. (2020). Soybean yield prediction from UAV using multimodal data fusion and deep learning. *Remote Sensing of Environment*, *237*, 111599.

Malambo, L., Popescu, S., Ku, N. W., Rooney, W., Zhou, T., & Moore, S. (2019). A deep learning semantic segmentation-based approach for field-level sorghum panicle counting. *Remote Sensing*, *11*(24), 2939.

Maman, N., Mason, S. C., Lyon, D. J., & Dhungana, P. (2004). Yield components of pearl millet and grain sorghum across environments in the Central Great Plains. *Crop Science*, *44*(6), 2138–2145.

Mayor, L., Demarco, P., Lira, S., Fang, Y., Abadie, T., Gambin, B. L., ... & Messina, C. (2023). Retrospective study in US commercial sorghum breeding: I. Genetic gain in relation to relative maturity. *Crop Science*, *63*(2), 501–510.

Miao, C., Guo, A., Thompson, A. M., Yang, J., Ge, Y., & Schnable, J. C. (2021). Automation of leaf counting in maize and sorghum using deep learning. *The Plant Phenome Journal*, *4*(1), e20022.

Miller, F. R., & Kebede, Y. (1984). Genetic contributions to yield gains in sorghum, 1950 to 1980. *Genetic Contributions to Yield Gains of Five Major Crop Plants*, *7*, 1–14.

Minderer, M., Gritsenko, A., Stone, A., Neumann, M., Weissenborn, D., Dosovitskiy, A., ... & Houlsby, N. (2022, October). Simple open-vocabulary object detection. In Avidan, S., Brostow, G., Cissé, M., Farinella, G.M. & Hassner, T. (eds.), *European Conference on Computer Vision* (pp. 728–755). Springer Nature Switzerland.

Taghavi Namin, S., Esmaeilzadeh, M., Najafi, M., Brown, T. B., & Borevitz, J. O. (2018). Deep phenotyping: Deep learning for temporal phenotype/genotype classification. *Plant Methods*, *14*, 1–14.

Nazir, S., Dickson, D. M., & Akram, M. U. (2023). Survey of explainable artificial intelligence techniques for biomedical imaging with deep neural networks. *Computers in Biology and Medicine*, *156*, 106668.

Ndlovu, E., Van Staden, J., & Maphosa, M. (2021). Morpho-physiological effects of moisture, heat and combined stresses on *Sorghum bicolor* [Moench (L.)] and its acclimation mechanisms. *Plant Stress*, *2*, 100018.

Nguyen, V. A. (2024). AnyLabeling – Effortless data labeling with AI support. https://github.com/vietanhdev/anylabeling. (accessed on October 05, 2024).

Nichol, A., Dhariwal, P., Ramesh, A., Shyam, P., Mishkin, P., McGrew, B., ... & Chen, M. (2021). Glide: Towards photorealistic image generation and editing with text-guided diffusion models. *arXiv preprint arXiv:2112.10741*.

Ninomiya, S. (2022). High-throughput field crop phenotyping: Current status and challenges. *Breeding Science*, *72*(1), 3–18.

Oh, M. H., Olsen, P., & Ramamurthy, K. N. (2019). Counting and segmenting sorghum heads. *arXiv preprint arXiv:1905.13291*.

Olsen, P. A., Ramamurthy, K. N., Ribera, J., Chen, Y., Thompson, A. M., Luss, R., ... & Abe, N. (2018, October). Detecting and counting panicles in sorghum images. In 2018 IEEE 5th International Conference on Data Science and Advanced Analytics (DSAA) (pp. 400–409). IEEE.

Onoro-Rubio, D., & López-Sastre, R. J. (2016). Towards perspective-free object counting with deep learning. In Computer Vision–ECCV 2016: 14th European Conference, Amsterdam, The Netherlands, October 11–14, 2016, Proceedings, Part VII 14 (pp. 615–629). Springer International Publishing.

Osborne, C. P., & Freckleton, R. P. (2009). Ecological selection pressures for C4 photosynthesis in the grasses. *Proceedings of the Royal Society B: Biological Sciences*, *276*(1663), 1753–1760.

Radford, A., Kim, J. W., Hallacy, C., Ramesh, A., Goh, G., Agarwal, S., ... & Sutskever, I. (2021, July). Learning transferable visual models from natural language supervision. In *International Conference On Machine Learning* (pp. 8748–8763). PMLR.

Rakshit, S., & Bellundagi, A. (2019). Conventional breeding techniques in sorghum. In Aruna, C., Visarada, K.B.R.S., Bhat, B.V. & Tonapi, V.A. (eds.), *Breeding Sorghum for Diverse End Uses* (pp. 77–91). Cambridge, MA: Woodhead.

Rakshit, S., Hariprasanna, K., Gomashe, S., Ganapathy, K. N., Das, I. K., Ramana, O. V., ... & Patil, J. V. (2014). Changes in area, yield gains, and yield stability of sorghum in major sorghum-producing countries, 1970 to 2009. *Crop Science*, *54*(4), 1571–1584.

Redmon, J. (2016). You only look once: Unified, real-time object detection. In *Proceedings of the IEEE Conference on Computer Vision and Pattern Recognition*. IEEE.

Redmon, J. (2018). Yolov3: An incremental improvement. *arXiv preprint arXiv:1804.02767*.

Ren, S., He, K., Girshick, R., & Sun, J. (2016). Faster R-CNN: Towards real-time object detection with region proposal networks. *IEEE Transactions on Pattern Analysis and Machine Intelligence*, *39*(6), 1137–1149.

Ribera, J., Chen, Y., Boomsma, C., & Delp, E. J. (2017, November). Counting plants using deep learning. In 2017 IEEE Global Conference on Signal and Information Processing (GlobalSIP) (pp. 1344–1348). IEEE.

Rombach, R., Blattmann, A., Lorenz, D., Esser, P., & Ommer, B. (2022). High-resolution image synthesis with latent diffusion models. In Proceedings of the IEEE/CVF Conference on Computer Vision and Pattern Recognition (pp. 10684–10695). IEEE.

Russell, B. C., Torralba, A., Murphy, K. P., & Freeman, W. T. (2008). LabelMe: A database and web-based tool for image annotation. *International Journal of Computer Vision*, *77*, 157–173.

Sakoda, K., Watanabe, T., Sukemura, S., Kobayashi, S., Nagasaki, Y., Tanaka, Y., & Shiraiwa, T. (2019). Genetic diversity in stomatal density among soybeans elucidated using high-throughput technique based on an algorithm for object detection. *Scientific Reports*, *9*(1), 7610.

Santiago, G. N., Cisdeli Magalhaes, P. H., Carcedo, A. J., Marziotte, L., Mayor, L., & Ciampitti, I. A. (2024). Deep learning methods using imagery from a smartphone for recognizing sorghum panicles and counting grains at a plant level. *Plant Phenomics*, *6*, 0234.

Singh, A., Ganapathysubramanian, B., Singh, A. K., & Sarkar, S. (2016). Machine learning for high-throughput stress phenotyping in plants. *Trends in Plant Science*, *21*(2), 110–124.

Staggenborg, S. A., Dhuyvetter, K. C., & Gordon, W. B. (2008). Grain sorghum and corn comparisons: Yield, economic, and environmental responses. *Agronomy Journal*, *100*(6), 1600–1604.

Tack, J., Lingenfelser, J., & Jagadish, S. V. K. (2017). Disaggregating sorghum yield reductions under warming scenarios exposes narrow genetic diversity in US breeding programs. *Proceedings of the National Academy of Sciences*, *114*(35), 9296–9301.

Ubbens, J. R., & Stavness, I. (2017). Deep plant phenomics: A deep learning platform for complex plant phenotyping tasks. *Frontiers in Plant Science*, *8*, 1190.

USDA. (2024). Sorghum production. https://fas.usda.gov/data/production/commodity/0459200 (accessed on September 23, 2024).

van Dijk, A. D. J., Kootstra, G., Kruijer, W., & de Ridder, D. (2021). Machine learning in plant science and plant breeding. *Iscience*, *24*(1), 101890.

Van Oosterom, E., & Hammer, G. (2008). Determination of grain number in sorghum. *Field Crops Research, 108*(3), 259–268.

Waliman, M., & Zakhor, A. (2020). Deep learning method for height estimation of sorghum in the field using LiDAR. *Electronic Imaging*, *32*, 1–7.

Walker, C. K., Ford, R., Muñoz-Amatriaín, M., & Panozzo, J. F. (2013). The detection of QTLs in barley associated with endosperm hardness, grain density, grain size and malting quality using rapid phenotyping tools. *Theoretical and Applied Genetics*, *126*, 2533–2551.

Wang, C. Y., Bochkovskiy, A., & Liao, H. Y. M. (2023). YOLOv7: Trainable bag-of-freebies sets new state-of-the-art for real-time object detectors. In Proceedings of the IEEE/CVF Conference on Computer Vision and Pattern Recognition (pp. 7464–7475). IEEE.

Wang, C. Y., Liao, H. Y. M., Wu, Y. H., Chen, P. Y., Hsieh, J. W., & Yeh, I. H. (2020). CSPNet: A new backbone that can enhance learning capability of CNN. In *Proceedings of the IEEE/CVF Conference on Computer Vision and Pattern Recognition Workshops* (pp. 390–391). IEEE.

Wang, C. Y., Liao, H. Y. M., & Yeh, I. H. (2022). Designing network design strategies through gradient path analysis. *arXiv preprint arXiv:2211.04800*.

Wang, C. Y., Yeh, I. H., & Mark Liao, H. Y. (2025). Yolov9: Learning what you want to learn using programmable gradient information. In Leonardis, A., Ricci, E., Roth, S., Russakovsky, O., Sattler, T. & Varol, G. (eds.), European Conference on Computer Vision (pp. 1–21). Cham, Switzerland: Springer.

Wang, X., Jin, J., Li, C., Tang, J., Zhang, C., & Wang, W. (2024). Pedestrian attribute recognition via clip based prompt vision-language fusion. *IEEE Transactions on Circuits and Systems for Video Technology*, *35*, 148–161.

Yang, Z., & Han, Y. (2020). A low-cost 3D phenotype measurement method of leafy vegetables using video recordings from smartphones. *Sensors*, *20*(21), 6068.

Zaidi, S. S. A., Ansari, M. S., Aslam, A., Kanwal, N., Asghar, M., & Lee, B. (2022). A survey of modern deep learning based object detection models. *Digital Signal Processing*, *126*, 103514.

Zannou, J. G. N., & Houndji, V. R. (2019, April). Sorghum yield prediction using machine learning. In 2019 3rd International Conference on Bio-engineering for Smart Technologies (BioSMART) (pp. 1–4). IEEE.

Zhang, Y., Zhou, D., Chen, S., Gao, S., & Ma, Y. (2016). Single-image crowd counting via multi-column convolutional neural network. In Proceedings of the IEEE Conference on Computer Vision and Pattern Recognition (pp. 589–597). IEEE.

Zhang, Z., Pless, R., Shakoor, N., Carnahan, A., & Stylianou, A. (2024). SG× P: A sorghum genotype× phenotype prediction dataset and benchmark. *Advances in Neural Information Processing Systems*, *36*.

Zhao, Z., Alzubaidi, L., Zhang, J., Duan, Y., & Gu, Y. (2024). A comparison review of transfer learning and self-supervised learning: Definitions, applications, advantages and limitations. *Expert Systems with Applications*, *242*, 122807.

Zoph, B., Cubuk, E. D., Ghiasi, G., Lin, T. Y., Shlens, J., & Le, Q. V. (2020). Learning data augmentation strategies for object detection. In Vedaldi, A., Bischof, H., Brox, T., Frahm, J.M. (eds.), Computer Vision–ECCV 2020: 16th European Conference, Glasgow, UK, August 23–28, 2020, Proceedings, Part XXVII 16 (pp. 566–583). Springer International Publishing.

15 Utilizing Crop Simulation Modeling to Address Climate Change Challenges in Sorghum Production

Opportunities and Challenges

Swarna Ronanki, Gangaiah Bandla, Srividhya S., Seva Nayak D., Deepika Cheruku, and Anuradha Narala

15.1 INTRODUCTION

Sorghum (*Sorghum bicolor* (L.) Moench) ranks fifth in global cereal production and generally thrives in drought-prone, semi-arid regions, making it vital for food security in areas with erratic rainfall (Enyew et al., 2022; Kamdi et al., 2023). With global cultivation spanning 41 million hectares and a production of 58.7 million tons, sorghum's average yield of 2.5 t/ha falls well below its potential of 6 t/ha due to low inputs, over-reliance on rainfall, and suboptimal management practices (Khalifa & Eltahir, 2023; Widodo et al., 2024). Climate change further exacerbates challenges, particularly in tropical arid and semi-arid regions, where rising temperatures and unpredictable rainfall adversely affect crop growth and yields (Parry et al., 2004). According to climate projections, global temperatures are expected to rise by approximately 1.5°C by 2050 and between 2 and 4°C by 2100, with the exact increase depending on future greenhouse gas emission levels; some regions may see significantly larger temperature increases than the global average (https://climatechange.chicago.gov/climate-change-science/future-climate-change).

In particular, climate change impacts sorghum through rising temperatures, altered precipitation patterns, and increased atmospheric CO_2 (Figure 15.1). Temperature directly influences sorghum yield by affecting biomass accumulation, photosynthesis, growth duration, seed setting, and viability (Prasad et al., 2021; 2015). The optimum temperature range for sorghum germination is 21–35°C, while the vegetative and reproductive periods have optimal ranges of 26–34°C and 25–28°C, respectively (Wawo et al., 2024; Prasad et al., 2015; Srivastava et al., 2010). High temperatures also accelerate Growing Degree Day (GDD) accumulation, shortening phenophase duration and adversely affecting yield and quality (Fatima et al., 2020). Furthermore, elevated temperatures hinder seed development by reducing seed-filling duration, seed size, and overall seed yields (Saravanakumar, 2018; Sari, 2023).

Rainfall, another critical climate factor, significantly impacts sorghum yields. Although sorghum is known for its stress tolerance, water stress during critical stages, such as panicle development, flowering, and grain filling, can cause substantial yield losses (Prasad et al., 2008; Prasad et al., 2021). Water stress impairs carbon assimilation, cell turgor, and stomatal conductance, reducing crop growth and development, and ultimately affecting yield (Earl & Davis, 2003; Prasad et al., 2019). While sorghum is a C_4 crop and generally shows limited responses to increased CO_2 concentrations, elevated CO_2 can positively affect yields under water stress by improving water-use

 DOI: 10.1201/9781003623717-15

efficiency through reduced stomatal conductance (Conley et al., 2001; Allen et al., 2011). However, some studies suggest that increased CO_2 may negatively affect sorghum yields by reducing nutrient availability, altering plant morphology, increasing temperature sensitivity, and shifting pest and disease dynamics (Lahlali et al., 2024; Skendžić et al., 2021).

Given the complex interactions among traits and the challenges posed by climate change, innovative technological solutions are critical for improving sorghum productivity. One such solution is crop simulation modeling, which has become a key tool for quantifying trade-offs and synergies between traits, helping adapt agricultural practices to a rapidly changing climate (Timlin et al., 2024). Crop simulation models (CSMs) integrate data on climate, soil types, genetics, and management practices to predict the outcomes of various scenarios, providing researchers, farmers and policymakers with valuable insights for informed decisions on crop management, resource allocation, and adaptation strategies. Crop models are mathematical representations of the soil-plant-environment system, predicting crop attributes across the life cycle by incorporating environmental (E), genetic (G), and management (M) factors. These models offer interpretive and predictive capabilities that go beyond traditional statistical methods. While crop modeling began in the 1960s (de Wit, 1969), its scientific legitimacy and application in crop management and genetic improvement grew with advancements in computational power and plant genotyping (Messina et al., 2009; Cooper et al., 2014a,b). Platforms like APSIM (Agricultural Production Systems sIMulator; Holzworth et al., 2014) and DSSAT (Decision Support System for Agrotechnology Transfer; Jones et al., 2003) played a pivotal role in model development, testing, and application, with APSIM facilitating model creation through structured templates (Brown et al., 2014). Sorghum modeling evolved similarly, beginning with early models in the 1970s and advancing with mechanistic models in the

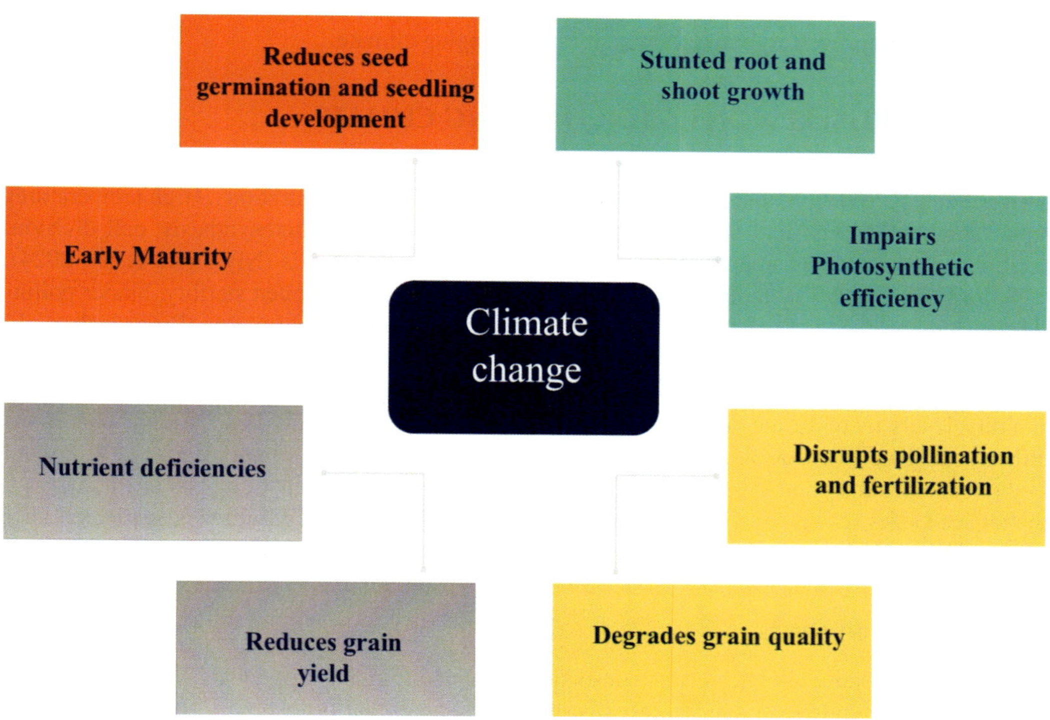

FIGURE 15.1 Illustration of climate change impacts on sorghum, highlighting how stresses like increased temperature, drought, and extreme weather disrupt physiological processes and affect agronomic and physiological traits.

1990s. These models, informed by plant growth and nitrogen dynamics, contributed to the development of the advanced sorghum model now part of APSIM (Verburg et al., 2025).

Early sorghum modeling sought to expand agronomic insights beyond traditional experimental constraints of location and season, where results often varied. Hammer et al. (1996) noted that, while average responses to sowing date were minimal, individual years showed significant yield advantages for early or late sowing. This emphasized the need for simulations to assess the risk implications of management decisions. Sorghum models were also applied to explore planting decisions in the U.S. Midwest (Heiniger et al., 1997), which eventually led to decision support tools for farmers (Carberry et al., 2009). In the 1990s, crop modeling was integrated into plant breeding strategies to analyze genotype-environment-management (G × E × M) interactions, thereby improving breeding decisions (Cooper & Hammer, 1996). These models quantified environmental stress and crop growth trajectories, allowing for simulations of trait manipulation and adaptation within the G × E × M framework (Hammer et al., 1996). As advancements in genotyping technologies were made, the integration of genetic regulation with crop modeling became more feasible (Yin et al., 2003). This approach was exemplified in maize studies (Messina et al., 2011), with an increasing potential to unravel crop complexity through the combination of genomics and simulation (Cooper et al., 2014a, b). Additionally, several studies have already employed crop models to assess the impact of climate change on sorghum production (Druille et al., 2020; Mohammed & Misganaw, 2022; Praveenkumar et al., 2024), underscoring the growing importance of simulation in addressing future climate risks.

This chapter examines responses of sorghum to drought, heat stress, and elevated CO_2, and provides an overview of CSMs, including their components and data requirements. The chapter explores how CSMs predict future impacts on sorghum yields, regional variability, and climate adaptation strategies. It also discusses how crop modeling can enhance breeding efficiency for climate resilience. Finally, the chapter outlines future prospects, challenges, and opportunities for using CSMs to improve sorghum productivity in a changing climate.

15.2 SORGHUM RESPONSES TO CHANGING CLIMATE

Climate change affects sorghum production through various factors, including seasonal and temperature variations, altered precipitation patterns, and increased CO_2 concentrations. High temperatures impair sorghum growth and development by negatively impacting photosynthesis, cell division, metabolism, and nutrient uptake, ultimately resulting in reduced yields (Niyibigira et al., 2024). In contrast, extremely low temperatures can delay flowering, induce pollen sterility, hinder pollination, and reduce seed count per panicle, leading to lower overall grain yields (Emendack et al., 2021). Additionally, excessive rainfall can saturate the soil, depriving roots of essential oxygen and reducing grain yields. Strong winds exacerbate the situation by causing crop lodging, while high humidity levels increase the risk of fungal infections (Sharma et al., 2019). Drought conditions further contribute to reduced productivity by negatively affecting germination, emergence, plant height, leaf area, and photosynthesis. These conditions often lead to premature plant death, lodging, smaller seed sizes, and significant yield losses (Abreha et al., 2022). The following sections briefly explore these factors.

15.2.1 Drought Stress

Drought stress, a major climate change consequence, threatens global crop yields. While considered drought-tolerant, sorghum remains highly susceptible to water scarcity, leading to severe yield losses (Figure 15.2; Keerthi et al., 2021). It requires 450–650 mm of water per season (FAO, 2002), with demand peaking at booting and flowering (7–10 mm/day), the most sensitive stages (Assefa et al., 2010). In drought-prone regions, insufficient rainfall exacerbates yield losses, particularly when stress coincides with critical growth phases. Early drought hampers establishment, while stress at booting and flowering drastically reduces yield (Begna et al., 2022). Post-flowering drought is especially damaging, decreasing grain number, accelerating senescence, and reducing seed size (Abreha et al., 2022). Yield losses vary by timing, with vegetative-stage drought reducing yield by 36% and reproductive-stage drought causing up to 55% loss (Abreha et al., 2022; Enyew et al., 2022; Assefa et al., 2010). When drought coincides with booting and flowering, losses can reach 87% (Craufurd & Peacock, 1993). Since reproductive processes are highly water-sensitive, stress during these stages severely impacts yield and quality.

15.2.2 Heat Stress

Despite its heat tolerance, sorghum remains vulnerable to heat stress, particularly during flowering and grain filling. High temperatures impair photosynthesis by reducing RuBisCO efficiency, limiting CO_2 fixation, and decreasing stomatal conductance (Ainsworth & Ort, 2010; Zahra et al., 2023). Increased transpiration accelerates soil moisture depletion, leading to dehydration, wilting, and stunted growth, especially in drought-prone areas (Prasad et al., 2008, 2021). Heat stress also induces oxidative damage by accumulating reactive oxygen species (ROS), causing membrane instability and cellular damage despite antioxidant defenses (Hasanuzzaman et al., 2013; Djanaguiraman et al., 2014). The reproductive phase is particularly sensitive, with heat stress reducing pollen viability, grain set, and starch accumulation, leading to smaller, lower-weight grains (Prasad et al., 2006; Kundu et al., 2024). Grain number can decline by up to 50% under severe heat (Singh et al., 2015). Temperatures above 35°C during grain filling further reduce grain weight and viability (Prasad et al., 2021). Additionally, heat stress accelerates senescence by promoting chlorophyll degradation, shortening the grain-filling period, and limiting carbohydrate synthesis, ultimately reducing yield. Enhancing sorghum's resilience to heat stress is essential for maintaining stable productivity under climate change.

15.2.3 Effect of Elevated CO_2

Elevated CO_2 reduces transpiration in sorghum by increasing stomatal resistance, conserving water, and enhancing drought tolerance in dryland conditions (van Bavel, 1974; Allen et al., 2011). It also promotes root mass expansion, improving access to deeper soil water and nutrients, particularly during drought (Chaudhuri et al., 1986; de Souza Rodrigues et al., 2023). This water-use efficiency supports critical growth stages like grain filling, ensuring sustained productivity despite reduced vegetative growth (Ottman et al., 2001). Elevated CO_2 boosts sorghum biomass by up to 30%, as reported by Torbert et al. (2004), enhancing carbon gain and yields, particularly under drought. It also improves grain quality; de Souza et al. (2015) noted a 60% rise in protein content and slight increases in fatty acids, with starch content largely unaffected. Elevated CO_2 raises the carbon-to-nitrogen (C:N) ratio, influencing physiological processes and nutritional quality (Torbert et al., 2004). However, prolonged exposure to high CO_2 may negatively affect sorghum, reducing stomatal conductance and potentially limiting photosynthesis under combined stresses like heat and drought (Silva et al., 2020; Al-Salman et al., 2023). After exploring sorghum's responses to drought, heat, and elevated CO_2, the next section examines the role of CSMs in optimizing sorghum production under climate change.

15.3 CROP SIMULATION MODELS – OVERVIEW AND COMPONENTS

A CSM is a computer-based program that uses mathematical equations to represent the growth and developmental stages of crops, taking into account weather variables, crop conditions, soil characteristics, and management practices. These models are valuable tools for addressing real-world

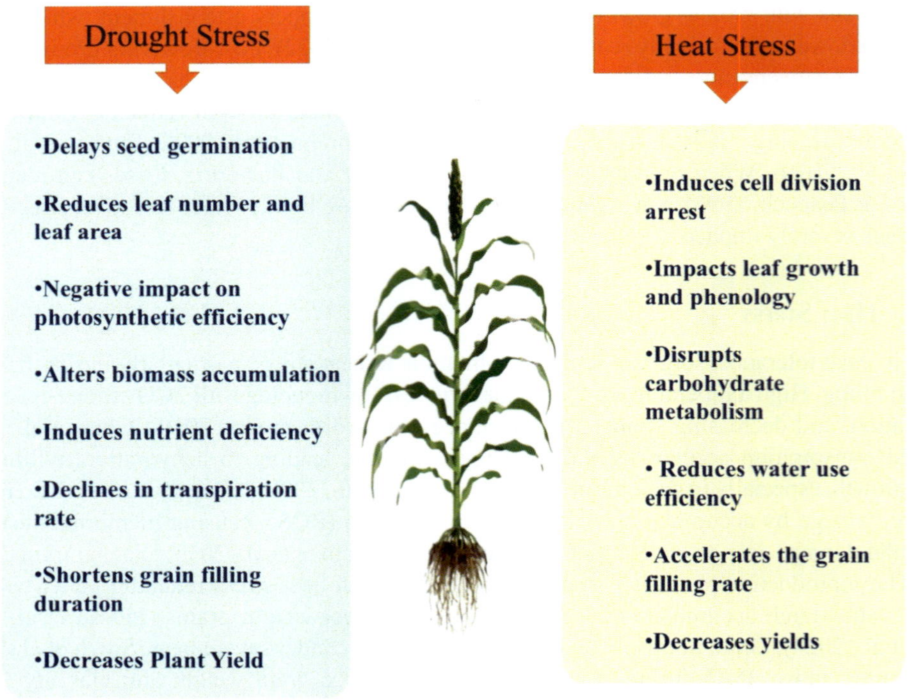

FIGURE 15.2 Depiction of the effects of drought and heat stress on sorghum, highlighting disruptions in physiological processes such as photosynthesis, transpiration, and nutrient uptake, as well as their impact on other key traits.

agricultural challenges, as they provide insights that assist in decision-making (Asseng et al., 2014). By simulating various crop conditions, crop models can enhance efficiency, saving both time and money. These models are particularly crucial for evaluating the impacts of climate change on crop productivity, as they can simulate plant growth, development, and yield under a range of environmental conditions. In recent years, significant progress has been made in the development of CSMs, leading to a wide array of models designed for almost all major crops (Timlin et al., 2024). A list of key crop simulation models is provided in Table 15.1.

15.3.1 Components of Crop Simulation Model

Crop models use mathematical equations to represent the various processes occurring within a plant and the interactions between the plant and its environment (Schulze et al., 2016). These models have become essential tools for predicting the effects of climate change on crop yields. By integrating data on weather, soil types, crop characteristics, and management practices, crop models predict outcomes related to crop growth and development parameters (Figure 15.3; Hodson & White, 2010). To perform crop simulations, the model incorporates data from different components, weather, soil, site, crop, and management, and generates output data necessary for model calibration, validation, and application (Table 15.2).

For accurate simulations, crop and site-specific details must be calibrated to align the model's predictions with actual field conditions. During calibration, the default parameters are adjusted to reflect the specific crop cultivars and field conditions. Some parameters are considered conservative and typically do not require adjustments, while others should be fine-tuned using relevant field

TABLE 15.1
Key crop simulation models for enhancing crop production and productivity

Crop model	Target trait(s)	References
DSSAT	Crop growth and development, soil and crop water and nutrient dynamics	Jones et al. (2003)
CERES- SORGHUM	Crop growth simulation based on soil water, evapo-transpiration, harvest index, and aboveground biomass	Alagarswamy & Ritchie, (1991)
APSIM	Growth, development and yield of crops, pastures and forests and their interactions with the soil	Keating et al. (2003)
InfoCrop	Growth and yield parameters, nitrogen uptake, greenhouse gas emissions, soil water and nitrogen balance and climate change studies	Agarwal et al. (2006)
AQUACROP	Crop yield response to water dynamics	Steduto et al. (2009)
RZWQM2	A model of root zone water quality that simulates the water, nutrient, and chemical fluxes in the root zone	Ma et al. (2007)
HERMES	A model to explain the dynamics of water and nitrogen in the soil-plant system and plant growth	Kersebaum, (2007)
WOFOST	For evaluating field crop development and yield in varied soil and climatic conditions	Hijmans et al. (1994)
CENTURY	Simulates C and nutrient dynamics for various ecosystem types, such as grasslands, agricultural areas, and forests	Parton et al. (1988)

data to ensure accuracy and reliability of prediction. During the calibration process, the model uses independent datasets, including phenological events, leaf area index (LAI), above-ground biomass and its partitioning to leaf, stem, and grain, as well as yield attributes and final yield. The simulated values are then compared with actual field data to validate the model's output. Once validated, the model's performance is assessed using various statistical parameters, such as Root Mean Square Error (RMSE), index of agreement, correlation coefficient (r), coefficient of determination (R^2), etc. (Kephe et al., 2021).

15.3.2 Data Requirement for Crop Simulation Models

Crop simulation models require a minimum dataset for effective operation. This includes data on the site where the model will be applied, weather data for the crop growth season, soil data at the start of the experiment, and management practices (Table 15.2).

15.4 CROP SIMULATION MODELING TO ADDRESS CLIMATE CHANGE CHALLENGES IN SORGHUM

The uncertainty of climate change highlights the need for reliable predictive tools, with crop simulation models crucial for assessing impacts and identifying strategies to mitigate yield losses, particularly for sorghum. Several crop models have been implemented within simulation frameworks to address the challenges faced by crop production, as highlighted in Table 15.1. These models vary in their description of plant physiological processes and soil interactions, leading to differences in their predictions. Comparing these models is essential for revealing uncertainties in crop growth and yield predictions (Palosuoa et al., 2011), particularly those related to model structure, a key source of uncertainty that is difficult to quantify (Chatfield, 1995; Akinseye et al., 2017). Model comparison also aids in identifying systematic errors, highlighting areas that need improvement (Adam et al., 2012).

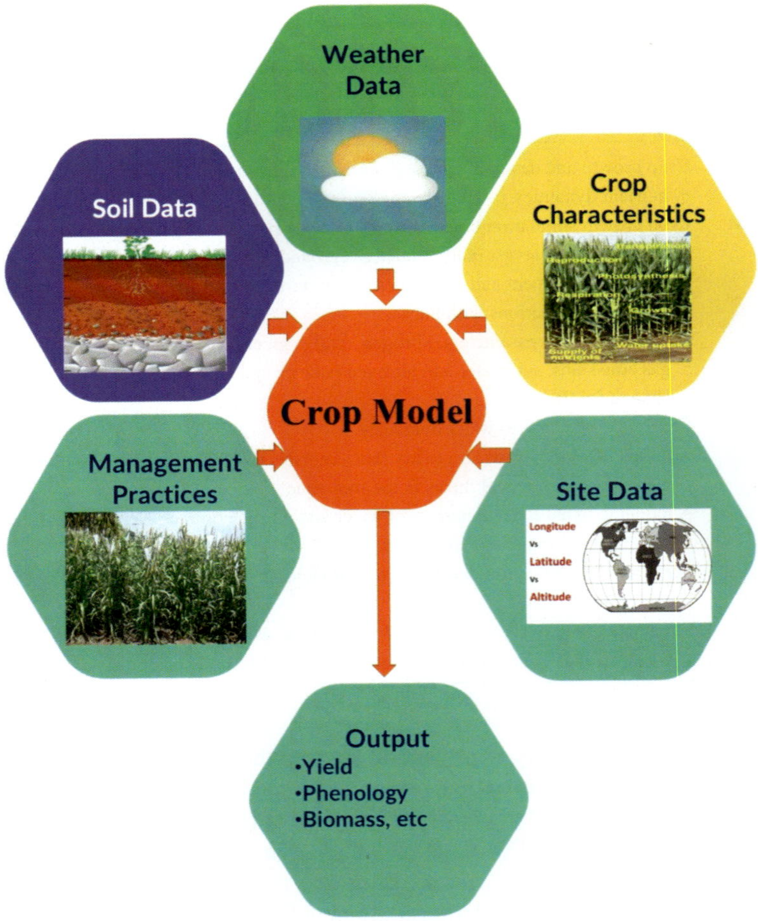

FIGURE 15.3 Illustration of the key components of a crop simulation model, highlighting its structural framework. The diagram showcases critical elements such as climate data inputs, soil properties, crop genetics, and management practices.

While there is a growing body of research comparing models across different crops (Battisti et al., 2017; Palosuo et al., 2011; Rötter et al., 2012; Salo et al., 2016), studies specific to sorghum remain limited. For instance, Perkins et al. (2012) compared five crop models (CropSyst, CERES-Sorghum, APSIM, ALMANAC, and SORKAM) for sweet sorghum biomass modeling, finding ALMANAC to be the most suitable for parameter development due to its superior performance. Akinseye et al. (2017) evaluated the performance of three crop models (APSIM, DSSAT, and Samara) in predicting the growth of diverse sorghum germplasm in the semi-arid region of West Africa. While all models accurately simulated phenology and morphology, APSIM and Samara provided the best results for phenology events, leaf area index, and total leaf number. However, predictions for grain yield and biomass were less accurate, with APSIM performing closest to observed values. In a similar study, Msongaleli et al. (2013) calibrated both DSSAT and APSIM to simulate sorghum under current and future climate scenarios in central Tanzania. Both models showed similar yield response trends, although their predictions varied slightly due to differences in their simulation frameworks. These comparisons highlight the usefulness of DSSAT and APSIM as tools for assessing climate change impacts on sorghum and provide valuable insights for developing adaptation strategies to ensure long-term productivity.

TABLE 15.2
Key parameters and required data for crop simulation models

	Parameter	Required Data
1.	Site	Geographic details: latitude, longitude, and elevation
		Average annual temperature
		Slope and aspect
2.	Soil	Physical properties: soil texture, depth, water-holding capacity, bulk density, and hydraulic conductivity
		Chemical properties: pH, organic carbon, and nitrogen content
3.	Weather	Daily weather during the growing season: global solar radiation, maximum and minimum temperatures, precipitation, and relative humidity
		Historical weather data (depending on the research objectives)
4.	Management	Crop details: cultivar name and type, planting date, depth, and method
		Field setup: row spacing and direction, plant population
		Water management: irrigation dates, methods, and quantities
		Fertilizer applications: inorganic fertilizers and inoculants
		Residue management: organic material type, incorporation depth, amounts, and nutrient content
		Chemical treatments: pesticides and application details
		Tillage practices
		Harvest schedule (indicating date or time frame within simulation when a crop is expected to be harvested)

15.4.1 Predicting Future Impacts on Sorghum Yields

Crop growth models, by analyzing future climate conditions, help identify potential yield trends, which is crucial for effective decision-making regarding resource management, breeding strategies, and climate adaptation. Several studies have examined how climate change may affect sorghum productivity, offering valuable insights. For instance, Sultan et al. (2019) used climate simulations from two crop models (SARRA-H and CYGMA) to assess the impact of historical climate change on sorghum and millet production in West Africa. They found that between 2000 and 2009, a 1°C temperature increase and more frequent climate extremes resulted in a 5–15% reduction in sorghum yields, leading to annual production losses of $0.73–2.17 billion. This historical perspective provides a crucial baseline for understanding the impact of climate change on sorghum, helping to relate past losses to future predictions and quantify potential further impacts.

In contrast, Mohammed & Misganaw (2022) utilized the CERES-Sorghum model to simulate future climate scenarios (RCP4.5 and RCP8.5), predicting significant yield reductions under both projections. However, their findings also highlighted that adopting early sowing, optimal nitrogen application, and supplemental irrigation could substantially improve yields under these challenging future conditions. Similarly, Srivastava et al. (2010) projected a reduction of up to 76% in monsoon sorghum grain yields across different regions of India by 2080, although adjustments in sowing dates and the use of climate-adapted varieties could mitigate these losses. Likewise, Praveenkumar et al. (2024) conducted crop simulations in Tamil Nadu's western zone, estimating a 30.5% decrease in sorghum yields by the century's end, with the most significant declines expected in the latter part of the century.

In Argentina, Druille et al. (2020) employed the ALMANAC model to evaluate the impact of climate warming on both forage and grain sorghum production, focusing on yield changes due to water and nitrogen stress. Their simulations, considering temperature increases of 1, 2, and 4°C, revealed that forage yield increased by 0.53 Mg ha^{-1} per °C, while grain yield decreased by 0.27 Mg ha^{-1} per °C. Forage yield was particularly sensitive to nitrogen stress, and grain yield was strongly negatively correlated with water stress. Similarly, Eggen et al. (2019) analyzed the sensitivity of sorghum to sub-seasonal rainfall variability under climate change in Ethiopia using crop models driven by 14 General Circulation Models (GCMs) representing ENSO and rainfall patterns. Despite projections of warmer and wetter seasons, they predicted that early-season sub-seasonal rainfall failures would negatively impact sorghum yields. Interviews conducted during the 2015 El Niño event highlighted that biophysical constraints severely limited farmers' ability to adapt to such extremes, further emphasizing the need for adaptation strategies to ensure sorghum productivity under climate change. These findings underscore the importance of proactive planning and investment in adaptation measures to safeguard sorghum production and ensure food security in the face of a changing climate.

15.4.2 Regional Variability in the Impact of Climate Change on Sorghum Production

Regional variability in the impact of climate change on sorghum production is influenced by local climate, soil conditions, and farming practices. While some regions may benefit from longer growing seasons or higher CO_2, others face challenges such as water scarcity, heat stress, and soil degradation, which threaten yields. Crop simulation models help assess these regional differences and guide adaptation strategies. For example, Msongaleli et al. (2014) found that in Central Tanzania, adjusting crop duration could mitigate the impacts of climate change, potentially even leading to yield increases in certain areas. In contrast, Fu et al. (2016) studied sorghum production in western North America and concluded that while elevated CO_2 levels could have a positive impact on grain yields, rising temperatures would likely offset this benefit, with temperature management emerging as a key focus for adaptation. Similarly, Ramirez-Villegas et al. (2013) projected vulnerabilities in sorghum production by 2030, particularly in already marginal areas such as the western Sahel, southern Africa, and parts of India. Traoré et al. (2021) assessed climate variability in Mali and found that suitable sorghum areas in the Sudanian zone could increase under RCP 4.5 but decline under more extreme scenarios like RCP 8.5, while the Sahelian zone could face significant reductions in suitable areas under both scenarios. These findings highlight the need for region-specific adaptation strategies to help farmers build resilience against the varying impacts of climate change.

15.4.3 Adaptation Strategies for Climate Resilience

The application of CSMs has proven essential for identifying effective adaptation strategies to enhance the resilience of crops to climate stress. Breeding for drought-responsive traits, such as limited transpiration, has shown promise in improving yields under water stress (Raymundo et al., 2024). Additionally, adjustments in crop management practices, such as leaf removal during grain filling, have been found to help maintain productivity in water-limited environments (Marziotte et al., 2025). In regions prone to drought, selecting drought-tolerant cultivars, especially during critical flowering periods, is key to adapting to climate variability (Diancoumba et al., 2024). Furthermore, studies by Carcedo et al. (2022) and Gohain et al. (2022) highlight the importance of optimizing sowing dates and irrigation strategies to mitigate water and heat stress, with tailored approaches for different environments.

Sowing date adjustments are also vital for enhancing yield resilience in response to changing climate patterns. Akinseye et al. (2020) found that adjusting sowing dates can lead to increased yields, particularly when synchronized with variety-specific growth patterns. Similarly, strategies like early planting in chilling-tolerant sorghum hybrids have been shown to extend the growing season and mitigate drought risks (Raymundo et al., 2021). Additionally, genetic traits such as photoperiod sensitivity are being explored to improve sorghum's adaptability to climate variability (Guitton et al., 2018). Water management strategies, such as optimizing irrigation for sorghum in rotation systems, have also been critical for climate adaptation (Kothari et al., 2019). These insights into climate-adaptive irrigation techniques ensure sustainable water use while maintaining yield potential under varying weather conditions. Finally, crop modeling studies like those by Akinseye et al. (2017) show that simulations of sorghum growth, accounting for genetic variability in photoperiod sensitivity, can support informed decision-making in selecting the most suitable cultivars and management practices for different environmental conditions. In conclusion, CSMs are crucial for refining adaptation strategies to enhance sorghum's resilience to climate stress. Ongoing research into genetic traits and climate-adaptive technologies strengthens efforts to mitigate climate variability, ensuring stable yields and long-term resilience.

15.5 MODELING-DRIVEN IDEOTYPE DEVELOPMENT IN SORGHUM

Implementing adaptation strategies can mitigate the negative impacts of climate change and even enhance crop productivity. Global crop-climate modeling assessments suggest that incremental adaptation strategies could boost yields by up to ~7% at any given warming level (Ramirez-Villegas et al., 2015). However, systemic and transformational strategies, such as ideotype development, are needed to maximize these opportunities. Ideotype breeding focuses on developing crop varieties with specific traits suited to future climates, enhancing food security and production stability. Recent advancements in crop modeling enable breeders to identify genotype-environment interactions across large spatial areas, helping to develop climate-resilient cultivars. Unlike conventional breeding, which eliminates undesirable traits one at a time, ideotype breeding proactively selects for an optimal combination of physiological and morphological traits (Donald, 1968). For instance, Anami et al. (2015) proposed super sweet sorghum ideotypes tailored for biofuel production, integrating traits such as flowering time, maturity, and biomass conversion efficiency to enhance yield under diverse climatic conditions.

Although idiotypic adaptations have been extensively studied in crops like maize, cotton, rice, and soybean, research on sorghum under projected climate scenarios remains limited. Addressing this gap, Getachew et al. (2023) evaluated 4,096 sorghum ideotypes under future climate conditions (SSP245 and SSP585) using four global circulation models from CMIP6. Their study identified genetic coefficients for a virtual ideotype capable of offsetting yield losses and achieving a national yield gain of up to 0.96 t/ha. Compared to the reference cultivar Teshale, this ideotype featured a shorter anthesis period (by up to eight days), an extended maturity period (by up to ten days), larger leaves, and a longer thermal time from emergence to grain filling. These traits improve heat stress tolerance and yield stability, highlighting key breeding targets for future sorghum adaptation strategies. In 2022, Texas Tech University, U.S.A. secured funding from the United Sorghum Checkoff Program (USCP) to integrate major U.S. public sorghum improvement programs (www.depts.ttu.edu/pss/newsroom/posts/2022/10/texas-tech-taking-lead-uscp-sorghum-project.php). This transdisciplinary initiative combines agronomy, crop physiology, breeding, machine learning, and crop-climate modeling to enhance sorghum productivity. APSIM-based modeling is being used to simulate trait-environment interactions, identifying optimal trait combinations for water-limited and favorable conditions. Historical yield datasets from Kansas and Texas are being used as benchmarks, while modeling enables the exploration of various water levels, management scenarios, and climate resilience strategies that would be impractical to test through field trials.

15.6 ENHANCING SORGHUM BREEDING EFFICIENCY THROUGH CROP MODELING AND SIMULATION

Despite advancements in crop breeding strategies, including quantitative genetics, marker-assisted selection, genomic selection, and improved trait-screening techniques, progress in developing improved cultivars has been slow due to the complex genotype-environment-management (G × E × M) interactions. Investigating these interactions *in vivo* requires extensive, costly, and time-consuming multi-location field trials, often covering too few relevant environments.

This challenge suggests that genotypes adaptable to a broad range of stress environments are unlikely, making it crucial for breeding strategies to focus on developing genotypes suited to specific environments. To address this, crop simulation modeling has emerged as a valuable tool for deciphering G × E × M interactions. By integrating mechanistic knowledge of crop growth, these models can estimate crop productivity across regions, thereby guiding precision breeding. This approach allows for the identification of environmental-restricting factors, such as drought stress types, and helps predict a crop's potential to face such stress. This methodology has been applied to characterize wheat and sorghum environments in Australia (Chenu et al., 2011; Hammer et al., 2014).

Additionally, modeling and simulation offer significant opportunities to enhance breeding efficiency by incorporating biological insights, improving the prediction of genotype-to-phenotype relationships crucial for plant breeding (Hammer et al., 2016, 2019a,b). Phenotype prediction based on genotype is vital for advancing yield in breeding. Traditionally, this is achieved through extensive field phenotyping of different genotypes, coupled with sophisticated statistical methods to analyze variance components from multi-environment trials. However, recent breakthroughs in genotyping and whole genome prediction techniques have transformed plant breeding approaches (Aziz & Masmoudi, 2024; Saini et al., 2024; Technow et al., 2015). While these advances have boosted the rate of genetic gain, they also increase the potential for crop modeling and biological understanding to make further contributions to breeding (Technow et al., 2015).

There are two key ways crop ecophysiology and modeling can enhance breeding efficiency (Hammer et al., 2016, 2019a, b). The first is by utilizing insights from dynamic models to refine phenotyping strategies. This approach breaks down complex traits into more manageable targets, addressing G × E × M interactions and improving connections to genetic regulation and predictive power. The second involves using crop growth and development models for trait assessment and phenotype prediction in target production regions, assisting in prioritizing efforts and evaluating breeding strategies. For instance, the stay-green trait in sorghum, which provides yield benefits in water-limited conditions (Borrell and Hammer, 2000), demonstrates both approaches. Stay-green, a drought-adaptation trait in sorghum, is associated with delayed leaf senescence during grain filling, which is largely a result of improved water availability post-anthesis (Borrell and Hammer, 2000). Van Oosterom et al. (2011) identified physiological processes that enhance grain yield under drought by improving post-anthesis water availability. Studies linked these processes to the stay-green trait and associated QTLs, showing that stay-green QTLs positively impact yield by altering water use patterns through changes in canopy development and water efficiency (Borrell et al., 2014a,b). Genetic regulation of tillering in sorghum (Alam et al., 2014) reveals QTLs overlapping with stay-green. Additionally, transpiration efficiency (Xin et al., 2009), root system architecture (Demissie et al., 2023), and leaf traits (van Oosterom et al., 2011; Borrell et al., 2014a, b) also contribute to improved water use during grain filling.

The ability of a crop model to simulate complex traits as emerging outcomes of the internal dynamics of processes within the model is crucial for effectively analyzing complex traits and establishing stronger connections to their underlying genetic control. To accurately address G × E × M interactions and improve the heritability of component traits, which in turn aids in identifying phenotyping targets, the model must exhibit enhanced biological realism in its structure (Hammer et al., 2006, 2010, 2019a, b). Studies have been conducted to explore the traits that influence stay-green in sorghum (Kholová et al., 2014; Hammer et al., 2019a). A simulation study using the APSIM-sorghum model was performed to analyze two component traits influencing staygreen, tillering, and transpiration efficiency (Hammer et al., 2019a). The tillering routine was modified to simulate a reduced tillering variety compared to a standard hybrid, and a maximum transpiration rate was introduced to limit water use during midday if potential transpiration exceeded the threshold. This required an hourly calculation and utilized a pipeline developed by Hammer et al. (2009), with transpiration limits set as per Sinclair et al. (2005). The simulations were carried out for Emerald, located in the sorghum-growing region of NE Australia, using long-term historical climate data and local soil information. The simulated results for a single year with a terminal moisture stress pattern demonstrated the model's ability to generate stay-green as an emergent trait in such conditions (Hammer et al., 2019a). Both the reduced tillering type and the variety with limited maximum transpiration maintained green foliage longer into the post-anthesis phase. This was linked to a smaller canopy size, which reduced pre-anthesis water use in the case of reduced tillering, or to improved transpiration efficiency, which lowered pre-anthesis water use despite similar canopy size. Both scenarios resulted in an increased simulated yield for the modified type (Hammer et al., 2019a). In the case of reduced tillering, the increase in yield was not associated with greater total biomass at maturity, but rather with a higher harvest index due to increased water availability and crop growth post-anthesis. In the case of limited maximum transpiration, both total and grain biomass increased due to the enhanced transpiration efficiency, which improved water use efficiency.

Over the long-term climate for this location, both reduced tillering and limited maximum transpiration provide significant yield advantages in seasons where the standard hybrid's yield is below approximately 4.5 t ha^{-1} (Hammer et al., 2019a). This aligns with findings from breeding trials on the stay-green trait (Jordan et al., 2012) and reflects the typical field conditions in NE Australia, where the long-term average yield is around 3 t ha^{-1}. However, in high-yielding seasons, the simulation scenario showed a yield reduction. This was due to the conservative growth (and resource use) of the adaptations simulated. These results emphasize the potential value of these component traits and suggest that targeting them in phenotyping strategies could be beneficial. This highlights the approach of breaking down a complex trait into component targets that may be more closely linked to genetic control (with higher heritability), offering the potential to improve genetic gain if effective phenotyping systems can be developed. In conclusion, crop modeling and simulation offer significant potential to enhance sorghum breeding efficiency by addressing complex G × E × M interactions. Through improved phenotyping strategies and the use of dynamic models to predict trait outcomes, these tools help identify key traits such as staygreen, which can enhance drought resilience and yield under stress conditions. By integrating biological insights with advanced genotyping and whole-genome prediction techniques, modeling approaches contribute to more precise, targeted breeding efforts, ultimately accelerating genetic gains and improving crop resilience in variable environments.

15.7 CHALLENGES IN USING CROP SIMULATION MODELS TO ADDRESS CLIMATE CHANGE

Despite their potential, CSMs face several challenges that limit their efficacy in addressing climate change impacts. A primary limitation lies in the accuracy of the models, which depends heavily on high-quality data regarding climate, soil, and crop management. Unfortunately, in many sorghum-growing regions, particularly in developing countries, access to reliable data is constrained. Incomplete or outdated datasets undermine the precision of model predictions, potentially leading to suboptimal decision-making (Kephe et al., 2021; Challinor et al., 2009).

Additionally, the predictive accuracy of these models is hampered by uncertainties inherent in climate projections. Models often rely on assumptions about crop responses to environmental factors like temperature, CO_2 levels, and precipitation. The interaction of these factors under future climate scenarios is difficult to predict accurately, even when employing ensemble modeling approaches. While ensemble models, which integrate outputs from multiple GCMs, can improve prediction robustness, their reliance on average outputs can obscure the full spectrum of potential impacts (Timlin et al., 2024). Crop responses to climate change are further complicated by the interplay of multiple physiological factors, including CO_2 fertilization effects, temperature thresholds, and water availability. Accurately simulating these responses requires highly detailed models capable of handling complex processes. However, increasing model complexity also heightens the risk of errors and reduces the models' applicability, particularly in practical decision-making contexts (Challinor et al., 2009).

Bridging the gap between the spatial and temporal scales of climate and crop models poses another significant challenge. While climate models operate on large spatial scales (e.g., hundreds of kilometers), crop models function at the field scale. Effective integration requires downscaling climate model outputs to match the resolution of crop models, a process that introduces additional uncertainties (Osborne et al., 2007). Moreover, the complexity of CSMs often demands specialized expertise for effective application. This creates a barrier for smallholder farmers and extension workers, who may lack the training or resources to utilize these tools effectively. Making CSMs more accessible and user-friendly is critical, particularly in resource-constrained settings. Finally, the inherent uncertainty of climate change projections complicates the formulation of adaptation strategies. While models can simulate a range of future scenarios, the variability in predictions makes it challenging to provide definitive guidance. Farmers and policymakers must therefore remain flexible, preparing for a variety of potential outcomes rather than relying on a single projection.

15.8 FUTURE PROSPECTS FOR SORGHUM CROP MODELING

Sorghum crop modeling has matured, demonstrating value in agronomy and breeding but with untapped potential. In agronomy, models addressing water and nitrogen dynamics have been used to assess production-risk trade-offs in developed systems, though challenges in developing systems persist. These include calibrating models for local varieties, landraces, and cultivation practices, as well as integrating soil and crop nutrition dynamics. Collaborative efforts between experimentalists and modelers are essential for refining models and addressing system-level constraints, such as pest and weed management. The broader integration of crop modeling within farming systems, such as through platforms like APSIM, offers additional opportunities to simulate the interplay between system-level dynamics and crop performance. Climate change applications further highlight the importance of sorghum modeling.

In breeding, crop modeling has proven to be a valuable support technology. It complements genomic prediction methods by integrating biological insights, improving genotype-phenotype predictions, and identifying adaptive traits under diverse conditions. Simulation approaches, such as the G × E × M framework, remain critical for exploring genotype-environment-management interactions and guiding breeding strategies. For instance, Hammer et al. (2014) identified promising sorghum genotypes and management strategies through simulations addressing productivity-risk trade-offs. To advance, crop modeling must balance biological detail with practical simplicity, emphasizing "the right answer for the right reason". Models tailored for breeding must connect genetic architecture with phenotypic outcomes while remaining accessible to practitioners. Continued transdisciplinary collaboration among experimentalists, theorists, and practitioners will ensure the relevance and impact of sorghum modeling in addressing agricultural challenges, driving innovation across agronomy and breeding domains.

15.9 CONCLUSIONS

Crop simulation models serve as powerful tools for addressing the challenges climate change poses to sorghum production. By integrating biophysical data with climate projections, these models provide critical insights into how varying climatic factors such as temperature, rainfall, and CO_2 concentrations affect crop growth, development, and yield. They help researchers and farmers identify climate-resilient sorghum varieties and optimize management practices tailored to specific agro-ecological zones. However, fully harnessing the potential of these models requires overcoming several challenges. Accurate data inputs, including soil properties, cultivar characteristics, and real-time weather data, are essential for generating reliable outputs. In many sorghum-growing regions, particularly in resource-limited settings, the scarcity or incompleteness of such data undermines model accuracy. Additionally, the complexity of climate systems necessitates continuous model updates to account for dynamic interactions between environmental factors. The inherent uncertainty of future climate scenarios further complicates predictions, limiting the precision of adaptation strategies. Despite these obstacles, advancements in data collection technologies, remote sensing, and machine learning are paving the way for more precise and accessible CSMs. Combining these innovations with local knowledge can empower stakeholders to make informed decisions, ensuring that sorghum production remains resilient and sustainable in a rapidly changing climate.

REFERENCES

Abreha, K. B., Enyew, M., Carlsson, A. S., Vetukuri, R. R., Feyissa, T., Motlhaodi, T., ... & Geleta, M. (2022). Sorghum in dryland: Morphological, physiological, and molecular responses of sorghum under drought stress. *Planta*, *255*, 1–23.

Adam, M., Corbeels, M., Leffelaar, P. A., Van Keulen, H., Wery, J., & Ewert, F. (2012). Building crop models within different crop modelling frameworks. *Agricultural Systems*, *113*, 57–63.

Aggarwal, P. K., Kalra, N., Chander, S., & Pathak, H. (2006). InfoCrop: A dynamic simulation model for the assessment of crop yields, losses due to pests, and environmental impact of agro-ecosystems in tropical environments. I. Model description. *Agricultural Systems*, *89*(1), 1–25.

Ainsworth, E. A., & Ort, D. R. (2010). How do we improve crop production in a warming world?. *Plant Physiology*, *154*(2), 526–530.

Akinseye, F. M., Adam, M., Agele, S. O., Hoffmann, M. P., Traore, P. C. S., & Whitbread, A. M. (2017). Assessing crop model improvements through comparison of sorghum (*Sorghum bicolor* L. moench) simulation models: A case study of West African varieties. *Field Crops Research*, *201*, 19–31.

Akinseye, F. M., Ajeigbe, H. A., Traore, P. C., Agele, S. O., Zemadim, B., & Whitbread, A. (2020). Improving sorghum productivity under changing climatic conditions: A modelling approach. *Field Crops Research*, *246*, 107685.

Alagarswamy, G., & Ritchie, J. T. (1991). Phasic development in CERES-Sorghum model. *Predicting Crop Phenology*, 143–152.

Alam, M. M., Hammer, G. L., Van Oosterom, E. J., Cruickshank, A. W., Hunt, C. H., & Jordan, D. R. (2014). A physiological framework to explain genetic and environmental regulation of tillering in sorghum. *New Phytologist*, *203*(1), 155–167.

Allen Jr, L. H., Kakani, V. G., Vu, J. C., & Boote, K. J. (2011). Elevated CO2 increases water use efficiency by sustaining photosynthesis of water-limited maize and sorghum. *Journal of Plant Physiology*, *168*(16), 1909–1918.

Al-Salman, Y., Ghannoum, O., & Cano, F. J. (2023). Elevated [CO2] negatively impacts C4 photosynthesis under heat and water stress without penalizing biomass. *Journal of Experimental Botany*, *74*(9), 2875–2890.

Anami, S. E., Zhang, L. M., Xia, Y., Zhang, Y. M., Liu, Z. Q., & Jing, H. C. (2015). Sweet sorghum ideotypes: Genetic improvement of stress tolerance. *Food and Energy Security*, *4*(1), 3–24.

Assefa, Y., Staggenborg, S. A., & Prasad, V. P. (2010). Grain sorghum water requirement and responses to drought stress: A review. *Crop Management*, *9*(1), 1–11.

Asseng, S., Zhu, Y., Basso, B., Wilson, T., & Cammarano, D. (2014). *Simulation Modeling: Applications in Cropping Systems*. Elsevier.

Aziz, M. A., & Masmoudi, K. (2024). Molecular breakthroughs in modern plant breeding techniques. *Horticultural Plant Journal, 11*(1), 15–41.

Battisti, R., Sentelhas, P. C., & Boote, K. J. (2017). Inter-comparison of performance of soybean crop simulation models and their ensemble in southern Brazil. *Field Crops Research*, *200*, 28–37.

Begna, T., Yali, W., & Gichile, H. (2022). impact of drought stress in sorghum (*Sorghum bicolor* (L.) Moench) production and drought genetic resistance mechanisms. *European Journal of Agriculture and Forestry Research*, *10*(2), 28–49.

Borrell, A. K., & Hammer, G. L. (2000). Nitrogen dynamics and the physiological basis of stay-green in sorghum. *Crop Science*, *40*(5), 1295–1307.

Borrell, A. K., Mullet, J. E., George-Jaeggli, B., van Oosterom, E. J., Hammer, G. L., Klein, P. E., & Jordan, D. R. (2014b). Drought adaptation of stay-green sorghum is associated with canopy development, leaf anatomy, root growth, and water uptake. *Journal of Experimental Botany*, *65*(21), 6251–6263.

Borrell, A. K., van Oosterom, E. J., Mullet, J. E., George-Jaeggli, B., Jordan, D. R., Klein, P. E., & Hammer, G. L. (2014a). Stay-green alleles individually enhance grain yield in sorghum under drought by modifying canopy development and water uptake patterns. *New Phytologist*, *203*(3), 817–830.

Brown, H. E., Huth, N. I., Holzworth, D. P., Teixeira, E. I., Zyskowski, R. F., Hargreaves, J. N., & Moot, D. J. (2014). Plant modelling framework: Software for building and running crop models on the APSIM platform. *Environmental Modelling & Software*, *62*, 385–398.

Carberry, P. S., Hochman, Z., Hunt, J. R., Dalgliesh, N. P., McCown, R. L., Whish, J. P. M., ... & Van Rees, H. (2009). Re-inventing model-based decision support with Australian dryland farmers. 3. Relevance of APSIM to commercial crops. *Crop and Pasture Science*, *60*(11), 1044–1056.

Carcedo, A. J., Mayor, L., Demarco, P., Morris, G. P., Lingenfelser, J., Messina, C. D., & Ciampitti, I. A. (2022). Environment characterization in sorghum (*Sorghum bicolor* L.) by modeling water-deficit and heat patterns in the Great Plains Region, United States. *Frontiers in Plant Science*, *13*, 768610.

Challinor, A. J., Ewert, F., Arnold, S., Simelton, E., & Fraser, E. (2009). Crops and climate change: Progress, trends, and challenges in simulating impacts and informing adaptation. *Journal of Experimental Botany*, *60*(10), 2775–2789.

Chatfield, C. (1995). Model uncertainty, data mining and statistical inference. *Journal of the Royal Statistical Society Series A: Statistics in Society*, *158*(3), 419–444.

Chaudhuri, U. N., Burnett, R. B., Kirkham, M. B., & Kanemasu, E. T. (1986). Effect of carbon dioxide on sorghum yield, root growth, and water use. *Agricultural and Forest Meteorology*, *37*(2), 109–122.

Chenu, K., Cooper, M., Hammer, G. L., Mathews, K. L., Dreccer, M. F., & Chapman, S. C. (2011). Environment characterization as an aid to wheat improvement: Interpreting genotype–environment interactions by modelling water-deficit patterns in North-Eastern Australia. *Journal of Experimental Botany*, *62*(6), 1743–1755.

Conley, M. M., Kimball, B. A., Brooks, T. J., Pinter Jr, P. J., Hunsaker, D. J., Wall, G. W., ... & Triggs, J. M. (2001). CO2 enrichment increases water-use efficiency in sorghum. *New Phytologist*, *151*(2), 407–412.

Cooper, M., & Hammer, G. L. (1996). *Synthesis of Strategies for Crop Improvement*. In M. Cooper & G.L. Hammer (eds.), Plant Adaptation and Crop Improvement (pp. 591–623). Wallingford, England: CAB International.

Cooper, M., Gho, C., Leafgren, R., Tang, T., & Messina, C. (2014a). Breeding drought-tolerant maize hybrids for the US corn-belt: Discovery to product. *Journal of Experimental Botany*, *65*(21), 6191–6204.

Cooper, M., Messina, C. D., Podlich, D., Totir, L. R., Baumgarten, A., Hausmann, N. J., ... & Graham, G. (2014b). Predicting the future of plant breeding: Complementing empirical evaluation with genetic prediction. *Crop and Pasture Science*, *65*(4), 311–336.

Craufurd, P. Q., & Peacock, J. M. (1993). Effect of heat and drought stress on sorghum (*Sorghum bicolor*). II. Grain yield. *Experimental Agriculture*, *29*(1), 77–86.

De Souza Rodrigues, T., Arge, L. W. P., de Freitas Guedes, F. A., Travassos-Lins, J., de Souza, A. P., Cocuron, J. C., ... & Alves-Ferreira, M. (2023). Elevated CO2 increases biomass of *Sorghum bicolor* green prop roots under drought conditions via soluble sugar accumulation and photosynthetic activity. *Physiologia Plantarum*, *175*(4), e13984.

De Souza, A. P., Cocuron, J. C., Garcia, A. C., Alonso, A. P., & Buckeridge, M. S. (2015). Changes in whole-plant metabolism during the grain-filling stage in sorghum grown under elevated CO2 and drought. *Plant Physiology*, *169*(3), 1755–1765.

De Wit, C. T. (1969). Dynamic concepts in biology. In Setlik, I. (ed.), *Prediction and Measurement of Photosynthetic Productivity. Proceedings of the IBP/PP Technical Meeting, Trebon, 1969* (pp. 17–23). Wageningen: Centre for Agricultural Publishing and Documentation.

Demissie, H. S., Mindaye, T. T., Teklu, D. N., & Kebede, F. G. (2023). Root system architecture analysis of sorghum genotypes and its effect on drought adaptation. *Rhizosphere*, *27*, 100772.

Diancoumba, M., Kholová, J., Adam, M., Famanta, M., Clerget, B., Traore, P. C., ... & Vadez, V. (2024). APSIM-based modeling approach to understand sorghum production environments in Mali. *Agronomy for Sustainable Development*, *44*(3), 25.

Djanaguiraman, M., Prasad, P. V., Murugan, M., Perumal, R., & Reddy, U. K. (2014). Physiological differences among sorghum (*Sorghum bicolor* L. Moench) genotypes under high temperature stress. *Environmental and Experimental Botany*, *100*, 43–54.

Donald, C. T. (1968). The breeding of crop ideotypes. *Euphytica*, *17*, 385–403.

Druille, M., Williams, A. S., Torrecillas, M., Kim, S., Meki, N., & Kiniry, J. R. (2020). Modeling climate warming impacts on grain and forage sorghum yields in Argentina. *Agronomy*, *10*(7), 964.

Earl, H. J., & Davis, R. F. (2003). Effect of drought stress on leaf and whole canopy radiation use efficiency and yield of maize. *Agronomy Journal*, *95*(3), 688–696.

Eggen, M., Ozdogan, M., Zaitchik, B., Ademe, D., Foltz, J., & Simane, B. (2019). Vulnerability of sorghum production to extreme, sub-seasonal weather under climate change. *Environmental Research Letters*, *14*(4), 045005.

Emendack, Y., Sanchez, J., Hayes, C., Nesbitt, M., Laza, H., & Burke, J. (2021). Seed-to-seed early-season cold resiliency in sorghum. *Scientific Reports*, *11*(1), 7801.

Enyew, M., Carlsson, A. S., Geleta, M., Tesfaye, K., Hammenhag, C., Seyoum, A., & Feyissa, T. (2022). Novel sources of drought tolerance in sorghum landraces revealed via the analyses of genotype-by-environment interactions. *Frontiers in Plant Science*, *13*, 1062984.

FAO. (2002). *Crop Water Management. Online*. AGLW Water Management Group, United Nations FAO.

Fatima, Z., Ahmed, M., Hussain, M., Abbas, G., Ul-Allah, S., Ahmad, S., ... & Hussain, S. (2020). The fingerprints of climate warming on cereal crops phenology and adaptation options. *Scientific Reports*, *10*(1), 18013.

Fu, T., Ko, J., Wall, G. W., Pinter, P. J., Kimball, B. A., Ottman, M. J., & Kim, H. Y. (2016). Simulation of climate change impacts on grain sorghum production grown under free air CO2 enrichment. *International Agrophysics*, *30*(3), 311–322.

Getachew, F., Bayabil, H. K., Hoogenboom, G., Kiker, G. A., Yu, Z., & Li, Y. (2023). Development of climate-smart sorghum ideotype for climate resilience in Ethiopia. *Field Crops Research*, *303*, 109135.

Gohain, G. B., Singh, K. K., Singh, R. S., Dakhore, K. K., & Ghosh, K. (2022). Application of CERES-sorghum crop simulation model DSSAT v4. 7 for determining crop water stress in crop phenological stages. *Modeling Earth Systems and Environment*, *8*(2), 1963–1975.

Guitton, B., Thera, K., Tékété, M. L., Pot, D., Kouressy, M., Teme, N., ... & Vaksmann, M. (2018). Integrating genetic analysis and crop modeling: A major QTL can finely adjust photoperiod-sensitive sorghum flowering. *Field Crops Research*, *221*, 7–18.

Hammer, G. L., Chapman, S. C., & Muchow, R. C. (1996). *Modelling Sorghum in Australia: The State of the Science and Its Role in the Pursuit of Improved Practices*. Proceedings of the Third Australian Sorghum Conference, Tamworth, 20–22 February 1996.

Hammer, G. L., Dong, Z., McLean, G., Doherty, A., Messina, C., Schussler, J., ... & Cooper, M. (2009). Can changes in canopy and/or root system architecture explain historical maize yield trends in the US corn belt?. *Crop Science*, *49*(1), 299–312.

Hammer, G. L., McLean, G., Chapman, S., Zheng, B., Doherty, A., Harrison, M. T., ... & Jordan, D. (2014). Crop design for specific adaptation in variable dryland production environments. *Crop and Pasture Science*, *65*(7), 614–626.

Hammer, G. L., van Oosterom, E., McLean, G., Chapman, S. C., Broad, I., Harland, P., & Muchow, R. C. (2010). Adapting APSIM to model the physiology and genetics of complex adaptive traits in field crops. *Journal of Experimental Botany*, *61*(8), 2185–2202.

Hammer, G., Cooper, M., Tardieu, F., Welch, S., Walsh, B., van Eeuwijk, F. & Podlich, D. (2006). Models for navigating biological complexity in breeding improved crop plants. *Trends in Plant Science*, *11*(12), 587–593.

Hammer, G., McLean, G., Doherty, A., van Oosterom, E., & Chapman, S. (2019a). Sorghum crop modeling and its utility in agronomy and breeding. *Sorghum: A State of the Art and Future Perspetives*, *58*, 215–239.

Hammer, G., Messina, C., van Oosterom, E., Chapman, S., Singh, V., Borrell, A., ... & Cooper, M. (2016). Molecular breeding for complex adaptive traits: How integrating crop ecophysiology and modelling can enhance efficiency. In: Yin, X. & Struik, (eds.), *Crop Systems Biology: Narrowing the Gaps Between Crop Modelling and Genetics* (pp. 147–162). Cham, Switzerland: Springer.

Hammer, G., Messina, C., Wu, A., & Cooper, M. (2019b). Biological reality and parsimony in crop models—why we need both in crop improvement!. In *Silico Plants*, *1*(1), diz010.

Hasanuzzaman, M., Nahar, K., Alam, M. M., Roychowdhury, R., & Fujita, M. (2013). Physiological, biochemical, and molecular mechanisms of heat stress tolerance in plants. *International Journal of Molecular Sciences*, *14*(5), 9643–9684.

Heiniger, R. W., Vanderlip, R. L., & Welch, S. M. (1997). Developing guidelines for replanting grain sorghum: I. Validation and sensitivity analysis of the SORKAM sorghum growth model. *Agronomy Journal*, *89*(1), 75–83.

Hijmans, R. J., Guiking-Lens, I. M., & Van Diepen, C. A. (1994). *WOFOST 6.0: User's Guide for the WOFOST 6.0 Crop Growth Simulation Model*. Wageningen: Winand Staring Centre.

Hodson, D., & White, J. (2010). GIS and crop simulation modelling applications in climate change research. *Climate Change and Crop Production*, *1*, 245–262.

Holzworth, D. P., Huth, N. I., deVoil, P. G., Zurcher, E. J., Herrmann, N. I., McLean, G., ... & Keating, B. A. (2014). APSIM–evolution towards a new generation of agricultural systems simulation. *Environmental Modelling & Software*, *62*, 327–350.

Jones, J. W., Hoogenboom, G., Porter, C. H., Boote, K. J., Batchelor, W. D., Hunt, L. A., ... & Ritchie, J. T. (2003). The DSSAT cropping system model. *European Journal of Agronomy*, *18*(3–4), 235–265.

Jordan, D. R., Hunt, C. H., Cruickshank, A. W., Borrell, A. K., & Henzell, R. G. (2012). The relationship between the stay-green trait and grain yield in elite sorghum hybrids grown in a range of environments. *Crop Science*, *52*(3), 1153–1161.

Kamdi, P. J., Swain, D. K., & Wani, S. P. (2023). Developing climate change agro-adaptation strategies through field experiments and simulation analyses for sustainable sorghum production in semi-arid tropics of India. *Agricultural Water Management*, *286*, 108399.

Keating, B. A., Carberry, P. S., Hammer, G. L., Probert, M. E., Robertson, M. J., Holzworth, D., ... & Smith, C. J. (2003). An overview of APSIM, a model designed for farming systems simulation. *European Journal of Agronomy*, *18*(3–4), 267–288.

Kephe, P. N., Ayisi, K. K., & Petja, B. M. (2021). Challenges and opportunities in crop simulation modelling under seasonal and projected climate change scenarios for crop production in South Africa. *Agriculture & Food Security*, *10*, 1–24.

Kersebaum, K. C. (2007). Modelling nitrogen dynamics in soil–crop systems with HERMES. *Nutrient Cycling in Agroecosystems*, *77*, 39–52.

Kerthi,C., Kumari, B. R., & Kumar, T. S. (2021). Sorghum mitigates climate variability and change on crop yield and quality. *Planta*, *253*(5), 113.

Khalifa, M., & Eltahir, E. A. (2023). Assessment of global sorghum production, tolerance, and climate risk. *Frontiers in Sustainable Food Systems*, *7*, 1184373.

Kholová, J., Murugesan, T., Kaliamoorthy, S., Malayee, S., Baddam, R., Hammer, G. L., ... & Vadez, V. (2014). Modelling the effect of plant water use traits on yield and stay-green expression in sorghum. *Functional Plant Biology*, *41*(11), 1019–1034.

Kothari, K., Ale, S., Bordovsky, J. P., Thorp, K. R., Porter, D. O., & Munster, C. L. (2019). Simulation of efficient irrigation management strategies for grain sorghum production over different climate variability classes. *Agricultural Systems*, *170*, 49–62.

Kundu, S., Saini, D. K., Meena, R. K., Bahuguna, R. N., & Jagadish, S. V. K. (2024). High-throughput phenotyping and AI technologies for deciphering crop resilience to heat stress. *Plant Physiology Reports*, *29*, 1–17.

Lahlali, R., Taoussi, M., Laasli, S. E., Gachara, G., Ezzouggari, R., Belabess, Z., ... & Barka, E. A. (2024). Effects of climate change on plant pathogens and host-pathogen interactions. *Crop and Environment*, *3*(3), 159–170.

Ma, L., Malone, R. W., Heilman, P., Jaynes, D. B., Ahuja, L. R., Saseendran, S. A., ... & Ascough Ii, J. C. (2007). RZWQM simulated effects of crop rotation, tillage, and controlled drainage on crop yield and nitrate-N loss in drain flow. *Geoderma*, *140*(3), 260–271.

Marziotte, L., Carcedo, A. J., Mayor, L., Prasad, P. V., Peraza, J. A., & Ciampitti, I. A. (2025). An in-silico approach exploring sorghum source: Sink balance across sorghum hybrids: How many leaves are enough?. *Crop Science*, *65*(1), e21449.

Messina, C. D., Podlich, D., Dong, Z., Samples, M., & Cooper, M. (2011). Yield–trait performance landscapes: From theory to application in breeding maize for drought tolerance. *Journal of Experimental Botany*, *62*(3), 855–868.

Messina, C., Hammer, G., Dong, Z., Podlich, D., & Cooper, M. (2009). Modelling crop improvement in a G× E× M framework via gene-trait-phenotype relationships. In Sadras, V. & Calderini D. (eds.), *Crop Physiology: Applications for Genetic Improvement and Agronomy* (pp. 235–265). The Netherlands: Elsevier.

Mohammed, A., & Misganaw, A. (2022). Modeling future climate change impacts on sorghum (*Sorghum bicolor*) production with best management options in Amhara Region, Ethiopia. *CABI Agriculture and Bioscience*, *3*(1), 22.

Msongaleli, B., Rwehumbiza, F. B. R., Tumbo, S. D., & Kihupi, N. (2013). *Investigation of Sorghum Yield Response to Variable and Changing Climatic Conditions in Semi-Arid Central Tanzania: Evaluating Crop Simulation Model Applications*. In: African crop science conference proceedings. African Crop Science Society, pp 899–909.

Msongaleli, B., Rwehumbiza, F., Tumbo, S. D., & Kihupi, N. (2014). Sorghum yield response to changing climatic conditions in semi-arid central Tanzania: Evaluating crop simulation model applicability. *Agricultural Sciences*, *5*(10), 822–833.

Niyibigira, T., Mohammed, W., Tana, T., Lemma Tefera, T., & Rukundo, P. (2024). Sorghum farmers' perceptions of climate change, its effects, temperature and precipitation trends, and determinants of adaptation strategies in the central plateau zone of Rwanda. *Cogent Food & Agriculture*, *10*(1), 2334999.

Osborne, T. M., Lawrence, D. M., Challinor, A. J., Slingo, J. M., & Wheeler, T. R. (2007). Development and assessment of a coupled crop–climate model. *Global Change Biology*, *13*(1), 169–183.

Ottman, M. J., Kimball, B. A., Pinter, P. J., Wall, G. W., Vanderlip, R. L., Leavitt, S. W., ... & Brooks, T. J. (2001). Elevated CO2 increases sorghum biomass under drought conditions. *New Phytologist*, *150*(2), 261–273.

Palosuo, T., Kersebaum, K. C., Angulo, C., Hlavinka, P., Moriondo, M., Olesen, J. E., ... & Rötter, R. (2011). Simulation of winter wheat yield and its variability in different climates of Europe: A comparison of eight crop growth models. *European Journal of Agronomy*, *35*(3), 103–114.

Parry, M. L., Rosenzweig, C., Iglesias, A., Livermore, M., & Fischer, G. (2004). Effects of climate change on global food production under SRES emissions and socio-economic scenarios. *Global Environmental Change*, *14*(1), 53–67.

Parton, W. J., Stewart, J. W., & Cole, C. V. (1988). Dynamics of C, N, P and S in grassland soils: A model. *Biogeochemistry*, *5*, 109–131.

Perkins, S. A. (2012). *Crop model review and sweet sorghum crop model parameter development* [Doctoral dissertation]. Kansas State University.

Prasad, P. V. V., Boote, K. J., & Allen, L. H. Jr. (2006). Adverse high dtemperature efects on pollen viability, seed-set, seed yield and harvest index of grain sorghum [*Sorghum bicolor* (L.) Moench] are more severe at elevated carbon dioxide due to higher tissue temperatures. *Agricultural and Forest Meteorology*, *139*, 237–251.

Prasad, P. V. V., Djanaguiraman, M., Jagadish, S. V. K., & Ciampitti, I. A. (2019). Drought and high temperature stress and traits associated with tolerance. *Sorghum: A State of the Art and Future Perspectives*, *58*, 241–265.

Prasad, P. V. V., Djanaguiraman, M., Perumal, R., & Ciampitti, I. A. (2015). Impact of high temperature stress on floret fertility and individual grain weight of grain sorghum: Sensitive stages and thresholds for temperature and duration. *Frontiers in Plant Science*, *6*, 820.

Prasad, P. V. V., Staggenborg, S. A., & Ristic, Z. (2008). Impacts of drought and/or heat stress on physiological, developmental, growth, and yield processes of crop plants. *Response of Crops to Limited Water: Understanding and Modeling Water Stress Effects on Plant Growth Processes*, *1*, 301–355.

Prasad, V. R., Govindaraj, M., Djanaguiraman, M., Djalovic, I., Shailani, A., Rawat, N., ... & Prasad, P. V. (2021). Drought and high temperature stress in sorghum: Physiological, genetic, and molecular insights and breeding approaches. *International Journal of Molecular Sciences*, *22*(18), 9826.

Praveenkumar, P., Sathyamoorthy, N., Dheebakaran, G., & Karthikeyan, R. (2024). Assessment of climate change impact on sorghum (*Sorghum bicolor*) production and adaptation strategies in the western zone of Tamil Nadu. *Journal of Agrometeorology*, *26*(2), 181–185.

Ramirez-Villegas, J., Jarvis, A., & Läderach, P. (2013). Empirical approaches for assessing impacts of climate change on agriculture: The EcoCrop model and a case study with grain sorghum. *Agricultural and Forest Meteorology*, *170*, 67–78.

Ramirez-Villegas, J., Watson, J., & Challinor, A. J. (2015). Identifying traits for genotypic adaptation using crop models. *Journal of Experimental Botany*, *66*(12), 3451–3462.

Raymundo, R., Mclean, G., Sexton-Bowser, S., Lipka, A. E., & Morris, G. P. (2024). Crop modeling suggests limited transpiration would increase yield of sorghum across drought-prone regions of the United States. *Frontiers in Plant Science*, *14*, 1283339.

Raymundo, R., SextonBowser, S., Ciampitti, I. A., & Morris, G. P. (2021). Crop modeling defines opportunities and challenges for drought escape, water capture, and yield increase using chilling-tolerant sorghum. *Plant Direct*, *5*(9), e349.

Rötter, R. P., Palosuo, T., Kersebaum, K. C., Angulo, C., Bindi, M., Ewert, F., ... & Trnka, M. (2012). Simulation of spring barley yield in different climatic zones of Northern and Central Europe: A comparison of nine crop models. *Field Crops Research*, *133*, 23–36.

Saini, D. K., Kumar, S., & Kaur, R. (2024). Applying genomics resources to accelerate the development of climate resilient crops. In Sheraz Mahdi, S., Singh, R. & Dhekale, B. (eds.), *Adapting to Climate Change in Agriculture-Theories and Practices: Approaches for Adapting to Climate Change in Agriculture in India* (pp. 43–120). Springer Nature Switzerland.

Salo, T. J., Palosuo, T., Kersebaum, K. C., Nendel, C., Angulo, C., Ewert, F., ... & Rötter, R. P. (2016). Comparing the performance of 11 crop simulation models in predicting yield response to nitrogen fertilization. *Journal of Agricultural Science*, *154*(7), 1218–1240.

Saravanakumar, V. (2018). Assessment of climate change impacts and adaptation strategies on Sorghum production in Tamil Nadu, India. In: 30th International conference of agricultural economists, Vancouver, British Columbia.

Sari, A. (2023). Germination characteristics of sorghum (Sorghum bicolor L.) affected by temperature variation. *Agronomy Research*, *21*(1), 347–356.

Schulze, R. E., Schütte, S., & Wallace, M. (2016). Wheat Production in South Africa and Climate Change. In: Schulze, R.E. (ed.), *Handbook for Farmers, Officials and Other Stakeholders on Adaptation to Climate Change in the Agricultural Sector within South Africa. Section C: Crops in South Africa and Climate Change C*, vol. 2.

Sharma, O. P., Kannan, N., Cook, S., Pokhrel, B. K., & McKenzie, C. (2019). Analysis of the effects of high precipitation in Texas on rainfed sorghum yields. *Water*, *11*(9), 1920.

Silva, R. G. D., Alves, R. D. C., & Zingaretti, S. M. (2020). Increased [CO2] causes changes in physiological and genetic responses in C4 crops: A brief review. *Plants*, *9*(11), 1567.

Sinclair, T. R., Hammer, G. L., & Van Oosterom, E. J. (2005). Potential yield and water-use efficiency benefits in sorghum from limited maximum transpiration rate. *Functional Plant Biology*, *32*(10), 945–952.

Singh, V., Nguyen, C. T., van Oosterom, E. J., Chapman, S. C., Jordan, D. R., & Hammer, G. L. (2015). Sorghum genotypes differ in high temperature responses for seed set. *Field Crops Research*, *171*, 32–40.

Skendžić, S., Zovko, M., Živković, I. P., Lešić, V., & Lemić, D. (2021). The impact of climate change on agricultural insect pests. *Insects*, *12*(5), 440.

Srivastava, A., Kumar, S. N., & Aggarwal, P. K. (2010). Assessment on vulnerability of sorghum to climate change in India. *Agriculture, Ecosystems & Environment*, *138*(3-4), 160–169.

Steduto, P., Hsiao, T. C., Raes, D., & Fereres, E. (2009). AquaCrop—The FAO crop model to simulate yield response to water: I. Concepts and underlying principles. *Agronomy Journal*, *101*(3), 426–437.

Sultan, B., Defrance, D., & Iizumi, T. (2019). Evidence of crop production losses in West Africa due to historical global warming in two crop models. *Scientific Reports*, *9*(1), 12834.

Technow, F., Messina, C. D., Totir, L. R., & Cooper, M. (2015). Integrating crop growth models with whole genome prediction through approximate Bayesian computation. *PloS One*, *10*(6), e0130855.

Timlin, D., Paff, K., & Han, E. (2024). The role of crop simulation modeling in assessing potential climate change impacts. *Agrosystems, Geosciences & Environment*, *7*(1), e20453.

Torbert, H. A., Prior, S. A., Rogers, H. H., & Runion, G. B. (2004). Elevated atmospheric CO2 effects on N fertilization in grain sorghum and soybean. *Field Crops Research*, *88*(1), 57–67.

Traoré, L., Bello, O. D., Chabi, F., Balogoun, I., Yabi, I., Issifou, M. Y., ... & Saïdou, A. (2021). Modelling of Sorghum (*Sorghum bicolor*) growing areas under current and future climate in the Sudanian and Sahelian zones of Mali. *American Journal of Climate Change*, *10*(2), 185–203.

Van Bavel, C. H. M. (1974). Antitranspirant action of carbon dioxide on intact sorghum plants. *Crop Science*, *14*(2), 208–212.

van Oosterom, E. J., Borrell, A. K., Deifel, K. S., & Hammer, G. L. (2011). Does increased leaf appearance rate enhance adaptation to postanthesis drought stress in sorghum?. *Crop Science*, *51*(6), 2728–2740.

Verburg, K., Pasley, H. R., Biggs, J. S., Vogeler, I., Wang, E., Mielenz, H., ... & Thorburn, P. J. (2025). Review of APSIM's soil nitrogen modelling capability for agricultural systems analyses. *Agricultural Systems*, *224*, 104213.

Wawo, A. H., Lestari, P., Setyowati, N., Syarif, F., Juhaeti, T., & Kartika, K. (2024). The influence of temperature on germination and seedling growth of sorghum. *Italian Journal of Agrometeorology*, *1*, 95–107.

Widodo, S., Purwaningsih, H., Pustika, A. B., Widyayanti, S., Muazam, A., Hanifa, A. P., ... & Yusuf, R. (2024). Sorghum productivity and its farming feasibility in dryland agriculture: Genotypic and planting distance insights. *Phyton (0031-9457)*, *93*(5), p1007.

Xin, Z., Aiken, R., & Burke, J. (2009). Genetic diversity of transpiration efficiency in sorghum. *Field Crops Research*, *111*(1–2), 74–80.

Yin, X., Stam, P., Kropff, M. J., & Schapendonk, A. H. (2003). Crop modeling, QTL mapping, and their complementary role in plant breeding. *Agronomy Journal*, *95*(1), 90–98.

Zahra, N., Hafeez, M. B., Ghaffar, A., Kausar, A., Al Zeidi, M., Siddique, K. H., & Farooq, M. (2023). Plant photosynthesis under heat stress: Effects and management. *Environmental and Experimental Botany*, *206*, 105178.

About the Book Editors

Dinesh Kumar Saini is a Research Associate in the Department of Plant and Soil Science at Texas Tech University, Texas, USA, with expertise in Plant Breeding, Genomics, and Crop Biotechnology. He holds a Bachelor of Science (Agriculture) and a Master of Science (Genetics and Plant Breeding) from Chaudhary Charan Singh University, Meerut, Uttar Pradesh (India). In addition to qualifying for multiple national-level exams in India, including ICAR-SRF, ASRB-NET, and GATE, he earned prestigious fellowships such as DST-INSPIRE and ICAR-SRF during his Ph.D. at Punjab Agricultural University, Ludhiana. A recipient of several honors, including the Chancellor Medal (twice), Vice-Chancellor Gold Medal (twice), Dr. M.S. Randhawa Medal, and PAU-ICAR Gold Medal, he has established a strong academic and research foundation. His current research explores physiological, molecular, and histological mechanisms driving climate resilience and yield potential in grain sorghum. By collaborating with breeders, physiologists, and machine learning experts, he develops advanced tools and approaches for high-throughput phenotyping and molecular breeding to enhance sorghum productivity. Dinesh has co-authored over 55 articles in peer-reviewed journals, along with 20 book chapters, and several abstracts. Actively engaged in the scientific community, he serves as a reviewer for several reputed journals and is on the editorial boards of some esteemed journals. He is a member of several professional societies, including the Crop Science Society of America, the American Society of Agronomy, Bioclues, Genome India International, etc.

S. V. Krishna Jagadish, a Professor and the Thornton Distinguished Chair of Plant & Soil Science at Texas Tech University, earned his Bachelor in Science (Agriculture) from the University of Agricultural Sciences, Bangalore and Master of Science (Agronomy) from the University of Agricultural Sciences, Dharwad, India, and his Ph.D. in Agriculture (Crop Physiology) from the University of Reading, UK, supported by the prestigious Felix Scholarship (2003–2007). Before joining Texas Tech, he served as Scientist I, II, and Deputy Division Head of the Crop and Environmental Sciences Division at the International Rice Research Institute, Philippines, for seven years. In 2015, he joined Kansas State University as a tenured Associate Professor, where he was later promoted to Professor in 2021. Dr. Jagadish's accolades include the Kansas State Gamma Sigma Delta Outstanding Research Award (2021), International Fellow of the Indian Society of Plant Physiology (2021), and the Outstanding Young Agricultural Scientist Award from the Association for Agricultural Scientists of Indian Origin (2015). He is a member of several professional societies, including the Crop Science Society of America, the American Society of Agronomy, and the North American Plant Phenotyping Network. During his tenure at Kansas State, Dr. Jagadish focused on enhancing resilience to abiotic stresses such as heat and drought in different crops including sorghum. His prolific research output includes more than 200 international peer-reviewed publications, over 16,500 citations, an H-index of 73, and more than $25 million in grant funding. At Texas Tech University, his research program focuses on exploring abiotic stress tolerance in sorghum, optimizing forage-based cropping systems, developing tools for water conservation, and improving soil health. Through the TeCSIS-TAWC program, his team integrates forage-crop-livestock systems to maximize economic benefits and promote environmental sustainability in the Southern High Plains.

Index

Note: Figures are indicated by *italics*. Tables are indicated by **bold**.

For Product Safety Concerns and Information please contact our EU representative GPSR@taylorandfrancis.com Taylor & Francis Verlag GmbH, Kaufingerstraße 24, 80331 München, Germany